# Mathematical Modeling and Simulation in Mechanics and Dynamic Systems

# Mathematical Modeling and Simulation in Mechanics and Dynamic Systems

Editors

**Maria Luminita Scutaru**
**Catalin I. Pruncu**

MDPI • Basel • Beijing • Wuhan • Barcelona • Belgrade • Manchester • Tokyo • Cluj • Tianjin

*Editors*
Maria Luminita Scutaru
Transilvania University of Brasov
Romania

Catalin I. Pruncu
University of Strathclyde
UK

*Editorial Office*
MDPI
St. Alban-Anlage 66
4052 Basel, Switzerland

This is a reprint of articles from the Special Issue published online in the open access journal *Mathematics* (ISSN 2227-7390) (available at: https://www.mdpi.com/journal/mathematics/special_issues/Math_Model_Simul_Mech_Dyn_Syst).

For citation purposes, cite each article independently as indicated on the article page online and as indicated below:

LastName, A.A.; LastName, B.B.; LastName, C.C. Article Title. *Journal Name* **Year**, *Volume Number*, Page Range.

**ISBN 978-3-0365-3276-9 (Hbk)**
**ISBN 978-3-0365-3277-6 (PDF)**

© 2022 by the authors. Articles in this book are Open Access and distributed under the Creative Commons Attribution (CC BY) license, which allows users to download, copy and build upon published articles, as long as the author and publisher are properly credited, which ensures maximum dissemination and a wider impact of our publications.

The book as a whole is distributed by MDPI under the terms and conditions of the Creative Commons license CC BY-NC-ND.

# Contents

**Maria Luminita Scutaru and Catalin-Iulian Pruncu**
Mathematical Modeling and Simulation in Mechanics and Dynamic Systems
Reprinted from: *Mathematics* **2022**, *10*, 448, doi:10.3390/math10030448 . . . . . . . . . . . . . . . 1

**K. Yakoubi, S. Montassir, Hassane Moustabchir, A. Elkhalfi, Catalin Iulian Pruncu, J. Arbaoui and Muhammad Umar Farooq**
An Extended Finite Element Method (XFEM) Study on the Elastic T-Stress Evaluations for a Notch in a Pipe Steel Exposed to Internal Pressure
Reprinted from: *Mathematics* **2021**, *9*, 507, doi:10.3390/math9050507 . . . . . . . . . . . . . . . . . 7

**Chukwuma Ogbonnaya, Chamil Abeykoon, Adel Nasser and Ali Turan**
A Computational Approach to Solve a System of Transcendental Equations with Multi-Functions and Multi-Variables
Reprinted from: *Mathematics* **2021**, *9*, 920, doi:10.3390/math9090920 . . . . . . . . . . . . . . . . . 17

**Yasser Zare and Kyong Yop Rhee**
Advanced Models for Modulus and Strength of Carbon-Nanotube-Filled Polymer Systems Assuming the Networks of Carbon Nanotubes and Interphase Section
Reprinted from: *Mathematics* **2021**, *9*, 990, doi:10.3390/math9090990 . . . . . . . . . . . . . . . . . 31

**Dongxu Li, Siwei Li, Zheshu Ma, Bing Xu, Zhanghao Lu, Yanju Li and Meng Zheng**
Ecological Performance Optimization of a High Temperature Proton Exchange Membrane Fuel Cell
Reprinted from: *Mathematics* **2021**, *9*, 1332, doi:10.3390/math9121332 . . . . . . . . . . . . . . . . 45

**Pau Fonseca i Casas, Joan Garcia i Subirana, Víctor García i Carrasco and Xavier Pi i Palomés**
SARS-CoV-2 Spread Forecast Dynamic Model Validation through Digital Twin Approach, Catalonia Case Study
Reprinted from: *Mathematics* **2021**, *9*, 1660, doi:10.3390/math9141660 . . . . . . . . . . . . . . . . 61

**Gregor Bánó, Jana Kubacková, Andrej Hovan, Alena Strejčková, Gergely T. Iványi, Gaszton Vizsnyiczai, Lóránd Kelemen;, Gabriel Žoldák, Zoltán Tomori andDenis Horvath**
Power Spectral Density Analysisof Nanowire-Anchored Fluctuating MicrobeadReveals a Double Lorentzian Distribution
Reprinted from: *Mathematics* **2021**, *9*, 1748, doi:10.3390/math9151748 . . . . . . . . . . . . . . . . 79

**Bing Xu, Dongxu Li, Zheshu Ma, Meng Zheng and Yanju Li**
Thermodynamic Optimization of a High Temperature Proton Exchange Membrane Fuel Cell for Fuel Cell Vehicle Applications
Reprinted from: *Mathematics* **2021**, *9*, 1792, doi:10.3390/math9151792 . . . . . . . . . . . . . . . . 97

**Botond-Pál Gálfi, Ioan Száva, Daniela Șova and Sorin Vlase**
Thermal Scaling of Transient Heat Transfer in a Round Cladded Rod with Modern Dimensional Analysis
Reprinted from: *Mathematics* **2021**, *9*, 1875, doi:10.3390/math9161875 . . . . . . . . . . . . . . . . 111

**Koldo Portal-Porras, Unai Fernandez-Gamiz, Ainara Ugarte-Anero, Ekaitz Zulueta and Asier Zulueta**
Alternative Artificial Neural Network Structures for Turbulent Flow Velocity Field Prediction
Reprinted from: *Mathematics* **2021**, *9*, 1939, doi:10.3390/math9161939 . . . . . . . . . . . . . . . . 135

**Mohamed Derbeli, Cristian Napole, Oscar Barambones**
Machine Learning Approach for Modeling and Control of a Commercial Heliocentris FC50 PEM Fuel Cell System
Reprinted from: *Mathematics* **2021**, *9*, 2068, doi:10.3390/math9172068 . . . . . . . . . . . . . . . . 157

**Alexandra Saviuc, Manuela Gîrțu, Liliana Topliceanu, Tudor-Cristian Petrescu and Maricel Agop**
"Holographic Implementations" in the Complex Fluid Dynamics through a Fractal Paradigm
Reprinted from: *Mathematics* **2021**, *9*, 2273, doi:10.3390/math9182273 . . . . . . . . . . . . . . . . 175

**Kiril Tenekedjiev, Simon Cooley, Boyan Mednikarov, Guixin Fan and Natalia Nikolova**
Reliability Simulation of Two Component Warm-Standby System with Repair, Switching, and Back-Switching Failures under Three Aging Assumptions
Reprinted from: *Mathematics* **2021**, *9*, 2547, doi:10.3390/math9202547 . . . . . . . . . . . . . . . . 195

**Alessandro Tarsi and Simone Fiori**
Lie-Group Modeling and Numerical Simulation of a Helicopter
Reprinted from: *Mathematics* **2021**, *9*, 2682, doi:10.3390/math9212682 . . . . . . . . . . . . . . . . 235

**Soufiane Montassir, Hassane Moustabchir, Ahmed Elkhalfi, Maria Luminita Scutaru and Sorin Vlase**
Fracture Modelling of a Cracked Pressurized Cylindrical Structure by Using Extended Iso-Geometric Analysis (X-IGA)
Reprinted from: *Mathematics* **2021**, *9*, 2990, doi:10.3390/math9232990 . . . . . . . . . . . . . . . . 269

**Gabriel Gavriluț, Liliana Topliceanu, Manuela Gîrțu, Ana Maria Rotundu, Stefan Andrei Irimiciuc and Maricel Agop**
Assessment of Complex System Dynamics via Harmonic Mapping in a Multifractal Paradigm
Reprinted from: *Mathematics* **2021**, *9*, 3298, doi:10.3390/math9243298 . . . . . . . . . . . . . . . . 291

**Asif Khan, Jun-Sik Kim and Heung Soo Kim**
Damage Detection and Isolation from Limited Experimental Data Using Simple Simulations and Knowledge Transfer
Reprinted from: *Mathematics* **2022**, *10*, 80, doi:10.3390/math10010080 . . . . . . . . . . . . . . . . 307

 *mathematics*

*Editorial*

# Mathematical Modeling and Simulation in Mechanics and Dynamic Systems

Maria Luminita Scutaru [1,*] and Catalin-Iulian Pruncu [2,*]

1. Department of Mechanical Engineering, Faculty of Mechanical Engineering, Transilvania University of Brașov, 500036 Brașov, Romania
2. Departimento di Meccanica, Matematica e Management, Politecnico di Bari, 70125 Bari, Italy
* Correspondence: lscutaru@unitbv.ro (M.L.S.); catalin.pruncu@gmail.com (C.-I.P.)

**Citation:** Scutaru, M.L.; Pruncu, C.-I. Mathematical Modeling and Simulation in Mechanics and Dynamic Systems. *Mathematics* **2022**, *10*, 448. https://doi.org/10.3390/math10030448

Received: 19 January 2022
Accepted: 20 January 2022
Published: 30 January 2022

**Publisher's Note:** MDPI stays neutral with regard to jurisdictional claims in published maps and institutional affiliations.

**Copyright:** © 2022 by the authors. Licensee MDPI, Basel, Switzerland. This article is an open access article distributed under the terms and conditions of the Creative Commons Attribution (CC BY) license (https://creativecommons.org/licenses/by/4.0/).

## 1. Introduction

Although it has previously been considered difficult to make further contributions in the field of mechanics, the spectacular evolution of technology and numerical calculation techniques has caused this opinion to be reconsidered and to the development of more and more sophisticated models that describe, as accurately as possible, the phenomena that take place in dynamic systems. Therefore, researchers have come to study mechanical systems with complicated behavior, observing them in experiments and computer models [1–3]. The key requirement in these studies is that the system must involve a nonlinearity. The impetus in mechanics and dynamical systems has come from many sources: computer simulation, experimental science, mathematics, and modeling [4–6]. There are a wide range of influences. Computer experiments change the way in which we analyze these systems. Topics of interest include, but are not limited to, modeling mechanical systems, new methods in dynamic systems, the behavior simulation of mechanical systems, nonlinear systems, multibody systems with elastic elements, multiple degrees of freedom, mechanical systems, experimental modal analyses, and the mechanics of materials.

## 2. Statistics of the Special Issue

The statistics of papers submitted to this Special Issue for both published and rejected items are as follows: 23 total submissions, of which 16 were published (69.6%) [7–23] and 7 rejected (30.4%). The authors' geographical distribution is shown in Table 1, where it can be seen that the 67 authors are from 13 different countries. Note that it is usual for a paper to be written by more than one author, and for authors to collaborate with authors with different affiliations or multiple affiliations.

**Table 1.** Geographic distribution of authors by country.

| Country | Number of Authors |
|---|---|
| Romania | 13 |
| China | 9 |
| Iran | 1 |
| Italy | 2 |
| Pakistan | 1 |
| UK | 3 |
| Morocco | 5 |
| Korea | 4 |
| Bulgaria | 1 |
| Australia | 4 |
| Spain | 12 |
| Slovakia | 7 |
| Hungary | 3 |

## 3. Authors of the Special Issue

The authors of this Special Issue and their main affiliations are summarized in Table 2; it can be seen that there are three authors on average per manuscript.

**Table 2.** Affiliations and bibliometric indicators for authors.

| Author | Affiliation | References |
|---|---|---|
| Gabriel Gavrilut | Faculty of Phisics, Alexandru Ioan Cuza University, Bulevardul Carol I nr. 11, 700506 Iași, Romania | [8] |
| Liliana Topliceanu | Faculty of Engineering, Vasile Alecsandri University of Bacau, 600115 Bacau, Romania | [8] |
| Manuela Girtu | Department of Mathematics and Informatics, Vasile Alecsandri University of Bacau, 600115 Bacau, Romania | [8] |
| Ana Maria Rotundu | Faculty of Phisics, Alexandru Ioan Cuza University, Bulevardul Carol I nr. 11, 700506 Iași, Romania | [8] |
| Stefan Andrei Irimiciuc | National Institute for Laser, Plasma and Radiation Physics, 409 Atomistilor Street, 077125 Bucharest, Romania | [8] |
| Maricel Agop | Department of Physics, "Gh. Asachi" Technical University of Iasi, 700050 Iasi, Romania | [8] |
| Ashif Khan | Department of Mechanical, Robotics and Energy Engineering, Dongguk University-Seoul, 30 Pildong-ro 1 Gil, Jung-gu, Seoul 04620, Korea | [9] |
| Jun-Sik Kim | Department of Mechanical System Engineering, Kumoh National Institute of Technology, Gumi-si 39177, Korea; junsik.kim@kumoh.ac.kr | [9] |
| Heung Soo Kim | Department of Mechanical, Robotics and Energy Engineering, Dongguk University-Seoul, 30 Pildong-ro 1 Gil, Jung-gu, Seoul 04620, Korea | [9] |
| Alessandro Tarsi | School of Automation Engineering, Alma Mater Studiorum—Università di Bologna, Viale del Risorgimento 2, I-40136 Bologna, Italy; | [10] |
| Simone Fiori | Department of Information Engineering, Marches Polytechnic University, Brecce Bianche Rd., I-60131 Ancona, Italy | [10] |
| Kiril Tenekedjiev | Australian Maritime College, University of Tasmania, 1 Maritime Way, Launceston, TAS 7250, Australia | [11] |
| Simon Cooley | Australian Maritime College, University of Tasmania, 1 Maritime Way, Launceston, TAS 7250, Australia | [11] |
| Boyan Mednikarov | Nikola Vaptsarov Naval Academy—Varna, 73 V. Drumev Str., 9002 Varna, Bulgaria | [11] |
| Guixin Fan | Australian Maritime College, University of Tasmania, 1 Maritime Way, Launceston, TAS 7250, Australia | [11] |
| Natalia Nikolova | Australian Maritime College, University of Tasmania, 1 Maritime Way, Launceston, TAS 7250, Australia | [11] |
| Mohamed Derbeli | System Engineering and Automation Department, Faculty of Engineering of Vitoria-Gasteiz, Basque Country University (UPV/EHU), 01006 Vitoria-Gasteiz, Spain | [14] |
| Cristian Napole | System Engineering and Automation Department, Faculty of Engineering of Vitoria-Gasteiz, Basque Country University (UPV/EHU), 01006 Vitoria-Gasteiz, Spain | [14] |
| Oscar Barambones | System Engineering and Automation Department, Faculty of Engineering of Vitoria-Gasteiz, Basque Country University (UPV/EHU), 01006 Vitoria-Gasteiz, Spain | [14] |
| Soufiane Montassir | Faculty of Science and Technology, Sidi Mohamed Ben Abdellah University, B.P. 2202 Route d'Imouzzer, Fez 30000, Morocco | [23] |
| Hassane Moustabchir | Laboratory of Science Engineering and Applications (LISA) National School of Applied Sciences, Sidi Mohamed Ben Abdellah University, BP 72 Route d'Imouzzer, Fez 30000, Morocco | [23] |
| Maria Luminita Scutaru | Department of Mechanical Engineering, Transilvania University of Brassov, B-dul Eroilor 20, 500036 Brassov, Romania | [3,5,6,9,12] |

Table 2. Cont.

| Author | Affiliation | References |
|---|---|---|
| Sorin Vlase | Department of Mechanical Engineering, Transilvania University of Brassov, B-dul Eroilor 20, 500036 Brassov, Romania | [1–6,9,16] |
| Alexandra Saviuc | Faculty of Physics, Alexandru Ioan Cuza University of Iasi, 700506 Iasi, Romania | [13] |
| Tudor-Cristian Petrescu | Department of Structural Mechanics, Gheorghe Asachi Technical University of Iasi, 700050 Iasi, Romania; | [13] |
| Botond-Pál Gálf | Autolive Romania, Brasov, Bucegi, Str. 8, 500053 Brasov, Romania | [16] |
| Ioan Száva | Department of Mechanical Engineering, Transilvania University of Brasov, B-dul Eroilor 20, 500036 Brassov, Romania | [16] |
| Daniela Sova | Department of Mechanical Engineering, Transilvania University of Brasov, B-dul Eroilor 20, 500036 Brassov, Romania | [16] |
| Koldo Portal-Porras | Nuclear Engineering and Fluid Mechanics Department, University of the Basque Country, UPV/EHU, Nieves Cano 12, Vitoria-Gasteiz, 01006 Araba, Spain | [15] |
| Unai Fernandez-Gamiz | Nuclear Engineering and Fluid Mechanics Department, University of the Basque Country, UPV/EHU, Nieves Cano 12, Vitoria-Gasteiz, 01006 Araba, Spain | [15] |
| Ainara Ugarte-Anero | Nuclear Engineering and Fluid Mechanics Department, University of the Basque Country, UPV/EHU, Nieves Cano 12, Vitoria-Gasteiz, 01006 Araba, Spain | [15] |
| Ekaitz Zulueta | System Engineering and Automation Control Department, University of the Basque Country, UPV/EHU, Nieves Cano 12, Vitoria-Gasteiz, 01006 Araba, Spain | [15] |
| Asier Zulueta | System Engineering and Automation Control Department, University of the Basque Country, UPV/EHU, Nieves Cano 12, Vitoria-Gasteiz, 01006 Araba, Spain | [15] |
| Dongxu Li | College of Automobile and Traffic Engineering, Nanjing Forestry University, Nanjing 210037, China | [17] |
| Bing Xu | College of Automobile and Traffic Engineering, Nanjing Forestry University, Nanjing 210037, China | [17] |
| Zheshu Ma | College of Automobile and Traffic Engineering, Nanjing Forestry University, Nanjing 210037, China | [17] |
| Yanju Li | College of Automobile and Traffic Engineering, Nanjing Forestry University, Nanjing 210037, China | [17] |
| Pau Fonseca i Casas | Department of Statistics and Operations Research, Universitat Politècnica de Catalunya, 08034 Barcelona, Spain | [19] |
| Joan Garcia i Subirana | Department of Statistics and Operations Research, Universitat Politècnica de Catalunya, 08034 Barcelona, Spain | [19] |
| Víctor García i Carrasco | Department of Statistics and Operations Research, Universitat Politècnica de Catalunya, 08034 Barcelona, Spain | [19] |
| Xavier Pi i Palomés | Open University of Catalonia, Computer Science, Multimedia and Telecommunications Studies, 08860 Barcelona, Spain | [19] |
| Gregor Bánó | Department of Biophysics, Faculty of Science, P. J. Šafárik University, Jesenná 5, 041 54 Košice, Slovakia | [18] |
| Jana Kubacková | Institute of Experimental Physics SAS, Department of Biophysics, Watsonova 47, 040 01 Košice, Slovakia | [18] |
| Andrej Hovan | Department of Biophysics, Faculty of Science, P. J. Šafárik University, Jesenná 5, 041 54 Košice, Slovakia | [18] |
| Alena Strejˇcková | Department of Chemistry, Biochemistry and Biophysics, University of Veterinary Medicine and Pharmacy, Komenského 73, 041 81 Košice, Slovakia | [18] |
| Gergely T. Iványi | Faculty of Science and Informatics, University of Szeged, Dugonics Square 13, 6720 Szeged, Hungary | [18] |
| Gaszton Vizsnyiczai | Biological Research Centre, Institute of Biophysics, Eötvös Loránd Research Network (ELKH), Temesvári krt. 62, 6726 Szeged, Hungary | [18] |

Table 2. Cont.

| Author | Affiliation | References |
|---|---|---|
| Lóránd Kelemen | Biological Research Centre, Institute of Biophysics, Eötvös Loránd Research Network (ELKH), Temesvári krt. 62, 6726 Szeged, Hungary | [18] |
| Gabriel Žoldák | Center for Interdisciplinary Biosciences, Technology and Innovation Park, P. J. Šafárik University, Jesenná 5, 041 54 Košice, Slovakia | [18] |
| Zoltán Tomori | Institute of Experimental Physics SAS, Department of Biophysics, Watsonova 47, 040 01 Košice, Slovakia | [18] |
| Denis Horvath | Center for Interdisciplinary Biosciences, Technology and Innovation Park, P. J. Šafárik University, Jesenná 5, 041 54 Košice, Slovakia | [18] |
| Dongxu Li | College of Automobile and Traffic Engineering, Nanjing Forestry University, Nanjing 210037, China | [20] |
| Siwei Li | College of Automobile and Traffic Engineering, Nanjing Forestry University, Nanjing 210037, China | [20] |
| Zhanghao Lu | College of Automobile and Traffic Engineering, Nanjing Forestry University, Nanjing 210037, China | [20] |
| Yanju Li | College of Automobile and Traffic Engineering, Nanjing Forestry University, Na College of Automobile and Traffic Engineering, Nanjing Forestry University, Nanjing 210037, China njing 210037, China | [20] |
| Meng Zheng | College of Automobile and Traffic Engineering, Nanjing Forestry University, Na College of Automobile and Traffic Engineering, Nanjing Forestry University, Nanjing 210037, China njing 210037, China | [20] |
| Chukwuma Ogbonnaya | Department of Mechanical, Aerospace and Civil Engineering, The University of Manchester, Manchester M13 9PL, UK | [22] |
| Chamil Abeykoon | Faculty of Engineering and Technology, Alex Ekwueme Federal University, Ndufu Alike Ikwo, Abakaliki PMB 1010, Nigeria | [22] |
| Adel Nasser | Faculty of Engineering and Technology, Alex Ekwueme Federal University, Ndufu Alike Ikwo, Abakaliki PMB 1010, Nigeria | [22] |
| Ali Turan | Independent Researcher, Manchester M22 4ES, Lancashire, UK | [22] |
| Yasser Zare | Breast Cancer Research Center, Biomaterials and Tissue Engineering Research Group, Department of Interdisciplinary Technologies, Motamed Cancer Institute, ACECR, Tehran 15179-64311, Iran | [21] |
| Kyongyop Rhee | Department of Mechanical Engineering (BK21 Four), College of Engineering, Kyung Hee University, Yongin 449-701, Gyeonggi, Korea | [21] |
| Khadija Yakoubi | Faculty of Science and Technology, Sidi Mohamed Ben Abdellah University, Fez 30000, Morocco | [23] |
| Ahmed Elkhalf | Faculty of Science and Technology, Sidi Mohamed Ben Abdellah University, Fez 30000, Morocco | [23] |
| Catalin Iulian Pruncu | Department of Mechanical Engineering, Imperial College London, Exhibition Rd., London SW7 2AK, UK | [23] |
| Jamal Arbaoui | National School of Applied Sciences of Safi, University Cadi Ayad, Marrakesh 40000, Morocco; | [23] |
| Muhammad Umar Farooq | Department of Industrial and Manufacturing Engineering, University of Engineering and Technology, Lahore 54890, Pakistan | [23] |

## 4. Brief Overview of the Contributions to the Special Issue

This analysis of topics identifies or summarizes the research undertaken. This section classifies the manuscripts according to the topics covered in this Special Issue. There are three topics that are dominant, namely: the modeling of the multibody systems with symmetries, symmetry in applied mathematics, and analytical methods in symmetric multibody systems.

**Author Contributions:** Conceptualization, M.L.S. and C.-I.P.; methodology, M.L.S. and C.-I.P.; software, M.L.S. and C.-I.P.; validation, M.L.S. and C.-I.P.; formal analysis, M.L.S. and C.-I.P.; investigation, M.L.S. and C.-I.P.; resources, M.L.S. and C.-I.P.; data curation, M.L.S. and C.-I.P.; writing—original draft preparation, M.L.S. and C.-I.P.; writing—review and editing, M.L.S. and C.-I.P.; visualization, M.L.S. and C.-I.P.; supervision, M.L.S. and C.-I.P.; project administration, M.L.S. and C.-I.P. All authors have read and agreed to the published version of the manuscript.

**Funding:** This research received no external funding.

**Institutional Review Board Statement:** Not applicable.

**Informed Consent Statement:** Not applicable.

**Data Availability Statement:** Not applicable.

**Conflicts of Interest:** The authors declare no conflict of interest.

## References

1. Vlase, S.; Năstac, C.; Marin, M.; Mihălcică, M. A Method for the Study of the Vibration of Mechanical Bars Systems with Symmetries. *Acta Technica Napocensis. Ser.-Appl. Math. Mech. Eng.* **2017**, *60*, 539–544.
2. Vlase, S. A Method of Eliminating Lagrangian Multipliers from the Equation of Motion of Interconnected Mechanical Systems. *J. Appl. Mech.* **1987**, *54*, 235–237. [CrossRef]
3. Scutaru, M.L.; Vlase, S.; Marin, M.; Modrea, A. New analytical method based on dynamic response of planar mechanical elastic systems. *Bound. Value Probl.* **2020**, *2020*, 104. [CrossRef]
4. Vlase, S.; Teodorescu, P.P.; Itu, C.; Scutaru, M.L. Elasto-Dynamics of a Solid with a General "Rigid" Motion using FEM Model. Part I. Theoretical Approach. *Rom. J. Phys.* **2013**, *58*, 872–881.
5. Vlase, S.; Marin, M.; Ochsner, A. Considerations of the transverse vibration of a mechanical system with two identical bars. *Proc. Inst. Mech. Eng. Part L J. Mater. Des. Appl.* **2019**, *233*, 1318–1323. [CrossRef]
6. Vlase, S.; Marin, M.; Scutaru, M.L.; Munteanu, R. Coupled transverse and torsional vibrations in a mechanical system with two identical beams. *AIP Adv.* **2017**, *7*, 065301. [CrossRef]
7. Khan, A.; Kim, J.-S.; Kim, H.S. Damage Detection and Isolation from Limited Experimental Data Using Simple Simulations and Knowledge Transfer. *Mathematics* **2021**, *10*, 80. [CrossRef]
8. Gavriluț, G.; Topliceanu, L.; Gîrțu, M.; Rotundu, A.M.; Irimiciuc, S.A.; Agop, M. Assessment of Complex System Dynamics via Harmonic Mapping in a Multifractal Paradigm. *Mathematics* **2021**, *9*, 3298. [CrossRef]
9. Montassir, S.; Moustabchir, H.; Elkhalfi, A.; Scutaru, M.L.; Vlase, S. Fracture Modelling of a Cracked Pressurized Cylindrical Structure by Using Extended Iso-Geometric Analysis (X-IGA). *Mathematics* **2021**, *9*, 2990. [CrossRef]
10. Tarsi, A.; Fiori, S. Lie-Group Modeling and Numerical Simulation of a Helicopter. *Mathematics* **2021**, *9*, 2682. [CrossRef]
11. Tenekedjiev, K.; Cooley, S.; Mednikarov, B.; Fan, G.; Nikolova, N. Reliability Simulation of Two Component Warm-Standby System with Repair, Switching, and Back-Switching Failures under Three Aging Assumptions. *Mathematics* **2021**, *9*, 2547. [CrossRef]
12. Teodorescu Draghicescu, H.; Scutaru, M.L.; Rosu, D.; Calin, M.R.; Grigore, P. New Advanced Sandwich Composite with twill weave carbon and EPS. *J. Optoelectron. Adv. Mater.* **2013**, *15*, 199–203.
13. Saviuc, A.; Gîrțu, M.; Topliceanu, L.; Petrescu, T.-C.; Agop, M. "Holographic Implementations" in the Complex Fluid Dynamics through a Fractal Paradigm. *Mathematics* **2021**, *9*, 2273. [CrossRef]
14. Derbeli, M.; Napole, C.; Barambones, O. Machine Learning Approach for Modeling and Control of a Commercial Heliocentris FC50 PEM Fuel Cell System. *Mathematics* **2021**, *9*, 2068. [CrossRef]
15. Portal-Porras, K.; Fernandez-Gamiz, U.; Ugarte-Anero, A.; Zulueta, E.; Zulueta, A. Alternative Artificial Neural Network Structures for Turbulent Flow Velocity Field Prediction. *Mathematics* **2021**, *9*, 1939. [CrossRef]
16. Gálfi, B.-P.; Száva, I.; Șova, D.; Vlase, S. Thermal Scaling of Transient Heat Transfer in a Round Cladded Rod with Modern Dimensional Analysis. *Mathematics* **2021**, *9*, 1875. [CrossRef]
17. Xu, B.; Li, D.; Ma, Z.; Zheng, M.; Li, Y. Thermodynamic Optimization of a High Temperature Proton Exchange Membrane Fuel Cell for Fuel Cell Vehicle Applications. *Mathematics* **2021**, *9*, 1792. [CrossRef]
18. Bánó, G.; Kubacková, J.; Hovan, A.; Strejčková, A.; Iványi, G.; Vizsnyiczai, G.; Kelemen, L.; Žoldák, G.; Tomori, Z.; Horvath, D. Power Spectral Density Analysis of Nanowire-Anchored Fluctuating Microbead Reveals a Double Lorentzian Distribution. *Mathematics* **2021**, *9*, 1748. [CrossRef]
19. Fonseca i Casas, P.; Garcia i Subirana, J.; Garcia i Carrasco, V.; Pi i Palomés, X. SARS-CoV-2 Spread Forecast Dynamic Model Validation through Digital Twin Approach, Catalonia Case Study. *Mathematics* **2021**, *9*, 1660. [CrossRef]
20. Li, D.; Li, S.; Ma, Z.; Xu, B.; Lu, Z.; Li, Y.; Zheng, M. Ecological Performance Optimization of a High Temperature Proton Exchange Membrane Fuel Cell. *Mathematics* **2021**, *9*, 1332. [CrossRef]
21. Zare, Y.; Rhee, K. Advanced Models for Modulus and Strength of Carbon-Nanotube-Filled Polymer Systems Assuming the Networks of Carbon Nanotubes and Interphase Section. *Mathematics* **2021**, *9*, 990. [CrossRef]

22. Ogbonnaya, C.; Abeykoon, C.; Nasser, A.; Turan, A. A Computational Approach to Solve a System of Transcendental Equations with Multi-Functions and Multi-Variables. *Mathematics* **2021**, *9*, 920. [CrossRef]
23. Yakoubi, K.; Montassir, S.; Moustabchir, H.; Elkhalfi, A.; Pruncu, C.; Arbaoui, J.; Farooq, M. An Extended Finite Element Method (XFEM) Study on the Elastic T-Stress Evaluations for a Notch in a Pipe Steel Exposed to Internal Pressure. *Mathematics* **2021**, *9*, 507. [CrossRef]

Article

# An Extended Finite Element Method (XFEM) Study on the Elastic T-Stress Evaluations for a Notch in a Pipe Steel Exposed to Internal Pressure

Khadija Yakoubi [1], Soufiane Montassir [1], Hassane Moustabchir [2], Ahmed Elkhalfi [1], Catalin Iulian Pruncu [3,4,*], Jamal Arbaoui [5] and Muhammad Umar Farooq [6,7]

[1] Faculty of Science and Technology, Sidi Mohamed Ben Abdellah University, Fez 30000, Morocco; khadija.yakoubi95@gmail.com (K.Y.); soufianemontassir@gmail.com (S.M.); aelkhalfi@gmail.com (A.E.)
[2] Laboratory of Systems Engineering and Applications (LISA), National School of Applied Sciences of Fez, Sidi Mohamed Ben Abdellah University, Fez 30000, Morocco; hmoustabchir@hotmail.com
[3] Department of Mechanical Engineering, Imperial College London, Exhibition Rd., London SW7 2AK, UK
[4] Design, Manufacturing & Engineering Management, University of Strathclyde, Glasgow G1 1XJ, UK
[5] National School of Applied Sciences of Safi, University Cadi Ayad, Marrakesh 40000, Morocco; jamal57010@yahoo.fr
[6] Department of Industrial and Manufacturing Engineering, University of Engineering and Technology, Lahore 54890, Pakistan; umarmuf0@gmail.com
[7] Department of Industrial and Systems Engineering, Advanced Institute of Science and Technology (KAIST), Daejeon 34141, Korea
* Correspondence: c.pruncu@imperial.ac.uk or Catalin.pruncu@strath.ac.uk

Citation: Yakoubi, K.; Montassir, S.; Moustabchir, H.; Elkhalfi, A.; Pruncu, C.I.; Arbaoui, J.; Farooq, M.U. An Extended Finite Element Method (XFEM) Study on the Elastic T-Stress Evaluations for a Notch in a Pipe Steel Exposed to Internal Pressure. *Mathematics* **2021**, *9*, 507. https://doi.org/10.3390/math9050507

Academic Editor: Krzysztof Kamil Żur

Received: 23 January 2021
Accepted: 22 February 2021
Published: 2 March 2021

**Publisher's Note:** MDPI stays neutral with regard to jurisdictional claims in published maps and institutional affiliations.

**Copyright:** © 2021 by the authors. Licensee MDPI, Basel, Switzerland. This article is an open access article distributed under the terms and conditions of the Creative Commons Attribution (CC BY) license (https://creativecommons.org/licenses/by/4.0/).

**Abstract:** The work investigates the importance of the K-T approach in the modelling of pressure cracked structures. T-stress is the constant in the second term of the Williams expression; it is often negligible, but recent literature has shown that there are cases where T-stress plays the role of opening the crack, also T-stress improves elastic modeling at the point of crack. In this research study, the most important effects of the T-stress are collected and analyzed. A numerical analysis was carried out by the extended finite element method (X-FEM) to analyze T-stress in an arc with external notch under internal pressure. The different stress method (SDM) is employed to calculate T-stress. Moreover, the influence of the geometry of the notch on the biaxiality is also examined. The biaxiality gave us a view on the initiation of the crack. The results are extended with a comparison to previous literature to validate the promising investigations.

**Keywords:** T-stress; X-FEM; notch; pipe; stress difference method (SDM)

## 1. Introduction

In the field of fracture mechanics especially linear elastics, the vicinity of notch tip is often symbolized by singular stress entities. Their resistance is measured through stress intensity factor (SIF). The SIF is mainly depended on the distance r from the tip of notch. In this field, parameter T is introduced to enrich the parameter K (SIF) to make the model better in the elastic stress field; this is the K-T approach [1].

In fracture mechanics body of knowledge, it is established that the same Stress Intensity Factor (SIF) is required for two cracks to propagate in the same way. Experiment [2] have shown that two plates with the same SIF and different crack length $a_1 > a_2$ show different the propagation of the cracks. The results have shown that the propagation speed of $a_2$ is higher than that of $a_1$. The study concluded that the first term of asymptotic development is not sufficient to predict crack behavior. Therefore, it is necessary to increase the order. The first term asymptotic development is the SIF that determines the initiation and propagation of the crack. In addition, the second term is constant and controls the stability of the propagation direction. It is the transverse component symbolized by T.

Several studies have shown the significance of the T-Stress, and its influence on different parameters of mechanics. Jayadevan [3] highlighted that plastic zone is manipulated by the variation of T-stress. It means, the plastic area escalates with the increase in absolute value of T-stress and changes its form. Sobotka et al. [4] demonstrated the alteration in the plastic wake with many T-stresses which depends on plastic wakes height (HPW).

T-Stress has been an essential contributing factor in the stability of the direction of the crack propagation. For instance, T negative gives a stable direction, and for T positive it is unstable [5]. Fayed et al. [6] explored the impact of T-stress on propagating crack direction by Maximum Tangential Stress (MTS). The principle of MTS is that the crack propagates in the trend of maximum tangential stress. They obtained that the directions of the overall crack no coincide with the initial direction of the crack. Many studies have concluded the behavior by the fact that the tangential stress is affected by the T-Stress. MTS becomes generalized maximum tangential stress (GMTS), which considers the constraint T in the expression of stress. In the same context Shahani [1] studied the result on the initiating angle of propagating crack of the stress T. The study has shown a negative T value declines the angle of crack initiation, and a positive T value enhances it. Nejati et al. [7] gauged the relationship between T-stress and material properties. Chen et al. [8] has shown that Graded Poisson's ratio affects the T-stress. Additionally, Toshio et al. [9] concluded that the Poison's ratio influences the T-stress on a three-dimensional edge-cracked plate.

Other important research that has shown the influences of the T-stress includes: Zhang et al. [10], which used numerical manifold method (NMM) is employed to calculate the T-stress for two-dimensional functionally graded material (FGM) having numerous cracks. Noritaka et al. [11] has resulted the T-stress might open micro-branches in the mist region. For the bending and tension load, Hancock [12] determined that T decreases with escalating crack length. Matvienko [13] explored the influence of T-stress in problems of the elastic and the elastic-plastic fracture mechanics.

Conventionally, the T-stress is often calculated at the crack tip, which is certainly not the case in this research. The research evaluates the T-stress at the tip of notch through extended finite element method (X-FEM).

The finite element method FE method is limited by the simple cases, as well as the presence of a singularity greatly degrades the convergence of the FEM. Belytschko and Blacken in 1999 added discontinuously enriching function in finite element approximation by respecting boundary conditions. Later, Moës et al. [14] developed the technique and called it as extended finite element method which is abbreviated as X-FEM. The efficiency of the X-FEM is well endorsed in the literature. To simulate the propagation of cracks in porous media, Wang et al. [15] integrated embedded discrete fracture method (EDFM) with X-FEM simulating fracture associated fluid and solid mechanics. Shu et al. [16] investigated the fatigue growth of 3-D multiple cracks by X-FEM. The implementation of X-FEM for composites resulted successfully [17–19]. X-FEM is also used for the calculation and analysis of failure mechanics parameters. Fakkoussi et al. [20] calculated stress intensity factor for mode one by X-FEM. Llavori et al. [21] studied the problems of contact fatigue by X-FEM.

The research study presents the use of X-FEM to calculate T-stress in the notch tip for an arc of the pipe of steel P264GH. Further, it demonstrates the benefits of using the X-FEM approach to compute K-T at the notch point in an arc under pression and revealing the possibility of detecting the crack initiation.

The remaining article is organized as, Section 2 talks about the K-T approach, X-FEM, and the geometry used. Section 3 deals with the numerical result obtained, compared against FEM conclusions, along with the discussion. Finally, the conclusion is in the other section.

## 2. Materials and Methods

### 2.1. K-T Approach

Stress Intensity Factor has been an essential parameter in the field of linear fracture mechanics, widely used for crack evaluation. SIF measures the strength of the singularity, and integrates various parameters such as load, geometry, and shape of crack. M. Hadj Meliani [22] suggested that a thin structure such as a thin pipe with a longitudinal crack, it is very difficult to characterize the stress field by a single parameter. For linear elasticity, the enrichment of the SIF with the T-stress is required to model the notch tip. In literature, studies show that T explains how geometry influences tenacity ($K_{IC}$) [22]. T-stress helps in approximating the level of stress at a crack or tip of the notch. The possibility of constructing the K(T) curve numerically has given the opportunity to predict the loading of a crack initiation [23,24]. Including T-stress in calculations, it improves the prediction of propagating crack under the control of mixed loading. The K-T was additionally calculated for through-wall-cracked pipes under various pressure conditions by three dimension-3D FE [25]. The importance of the T-Stress is established in many works.

- T-stress enhances the possibility of crack opening stresses in the context of small crack [26];

$$\sigma_{xx}(r,\theta) = \frac{K_1}{\sqrt{2\pi r}} f_{xx}(\theta) + T \tag{1}$$

Taking $\sigma_{xx}(r,\theta) = \sigma_{cr}$, for the crack propagation, the first term tends to $\frac{K_1}{\sqrt{2\pi r}} f_{xx}(\theta)$ to a because $K_1 = S_{yy}\sqrt{\pi a}$, and if a tends to zero; the first term becomes negligible intheface to T at the crack point.

$$\lim_{a \to 0} \sigma_{xx}(r,\theta) = \sigma_{cr} = T \tag{2}$$

In this case, the T-stress cannot be ignored, T play the role of crack opening, and therefore the importance of T varies with the size of the cracks.

- T-stress influences the plastic zone. Jayadevan [3] highlighted that plastic zone is affected by the variation of T-stress. The plastic area escalates with the upsurge in the absolute value of T-stress and changes its form. Sobotka et al. [4] demonstrated the deviation in the plastic wake with various T-stress which depends on plastic wakes height (hpw).
- Propagation's direction: Fayed et al. [6] analyzed the impact of T-stress on the propagation's direction by Maximum Tangential Stress (MTS). The principle of MTS is in the crack propagation direction which is in line with that of maximum tangential stress. MTS only considers the term singularity. Therefore, the direction of crack does not coincide with the initial directions of the crack. This behavior shows that the tangential stress is affected by the T-Stress. MTS becomes generalized maximum tangential stress (GMTS), which considers the T-stress in the expression of stress.
- T-stress has an impact on crack initiation angle. Shahani [1] analyzed the consequence of the T-stress on the angle of initiation of crack propagation. The study has shown that a negative T value declines the angle of crack initiation, and a positive T value enhances it.
- T-Stress is of prime significance when ensuring the stability of the direction of crack propagation, such as T negative gives a stable direction, and for T positive it is unstable [5].

T-Stress could be computed through numerous techniques. Weight Function Method has shown its efficiency in several problems cracking-related such as edge-cracked rectangular plate, circular disk [27]. Kfouri [28] developed a technique for evaluating the T-stress. The method uses the attributes of the path-independent J-integral and is called the Esheby–Integral method.

The stress different method (SDM) has been proposed by Yang [29]. The idea of this method is the errors of the numerical values of $\sigma_{11}$ and $\sigma_{22}$ near a crack point progress with r in the same way, and the variation must effectively eliminate errors.

$$T = \sigma_{yy} - \sigma_{xx} \qquad (3)$$

Biaxiality is a parameter that relates the SIF and T-stress:

$$\beta = \frac{T\sqrt{\pi a}}{K} \qquad (4)$$

### 2.2. Extended Finite Elements

One of the uses of the FE method is the study of crack propagation but is limited for simple cases. If the mesh size does not conform to the crack, the FE method does not treat the propagation, and the presence of a singularity degrades the convergence of the FE method.

The solution is to add enrichment function to the FE approximation (see Figure 1); this is the extended finite element method.

$$U = \sum_{1}^{N} N_i u_i + \sum_{i}^{N_{saut}} N_i H(x) a_i + \sum_{i}^{N_{sing}} \sum_{j} N(x)_i F(x)_j b_i^j \qquad (5)$$

where:

- $H(x)$: The Heaviside enrichment function, $H(x) = \begin{cases} -1, & x > 0 \\ +1, & x < 0 \end{cases}$
- $F(x)$: Enrichment functions near the crack front.
- N: Interpolation function of finite element. $N_{saut}$: Number of nodes enriched with Heaviside function. $N_{sing}$: Number of knots enriched near the crack front.

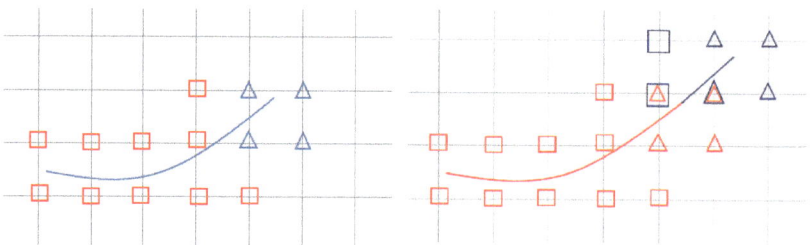

**Figure 1.** Step of enrichment methods.

The X-FEM requires operation to confirm the enrichment status of a knot according to its position in reference to the crack and to evaluate the functions H(x) and F(x). The position (r, θ) in relation to the notch point is calculated herein to know if x is above or below the crack. These operations are carried out using the level sets method. The technique for describing crack is known as level set method. In the X-FEM, it determines the location of the crack and the crack point, and the position to apply discontinuous enrichment and enrichment to the crack point Figure 1. Most importantly, the level set provides an instant result that helps track crack propagation, i.e., as the crack propagates, enrichment at the crack front becomes discontinuous enrichment, and nodes (not enriched) become enriched. There are two-level functions, and the first describes the crack surface (φ), second gives the crack front (ψ) [30].

The X-FEM method was applied in several studies. Yousheng Xie et al. [31] have implemented the X-FEM method in the study of propagating crack in mixed mode, and evaluated the crack initiation angle. Reference [32] has shown the performance of the method. The study applied X-FEM to calculate SIF for 3D crack propagation problems for a Compact Tension C-T specimen. X-FEM was also implemented in the analysis of bi-material interfaces, calculating service life and fatigue resistance [33]. An integration between the X-FEM and embedded discrete fracture method (EDFM) is established for simulation of the process of fluid fracture propagation in porous media [15]. Savenkov et al. [34] employed the X-FEM to represent the central surface of the crack. The application of X-FEM for composite models is also carried out supporting current investigation [35,36].

X-FEM was used to predict components failure from a different form of notch [37]. Patria et al. [38] adopted X-FEM to study the mechanical attributes and fracture behavior of (Reinforced Polymeric Composites) RPC materials with single edge notch three-point bending.

## 2.3. Geometry

An arc of pipe containing a notch under pression was numerically analyzed using X-FEM in ABAQUS software. The material used is a steel P264GH. The arc characterized by an inner radius Ri = 219.55 mm and thickness t = 6.1 mm. More details on geometry, the shape of the notch, and boundary requirements used are illustrated in Figure 2 and Table 1.

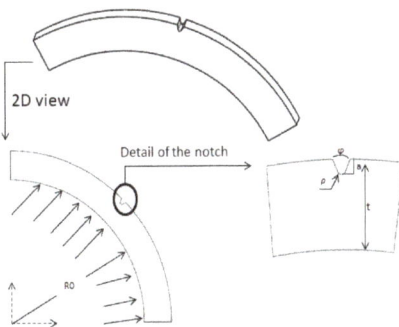

**Figure 2.** Details of the geometry and notch study, with boundary conditions.

**Table 1.** Geometry properties and load.

| Ri [mm] | P [MPa] | φ [deg] | a [mm] | t [mm] | ρ [mm] |
|---|---|---|---|---|---|
| 213.45 | 15 | 45 | 1.22–4.88 | 6.1 | 0.15 |

The mechanical attributes of the material used are presented in Table 2, and the chemical composition of the material are included in Table 3.

**Table 2.** Mechanical characteristics of P264GH.

| Young's Modulus | Poisson's Ratio | Yield Stress | Elongation to Fracture |
|---|---|---|---|
| 207,000 MPa | 0.3 | 430 MPa | 35% |

**Table 3.** Chemical composition of P264GH.

| Material | C | Mn | S | Si | P | Al |
|---|---|---|---|---|---|---|
| Tested | 0.135 | 0.665 | 0.002 | 0.195 | 0.027 | 0.027 |
| Steel P264GH (Standard EN10028.2–92) | 0.18 | 1 | 0.015 | 0.4 | 0.025 | 0.02 |

## 3. Results

This section presented the results of SIF and T-stress given by X-FEM via an user element UEL subroutine, the calculation was executed by ABAQUS software, we used quadratic element C3D20 in the mesh Figure 3, with a size of 0.5 mm. The number of elements is 246117. The number of nodes is 363256.

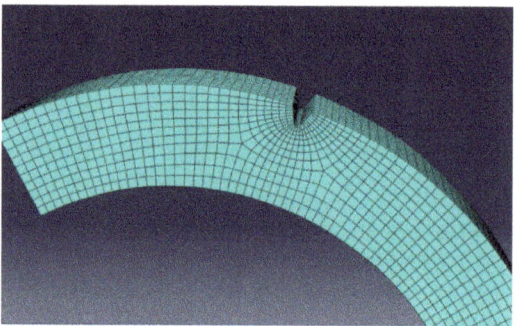

**Figure 3.** The mesh of 3D arc, C3D10 with 0.5 mm of size.

The T-stress is calculated through stress different method, and normalization is done for effect of the T-stress relative to the stress intensity factor by a parameter dimensionless termed biaxiality.

$$\beta = \frac{T\sqrt{\pi a}}{K}$$

K is the value of stress intensity factor, and a is the notch length.

Figure 4 illustrates the difference of the SIF in mode 1 as a function of r, for a/t = 0.2, by extended finite element.

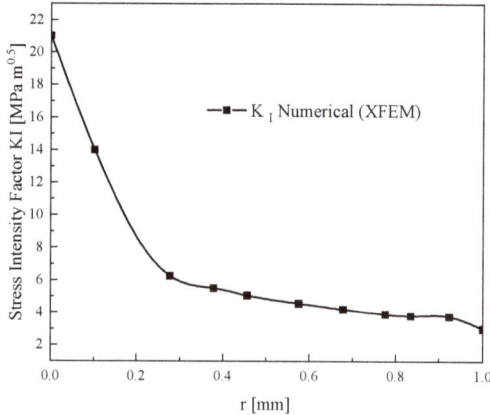

**Figure 4.** Distribution stress intensity factor SIF at the notch tip–SIF and r for a/t = 0.2.

The elastic SIF distribution at the notch tip decreases with distance from the tip of notch, for a/t = 0.2 the maximum value of SIF is 21 MPa$\sqrt{m}$ at notch tip, i.e., $r = 0$ (see Figure 4). Near the notch tip SIF decreases rapidly to 6 MPa$\sqrt{m}$ at r = 0.3 mm, then its variation becomes slower.

Figure 5 shows the variation of stress $\sigma_{xx}, \sigma_{yy}$ and T. T-stress increases with increasing r up to r = 0.43, and after that it starts to stabilize. The numerical calculation of the stress $\sigma_{xx}, \sigma_{yy}$ by X-FEM is executed by ABAQUS software. Figure 6 gives the Von Mises stress obtained by the Abaqus software.

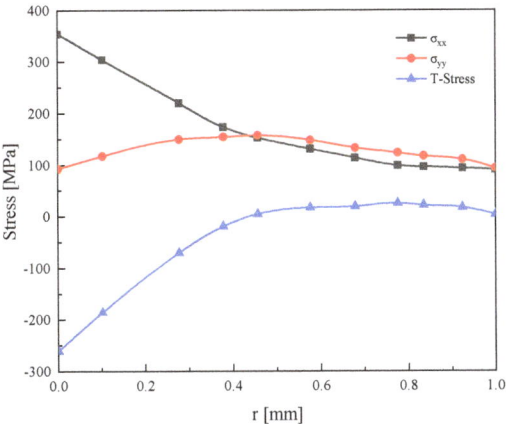

**Figure 5.** T-stress by stress different method.

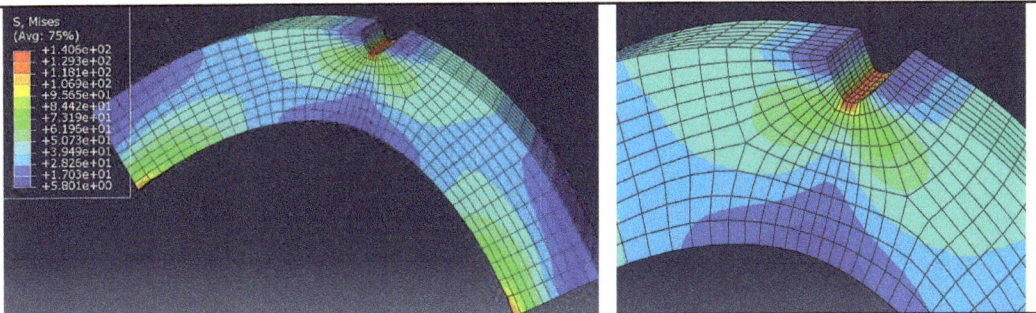

**Figure 6.** Distributions of Von Mises stress near the notch.

The result obtained of biaxiality are compared with H. Moustabchir [39], who calculated the biaxiality through the finite element method, for the same geometry which is used in this study. Figure 7 gives the variation of biaxiality as a function of a/t, by X-FEM and FEM.

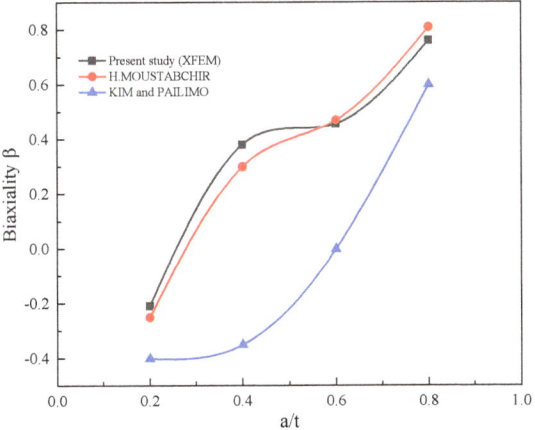

**Figure 7.** Variation of the biaxiality with a/t at the notch.

The biaxiality levels up with the increase a/t, The results are identical with the H. Moustabchir [39] recommendations. Kim and Paulino [40] has also got the same variation. By X-FEM the biaxiality varies from $-0.21$ to $0.76$ for $a/t = 0.2$ and $0.8$, respectively, and it goes from a positive to a negative value at $a/t = 0.3$. The difference between the results given by X-FEM and FEM is 0.013.

## 4. Discussion

SIF measures the strength of the singularity, which explains the rapid variation of the SIF obtained for $r = [0, 0.3]$. The more approach is made towards the notch tip, the more the stress concentration increases, and therefore, SIF increases. Moustabchir et al. [39] used the volumetric approach to calculate SIF, in the same condition that this research studied. Moustabchir et al. obtained for mode I at tip notch $K_1 = 21.6$ MPa$\sqrt{m}$, which differs from our result by 0.5. The T-stress can be analyzed by the SDM. The biaxiality increases with the rise in a/t, and the notch depth affects the value of T-stress. The same variation was obtained in other investigations on various materials. Bouchard et al. [41] have shown that T increases with increasing depth for a mono silicon. In [42], Sherry et al. obtained an increase in T in absolute value with the variation of the crack size over the width of a plate. In addition, Ayatollahi et al. [43] obtained for mode I, an increase in T-stress as a function of the depth of the crack for a single edge notched.

If
$$\beta = 1$$
and
$$\beta = \frac{T\sqrt{\pi a}}{K}$$
So
$$K = T\sqrt{\pi a} \text{ i.e., } T = \sigma$$

Which is not the case for this study $\beta_{max} = 0.78$, so $T \neq \sigma$, which implies that T has no influence on the notch in our case and our condition.

Many studies have resulted that the influence of the T-stress is remarkable and significant when T is negative [1,3,6], however, in this study for $r < 0.43$, negative T-stress causes an increase in the plastic zone [27]. This will cause a crack to initiate. In the presence of a crack, negative T can change the direction of propagating crack and decreases the crack growth initiation angle [31].

The K-T approach is an integration between the SIF and the T-stress to improve modelling the elastic stress at the point of the crack. The importance of T-stress has been highlighted, and it takes the place of short crack opening stress. Besides, the importance of the cooperation of SIF and T-stress, such as the K(T) curve was elaborated which gives a prediction of the stress of crack initiation [44]. Neggaz et al. [44] studied the influences of the reinforcements in the structure of composites, with the aim of reducing constraints at notch-tip. Moreover, authors evaluated the effective stress intensity factors in the regard of propagating crack in thin and thick panels. Therefore, an Extended Finite Element Method (XFEM) is novel and improved technique on the elastic T-stress evaluations for a notch in a pipe steel exposed to internal pressure.

## 5. Conclusions

Three-dimensional Extended Finite Element (X-FEM) analysis is applied to evaluate the stress intensity factor and the T-stress for an arc of pipe with external notch under internal pressure. The results are presented below:

- To study the influence of geometry and notch size on the T-stress, authors have approached the biaxiality as a function of a/t. The evaluation endorsed that biaxiality $\beta$ increases with the increasing a/t which is in accord with president results.
- The integration of biaxiality allowed us to determine the state of the crack initiation, and we can say that the pipe is safe in the used conditions.

- The SIF alone does not characterize the behavior of notches. T-stress is obtained by Stress Difference Method (SDM) along with the notch in mode I. SDM is an efficient and simple method to calculate the fracture parameters.
- With ABAQUS-based investigations, the numerical results achieved by X-FEM are in good agreement with the Moustabchir result [39]. The implementation of X-FEM in the presence of a notch corrected the problems of the standard finite element method. The advantage of the X-FEM is that the mesh is independent of the notch.

For more precision, the future objective is to calculate the parameters of fracture mechanics by iso-geometrical analysis.

**Author Contributions:** Data curation, K.Y. and J.A., formal analysis, S.M. and A.E., investigation, H.M. and C.I.P., writing and review of original draft, M.U.F., H.M., and C.I.P. All authors have read and agreed to the published version of the manuscript.

**Funding:** This research received no external funding.

**Institutional Review Board Statement:** Not applicable.

**Informed Consent Statement:** Not applicable.

**Data Availability Statement:** The data that support the findings of this study are available on request from the corresponding author.

**Acknowledgments:** Authors are thankful to Aqib Mashood Khan for his constructive feedback.

**Conflicts of Interest:** The authors declare no conflict of interest.

## References

1. Shahani, A.; Tabatabaei, S. Effect of T-stress on the fracture of a four point bend specimen. *Mater. Des.* **2009**, *30*, 2630–2635. [CrossRef]
2. Hamam, R.; Pommier, S.; Bumbieler, F. Mode I fatigue crack growth under biaxial loading. *Int. J. Fatigue* **2005**, *27*, 1342–1346. [CrossRef]
3. Jayadevan, K.; Narasimhan, R.; Ramamurthy, T.; Dattaguru, B. Effect of T-stress and loading rate on crack initiation in rate sensitive plastic materials. *Int. J. Solids Struct.* **2002**, *39*, 1757–1775. [CrossRef]
4. Sobotka, J.; Dodds, R. T-stress effects on steady crack growth in a thin, ductile plate under small-scale yielding conditions: Three-dimensional modeling. *Eng. Fract. Mech.* **2011**, *78*, 1182–1200. [CrossRef]
5. Cotterell, B.; Rice, J. Slightly curved or kinked cracks. *Int. J. Fract.* **1980**, *16*, 155–169. [CrossRef]
6. Fayed, A.S. Numerical Analysis of Crack Initiation Direction in Quasi-brittle Materials: Effect of T-Stress. *Arab. J. Sci. Eng.* **2019**, *44*, 7667–7676. [CrossRef]
7. Nejati, M.; Ghouli, S.; Ayatollahi, M.R. Crack tip asymptotic fields in anisotropic planes: Importance of higher order terms. *Appl. Math. Model.* **2021**, *91*, 837–862. [CrossRef]
8. Chen, X.; Yue, Z. Mode-I pressurized axisymmetric penny-shaped crack in graded interfacial zone with variable modulus and Poisson's ratio. *Eng. Fract. Mech.* **2020**, *235*, 107164. [CrossRef]
9. Toshio, N.; Parks, D.M. Determination of elastic T-stress along three-dimensional crack fronts using an interaction integral. *Int. J. Solids Struct.* **1992**, *29*, 1597–1611. [CrossRef]
10. Zhang, H.; Liu, S.; Han, S.; Fan, L. T-stress evaluation for multiple cracks in FGMs by the numerical manifold method and the interaction integral. *Theor. Appl. Fract. Mech.* **2020**, *105*, 102436. [CrossRef]
11. Nakamura, N.; Kawabata, T.; Takashima, Y.; Yanagimoto, F. Effect of the stress field on crack branching in brittle material. *Theor. Appl. Fract. Mech.* **2020**, *108*, 102583. [CrossRef]
12. Hancock, J.W.; Renter, W.G.; Parks, D.M. Constraint and Toughness Parameterized by T. In *Constraint Effects in Fracture*; Hackett, E., Schwalbe, K., Dodds, R., Eds.; ASTM International: West Concord, PA, USA, 1993; pp. 21–40.
13. Matvienko, Y. The effect of crack-tip constraint in some problems of fracture mechanics. *Eng. Fail. Anal.* **2020**, *110*, 104413. [CrossRef]
14. Moes, J.N.; Dolbow, T.B. *A Finite Element Method for Crack Growth without Remeshing*; John Wiley & Sons, Ltd.: Hoboken, NJ, USA, 1999. [CrossRef]
15. Wang, C.; Huang, Z.; Wu, Y.-S. Coupled numerical approach combining X-FEM and the embedded discrete fracture method for the fluid-driven fracture propagation process in porous media. *Int. J. Rock Mech. Min. Sci.* **2020**, *130*, 104315. [CrossRef]
16. Shu, Y.; Li, Y.; Duan, M.; Yang, F. An X-FEM approach for simulation of 3-D multiple fatigue cracks and application to double surface crack problems. *Int. J. Mech. Sci.* **2017**, *130*, 331–349. [CrossRef]
17. Liang, Y.-J.; McQuien, J.S.; Iarve, E.V. Implementation of the regularized extended finite element method in Abaqus framework for fracture modeling in laminated composites. *Eng. Fract. Mech.* **2020**, *230*, 106989. [CrossRef]

18. Akhondzadeh, S.; Khoei, A.; Broumand, P. An efficient enrichment strategy for modeling stress singularities in isotropic composite materials with X-FEM technique. *Eng. Fract. Mech.* **2017**, *169*, 201–225. [CrossRef]
19. Nagashima, T.; Suemasu, H. X-FEM analyses of a thin-walled composite shell structure with a delamination. *Comput. Struct.* **2010**, *88*, 549–557. [CrossRef]
20. El Fakkoussi, S.; Moustabchir, H.; Elkhalfi, A.; Pruncu, C.I. Computation of the stress intensity factor KI for external longitudinal semi-elliptic cracks in the pipelines by FEM and XFEM methods. *Int. J. Interact. Des. Manuf.* **2018**, *13*, 545–555. [CrossRef]
21. Llavori, I. A coupled crack initiation and propagation numerical procedure for combined fretting wear and fretting fa-tigue lifetime assessment. *Theor. Appl. Fract. Mech.* **2019**, *101*, 294–305. [CrossRef]
22. Meliani, H.M. *Mécanique de la Rupture d'Entaille par l'Approche Globale: Estimation des Contraintes de Confinements dans des Structures Portant des Entailles*; Editions Universitaires Européennes: Saarbrücken, Germany, 2010.
23. Anderson, T.L. *Fracture Mechanics: Fundamentals and Applications*; CRC Press LLC: Boca Raton, FL, USA, 2017.
24. Ravera, R.J.; Sih, G.C. Transient Analysis of Stress Waves around Cracks under Antiplane Strain. *J. Acoust. Soc. Am.* **1970**, *47*, 875–881. [CrossRef]
25. Yu, P.; Wang, Q.; Zhang, C.; Zhao, J. Elastic T-stress and I-II mixed mode stress intensity factors for a through-wall crack in an inner-pressured pipe. *Int. J. Press. Vessel. Pip.* **2018**, *159*, 67–72. [CrossRef]
26. Brugier, F. *Modèle Condensé de Plasticité Pour la Fissuration et Influence de la Contrainte T*; Université Paris-Saclay (ComUE): Saint-Aubin, France, 2017; Available online: https://www.theses.fr/2017SACLN028 (accessed on 20 January 2021).
27. Gupta, M.; Alderliesten, R.; Benedictus, R. A review of T-stress and its effects in fracture mechanics. *Eng. Fract. Mech.* **2015**, *134*, 218–241. [CrossRef]
28. Kfouri, A.P. Some evaluations of the elastic T-term using Eshelby's method. *Int. J. Fract.* **1986**, *30*, 301–315. [CrossRef]
29. Yang, B.; Ravi-Chandar, K. Evaluation of elastic T-stress by the stress difference method. *Eng. Fract. Mech.* **1999**, *64*, 589–605. [CrossRef]
30. Du, Z. *eXtended Finite Element Method (XFEM) in Abaqus*; Simulia: Jhonston, RI, USA, 2009.
31. Xie, Y.; Cao, P.; Jin, J.; Wang, M. Mixed mode fracture analysis of semi-circular bend (SCB) specimen: A numerical study based on extended finite element method. *Comput. Geotech.* **2017**, *82*, 157–172. [CrossRef]
32. Yixiu, S.; Yazhi, L. A Simple and Efficient X-FEM Approach for Non-planar Fatigue Crack Propagation. *Procedia Struct. Integr.* **2016**, *2*, 2550–2557. [CrossRef]
33. Nasri, K.; Zenasni, M. Fatigue crack growth simulation in coated materials using X-FEM. *Comptes Rendus Mécanique* **2017**, *345*, 271–280. [CrossRef]
34. Savenkov, E.B.; Borisov, V.E.; Kritskiy, B.V. Surface Representation with Closest Point Projection in the X-FEM. *Math. Model. Comput. Simul.* **2020**, *12*, 36–52. [CrossRef]
35. Angioni, S.; Visrolia, A.; Meo, M. Combining X-FEM and a multilevel mesh superposition method for the analysis of thick composite structures. *Compos. Part B Eng.* **2012**, *43*, 559–568. [CrossRef]
36. Koutsawa, Y.; Belouettar, S.; Makradi, A.; Tiem, S. X-FEM implementation of VAMUCH: Application to active structural fiber multi-functional composite materials. *Compos. Struct.* **2012**, *94*, 1297–1304. [CrossRef]
37. Schiavone, A.; Abeygunawardana-Arachchige, G.; Silberschmidt, V.V. Crack initiation and propagation in ductile specimens with notches: Experimental and numerical study. *Acta Mech.* **2015**, *227*, 203–215. [CrossRef]
38. Patria, K.; Bambang, B.; Muhammad, F. *XFEM Based Fracture Analysis of Single Notch Reactive Powder Concrete Specimen Subjected to Three Point Bending Test*; Web of Conferences; EDP Sciences: Paris, France, 2020; p. 05027. [CrossRef]
39. Moustabchir, H.; Arbaoui, J.; Zitouni, A.; Hariri, S.; Dmytrakh, I. Numerical analysis of stress intensity factor and T-stress in pipeline of steel P264GH submitted to loading conditions. *J. Theor. Appl. Mech.* **2015**, *53*, 665–672. [CrossRef]
40. Kim, J.-H.; Paulino, G.H. T-stress, mixed-mode stress intensity factors, and crack initiation angles in functionally graded materials: A unified approach using the interaction integral method. *Comput. Methods Appl. Mech. Eng.* **2003**, *192*, 1463–1494. [CrossRef]
41. Bouchard, P.-O.; Bernacki, M.; Parks, D.M. Analysis of stress intensity factors and T-stress to control crack propagation for kerf-less spalling of single crystal silicon foils. *Comput. Mater. Sci.* **2013**, *69*, 243–250. [CrossRef]
42. Sherry, A.H.; France, C.C.; Goldthorpe, M.R. Compendium of t-stress solutions for two and three dimensional cracked geometries. *Fatigue Fract. Eng. Mater. Struct.* **1995**, *18*, 141–155. [CrossRef]
43. Ayatollahi, M.; Pavier, M.; Smith, D. Determination of T-stress from finite element analysis for mode I and mixed mode I/II loading. *Int. J. Fract.* **1998**, *91*, 283–298. [CrossRef]
44. Bouledroua, O.; Meliani, M.H.; Pluvinage, G. A Review of T-Stress Calculation Methods in Fracture Mechanics Computation. *Nat. Technol.* **2016**, *11*, 20.

Article

# A Computational Approach to Solve a System of Transcendental Equations with Multi-Functions and Multi-Variables

Chukwuma Ogbonnaya [1,2,*], Chamil Abeykoon [3], Adel Nasser [1] and Ali Turan [4]

[1] Department of Mechanical, Aerospace and Civil Engineering, The University of Manchester, Manchester M13 9PL, UK; a.g.nasser@manchester.ac.uk
[2] Faculty of Engineering and Technology, Alex Ekwueme Federal University, Ndufu Alike Ikwo, Abakaliki PMB 1010, Nigeria
[3] Aerospace Research Institute and Northwest Composites Centre, School of Materials, The University of Manchester, Manchester M13 9PL, UK; chamil.abeykoon@manchester.ac.uk
[4] Independent Researcher, Manchester M22 4ES, Lancashire, UK; a.turan@ntlworld.com
* Correspondence: chukwuma.ogbonnaya@manchester.ac.uk; Tel.: +44-(0)74-3850-3799

**Citation:** Ogbonnaya, C.; Abeykoon, C.; Nasser, A.; Turan, A. A Computational Approach to Solve a System of Transcendental Equations with Multi-Functions and Multi-Variables. *Mathematics* **2021**, *9*, 920. https://doi.org/10.3390/math9090920

Academic Editor: Maria Luminița Scutaru

Received: 2 April 2021
Accepted: 19 April 2021
Published: 21 April 2021

**Publisher's Note:** MDPI stays neutral with regard to jurisdictional claims in published maps and institutional affiliations.

**Copyright:** © 2021 by the authors. Licensee MDPI, Basel, Switzerland. This article is an open access article distributed under the terms and conditions of the Creative Commons Attribution (CC BY) license (https://creativecommons.org/licenses/by/4.0/).

**Abstract:** A system of transcendental equations (SoTE) is a set of simultaneous equations containing at least a transcendental function. Solutions involving transcendental equations are often problematic, particularly in the form of a system of equations. This challenge has limited the number of equations, with inter-related multi-functions and multi-variables, often included in the mathematical modelling of physical systems during problem formulation. Here, we presented detailed steps for using a code-based modelling approach for solving SoTEs that may be encountered in science and engineering problems. A SoTE comprising six functions, including Sine-Gordon wave functions, was used to illustrate the steps. Parametric studies were performed to visualize how a change in the variables affected the superposition of the waves as the independent variable varies from $x_1 = 1:0.0005:100$ to $x_1 = 1:5:100$. The application of the proposed approach in modelling and simulation of photovoltaic and thermophotovoltaic systems were also highlighted. Overall, solutions to SoTEs present new opportunities for including more functions and variables in numerical models of systems, which will ultimately lead to a more robust representation of physical systems.

**Keywords:** system of transcendental equation; computational solutions; code-based modelling approach; numerical analysis; Sine-Gordon equations; photovoltaics; thermophotovoltaics; solar energy

## 1. Introduction

The advent of the computer has made explicit solution and visualization of transcendental equations (TE) easier [1]. Computing has, indeed, expanded the possibilities of modelling and simulation of complex phenomena, processes, and systems [2]. However, encountering non-zero TE of the form f(x) = g(x) in science and engineering poses challenges, particularly when the TE is included in a system of equations to create a system of transcendental equations (SoTE). A TE may have many roots which may require explicit method to find their roots using Cauchy's integral theorem [3]. The computational solution to SoTE may result in a single output in a case where some functions act as functions of the output function. The output function can be represented graphically to visualize and analyze how it changes with respect to some system variables or functions in the SoTE.

In order to increase the number of variables and parameters of a physical system captured during numerical modelling, multi-functions may be required to be solved simultaneously. Consequently, more methods/techniques for solving SoTEs are required to facilitate numerical solutions of physical systems involving TE. Over the years, the need to solve problems involving TEs or SoTEs caused scientists and engineers to use different methods/techniques to find solutions to them [4]. For instance, Artificial Neural Network (ANN) has been proposed for solving SoTEs [5]. A Chebyshev series has been added to a transcendental equation to convert it into a polynomial equation so that it can be truncated and solved [6]. A decomposition technique has also been applied to solve TEs [7]. Lagrange inversion theorem and Pade approximation were used by Luo [8] to solve TEs encountered in physics. Furthermore, Ruggiero [9] adopted a computational iteration procedure to solve a TE involving wave propagations in elastic plates. The iterative process has also been applied to solve a fourth order transcendental nonlinear equation of the form f(x) = 0 [10]. There have been some specific attempts to solve physical problems involving SoTEs. For instance, Falnes [11] demonstrated that the electrical impedance of a semiconductor supporting two waves contains an entire transcendental function of the form f(z) = exp(−z) − 1 − cz. Danhua et al. [12] studied a perturbed Sine-Gordon equation with impulsive forcing to describe non-linear oscillations. They highlighted the various application of the Sine-Gordon equations in science and engineering.

Computational methods have been applied to solve non-linear wave equations [13]. Recently, a code-based modelling (CBM) approach is an example of computational approach proposed for solving SoTEs applicable to photovoltaic and thermophotovoltaic systems [14–16]. Although the CBM approach appears to be robust in achieving numerical solutions to SoTEs, there are no clear steps for formulating and solving of scientific and engineering problems involving SoTE. Therefore, the aim of this paper is to present detailed steps of how CBM approach can be implemented to solve SoTEs. To achieve this aim, the specific objectives are to:

1. Describe the steps for using the CBM approach for solving SoTEs.
2. Demonstrate how the CBM approach is used to solve a hypothetical SoTE including Sine-Gordon equations.
3. Perform parametric analysis of wavelength and amplitude in Objective 2.
4. Discuss the application of the CBM approach for modelling and simulation of photovoltaic and thermophotovoltaic systems.

The originality of this study is realized in being the first paper to present detailed steps for applying the CBM approach to facilitate numerical/computational solutions to SoTEs. Although the steps are proposed for problems that may be encountered in science and engineering, there is no doubt that any researcher from any field can adopt/adapt the steps. The major contribution of this paper is to demonstrate how the CBM approach can allow scientists and engineers more degrees of freedom to overcome the limitations of including multi-functions and multi-variables during model representation of physical systems involving SoTEs. Henceforth, Section 2 presents detailed steps for formulating SoTEs including Sine-Gordon functions. Section 3 presents the results generated from the simulations of the SoTE formulated in Section 2. Then, Section 4 discusses the application of the CBM approach for solving SoTE related to photovoltaics and thermophotovoltaics, while Section 5 concludes the study.

## 2. Detailed Steps for Implementing the CBM Approach

The mathematical model of a system facilitates the predictable physical behaviors which can allow scientists or engineers to investigate the system using simulations. The functions describing the system may include linear or/and non-linear equations and can be solved as a system of equations. This means that the mathematical formulation of physical problems is critically important for capturing the crucial parameters and variables of the system under study [17]. Since any parameter excluded from the model cannot be accounted for, robust formulation becomes a necessary step in accurate representation of any physical system or phenomena. Once the problem is adequately formulated with all the possible dependent and independent variables, computing can facilitate solutions faster and accurately. As algorithms continue to facilitate the application of the computer in different facets of human existence [18], CBM approach appears compatible with code-based algorithms for solutions to SoTEs.

Figure 1 summarizes the steps for solving SoTEs using the CBM approach. Although the steps may vary depending on the nature and complexity of the problem, the steps in the flowchart are further described to show how the approach can be adopted for finding numerical solutions to SoTEs. Where applicable, illustrations were used to explain the applicability of the steps.

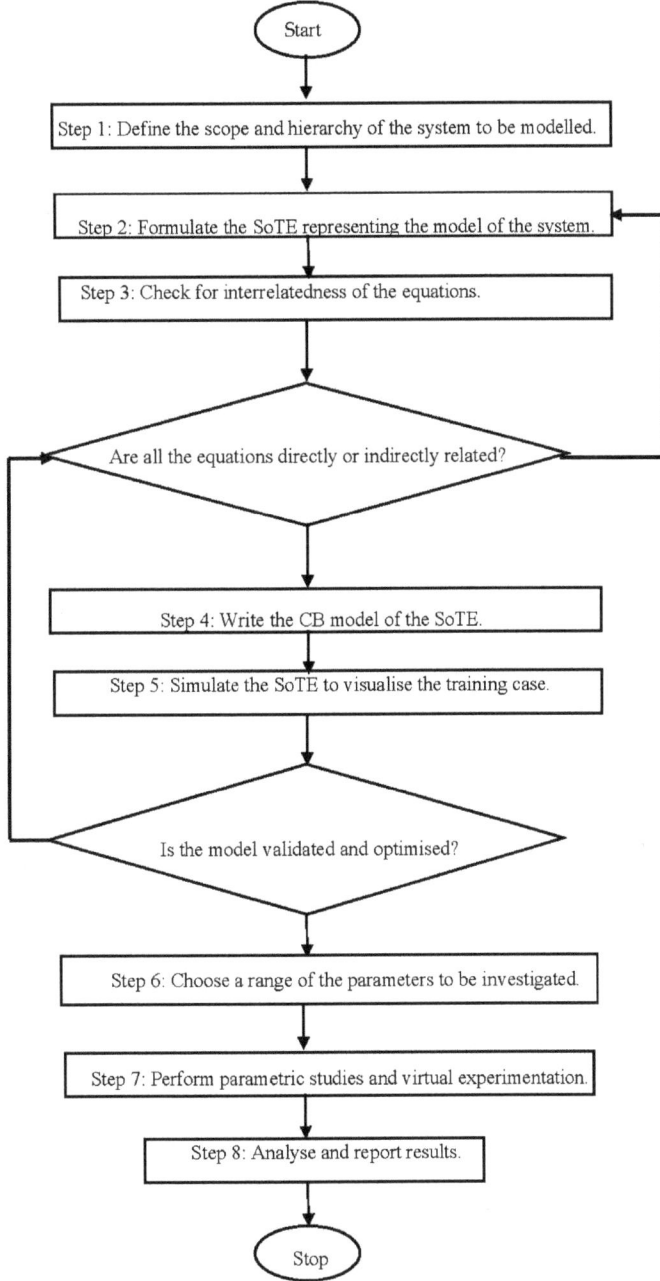

**Figure 1.** Flowchart for implementing the CBM approach for solving SoTEs.

*Step 1: Define the scope and hierarchy of the system to be modelled.* It is not always possible to capture all the parameters and variables affecting a system in a single equation. This step is crucial for defining the aspects of the system that would be included in the CB model. It might be helpful to recognize that more parameters and functions can be added once the basic model is established so that the decision space can be expanded. The idea is to start from a simple model, and then increase the complexity of the model to capture more parameters and variables.

*Step 2: Formulate the SoTE representing the model of the system.* This step may not require entirely new equations. Established equations in the field of study can be used. As an example, solar exergy equation by Petela [19] was used as solar exergy input into a numerical integration of solar, thermal, and electrical exergies of photovoltaic module [14]. Since electrical exergy involves a SoTE, the integrated model remained a SOTE. However, for situations where there are no extant equations, a formulation of new equations from the first principle, statistical modelling, or through experimental study can be considered. For instance, in order to determine the optimal location for a large-scale photovoltaic power generation, new thermodynamic indices were formulated and combined with a SoTE [15]. In formulating a SoTE, inter-relationships among the equations in the system of equation is fundamental in order to reach a point of convergence. Depending on the scope of the modelling, a SoTE may include as many functions as may be required.

*Step 3: Check for inter-relatedness of the equations.* Without an inter-relationship between the equations, the requirement for solving a SoTE simultaneously may prove elusive. Therefore, any convergence reached when the equations are not inter-related does not exactly represent a solution to a SoTE. Thus, properly linked equations to interact simultaneously directly or indirectly during computational iteration is crucial in solving SoTEs. To achieve direct or indirect inter-relationships, an equation may be rearranged to make the required dependent variable the subject of the equation. In direct inter-relationship, the total output of a function is substituted into another function, thereby creating a function of a function relationship. A function can be decomposed during coding, where possible, to make the algorithm easier to implement. On the other hand, indirect inter-relationship exists when two or more functions share the same parameter or variable. The shared parameter or variable by two or more functions in a SoTE may affect the output of the functions differently. For instance, in formulating the net temperature of a body undergoing heating and cooling simultaneously, the temperature of the body will exist in the heating and cooling functions. Nonetheless, whilst heating tends to increase the temperature of the body, cooling tends to reduce it.

*Step 4: Write the CB model of the SoTE.* In this step, each function is written as a code in accordance with the syntax/structure of the software used. The codes can be written and tested step-by-step instead of attempting to run the SoTEs after integrating them. By testing preceding codes before integrating more functions, troubleshooting, or debugging of the CB model would be enhanced because errors can be traced from the latest step. The algorithm used for implementing SoTE codes are important because it determines how many results that can be generated from a SoTE. CBM approach can be implemented in software such as MATLAB, Python, Mathematica, etc. MATLAB [16] appears to be very useful for creating CB models because it is easy to integrate the functions and visualize the effects of the change of the independent variable on the dependent variables. In MATLAB, an input function is presented before the output function. The algorithm is also designed to generate visualizations of the outputs from the SoTE.

*Step 5: Simulate the SoTE to visualize the training case.* Testing of the CB model involves validation tests against experimental results and training cases. The model should predict the training cases with a reasonable accuracy and significant precision for the model to be applied further. The model can be optimized at this stage if the accuracy of prediction is unacceptable.

*Step 6: Choose a range of the parameter to be investigated.* The nature of non-zero TEs (i.e., $f(x) = g(x)$) means that it cannot be solved explicitly like linear, quadratic, or polynomial equations with roots when f(x) = 0. For non-zero SoTEs, a range of value of a parameter or variable can be simulated. The solution may require an iterative process [20] with a range of values of the parameters and variables of the system. The parameters or variables chosen depend on the aspect of the system under investigation.

*Step 7: Perform parametric studies and virtual experimentation.* Parametric studies are important during model-based studies because they allow different scenarios that can affect the system to be simulated and analyzed. It also helps to investigate optimal solutions as well as carryout "what if" analysis. In a study [16], after validating the CB model of a PV module, parametric studies were used to study the effect of solar radiation, temperature, ideality factor, number of solar cells, and number of modules in parallel on the maximum power point. Model-based parametric study is useful for gaining deep insights without necessarily committing excessive resources in experimental studies. Yet, results from parametric studies may inform the ultimate design of experimental studies.

Virtual experimentation is a novel computational approach for gaining deeper insights into direct and indirect relationships between variables or a variable and other functions in the SoTE. This is an advanced application of CBM approach. Virtual experimentation, as the name implies, is a virtual implementation of steps similar to the steps performed in the laboratory. It allows some parameters of the system to be kept constant while other variables change. The effect of the changes on the system are then analyzed. There are studies that have discussed how virtual experimentation can be implemented [14,16]. For instance, if solar radiation increases, it may be of interest to investigate how power and heat generation evolve in a PV module [14]. However, this may require incorporating an additional user defined function and/or algorithm to the CB model. For instance, to model solar photovoltaic and thermophotovoltaics, the input radiation function is the solar radiation function in the case of solar photovoltaic systems, whilst the input radiation function is the radiative heat flux in the case of thermophotovoltaic system. Although solar radiation or thermal heat flux can cause the photovoltaic process in PV cells, parametric study may reveal more insights into how they specifically differ during power generation.

*Step 8: Analyze and report results.* This step involves a critical analysis of the results from Step 7 and reporting them in the required format. Reporting may encompass generating internal reports for decision-making as well as reporting in scholarly publications. The detailed steps for using CBM approach for solving SoTEs have satisfied the first objective of this study.

*A Hypothetical SoTE Including a Sine-Gordon Equation*

Sine wave functions are applied in physics and engineering, particularly in oscillations, vibrations, and signal processing. As an example, TE is encountered in transverse and longitudinal wave diffraction [21]. A study by Sun [22] proposed an exact solution to Sine-Gordon equations with transcendental characteristics. Here, hypothetical Sine-Gordon equations ($u_{tt} = u_{xx} + Sin(u)$) [22,23], linear, and quadratic functions are solved simultaneously. The SoTE including Sine-Gordon equations simulates how interferences affect the output wavelength, frequency, and amplitude of the resultant wave function. The SoTE expressed in Equation (1) composed of the six functions expressed in $f_1$ to $f_6$ where $f_6$ represents the output function, while $f_1$ to $f_5$ represent the input functions. Since $f_1$, a linear function, is a TE (i.e., f(x) = g(x)), solving it alongside other equations creates a SoTE. $f_1$ can be transmitted through the system and visualized through $f_6$. Likewise, a change in the parameters in $f_1$ to $f_5$ can be visualized in the output function. Here, the functions are formulated to have inter-relationships in order to facilitate convergence so that the output function ($f_6$) can predict the behavior of the wave as a function of the input functions, parameters, and variables. The CB model of the SoTE represented in $f_1$ to $f_6$ is written with MATLAB codes. The output wave ($f_6$) is simulated for $x_1 \in \mathbb{R}_+$ between 1 and 100 within which the input and output functions are visualized.

$$\begin{cases} f_1 = x_1 = v_1 x_1. \\ f_2 = x_2 = k_1 x_1^2. \\ f_3 = x_3 = -100 x_2 + \frac{K_2}{K_3}. \\ f_4 = x_4 = k_4 + v_2 \text{Sin}(x_1). \\ f_5 = x_5 = k_5 - v_3 \text{Sin}(x_4). \\ f_6 = x_6 = x_2\, x_4. \end{cases} \quad (1)$$

where $v_1$, $v_2$ and $v_3$ are variables, $k_1$ to $k_5$ are constants and $x_1$ to $x_6$ represents the functions $f_1$ to $f_6$, respectively.

## 3. Results and Discussions

The results from simulating the CB model of the SoTE are presented in this section. The variables $v_1 = 10$, $v_2 = 50$, $v_3 = 20,000$, and constants $k_1 = 2000$, $k_2 = 15,000$, $k_3 = 10$, $k_4 = 20$, and $k_5 = 20$ for $x_1$ = 1:0.5:100 was simulated so that individual functions can be observed, as well as their overall effect on the output function. Figure 2a–e shows the relationship between $x_1$ with $x_2$, $x_3$, $x_4$, $x_5$, and $x_6$. Considering that $f_1$ is a TE, a possible solution to it is that $v_1 = 1$. However, suppose that the mathematical model represents a process in which the parameter $x_1$ undergoes a process change but it is expected that its index should remain as unity. Then the input value of $x_1$ into the process must be equal to the output value of $x_1$. $f_1$ is satisfied if the value of $x_1$, before and after the process change remains the same. By this condition, the value of $x_1$ can be an integer, decimal, or indices since it will yield an index of 1 (i.e., $v_1 = \frac{x_1}{x_1} = 1$). Later, it will be shown that changing $v_1$ in $f_1$ affects the frequency of the Sine-Gordon wave $f_4$.

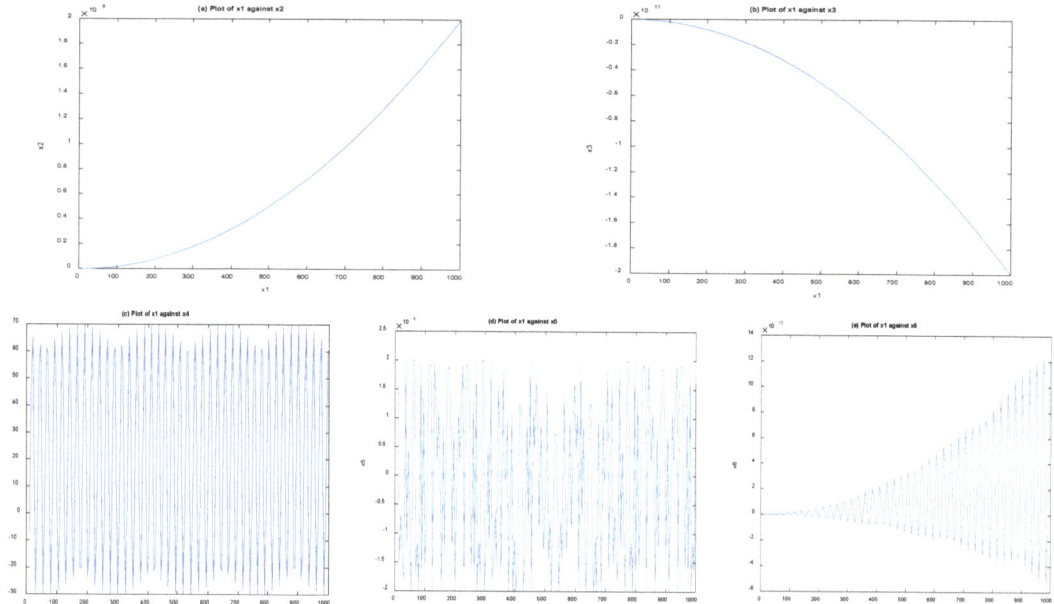

**Figure 2.** Simulation of the CB model of SoTE for $x_1$ = 1:0.5:100. (**a**) input quadratic function, (**b**) input quadratic function, (**c**) resultant polynomial function, (**d**) induced interference (**e**) output wave function.

A parametric study is performed to observe how a change in the variables affect the output of the Sine-Gordon wave model. The variables $v_1 = 10$, $v_2 = 50$, $v_3 = 20,000$ and constants $k_1 = 2000$, $k_2 = 15,000$, $k_3 = 10$, $k_4 = 20$, and $k_5 = 20$ for $x_1 = 1:0.5:100$ were maintained while the division of the scale of the wavelength was reduced from $x_1 = 1:0.5:100$ to $x_1 = 1:5:100$ to observe how the characteristics of the output wave would change. Based on the result of the simulations, the outputs of $f_4$, $f_5$, and $f_6$ significantly changed, as shown in Figure 3c–e. $f_5$ appears to have introduced the highest interference, as shown in Figure 3d, although it was superimposed at the output, as shown in Figure 3e.

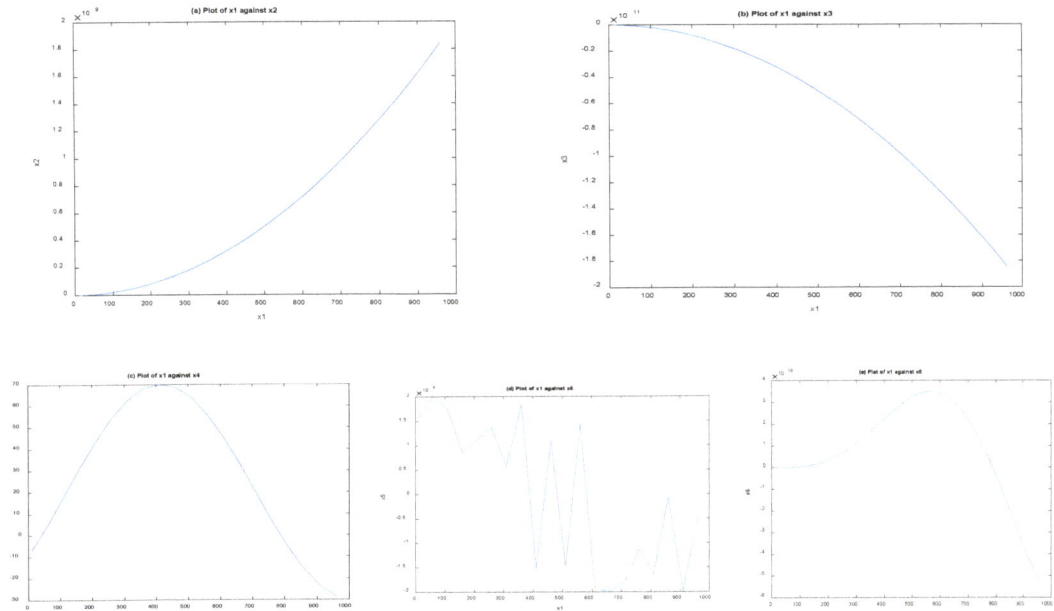

**Figure 3.** Simulation of the CB model of SoTE for $x_1 = 1:5:100$. (**a**) input quadratic function, (**b**) input quadratic function, (**c**) resultant polynomial function, (**d**) induced interference (**e**) output wave function.

Again, the variables $v_1 = 10$, $v_2 = 50$, $v_3 = 20,000$ and constants $k_1 = 2000$, $k_2 = 15,000$, $k_3 = 10$, $k_4 = 20$, and $k_5 = 20$ for $x_1 = 1:0.5:100$ were maintained while the scale of the divisions of the wavelength was increased from $x_1 = 1:0.5:100$ to $x_1 = 1:0.0005:100$ to observe how the characteristics of the output wave would change over a larger scope. There was a significant increase in the frequency of the wave, as shown in the outputs of $f_4$, $f_5$, and $f_6$ as visualized in Figure 4c–e. Still, the amplitude of the wave was virtually bounded by the two quadratic functions (Figure 4a,b) with the resultant amplitude increasing progressively, as shown in Figure 4e. This implies that the resultant discontinuity of the two quadratic functions, when visualized from the relationship between $x_1$ and $x_6$, did not eliminate their effects as they constrained the amplitude of the output wave even in their discontinuous states.

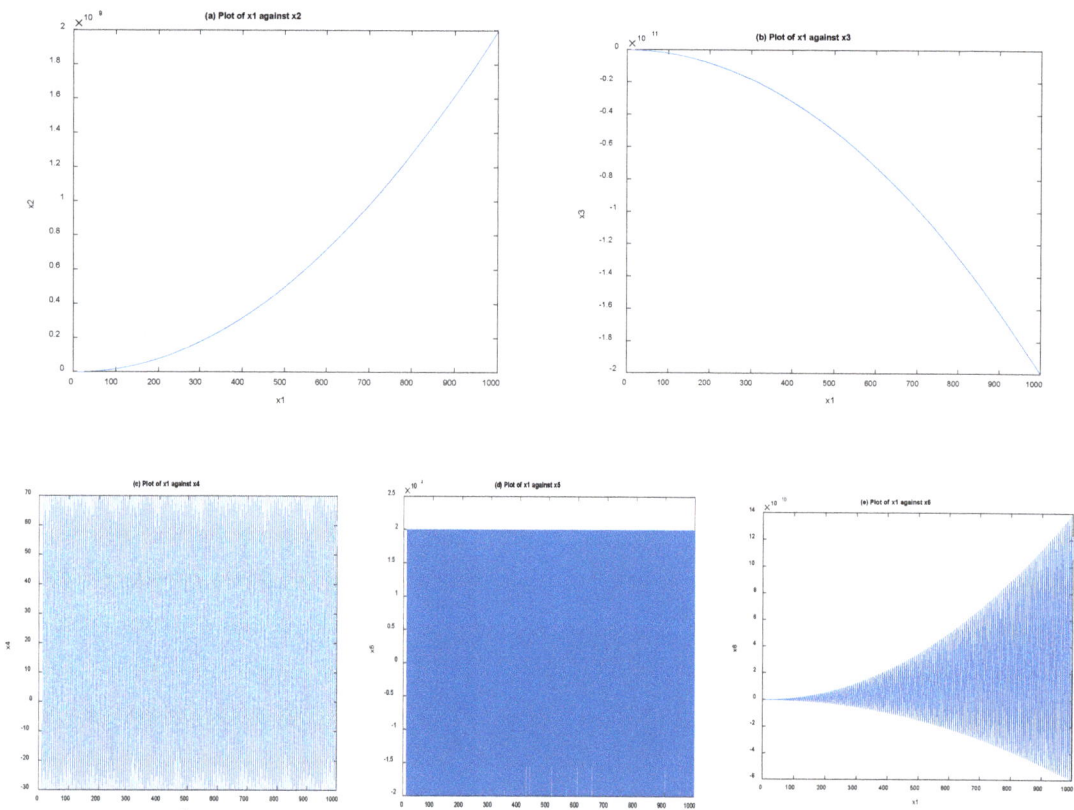

**Figure 4.** Simulation of the CB model of SoTE for $x_1$ = 1:0.0005:100. (**a**) input quadratic function, (**b**) input quadratic function, (**c**) resultant polynomial function, (**d**) induced interference (**e**) output wave function.

In addition to changing the wavelength of the sine functions over the range of $x_1$, other parameters can be adjusted so that their effects can be visualized. With $x_1$ = 1:0.5:100, $v_1 = 100$, $v_2 = 5000$, $v_3 = 20,000$, and $k_1 = 2000$, $k_2 = 15,000$, $k_3 = 10$, $k_4 = 20$, $k_5 = 20$, the effect of the change in $v_1$ and $v_2$ in the characteristics of the wave was visualized. Although the scale of the wavelength in Figures 2 and 5 were the same, the output waves in Figures 2e and 5e differ significantly in their frequency when $v_1$ and $v_2$ changed. Based on the analysis of the outputs, the reduced frequency was caused by the effects seen in Figures 2c and 5c, due to $f_4$. In all the cases, there was an increasing amplitude because of the two quadratic input functions.

From the foregoing analysis in this section, the CBM approach has been used to demonstrate how a hypothetical SoTE including the Sine-Gordon equation was solved to the satisfaction of research objective 2. Also, parametric analysis, which showed how the wavelength and the amplitude responds to the change in the variables, satisfy the research objective 3.

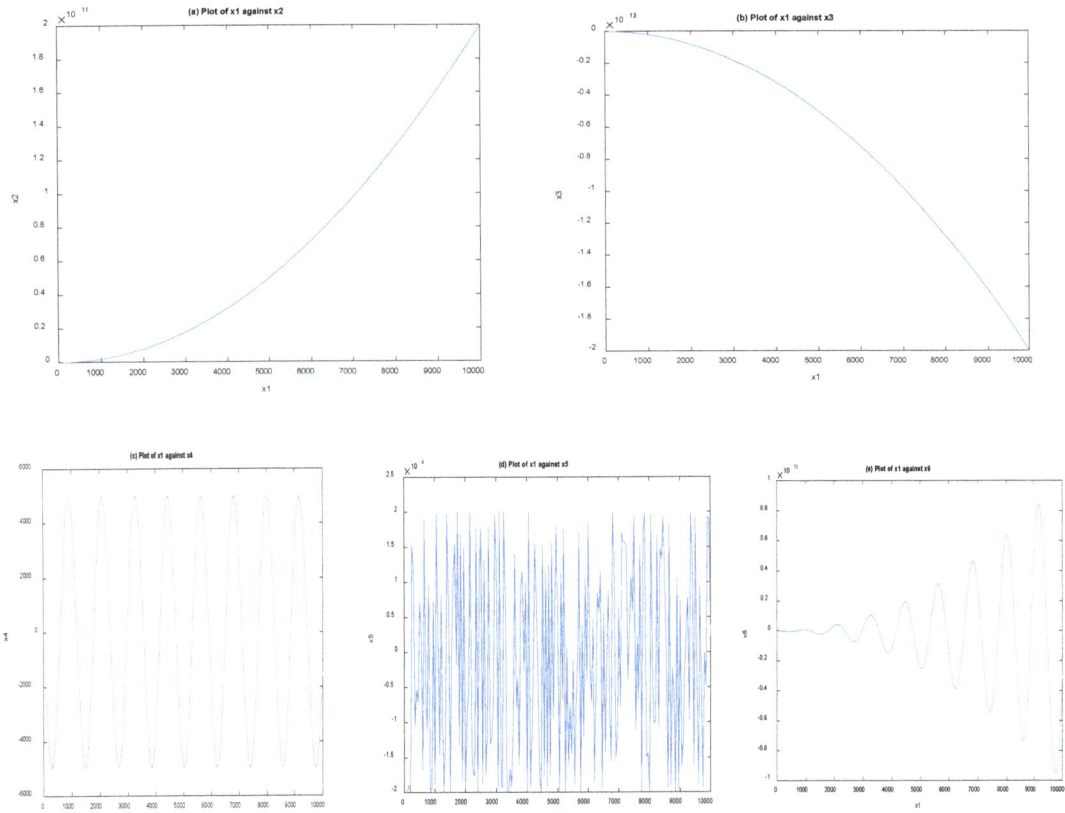

**Figure 5.** Simulation of the CB model of SoTE for $x_1 = 1:0.5:100$ with variation in $v_1$ and $v_2$. (**a**) input quadratic function, (**b**) input quadratic function, (**c**) resultant polynomial function, (**d**) induced interference (**e**) output wave function.

## 4. Application of SoTE in Photovoltaic and Thermophotovoltaic Modelling and Simulation

This section presents the application of the CBM approach for modelling and simulation of photovoltaic and thermophotovoltaic systems, pursuant the achievement of objective 4. The photovoltaic and thermophotovoltaic modelling and simulation involves a SoTE because the computation of the output voltage of the PV cells involves a transcendental function [16,24,25]. Equation (2) presents the functions that have been used to create a predictive model for power generation characteristics of the PV module. The interrelationships between the equations are further highlighted. Bandgap function ($E_g$) [26] is an input function in calculating the saturation current function ($I_s$) [27]. The photocurrent function ($I_{ph}$) [28] is an input function for calculating the output current of the PV ($I_0$). The typical problem that qualifies this to be a SoTE is that the output voltage ($V_0$) is within the function $I_0$ which is already defining the output current [29]. In order to compute the output power of PV ($P_o$), the function is iterated over a range of the voltage ($V_0$) [16].

$$\begin{cases} E_g = E_g(0) - \frac{\alpha T^2}{T+\beta}. \\ I_{ph} = (I_{sc} + K_I (T_{cell} - T_{ref})) \times \frac{G}{G_{ref}}. \\ I_s = I_{s,ref} \left[\frac{T_{cell}}{T_{ref}}\right]^3 \exp\left[\frac{1}{k}\left(\frac{E_g}{T_{ref}} - \frac{E_g}{T_{cell}}\right)\right]. \\ I_0 = I_{ph} N_p - I_s N_p \left[\exp\left(\frac{qV_0}{AN_s kT}\right) - 1\right]. \\ P_o = I_0 \times V_0. \end{cases} \quad (2)$$

As an example, the effect of increasing number of solar cells in series ($N_s$) from 36 to 72 cells is simulated and presented in Figure 6. The open circuit voltage increased as the number of cells increases leading to an increased maximum power point of the system.

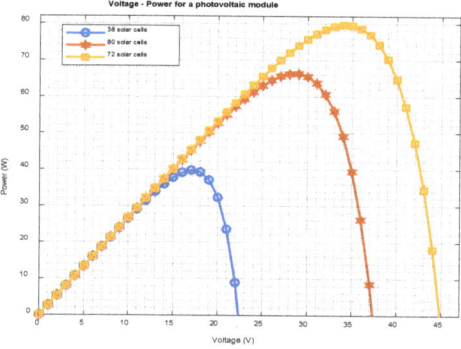

**Figure 6.** CB model of photovoltaic module to predict maximum power points as the number of solar cells changes.

Furthermore, the SoTE in Equation (2) which represents solar photovoltaic generation can be utilized to create a thermophotovoltaic model [14]. The insolation (G) can be replaced by Steffan–Boltzmman's radiation, where the thermal heat flux is from an artificial source with unique radiation surface characteristics. The utility of the CBM approach is the opportunity to adapt the algorithm for implementing SoTE to the specific problem under investigation. In the instant case of the application of CBM approach to implement a numerical solution for SoTE encountered in photovoltaics, two illustrations are hereby highlighted. Equation (2) acted as the power output in the integration of solar, thermal, and electrical exergies of a photovoltaic module, as shown in Equation (3).

$$\dot{Q}_{loss} = \left[G \times A_{cell} \times \tau_{glass}\left(1 - \frac{4}{3}\frac{T}{T_{sun}} + \frac{1}{3}\left(\frac{T}{T_{sun}}\right)^4\right)\right] - \left(I_{ph} N_p - I_s N_p \left[\exp\left(\frac{qV_{pv}}{AN_s kT}\right) - 1\right]\right) \times V_{pv}. \quad (3)$$

Apart from using solar radiation to generate excitation in PV cells, there are increasing research efforts to use sources of heat to generate radiative heat transfer that can cause excitations in PV cells. Regardless, thermophotovoltaic systems still depend on the physics of photovoltaic power generation except that thermophotovoltaic systems are not limited by the risks associated with the intermittency of solar radiation [30]. Equation (4) shows a numerical integration of radiative heat transfer, power density output, and thermal losses in the core of a thermophotovoltaic system [31].

$$\dot{Q}_{losses} = \left[n^2 \varepsilon \sigma F A_R \left(T_{rad}^4 - T_{pv}^4\right)\right] - \left(I_{ph} N_p - I_s N_p \left[\exp\left(\frac{qV_{pv}}{AN_s kT_{pv}}\right) - 1\right]\right) \times V_{pv}. \quad (4)$$

Both Equations (3) and (4) are SoTEs and their solutions were facilitated by using CBM approach. The two models are crucial models for investigating the thermodynamics of photovoltaics and thermophotovoltaics.

## 5. Conclusions

This study provides detailed steps on how the CBM approach can be used to solve SoTE with multi-functions and multi-variables. To formulate the steps, a hypothetical SoTE including Sine-Gordon equations was used to illustrate the steps for solving a SoTE. Also, a parametric analysis was performed to investigate how a change in the variables affected the superposition of the waves, the wavelength, and the amplitude. From the results of the simulations, the amplitude, wavelength, and frequency of the output wave reflects the changes in the parameters and variables of the SoTE. This means that the properties of the Sine-Gordon wave were altered when the variables and parameters in the CB model of the waves were adjusted. The application of the CBM approach in the modelling and simulation of photovoltaic and thermophotovoltaic systems was presented as practical application of CBM approach in solving a complex SoTE. In conclusion, more functions and variables of physical systems or phenomena can be added during mathematical modelling of problems exhibiting the characteristics of a SoTE, using the steps outlined in this study.

**Author Contributions:** Conceptualization, C.O.; methodology, C.O., C.A., and A.N.; software, C.O.; validation, C.O., C.A., and A.T.; formal analysis, C.O.; investigation, C.O.; resources, C.A. and A.N.; data curation, C.O.; writing—original draft preparation, C.O. and C.A.; writing—review and editing, C.A. and A.N.; visualization, C.O.; supervision, A.T., C.A. and A.N.; project administration, C.O.; funding acquisition, C.O. All authors have read and agreed to the published version of the manuscript.

**Funding:** This research was funded by the Petroleum Technology Development Fund (PTDF) Nigeria number PTDF/ED/PHD/OC/1078/17.

**Institutional Review Board Statement:** Not applicable.

**Informed Consent Statement:** Not applicable.

**Data Availability Statement:** Data associated with this research is available at request.

**Conflicts of Interest:** The authors declare no conflict of interest.

## Nomenclature

| | |
|---|---|
| A | ideality constant |
| CBM | code-based modelling |
| CB | code-based |
| $E_g$ | bandgap energy |
| $I_0$ | output current of PV module |
| $I_{ph}$ | photocurrent |
| $I_s$ | saturation current of PV module |
| $I_{sc}$ | short circuit current of PV module |
| k | Boltzmann's const. ($1.38 \times 10^{-23}$ J/K) |
| MPP | maximum power point |
| $N_s$ | number of solar cells in series |
| $N_p$ | number of solar cells in parallel |
| $P_0$ | output power of PV module |
| PV | photovoltaic |
| q | electron charge ($1.602 \times 10^{-19}$ C) |
| SoTE | system of transcendental equations |
| STC | standard test condition (25 °C, 1000 W/m$^2$, AM 1.5) |
| T | temperature |
| TE | transcendental equation |
| $V_{oc}$ | open circuit voltage |

**Greek symbols**

α  solar cell material constant
β  solar cell material constant

**Subscripts**

cell  solar cell
ph  photon
pv  photovoltaic
ref  reference

## References

1. Luck, R.; Stevens, J.W. Explicit Solutions for Transcendental Equations. *SIAM Rev.* **2002**, *44*, 227–233. [CrossRef]
2. Zhang, S.; Zhu, C.; Gao, Q. Numerical Solution of High-Dimensional Shockwave Equations by Bivariate Multi-Quadric Quasi-Interpolation. *Mathematics* **2019**, *7*, 734. [CrossRef]
3. Luck, R.; Zdaniuk, G.J.; Cho, H. An Efficient Method to Find Solutions for Transcendental Equations with Several Roots. *Int. J. Eng. Math.* **2015**, *2015*, 1–4. [CrossRef]
4. Zaguskin, V.L. *Handbook of Numerical Methods for the Solution of Algebraic and Transcendental Equations*; Pergamon Press: Oxford, UK, 1961.
5. Jeswal, S.; Chakraverty, S. Solving Transcendental Equation Using Artificial Neural Network. *Appl. Soft Comput.* **2018**, *73*, 562–571. [CrossRef]
6. Boyd, J.P. Computing the Zeros, Maxima and Inflection Points of Chebyshev, Legendre and Fourier series: Solving Tran-Scendental Equations by Spectral Interpolation and Polynomial Rootfinding. *J. Eng. Math.* **2006**, *56*, 203–219. [CrossRef]
7. Ruan, S.; Wei, J. On the Zeros of Transcendental Functions with Applications to Stability of Delay Differential Equations with Two Delays. *Dyn. Contin. Discret. Impuls. Syst. Ser. A Math. Anal.* **2003**, *6*, 863–874.
8. Luo, Q.; Wang, Z.; Han, J. A Padé Approximant Approach to Two Kinds of Transcendental Equations with Applications in Physics. *Eur. J. Phys.* **2015**, *36*, 35030. [CrossRef]
9. Ruggiero, C.M. *Solution of Transcendental and Algebraic Equations with Applications to Wave Propagation in Elastic Plates*; Naval Research Lab: Washington, DC, USA, 1981.
10. Maheshwari, A.K. A Fourth Order Iterative Method for Solving Nonlinear Equations. *Appl. Math. Comput.* **2009**, *211*, 383–391. [CrossRef]
11. Falnes, J. Complex Zeros of a Transcendental Impedance Function. *J. Eng. Math.* **1968**, *2*, 389–401. [CrossRef]
12. Wang, D.; Jung, J.-H.; Biondini, G. Detailed Comparison of Numerical Methods for the Perturbed Sine-Gordon Equation with Impulsive Forcing. *J. Eng. Math.* **2014**, *87*, 167–186. [CrossRef]
13. Cao, Q.; Djidjeli, K.; Price, W.G.; Twizell, E.H. Computational Methods for Some Non-Linear Wave Equations. *J. Eng. Math.* **1999**, *35*, 323–338. [CrossRef]
14. Ogbonnaya, C.; Turan, A.; Abeykoon, C. Numerical Integration of Solar, Electrical and Thermal Exergies of Photovoltaic Module: A Novel Thermophotovoltaic Model. *Sol. Energy* **2019**, *185*, 298–306. [CrossRef]
15. Ogbonnaya, C.; Turan, A.; Abeykoon, C. Novel Thermodynamic Efficiency Indices for Choosing an Optimal Location for Large-Scale Photovoltaic Power Generation. *J. Clean. Prod.* **2020**, *249*, 119405. [CrossRef]
16. Ogbonnaya, C.; Turan, A.; Abeykoon, C. Robust Code-Based Modeling Approach for Advanced Photovoltaics of the Future. *Sol. Energy* **2020**, *199*, 521–529. [CrossRef]
17. Papalambros, P.Y.; Wilde, D.J. Principles of Optimal Design—Modeling and Computation. *Math. Comput.* **1992**, *59*, 726. [CrossRef]
18. Slowik, A.; Kwasnicka, H. Evolutionary Algorithms and their Applications to Engineering Problems. *Neural Comput. Appl.* **2020**, *32*, 12363–12379. [CrossRef]
19. Petela, R. Exergy of Undiluted Thermal Radiation. *Sol. Energy* **2003**, *74*, 469–488. [CrossRef]
20. Abed, S.S.; Taresh, N.S. Abed on Stability of Iterative Sequences with Error. *Mathematics* **2019**, *7*, 765. [CrossRef]
21. Aleshin, N.; Kamenskii, V.; Mogil'Ner, L. Solution of a Transcendental Equation Encountered in Diffraction Problems. *J. Appl. Math. Mech.* **1983**, *47*, 139–141. [CrossRef]
22. Sun, Y. New Exact Traveling Wave Solutions for Double Sine–Gordon Equation. *Appl. Math. Comput.* **2015**, *258*, 100–104. [CrossRef]
23. Porubov, A.; Fradkov, A.; Andrievsky, B. Feedback Control for some Solutions of the Sine-Gordon Equation. *Appl. Math. Comput.* **2015**, *269*, 17–22. [CrossRef]
24. Ogbonnaya, C.; Abeykoon, C.; Damo, U.; Turan, A. The Current and Emerging Renewable Energy Technologies for Power Generation in Nigeria: A Review. *Therm. Sci. Eng. Prog.* **2019**, *13*, 100390. [CrossRef]
25. Ogbonnaya, C.; Turan, A.; Abeykoon, C. Energy and Exergy Efficiencies Enhancement Analysis of Integrated Photovoltaic-Based Energy Systems. *J. Energy Storage* **2019**, *26*, 101029. [CrossRef]
26. Ünlü, H. A Thermodynamic Model for Determining Pressure and Temperature Effects on the Bandgap Energies and other Properties of Some Semiconductors. *Solid-State Electron.* **1992**, *35*, 1343–1352. [CrossRef]

27. Muhammad, F.F.; Yahya, M.Y.; Hameed, S.S.; Aziz, F.; Sulaiman, K.; Rasheed, M.A.; Ahmad, Z. Employment of Single-Diode Model to Elucidate the Variations in Photovoltaic Parameters Under Different Electrical and Thermal Conditions. *PLoS ONE* **2017**, *12*, e0182925. [CrossRef] [PubMed]
28. Bellia, H.; Youcef, R.; Fatima, M. A Detailed Modeling of Photovoltaic Module Using MATLAB. *NRIAG J. Astron. Geophys.* **2014**, *3*, 53–61. [CrossRef]
29. Zeitouny, J.; Katz, E.A.; Dollet, A.; Vossier, A. Band Gap Engineering of Multi-Junction Solar Cells: Effects of Series Resistances and Solar Concentration. *Sci. Rep.* **2017**, *7*, 1–9. [CrossRef]
30. Ogbonnaya, C.; Abeykoon, C.; Nasser, A.; Ume, C.; Damo, U.; Turan, A. Engineering Risk Assessment of Photovoltaic-Thermal-Fuel Cell System Using Classical Failure Modes, Effects and Criticality Analyses. *Clean. Environ. Syst.* **2021**, *2*, 100021. [CrossRef]
31. Ogbonnaya, C.; Abeykoon, C.; Nasser, A.; Turan, A. Radiation-Thermodynamic Modelling and Simulating the Core of a Thermophotovoltaic System. *Energies* **2020**, *13*, 6157. [CrossRef]

Article

# Advanced Models for Modulus and Strength of Carbon-Nanotube-Filled Polymer Systems Assuming the Networks of Carbon Nanotubes and Interphase Section

Yasser Zare [1] and Kyongyop Rhee [2,*]

[1] Breast Cancer Research Center, Biomaterials and Tissue Engineering Research Group, Department of Interdisciplinary Technologies, Motamed Cancer Institute, ACECR, Tehran 15179-64311, Iran; y.zare@aut.ac.ir
[2] Department of Mechanical Engineering (BK21 Four), College of Engineering, Kyung Hee University, Yongin 449-701, Gyeonggi, Korea
* Correspondence: rheeky@khu.ac.kr; Tel.: +82-31-201-2565; Fax: +82-31-202-6693

**Abstract:** This study focuses on the simultaneous stiffening and percolating characteristics of the interphase section in polymer carbon nanotubes (CNTs) systems (PCNTs) using two advanced models of tensile modulus and strength. The interphase, as a third part around the nanoparticles, influences the mechanical features of such systems. The forecasts agree well with the tentative results, thus validating the advanced models. A CNT radius of >40 nm and CNT length of <5 μm marginally improve the modulus by 70%, while the highest modulus development of 350% is achieved with the thinnest nanoparticles. Furthermore, the highest improvement in nanocomposite's strength (350%) is achieved with the CNT length of 12 μm and interfacial shear strength of 8 MPa. Generally, the highest ranges of the CNT length, interphase thickness, interphase modulus and interfacial shear strength lead to the most desirable mechanical features.

**Keywords:** polymer CNTs systems; interphase section; percolation onset; mechanics

Citation: Zare, Y.; Rhee, K. Advanced Models for Modulus and Strength of Carbon-Nanotube-Filled Polymer Systems Assuming the Networks of Carbon Nanotubes and Interphase Section. *Mathematics* 2021, 9, 990. https://doi.org/10.3390/math9090990

Academic Editor: Maria Luminiţa Scutaru

Received: 30 March 2021
Accepted: 26 April 2021
Published: 28 April 2021

**Publisher's Note:** MDPI stays neutral with regard to jurisdictional claims in published maps and institutional affiliations.

**Copyright:** © 2021 by the authors. Licensee MDPI, Basel, Switzerland. This article is an open access article distributed under the terms and conditions of the Creative Commons Attribution (CC BY) license (https://creativecommons.org/licenses/by/4.0/).

## 1. Introduction

Carbon nanotubes (CNTs) have attracted considerable interest due to their outstanding physical and mechanical features [1–7]. Since the CNTs have an exceptionally high Young's modulus and tensile strength, they are used as fortifications in polymers to form polymer nanocomposites (PCNTs) [8–12]. Nevertheless, the CNTs incline to form agglomerates owing to van der Waals attraction, which reduces their surface area, disturbs net formation and eventually, weakens the mechanical features of nanocomposites [13]. Therefore, a satisfactory CNT dispersion is essential to exploit the potential of nanoparticles as reinforcing agents.

The joined net of the CNTs is produced after the percolation onset [14–16]. In fact, the percolation onset is the minimum filler concentration that can lead to net formation in a medium. Moreover, the electrical conductivity of the system increases prominently after the percolation onset. Studies have attempted to obtain a low percolation onset by altering material- and fabrication-related factors [16,17]. The percolation onset can affect the mechanical features of systems [18–20]. Favier et al. [21] correlated the high shear modulus of films composed of cellulose whiskers to the percolation onset and formation of the net. Accordingly, development of the percolated microstructures significantly enhances the mechanical features of such materials. Most models for estimating the percolation onset have the functional form of a power law [22]. These models, which fairly predict the electrical conductivity of composites, have been used to model the mechanical features of composites since percolation of electrical conductivity tends to occur along with mechanical percolation. However, they do not consider microstructural mechanisms other than connectivity in the modeling process.

The interface/interphase section in the system originates from a perturbation of media in the presence of nanoparticles (interfacial bonding between phases) or interference in the mobility of long polymer chains [23–25]. Actually, the extremely large surface area per unit volume of nanoparticles and the robust interfacial connections lead to formation of a significant third phase as an interphase section in the nanocomposites [26–30]. The interphase is created between the polymer matrix and nanoparticles, which is different from both the polymer and nanoparticles. The interphase section is tougher than the polymer matrix reinforcing the nanocomposites. It was reported that the viscoelastic behavior of the PCNTs depends on the interface state and consequently, the functionalization of CNT surfaces is essential for using the CNTs [31]. Hence, the interphase should be considered to realize unforeseen trends in the features of a nanocomposite. Figure 1 schematically depicts a CNT and the surrounding interphase in a polymer system.

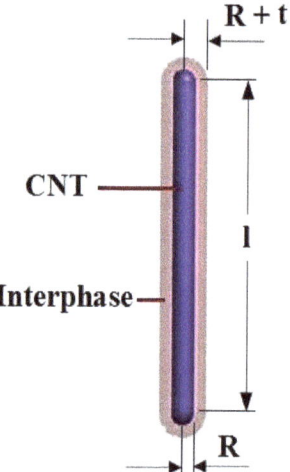

**Figure 1.** Schematic diagram of a CNT and the surrounding interphase in systems.

Many researchers have attempted to characterize the interface/interphase features. Generally, it has been reported that the dimensions and stiffness of the interphase are the main factors affecting the mechanical behavior of systems. Therefore, the interphase plays a reinforcing role in the mechanical testing of systems. The percolation onset may occur in the interphase section since it accelerates formation of a connected structure before the physical assembly of nanoparticles. The positive effect of interfacial interaction on the percolation onset has been confirmed and it is indicative of interphase percolation [32]. Interphase percolation has been studied based on the extension of the filler-excluded volume [33,34].

Although existing studies have considered the reinforcing feature of the interphase, percolation of the interphase section has not been clarified adequately. The interphase section can accelerate the percolation onset in PCNTs, resulting in a new approach for the formation of the network structure and significant improvement in the mechanical features. In this work, two advanced models of the tensile modulus and strength of systems are proposed to express the stiffening and percolating characteristics of the interphase in PCNTs. Likewise, the excluded volume of nanoparticles assumes the role of the interphase section in the percolation onset. The predictions of the proposed advanced models are compared to the tentative ranks of several examples. Finally, the influences of various factors on the mechanical features of systems are plotted considering the strengthening and percolating roles of the interphase section. The advanced models are helpful and valuable to predict and optimize the tensile modulus and strength of the PCNTs assuming the stiffening and percolating characteristics of the interphase section.

## 2. Upgrading of Models and Equations

The Halpin–Tsai model supposes the perfect stress transference between the polymer medium and a filler and the random three-dimensional (3D) arrangement of the filler [35] as:

$$E_L = E_m \left( \frac{1 + 2\alpha\eta\varphi_f}{1 - \eta\varphi_f} \right) \tag{1}$$

$$E_T = E_m \left( \frac{1 + 0.5\varphi_f}{1 - \varphi_f} \right) \tag{2}$$

$$\eta = \frac{E_f/E_m - 1}{E_f/E_m + 2\alpha} \tag{3}$$

$$\alpha = \frac{l}{d} \tag{4}$$

where $E_L$ and $E_T$ are the moduli in the longitudinal and transverse directions, respectively. Moreover, $E_m$ and $E_f$ denote the Young's moduli of the polymer media and the filler, respectively and $\varphi_f$ denotes the filler volume fraction. In addition, $\alpha$ is the aspect ratio of the filler and $l$ and $d$ denote the length and diameter of the particles, respectively. Moreover, $E_R$ the relative modulus (nanocomposite's modulus per media modulus) for a random 3D arrangement of the fillers is as follows:

$$E_R = \frac{1}{5}\frac{E_L}{E_m} + \frac{4}{5}\frac{E_T}{E_m} \tag{5}$$

This model does not consider the reinforcing effect of the interphase section. Therefore, it incorrectly predicts the modulus of systems. The above equations can be improved by assuming the interphase section. Accordingly, the interphase is considered as a separate phase that reinforces the systems apart from the nanoparticles. In fact, the dimensions and concentration of the interphase are assumed to be similar to those of the nanoparticles since both the nanoparticles and the interphase section reinforce a system simultaneously. The advanced equations are given as:

$$E_L = E_m \left( \frac{1 + 2\alpha_f\eta_f\varphi_f + 2\alpha_i\eta_i\varphi_i}{1 - \eta_f\varphi_f - \eta_i\varphi_i} \right) \tag{6}$$

$$E_T = E_m \left( \frac{1 + 0.5\varphi_f + +0.5\varphi_i}{1 - \varphi_f - \varphi_i} \right) \tag{7}$$

$$\eta_f = \frac{E_f/E_m - 1}{E_f/E_m + 2\alpha_f} \tag{8}$$

$$\eta_i = \frac{E_i/E_m - 1}{E_i/E_m + 2\alpha_i} \tag{9}$$

$$\alpha_i = \frac{l}{t} \tag{10}$$

where the subscripts $f$ and $i$ denote the filler and the interphase, respectively. Moreover, $\alpha_i$ denotes the aspect ratio of the interphase around the nanoparticles and t denotes the thickness of the interphase section. These equations express the reinforcing effect of the interphase in systems. The interphase modulus ($E_i$) is an intermediate quantity between $E_f$ and $E_m$ and its value provides information on the quality of the interphase section. The modulus of a nanocomposite, calculated using this model, depends on $\varphi_f$, R, l, $E_f$ and $E_m$, which are the factors of the classical Halpin–Tsai model, as well as on the additional factors due to the interphase section, namely, t and $E_i$. The advanced model assuming the interphase area does not consider the perfect stress transfer between the polymer and the filler since it correlates the extent of stress transferred to the interphase features.

For systems containing cylindrical fillers, the volume portion of the interphase $\varphi_i$ is defined [36] as:

$$\varphi_i = \varphi_f[(1+\frac{t}{R})^2 - 1] \quad (11)$$

where $R$ is the radius of the nanotubes ($R = d/2$). By substituting Equation (11) into Equations (6) and (7), the modulus can be correlated to the characteristics of the polymer media, nanoparticles and interphase section.

Chatterjee [37] proposed a relationship between the percolation onset and the aspect ratio of CNTs as follows:

$$\varphi_p \approx \frac{1}{\alpha_f} \quad (12)$$

By substituting $\alpha_f$ (the aspect ratio of the CNT as $l/d$) from Equation (12) into Equation (6), the modulus can be expressed in terms of the percolation onset. However, the interphase section swells the CNT nets and accelerates the percolation onset. Assuming the existence of interphase section, the percolation onset of nanoparticles in the PCNTs can be expressed in terms of the CNT volume ($V$) and excluded volume around CNTs ($V_{ex}$) [38] as:

$$\varphi_p = \frac{V}{V_{ex}} = \frac{\pi R^2 l + (4/3)\pi R^3}{\frac{32}{3}\pi(R+t)^3[1+\frac{3}{4}(\frac{l}{R+t})+\frac{3}{32}(\frac{l}{R+t})^2]} \quad (13)$$

Clearly, the CNT volume does not change due to the interphase part, while the excluded volume is assumed to increase to $R + t$ due to the interphase section. When this equation is compared to the tentative rank of percolation onset, the value of $t$ is obtained, whereas Equation (12) does not consider the interphase section. Moreover, it is possible to calculate the percolation onset in a sample using the CNT dimensions and interphase thickness. It is necessary to precisely determine the percolation onset to compute the minimum CNT concentration required for the formation of CNT nets in systems. We added the interphase part to the terms $E_L$, $E_T$ and $\varphi_p$ and developed the terms $\eta_i$, $\alpha_i$ and $\varphi_i$ by assuming the interphase section. All of these terms assume the role of the interphase section in the stiffening of systems. By substituting $\varphi_p$ from the above equation into Equations (6) and (8) ($\alpha_f = 1/\varphi_p$), the modulus of PCNTs can be expressed by assuming the reinforcing and networking effects of the interphase section.

We proposed a model to determine the tensile strength of the PCNTs by assuming the interphase features in our previous work [39]:

$$\sigma_c = \eta_0 \alpha_f \tau (1+\frac{t}{R})\varphi_f + \sigma_m[1-(1+\frac{t}{R})^2 \varphi_f] \quad (14)$$

where $\sigma_m$ denotes the media strength. In addition, $\eta_o$ is an orientation factor (1 for full filler arrangement, 3/8 for arbitrary in-plane 2D location and 1/5 for haphazard 3D filler organization). Moreover, $\tau$ denotes the interfacial shear strength. This equation considers the reinforcing effect of the interphase in PCNTs.

By substituting $\alpha_f$ ($\alpha_f = 1/\varphi_p$) from Equation (13) into the above model, the tensile strength can be expressed by simultaneously considering the stiffening and percolating influences of the interphase section. The relative strength (nanocomposite's strength per media strength) can be computed by rearranging Equation (14) as:

$$\sigma_R = 1 + \frac{\eta_0 \alpha_f \tau (1+\frac{t}{R})\varphi_f}{\sigma_m} - (1+\frac{t}{R})^2 \varphi_f \quad (15)$$

Moreover, it was suggested that $\tau$ is correlated to the interfacial factor, $B$ in the Pukanszky model [39] as:

$$\tau = \frac{\sigma_m(B-2.04)}{\eta_0 \alpha_f} \quad (16)$$

By substituting $\tau$ from Equation (16) into Equation (15), one can calculate the relative strength as follows:

$$\sigma_R = 1 + (B - 2.04)(1 + \frac{t}{R})\varphi_f - (1 + \frac{t}{R})^2 \varphi_f \qquad (17)$$

Pukanszky [40] recommended the following model to determine the strength of a system by:

$$\sigma_R = \frac{1 - \varphi_f}{1 + 2.5\varphi_f} \exp(B\varphi_f) \qquad (18)$$

where $B$ displays a measurable rank for filler–media interaction/linkage as:

$$B = (1 + A_c d_f t) \ln(\frac{\sigma_i}{\sigma_m}) \qquad (19)$$

where $A_c$ and $d_f$ are the specific surface area and density of the filler, respectively. Similarly, $\sigma_i$ is the interphase strength. The Pukanszky model can be restructured as follows:

$$\ln(\sigma_R \frac{1 + 2.5\varphi_f}{1 - \varphi_f}) = B\varphi_f \qquad (20)$$

One can calculate $B$ based on the linear association between $\ln(\sigma_R \frac{1+2.5\varphi_f}{1-\varphi_f})$ and $\varphi_f$. The value of $B$ is calculated using the tentative ranks of strength determined using the latter relationship. This rank is then used in Equation (17) to express the tensile strength in terms of the interphase thickness and other material factors. There are many types of CNTs such as single-walled nanotubes (SWNTs) and multi-walled nanotubes (MWNTs). The improved models can be applied for all types of long CNTs.

## 3. Results and Discussion

### 3.1. Confirming Views

The concurrent stiffening and percolating features of the interphase section in the PCNTs are described in this section using the proposed advanced models. The models are first applied to estimate the modulus and strength of some specimens described in the literature. The tentative ranks of samples are fitted to the proposed models. Based on the fitting results, the values of all of the interphase terms are calculated. The ranks of $t$ and $E_i$ are obtained by fitting Equations (6)–(10) to the tentative ranks. Additionally, $\tau$ and $B$ are calculated by fitting Equations (16) and (17) to the tentative ranks of strength. These equations may yield dissimilar values of each interphase term, but we report the average and reasonable ranks for the examples considered herein. Moreover, the models are used to demonstrate influences of all the factors on the modulus and strength of PCNTs.

### 3.2. Tensile Modulus

Figure 2 shows the tentative points of the relative modulus and the forecasts made using the advanced model (Equations (6)–(10)) for the polyamide 6 (PA6)/multi-walled carbon nanotubes (MWCNTs) [41], epoxy/MWCNTs [42], phenolic/MWCNTs [43] and PA6/MWCNTs-NH$_2$ [44] samples. The forecasted values agree well with the tentative ranges, thus validating the ability of the advanced model to calculate the modulus. This model usually provides superior forecasts at low volume proportions of nanoparticles since the deficient dispersion of nanoparticles at high filler percentages may lead to deviations from the forecasts.

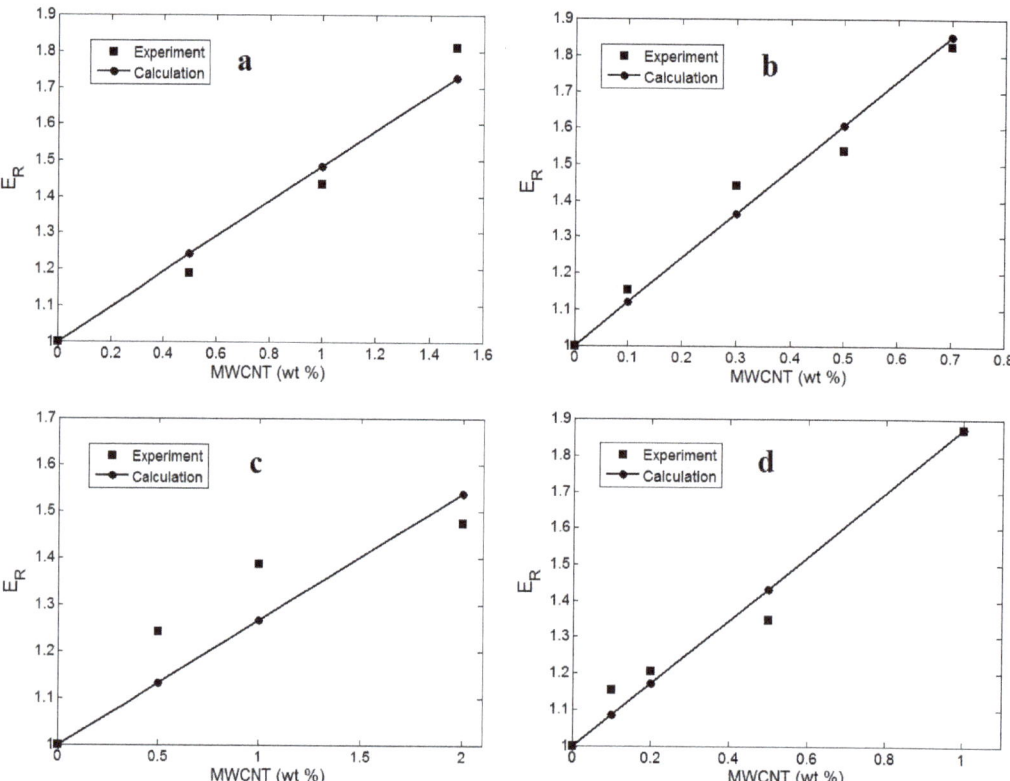

**Figure 2.** Tentative and theoretical ranks of relative modulus computed using the advanced model for (**a**) PA6/MWCNTs [41], (**b**) epoxy/MWCNTs [42], (**c**) phenolic/MWCNTs [43] and (**d**) PA6/MWCNTs-NH$_2$ [44] samples.

According to the expressed equations, the advanced model can forecast the modulus using suitable values of several factors. The material factors, such as $R$, $E_m$ and $l$ can be found in the original references reporting the mentioned samples. Moreover, the rank of $E_f$ is reflected as 1000 GPa according to [45]. Consequently, the values of the interphase factors, including $t$ and $E_i$, can be computed by applying the tentative ranks to the advanced model. This technique forecasts the interphase features using the tentative outputs of the modulus.

In the case of the PA6/MWCNTs, $t$ and $E_i$ are calculated as 4 nm and 10 GPa, respectively. Moreover, in the case of the epoxy/MWCNTs sample, $t$ = 14 nm and $E_i$ = 70 GPa, based on the fitting of the tentative data to the advanced model. Moreover, the computed values of t and $E_i$ for the phenolic/MWCNTs system are 25 nm and 130 GPa, respectively, while those for the PA6/MWCNTs-NH$_2$ system are 25 nm and 50 GPa. As identified, the interphase thickness cannot be higher than the radius of gyration of the macromolecules. Moreover, the interphase modulus varies between the values of the media modulus and the filler stiffness, that is, $E_m < E_i < E_f$. According to these criteria, the calculations of the interphase features are correct. Therefore, the advanced model provides correct ranks for the interphase attributes of the PCNTs.

The calculations indicate different ranks of the interphase within the reported samples. The best interphase is formed in the case of the phenolic/MWCNTs system, while the poorest interphase is formed in the case of the PA6/MWCNTs sample. The interphase thickness and modulus depend on the interfacial interaction/attachment between the polymer media and the nanoparticles [46,47]. Researchers have applied different methods to improve the interfacial features such as modification of the nanofiller surface or use of compatibilizers [48]. According to the above calculations, the best and the poorest interfacial features are obtained in the case of the phenolic/MWCNTs and PA6/MWCNTs samples, respectively.

Figure 3 reveals the effects of R and l on the modulus, computed using the advanced model at the normal ranks of $E_m$ = 2 GPa, $\varphi_f$ = 0.02, $E_f$ = 1000 GPa, $E_i$ = 100 GPa and t = 10 nm. The modulus is low when R is large and l is small, while the thinnest nanotubes yield the highest modulus. These comments indicate the undesirable effects of thick and small nanotubes and the desirable effects of thin nanoparticles on the modulus of PCNTs. $E_R$ improves by approximately 70 % upon the addition of nanotubes with R > 40 nm and l < 5 μm, while we get the best relative modulus of 4.5 when R = 5 nm. Therefore, it is significant to use thin nanoparticles when producing these systems. However, the nanoparticles tend to aggregate/agglomerate [49,50], which increases their thickness. Thus, the aggregation/agglomeration of nanoparticles should be prevented to control their thickness.

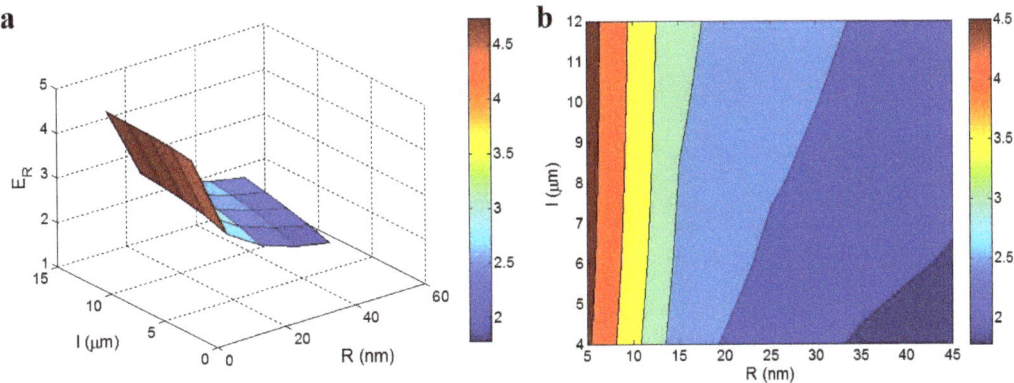

**Figure 3.** (a) The 3D and (b) contour plans of the influences of R and l on the relative modulus as predicted by the advanced model when $E_m$ = 2 GPa, $\varphi_f$ = 0.02, $E_f$ = 1000 GPa, $E_i$ = 100 GPa and t = 10 nm.

A low rank of R increases the extent of the surface area of nanoparticles, which increases the interfacial area/interaction, since smaller nanoparticles induce stronger interfacial contact with the polymer media owing to the analogous sizes of the nanoparticles and the macromolecules, which is also called the nano-effect [51]. Therefore, smaller nanoparticles increase the interfacial/interphase area and strengthen the interfacial communication. Since a larger and stronger interphase leads to stronger reinforcement in such systems, the advanced model properly predicts the influence of nanoparticle size on the modulus.

Figure 4 additionally illustrates the roles of the interphase factors in the relative modulus, as determined using the advanced model at $E_m$ = 2 GPa, $\varphi_f$ = 0.02, $E_f$ = 1000 GPa, l = 5 μm and R = 10 nm. The maximum modulus is observed at the uppermost ranks of t and $E_i$. The relative modulus of 7.5 is obtained with t = 25 nm and $E_i$ = 250 GPa. However, the modulus decreases as the values of the interphase features decrease, for example, t = 5 nm and $E_i$ = 50 GPa lead to $E_R$ = 2.7. Therefore, the interphase features directly affect the modulus according to the advanced model. These ranges are commonsensical since the interphase acts as a reinforcing agent in the systems.

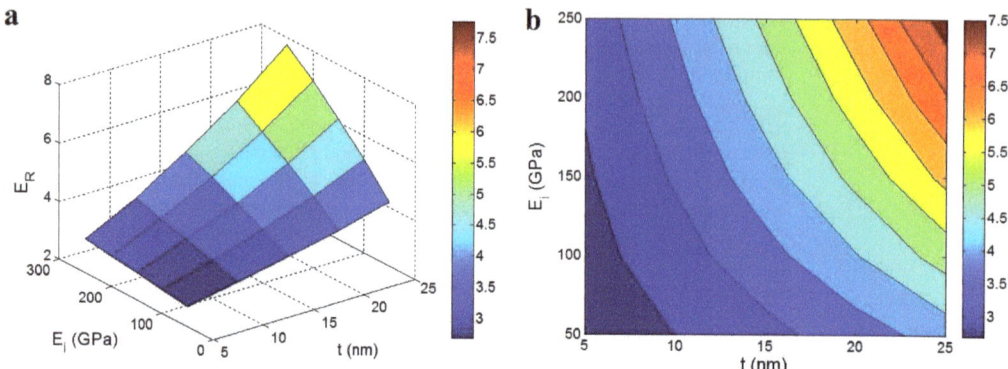

**Figure 4.** Relative modulus assuming interphase factors according to the advanced model at $E_m = 2$ GPa, $\varphi_f = 0.02$, $E_f = 1000$ GPa, $l = 5$ μm and $R = 10$ nm: (**a**) 3D and (**b**) contour plans.

A thicker interphase produces a stronger nanocomposite, since a thicker interphase shows the stronger connections between the polymer matrix and nanoparticles, which can bear and transfer more stress during loading [52,53]. In fact, a thick interphase reveals the strong interfacial attachments reinforcing the samples. Moreover, the interphase section helps accelerate the percolation onset in the PCNTs owing to the connections within the section. Certainly, a denser interphase accelerates the formation of connections in the interphase area, which advances the percolation onset. Therefore, the effect of interphase thickness on the modulus is reasonable. Moreover, a strong interphase stops the separation of nanoparticles from the polymer media in the loading process. Therefore, a greater amount of stress can be transferred from the polymer media to the nanoparticles without debonding, thus leading to a high range of modulus.

*3.3. Tensile Strength*

Figure 5 compares the tentative ranks of the relative strength with the forecasts obtained using the advanced model (Equation (17)) for the poly (vinyl alcohol) (PVA)/MWCNTs [54], polysilsesquioxane (PSE)/MWCNTs [55], chitosan/MWCNTs [56] and poly (phenylene sulfide) (PPS)/MWCNTs [57] systems. The forecasts follow the tentative ranges for all the systems. Therefore, the advanced model can appropriately forecast the tensile strength.

By comparing the tentative ranks to the forecasts obtained using the proposed model, the value of $t$ can be calculated for the samples. To calculate $\tau$ from Equation (16), $B$ should be determined using the tentative ranks of strength and the Pukanszky model. $B$ is calculated as 17.18, 145.5, 118.3 and 26.43, respectively, for the PVA/MWCNTs, PSE/MWCNTs, chitosan/MWCNTs and PPS/MWCNTs systems. With the abovementioned values, the values of $\tau$ are calculated to be 3.8, 3.54, 12.9 and 1.18 MPa, respectively for the mentioned samples. Based on the values of $B$ and those of other factors, such as $R$, $l$ and $\sigma_m$ reported in the original references, the values of $t$ are predicted as 4, 7, 8 and 10 nm, respectively, for the PVA/MWCNTs, PSE/MWCNTs, chitosan/MWCNTs and PPS/MWCNTs systems. Hence, the advanced model can be used to compute $t$ by assuming the fortifying and percolating characteristics of the interphase section. Moreover, the advanced model can be used to compare the interphase conditions of several samples. The interphase condition is illustrative of interfacial interaction/attachment, which controls the mechanical features of a system. Among the samples, the PPS/MWCNTs system exhibits the best interfacial/interphase features, while the PVA/MWCNTs system shows the poorest interphase section.

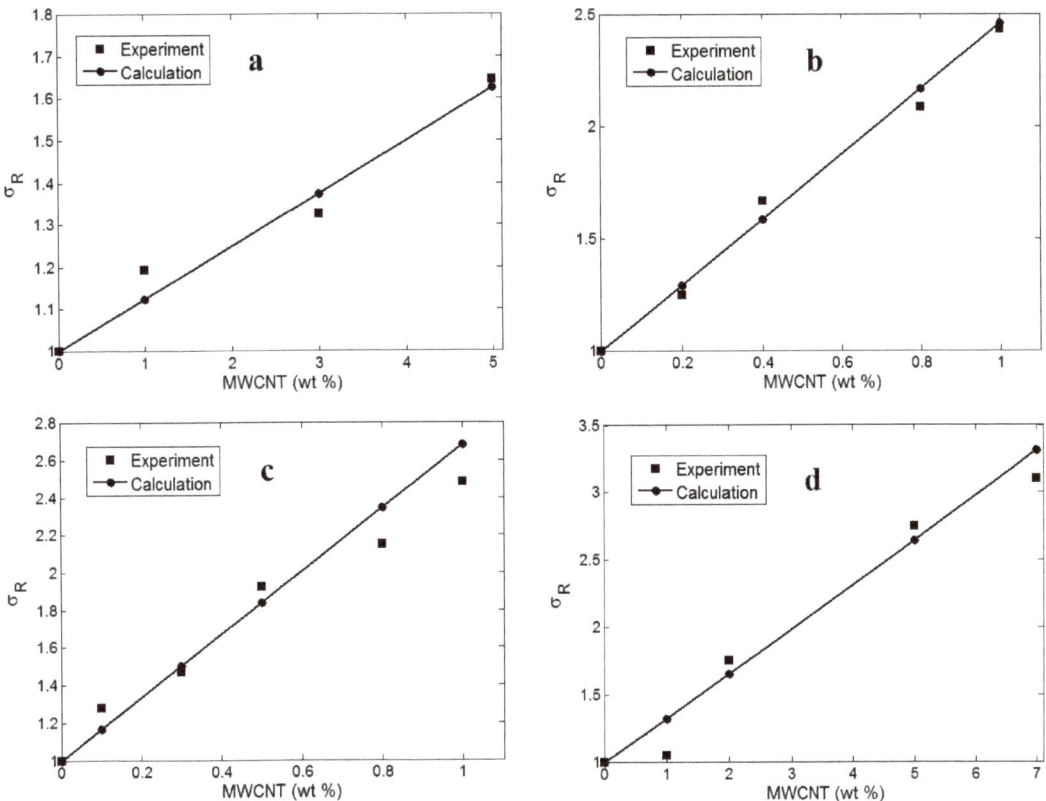

**Figure 5.** Comparison between the tentative and theoretical (Equation (17)) ranks of relative strength for (**a**) PVA/MWCNTs [54], (**b**) PSE/MWCNTs [55], (**c**) chitosan/MWCNTs [56] and (**d**) PPS/MWCNTs [57] systems.

Figure 6 shows the influences of $R$ and $\varphi_f$ on the relative strength using Equation (17) with $\sigma_m = 40$ MPa, $l = 10$ μm, $B = 30$ and $t = 10$ nm. As expected, a high volume proportion of thin nanoparticles increase the strength greatly. The $\sigma_R$ rank of 4.5 is achieved with $R = 5$ nm and $\varphi_f = 0.05$, whereas $\sigma_R = 1.3$ is obtained for $R > 25$ nm and $\varphi_f < 0.014$. Accordingly, the size and concentration of the CNTs significantly alter the strength of systems. Small nanoparticles generate a large surface area, thus intensifying the interfacial interaction (nano-effect) [51]. The interphase power depends on the interfacial communication/linkage and therefore, thin CNTs strengthen the interphase area in systems. Consequently, thin nanotubes increase the strength of systems owing to the higher ranks of interfacial range/contacts/bonding between thinner CNTs and polymer media.

The overall strength of a sample depends on the interfacial/interphase features between the polymer media and the nanofillers [58,59] since the stress is transferred through the interphase section. Moreover, a high nanoparticle content leads to a significant increase in strength since the strength of the CNTs is considerably higher (11–50 GPa) than that of polymer media (up to 60 MPa). However, good dispersion of nanoparticles is assumed in the advanced model and the aggregation/agglomeration of nanoparticles at high $\varphi_f$ may reduce the strength of a system. Generally, the advanced model shows the reasonable effects of $R$ and $\varphi_f$ on the strength by considering the stiffening and percolation of the interphase section.

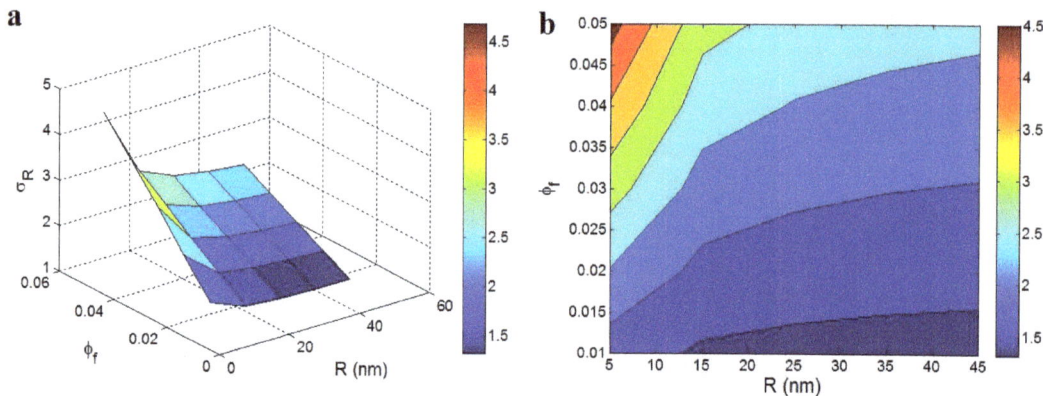

**Figure 6.** Influences of $R$ and $\varphi_f$ on the relative strength as determined using the advanced model: (**a**) 3D and (**b**) contour plots.

Figure 7 shows the influences of $l$ and $\tau$ on the relative strength as determined using the advanced model at $\sigma_m = 40$ MPa, $\varphi_f = 0.02$, $R = 10$ nm and $t = 10$ nm. The highest relative strength of 4.5 is expectedly observed at the maximum ranks of $l$ and $\tau$, that is, $l = 12$ μm and $\tau = 8$ MPa. The poorest relative strength of 1.25 is observed at the least ranks of $l = 4$ μm and $\tau = 2$ MPa. Consequently, $l$ and $\tau$ positively influence the strength of the systems.

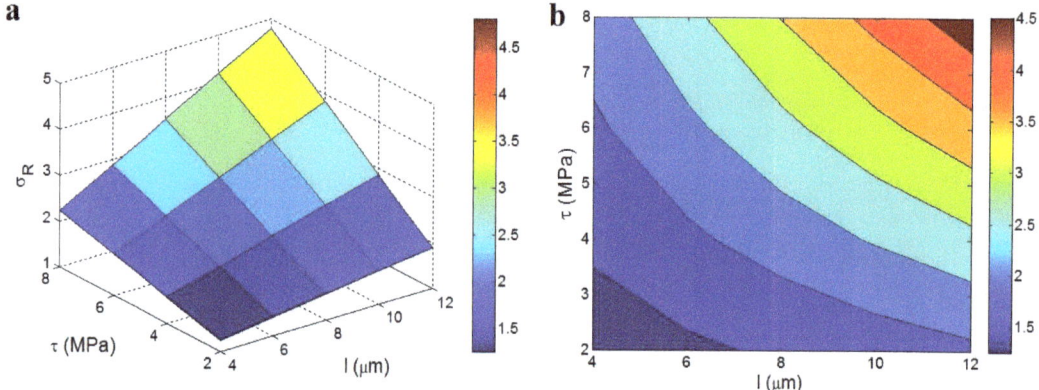

**Figure 7.** Influences of $l$ and $\tau$ on the relative strength (Equation (17)) at $\sigma_m = 40$ MPa, $\varphi_f = 0.02$, $R = 10$ nm and $t = 10$ nm: (**a**) 3D and (**b**) contour plots.

A long nanotube can promote the interfacial extent, which enhances the mechanical involvement between the polymer media and the nanoparticles. Moreover, it increases the aspect ratio, which increases the overall system's strength [60]. Furethermore, a high rank of the interfacial shear strength $\tau$ indicates the strong interfacial interaction/adhesion, which leads to formation of a strong interphase. As stated, a strong interphase mainly strengthens the system due to the fortifying efficiency of the interphase as the third phase in addition to the polymer media and the nanoparticles. The reinforcing effect of the interphase in systems has been extensively validated in tentative and theoretical studies [61,62].

Figure 8 additionally shows the relative strength determined using the advanced model in terms of the interfacial/interphase factors, $t$ and $B$ at $\sigma_m = 40$ MPa, $\varphi_f = 0.02$, $l = 10$ μm and $R = 10$ nm. The worst outputs are obtained for the lowest ranges of these factors, while the best outputs are obtained for the highest values of $t$ and $B$. Consequently, these factors directly affect the tensile strength and a significant strength enhancement is achieved with a thick interphase and high B, that is, substantial interfacial features. The maximum $\sigma_R$ of 6.5 is observed at $t = 25$ nm and $B = 90$, whereas a slight improvement in strength is achieved with the lowest ranks of $t$ and $B$. This trend is sensible since the high ranks of these factors promote the reinforcing and percolation effects of the interphase section. A denser interphase leads to a greater interphase concentration in systems, which increases the strength.

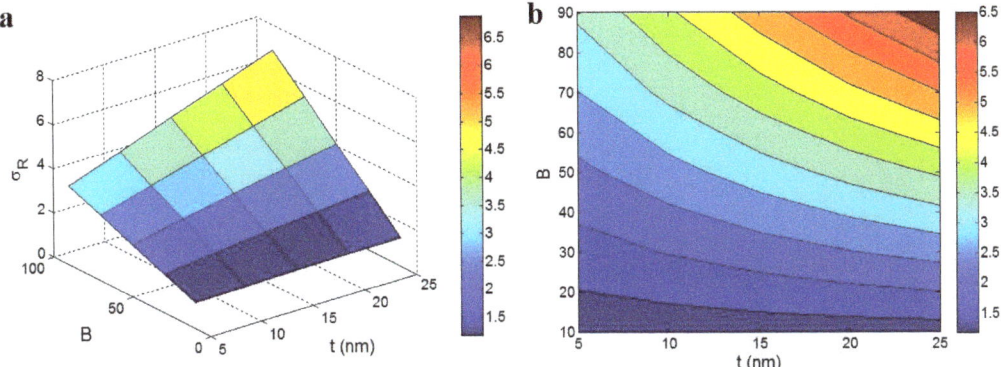

**Figure 8.** (a) The 3D and (b) contour plots of the relative strength in terms of $t$ and $B$ as determined using the advanced model at $\sigma_m = 40$ MPa, $\varphi_f = 0.02$, $l = 10$ μm and $R = 10$ nm.

The positive influence of a thick interphase on the mechanical features of systems has been established in the literature [35,63]. Likewise, a thick interphase easily percolates in a system, which accelerates the percolation onset. The networking of nanoparticles improves the mechanical features considerably. Consequently, it is expected that a thicker interphase will increase the strength. The direct relationship between the strength of a nanocomposite and $B$ is common in systems, according to the Pukanszky model (Equation (18)). The rank of $B$ reveals the features of the interphase section, such as $t$ and $\sigma_i$ (Equation (19)). A higher value of $B$ indicates a thicker and stronger interphase section, which leads to a stronger system based on the strengthening and percolating influences of the interphase section. Hence, the advanced equations accurately illustrate the effects of $t$ and $B$ on the strength of a nanocomposite.

## 4. Conclusions

Two models were improved for determining the modulus and strength of systems and used to investigate the fortifying and percolating characteristics of the interphase section in PCNTs. The forecasts generated by both models exhibited a good match with the tentative ranges, which validated the modulus and strength values calculated using the advanced models. In addition, the tentative ranges and the advanced models were used to determine the interphase features. The undesirable effects of thick and small nanotubes and the desirable effects of thin nanoparticles on the modulus were observed. The $E_R$ improved by only approximately 70% when $R > 40$ nm and $l < 5$ μm, but the maximum relative modulus of 4.5 was obtained with the lowest value of $R$ of 5 nm. Likewise, the best modulus was detected with the highest ranges of $t$ and $E_i$. The highest relative modulus of 7.5 was achieved with $t = 25$ nm and $E_i = 250$ GPa. Consequently, $t$, $E_i$ and $l$ positively influenced the stiffness of systems. A high concentration of thin nanoparticles led to considerable increase in strength, as determined using the advanced model. A $\sigma_R$ grade of 4.5 was obtained with $R = 5$ nm and $\varphi_f = 0.05$, whereas $\sigma_R = 1.3$ was obtained with $R > 25$ nm and $\varphi_f < 0.014$. The greatest relative strength of 4.5 was acquired with the maximum ranges of $l$ and $\tau$. However, the poorest relative strength of 1.25 was observed at the minimum series of $l = 4$ μm and $\tau = 2$ MPa. Moreover, the best grades of the advanced model for strength were obtained with the highest ranks of $t$ and $B$ since they indicate the magnitude of the interfacial/interphase features. Generally, the strength of a nanocomposite meaningfully improves as the values of $\varphi_f$, $l$, $\tau$, $t$ and $B$ increase largely, although a high value of $R$ (thick CNTs) decreases the strength.

**Author Contributions:** Formal analysis, Y.Z.; Funding acquisition, K.R.; Methodology, Y.Z. and K.R.; Validation, K.R.; Writing—original draft, Y.Z.; Writing—review & editing, K.R. All authors have read and agreed to the published version of the manuscript.

**Funding:** This research received no external funding.

**Institutional Review Board Statement:** Not applicable.

**Informed Consent Statement:** Not applicable.

**Data Availability Statement:** Not applicable.

**Acknowledgments:** This work was supported by the Basic Science Research Program through the National Research Foundation of Korea (NRF) funded by the Ministry of Education, Science and Technology (project number: 2020R1A2B5B02002203).

**Conflicts of Interest:** The authors declare no conflict of interest.

## References

1. Khan, A.A.P.; Bazan, G.C.; Alhogbi, B.G.; Marwani, H.M.; Khan, A.; Alam, M.; Rahman, M.M.; Asiri, A.M. Nanocomposite cross-linked conjugated polyelectrolyte/MWCNT/poly (pyrrole) for enhanced Mg2+ ion sensing and environmental remediation in real samples. *J. Mater. Res. Technol.* **2020**, *9*, 9667–9674. [CrossRef]
2. Kim, S.; Zare, Y.; Garmabi, H.; Rhee, K.Y. Variations of tunneling properties in poly (lactic acid)(PLA)/poly (ethylene oxide)(PEO)/carbon nanotubes (CNT) nanocomposites during hydrolytic degradation. *Sens. Actuators A Phys.* **2018**, *274*, 28–36. [CrossRef]
3. Zare, Y.; Rhee, K.Y. Expression of normal stress difference and relaxation modulus for ternary nanocomposites containing biodegradable polymers and carbon nanotubes by storage and loss modulus data. *Compos. Part B Eng.* **2018**, *158*, 162–168. [CrossRef]
4. El Sayed, A.M. Synthesis, optical, thermal, electric properties and impedance spectroscopy studies on P (VC-MMA) of optimized thickness and reinforced with MWCNTs. *Results Phys.* **2020**, *17*, 103025. [CrossRef]
5. Zare, Y.; Rhee, K.Y. Following the morphological and thermal properties of PLA/PEO blends containing carbon nanotubes (CNTs) during hydrolytic degradation. *Compos. Part B Eng.* **2019**, *175*, 107132. [CrossRef]
6. Zare, Y.; Rhee, K.Y. A power model to predict the electrical conductivity of CNT reinforced nanocomposites by considering interphase, networks and tunneling condition. *Compos. Part B Eng.* **2018**, *155*, 11–18. [CrossRef]
7. Behdinan, K.; Moradi-Dastjerdi, R.; Safaei, B.; Qin, Z.; Chu, F.; Hui, D. Graphene and CNT impact on heat transfer response of nanocomposite cylinders. *Nanotechnol. Rev.* **2020**, *9*, 41–52. [CrossRef]
8. Fard, M.Y.; Raji, B.; Pankretz, H. Correlation of nanoscale interface debonding and multimode fracture in polymer carbon composites with long-term hygrothermal effects. *Mech. Mater.* **2020**, *150*, 103601. [CrossRef]
9. Yazik, M.M.; Sultan, M.; Mazlan, N.; Talib, A.A.; Naveen, J.; Shah, A.; Safri, S. Effect of hybrid multi-walled carbon nanotube and montmorillonite nanoclay content on mechanical properties of shape memory epoxy nanocomposite. *J. Mater. Res. Technol.* **2020**, *9*, 6085–6100. [CrossRef]
10. Chen, J.; Han, J. Effect of hydroxylated carbon nanotubes on the thermal and electrical properties of derived epoxy composite materials. *Results Phys.* **2020**, *18*, 103246. [CrossRef]
11. Mahmoodi, M.; Rajabi, Y.; Khodaiepour, B. Electro-thermo-mechanical responses of laminated smart nanocomposite moderately thick plates containing carbon nanotube—A multi-scale modeling. *Mech. Mater.* **2020**, *141*, 103247. [CrossRef]
12. Ajori, S.; Parsapour, H.; Ansari, R. Structural properties and buckling behavior of non-covalently functionalized single-and double-walled carbon nanotubes with pyrene-linked polyamide in aqueous environment using molecular dynamics simulations. *J. Phys. Chem. Solids* **2019**, *131*, 79–85. [CrossRef]
13. Heller, D.A.; Barone, P.W.; Swanson, J.P.; Mayrhofer, R.M.; Strano, M.S. Using Raman spectroscopy to elucidate the aggregation state of single-walled carbon nanotubes. *J. Phys. Chem. B* **2004**, *108*, 6905–6909. [CrossRef]
14. Haghgoo, M.; Ansari, R.; Hassanzadeh-Aghdam, M.; Nankali, M. Analytical formulation for electrical conductivity and percolation threshold of epoxy multiscale nanocomposites reinforced with chopped carbon fibers and wavy carbon nanotubes considering tunneling resistivity. *Compos. Part A Appl. Sci. Manuf.* **2019**, *126*, 105616. [CrossRef]
15. Maiti, S.; Bera, R.; Karan, S.K.; Paria, S.; De, A.; Khatua, B.B. PVC bead assisted selective dispersion of MWCNT for designing efficient electromagnetic interference shielding PVC/MWCNT nanocomposite with very low percolation threshold. *Compos. Part B Eng.* **2019**, *167*, 377–386. [CrossRef]

16. Poothanari, M.A.; Xavier, P.; Bose, S.; Kalarikkal, N.; Komalan, C.; Thomas, S. Compatibilising action of multiwalled carbon nanotubes in polycarbonate/polypropylene (PC/PP) blends: Phase morphology, viscoelastic phase separation, rheology and percolation. *J. Polym. Res.* **2019**, *26*, 178. [CrossRef]
17. Haghgoo, M.; Hassanzadeh-Aghdam, M.; Ansari, R. A comprehensive evaluation of piezoresistive response and percolation behavior of multiscale polymer-based nanocomposites. *Compos. Part A Appl. Sci. Manuf.* **2020**, *130*, 105735. [CrossRef]
18. Zare, Y. An approach to study the roles of percolation threshold and interphase in tensile modulus of polymer/clay nanocomposites. *J. Colloid Interface Sci.* **2017**, *486*, 249–254. [CrossRef]
19. Li, H.-X.; Zare, Y.; Rhee, K.Y. The percolation threshold for tensile strength of polymer/CNT nanocomposites assuming filler network and interphase regions. *Mater. Chem. Phys.* **2018**, *207*, 76–83. [CrossRef]
20. Zare, Y.; Rhee, K.Y. The mechanical behavior of CNT reinforced nanocomposites assuming imperfect interfacial bonding between matrix and nanoparticles and percolation of interphase regions. *Compos. Sci. Technol.* **2017**, *144*, 18–25. [CrossRef]
21. Favier, V.; Chanzy, H.; Cavaille, J. Polymer nanocomposites reinforced by cellulose whiskers. *Macromolecules* **1995**, *28*, 6365–6367. [CrossRef]
22. Ouali, N.; Cavaillé, J.; Perez, J. Elastic, viscoelastic and plastic behavior of multiphase polymer blends. *Plast. Rubber Compos. Process. Appl.* **1991**, *16*, 55–60.
23. Zare, Y.; Garmabi, H. Modeling of interfacial bonding between two nanofillers (montmorillonite and CaCO3) and a polymer matrix (PP) in a ternary polymer nanocomposite. *Appl. Surf. Sci.* **2014**, *321*, 219–225. [CrossRef]
24. Wan, C.; Chen, B. Reinforcement and interphase of polymer/graphene oxide nanocomposites. *J. Mater. Chem.* **2012**, *22*, 3637–3646. [CrossRef]
25. Herasati, S.; Zhang, L.; Ruan, H. A new method for characterizing the interphase regions of carbon nanotube composites. *Int. J. Solids Struct.* **2014**, *51*, 1781–1791. [CrossRef]
26. Zare, Y.; Rhee, K.Y. Development of a conventional model to predict the electrical conductivity of polymer/carbon nanotubes nanocomposites by interphase, waviness and contact effects. *Compos. Part A Appl. Sci. Manuf.* **2017**, *100*, 305–312. [CrossRef]
27. Zare, Y. A simple technique for determination of interphase properties in polymer nanocomposites reinforced with spherical nanoparticles. *Polymer* **2015**, *72*, 93–97. [CrossRef]
28. Zare, Y. Assumption of interphase properties in classical Christensen–Lo model for Young's modulus of polymer nanocomposites reinforced with spherical nanoparticles. *RSC Adv.* **2015**, *5*, 95532–95538. [CrossRef]
29. Hassanzadeh-Aghdam, M.K.; Mahmoodi, M.J.; Ansari, R. Creep performance of CNT polymer nanocomposites-An emphasis on viscoelastic interphase and CNT agglomeration. *Compos. Part B Eng.* **2019**, *168*, 274–281. [CrossRef]
30. Zare, Y. Estimation of material and interfacial/interphase properties in clay/polymer nanocomposites by yield strength data. *Appl. Clay Sci.* **2015**, *115*, 61–66. [CrossRef]
31. Pan, Y.; Weng, G.; Meguid, S.; Bao, W.; Zhu, Z.; Hamouda, A. Interface effects on the viscoelastic characteristics of carbon nanotube polymer matrix composites. *Mech. Mater.* **2013**, *58*, 1–11. [CrossRef]
32. Celzard, A.; McRae, E.; Deleuze, C.; Dufort, M.; Furdin, G.; Marêché, J. Critical concentration in percolating systems containing a high-aspect-ratio filler. *Phys. Rev. B* **1996**, *53*, 6209. [CrossRef]
33. Pontefisso, A.; Zappalorto, M.; Quaresimin, M. Influence of interphase and filler distribution on the elastic properties of nanoparticle filled polymers. *Mech. Res. Commun.* **2013**, *52*, 92–94. [CrossRef]
34. Dominkovics, Z.; Hári, J.; Kovács, J.; Fekete, E.; Pukánszky, B. Estimation of interphase thickness and properties in PP/layered silicate nanocomposites. *Eur. Polym. J.* **2011**, *47*, 1765–1774. [CrossRef]
35. Saatchi, M.; Shojaei, A. Mechanical performance of styrene-butadiene-rubber filled with carbon nanoparticles prepared by mechanical mixing. *Mater. Sci. Eng. A* **2011**, *528*, 7161–7172. [CrossRef]
36. Ji, X.L.; Jiao, K.J.; Jiang, W.; Jiang, B.Z. Tensile modulus of polymer nanocomposites. *Polym. Eng. Sci.* **2002**, *42*, 983. [CrossRef]
37. Chatterjee, A.P. A model for the elastic moduli of three-dimensional fiber networks and nanocomposites. *J. Appl. Phys.* **2006**, *100*, 054302. [CrossRef]
38. Zare, Y.; Rhee, K.Y.; Park, S.-J. Modeling the roles of carbon nanotubes and interphase dimensions in the conductivity of nanocomposites. *Results Phys.* **2019**, *15*, 102562. [CrossRef]
39. Zare, Y. Effects of interphase on tensile strength of polymer/CNT nanocomposites by Kelly–Tyson theory. *Mech. Mater.* **2015**, *85*, 1–6. [CrossRef]
40. Pukanszky, B. Influence of interface interaction on the ultimate tensile properties of polymer composites. *Composites* **1990**, *21*, 255–262. [CrossRef]
41. Shao, W.; Wang, Q.; Wang, F.; Chen, Y. The cutting of multi-walled carbon nanotubes and their strong interfacial interaction with polyamide 6 in the solid state. *Carbon* **2006**, *44*, 2708–2714. [CrossRef]
42. Zou, W.; Du, Z.-J.; Liu, Y.-X.; Yang, X.; Li, H.-Q.; Zhang, C. Functionalization of MWNTs using polyacryloyl chloride and the properties of CNT–epoxy matrix nanocomposites. *Compos. Sci. Technol.* **2008**, *68*, 3259–3264. [CrossRef]
43. Yeh, M.-K.; Tai, N.-H.; Lin, Y.-J. Mechanical properties of phenolic-based nanocomposites reinforced by multi-walled carbon nanotubes and carbon fibers. *Compos. Part A Appl. Sci. Manuf.* **2008**, *39*, 677–684. [CrossRef]
44. Chen, G.-X.; Kim, H.-S.; Park, B.H.; Yoon, J.-S. Multi-walled carbon nanotubes reinforced nylon 6 composites. *Polymer* **2006**, *47*, 4760–4767. [CrossRef]

45. Bhuiyan, M.A.; Pucha, R.V.; Worthy, J.; Karevan, M.; Kalaitzidou, K. Understanding the effect of CNT characteristics on the tensile modulus of CNT reinforced polypropylene using finite element analysis. *Comput. Mater. Sci.* **2013**, *79*, 368–376. [CrossRef]
46. Li, Y.; Waas, A.M.; Arruda, E.M. The effects of the interphase and strain gradients on the elasticity of layer by layer (LBL) polymer/clay nanocomposites. *Int. J. Solids Struct.* **2011**, *48*, 1044–1053. [CrossRef]
47. Lu, P.; Leong, Y.; Pallathadka, P.; He, C. Effective moduli of nanoparticle reinforced composites considering interphase effect by extended double-inclusion model–Theory and explicit expressions. *Int. J. Eng. Sci.* **2013**, *73*, 33–55. [CrossRef]
48. Nesterov, A.; Lipatov, Y. Compatibilizing effect of a filler in binary polymer mixtures. *Polymer* **1999**, *40*, 1347–1349. [CrossRef]
49. Shokri-Oojghaz, R.; Moradi-Dastjerdi, R.; Mohammadi, H.; Behdinan, K. Stress distributions in nanocomposite sandwich cylinders reinforced by aggregated carbon nanotube. *Polymer Compos.* **2019**, *40*, E1918–E1927. [CrossRef]
50. Daghigh, H.; Daghigh, V. Free vibration of size and temperature-dependent carbon nanotube (CNT)-reinforced composite nanoplates with CNT agglomeration. *Polym. Compos.* **2019**, *40*, E1479–E1494. [CrossRef]
51. Crosby, A.J.; Lee, J.Y. Polymer nanocomposites: The "nano" effect on mechanical properties. *Polym. Rev.* **2007**, *47*, 217–229. [CrossRef]
52. Zare, Y.; Fasihi, M.; Rhee, K.Y. Efficiency of stress transfer between polymer matrix and nanoplatelets in clay/polymer nanocomposites. *Appl. Clay Sci.* **2017**, *143*, 265–272. [CrossRef]
53. Zare, Y. Effects of imperfect interfacial adhesion between polymer and nanoparticles on the tensile modulus of clay/polymer nanocomposites. *Appl. Clay Sci.* **2016**, *129*, 65–70. [CrossRef]
54. Mi, Y.; Zhang, X.; Zhou, S.; Cheng, J.; Liu, F.; Zhu, H.; Dong, X.; Jiao, Z. Morphological and mechanical properties of bile salt modified multi-walled carbon nanotube/poly (vinyl alcohol) nanocomposites. *Compos. Part A Appl. Sci. Manuf.* **2007**, *38*, 2041–2046. [CrossRef]
55. Yuen, S.M.; Ma, C.C.M. Morphological, electrical, and mechanical properties of multiwall carbon nanotube/polysilsesquioxane composite. *J. Appl. Polym. Sci.* **2008**, *109*, 2000–2007. [CrossRef]
56. Cao, X.; Dong, H.; Li, C.M.; Lucia, L.A. The enhanced mechanical properties of a covalently bound chitosan-multiwalled carbon nanotube nanocomposite. *J. Appl. Polym. Sci.* **2009**, *113*, 466–472. [CrossRef]
57. Yu, S.; Wong, W.M.; Hu, X.; Juay, Y.K. The characteristics of carbon nanotube-reinforced poly (phenylene sulfide) nanocomposites. *J. Appl. Polym. Sci.* **2009**, *113*, 3477–3483. [CrossRef]
58. Tserpes, K.; Chanteli, A.; Floros, I. Prediction of yield strength of MWCNT/PP nanocomposite considering the interphase and agglomeration. *Compos. Struct.* **2017**, *168*, 657–662. [CrossRef]
59. Lazzeri, A.; Phuong, V.T. Dependence of the Pukánszky's interaction parameter B on the interface shear strength (IFSS) of nanofiller-and short fiber-reinforced polymer composites. *Compos. Sci. Technol.* **2014**, *93*, 106–113. [CrossRef]
60. Ghasemi, A.; Mohammadi, M.; Mohandes, M. The role of carbon nanofibers on thermo-mechanical properties of polymer matrix composites and their effect on reduction of residual stresses. *Compos. Part B Eng.* **2015**, *77*, 519–527. [CrossRef]
61. Zare, Y.; Rhim, S.; Garmabi, H.; Rhee, K.Y. A simple model for constant storage modulus of poly (lactic acid)/poly (ethylene oxide)/carbon nanotubes nanocomposites at low frequencies assuming the properties of interphase regions and networks. *J. Mech. Behav. Biomed. Mater.* **2018**, *80*, 164–170. [CrossRef]
62. Zare, Y. Modeling the yield strength of polymer nanocomposites based upon nanoparticle agglomeration and polymer–filler interphase. *J. Colloid Interface Sci.* **2016**, *467*, 165–169. [CrossRef]
63. Joshi, P.; Upadhyay, S. Effect of interphase on elastic behavior of multiwalled carbon nanotube reinforced composite. *Comput. Mater. Sci.* **2014**, *87*, 267–273. [CrossRef]

# Article
# Ecological Performance Optimization of a High Temperature Proton Exchange Membrane Fuel Cell

Dongxu Li, Siwei Li, Zheshu Ma *, Bing Xu, Zhanghao Lu, Yanju Li and Meng Zheng

College of Automobile and Traffic Engineering, Nanjing Forestry University, Nanjing 210037, China; Ldx961203@163.com (D.L.); wesly.li@outlook.com (S.L.); xb18260078388@163.com (B.X.); L18252038553@163.com (Z.L.); njfulyj@163.com (Y.L.); mengzai19950929@163.com (M.Z.)
* Correspondence: mazheshu@njfu.edu.cn; Tel.: +86-137-7665-9269

**Abstract:** According to finite-time thermodynamics, an irreversible high temperature proton exchange membrane fuel cell (HT-PEMFC) model is established, and the mathematical expressions of the output power, energy efficiency, exergy efficiency and ecological coefficient of performance (ECOP) of HT-PEMFC are deduced. The ECOP is a step forward in optimizing the relationship between power and power dissipation, which is more in line with the principle of ecology. Based on the established HT-PEMFC model, the maximum power density is obtained under different parameters that include operating temperature, operating pressure, phosphoric acid doping level and relative humidity. At the same time, the energy efficiency, exergy efficiency and ECOP corresponding to the maximum power density are acquired so as to determine the optimal value of each index under the maximum power density. The results show that the higher the operating temperature and the doping level, the better the performance of HT-PEMFC is. However, the increase of operating pressure and relative humidity has little effect on HT-PEMFC performance.

**Keywords:** high temperature proton exchange membrane fuel cell; exergy analysis; ecological analysis; ecological coefficient of performance

## 1. Introduction

In recent years, proton exchange membrane fuel cells (PEMFCs) have been considered efficient and clean energy conversion devices. PEMFCs have been widely used in home equipment and automobiles [1] due to the advantage of higher power density, lower emission and noise. According to the operating temperature, PEMFC can be divided into low temperature proton exchange membrane fuel cell (70–95 °C) and high temperature proton exchange membrane fuel cell (120–200 °C). HT-PEMFC has the superiority of the accelerated kinetics of electrode reaction [2], higher CO tolerance [3], and simpler water and heat management systems [4,5].

At present, the research on HT-PEMFC mainly includes materials [6–8] and preparation methods [9,10]. Few people have used the first and second laws of thermodynamics to analyze and optimize the performance of HT-PEMFC. However, thermodynamic analysis and optimization of LT-PEMFC have been mature. Miansari et al. [11], Ozen et al. [12] and Esfeh et al. [13] verified that the operating temperature has a significant effect on the performance improvement of PEMFC. Li et al. [14,15] established the finite-time thermodynamic model of irreversible PEMFC, which considered polarization loss and leakage current. The effects of operating temperature, operating pressure and proton exchange membrane water content on the optimal performance of the irreversible proton exchange membrane fuel cell were numerically studied. Wei et al. [16] took the entropy production rate and ecological coefficient of performance of PEMFC as the objective function for numerical analysis, while optimal current density ranges were determined by different optimization objectives. Midilli et al. [17] found that higher current density and proton film thickness leads to a decrease in exergy efficiency of PEMFC. If the film thickness was the same, the exergy efficiency of PEMFC is improved with the increase of operating pressure and the decrease of current density. Xu et al. [18] investigated exergetic sustainability indicators (ESI) of PEMFC under different parameters. Increasing the operating temperature and pressure decreases the irreversibility of PEMFC and increases exergetic sustainability indicators.

Therefore, the study of parameters based on thermodynamics is very important to boost the performance of PEMFC [11–23], thus, it has been applied to the irreversibility analysis and optimization of HT-PEMFC. Barati et al. [24] studied the influence of air and hydrogen flow rate, operating temperature and doping level of phosphoric acid on HT-PEMFC performance. The doping level has a significant effect on the performance improvement, mainly because the doping level of the membrane affected the proton conductivity of the membrane.

Lu et al. [25] established a mathematical model of HT-PEMFC and analyzed the exergy performance of the HT-PEMFC power generation system. Consequently, an improved farmland fertility optimization design method was put forward for optimizing the exergy, irreversibility and output power. Compared with the original design method and genetic algorithm, the number of iterations in the improved method was less, optimization speed was faster, and the output power density increased by 5.2 and 2.9%, respectively.

Xia et al. [26] investigated the effects of catalyst layer thickness, operating temperature, and proton exchange membrane thickness on HT-PEMFC performance. The results showed that the operating temperature has a significant effect on the performance. Operating temperature at 160–180 °C not only ensured the fuel cell performance, but also reduced maintenance costs at high temperatures. The thinner thickness of the catalyst layer and proton exchange membrane had a positive influence on the performance of HT-PEMFC, but it was easily damaged.

Guo et al. [19] analyzed the energetic, exergetic and ecological performance of HT-PEMFC and mathematical models of power density, entropy production rate and ecological coefficient of performance were established based on finite time thermodynamics theory. The results showed that the operating temperature and doping level have significant effects on the performance of HT-PEMFC. According to the optimization criterion of maximum power density, the optimization interval of current density is found to be the left of the current density corresponding to the maximum power density.

Lin et al. [27] investigated the exergy efficiency of HT-PEMFC using the meta-heuristic technique, and an improved collective animal behavior algorithm was utilized to evaluate and optimize the thermodynamic irreversibility, exergy efficiency and output power. Compared with the standard collective animal behavior algorithm and genetic algorithm, the proposed improved collective animal behavior algorithm increased the output power density by 1.2 and 12.1% and the exergy efficiency increased by 22.9%.

In this paper, firstly, a finite-time thermodynamic was introduced to analyze the irreversibility of HT-PEMFC, and a mathematical model which took irreversible losses and leakage current into consideration was established. Secondly, according to the maximum power density criterion, the optimization interval of current density was obtained and the optimal output efficiency, exergy efficiency and ecological coefficient of performance corresponding to the maximum power density were achieved. At the same time, the effects of operating temperature, operating pressure, relative humidity and doping level on the performance of HT-PEMFC were studied.

## 2. Thermodynamic Model
### 2.1. Working Principle of HT-PEMFC

As shown in Figure 1, HT-PEMFC can directly convert the chemical energy containing hydrogen and oxygen into electrical energy and heat energy. The whole system mainly includes a cathode, an anode and electrolyte. The reaction of anode and cathode show as follows:

$$\text{Anode reaction}: H_2 \rightarrow 2H^+ + 2e^- \tag{1}$$

$$\text{Cathodic reaction}: 2H^+ + \frac{1}{2}O_2 + 2e^- \rightarrow H_2O + heat \tag{2}$$

$$\text{Total reaction}: H_2 + \frac{1}{2}O_2 \rightarrow H_2O + heat + electricity \tag{3}$$

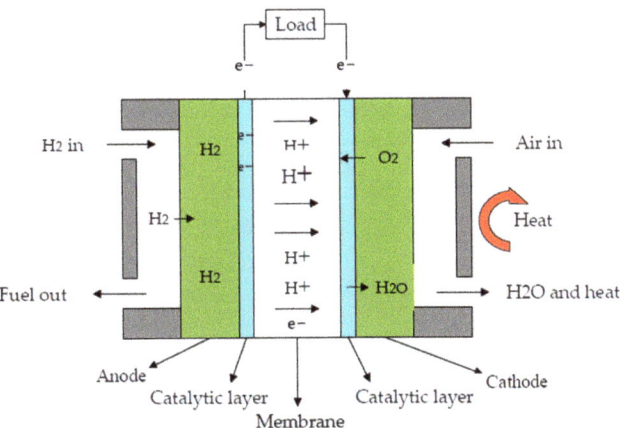

**Figure 1.** Working principle of an HT-PEMFC system fueled with $H_2$ and $O_2$.

*2.2. Reversible Potential of HT-PEMFC*

For HT-PEMFC, reversible potential [28,29] shows as follows:

$$E_r = E_r^0 + \frac{\Delta S}{nF}(T - 298.15) + \frac{RT}{nF}\ln(\frac{p_{H_2}p_{O_2}^{0.5}}{p_{H_2O}}) \tag{4}$$

in Equation (4), $E_r^0$ is the ideal standard potential which value is 1.185 V, $\Delta S$ is the change of standard molar entropy, $T$ is the operating temperature of HT-PEMFC, $R$ is the gas constant, $p_{H_2}$, $p_{O_2}$ and $p_{H_2O}$ are partial pressures of $H_2$, $O_2$ and $H_2O$, respectively.

$$\frac{\Delta S}{n} = -18.449 - 0.01283 \cdot T \tag{5}$$

where $\Delta S$ is related to the operating temperature.

*2.3. Overpotential of HT-PEMFC*

For HT-PEMFC, due to three types of overpotential containing activation overpotential, concentration overpotential and ohmic overpotential, its actual output voltage is generally less than the reversible potential.

- Activation overpotential [29,30];

$$E_{act} = \frac{RT}{n\alpha F}\ln(\frac{j + j_{leak}}{j_0}) \tag{6}$$

$$\ln j_{leak} = \left(-2342.9\frac{1}{T} + 9.0877\right) \tag{7}$$

where $\alpha$ is charge transfer coefficient, $j_{leak}$ is leakage current density, $j_0$ is exchange current density.

- Concentration overpotential [28];

$$E_{con} = \left(1 + \frac{1}{\alpha}\right)\frac{RT}{nF}\ln\left(\frac{j_L}{j_L - j}\right) \quad (8)$$

where $j$ is operating current density, $j_L$ is limiting current density [29].
- Ohmic overpotential [31];

$$E_{ohm} = j\frac{t_{mem}}{\sigma_{mem}} \quad (9)$$

where $t_{mem}$ is the thickness of the electrolyte, $\sigma_{mem}$ is the proton conductivity of the electrolyte [31].

$$\sigma_{mem} = \frac{A_0 B}{T} e^{\frac{-b_{act}}{RT}} \quad (10)$$

$$A_0 = 68DL^3 - 6324DL^2 + 65750DL + 8460 \quad (11)$$

$$B = \begin{cases} 1 + (0.01704T - 4.767)RH & 373.15K \leq T \leq 413.15 \\ 1 + (0.1432T - 56.89)RH & 413.15K < T \leq 453.15 \\ 1 + (0.7T - 309.2)RH & 453.15 < T \leq 473.15 \end{cases} \quad (12)$$

$$b_{act} = -619.6DL + 21750 \quad (13)$$

noindent where $DL$ is the doping level of the electrolyte, $RH$ is the relative humidity of the electrolyte [29].
- Output voltage;

According to Equations (1)–(13), the output voltage [32] of the HT-PEMFC can be derived as follows:

$$U = E_{rev} - E_{con} - E_{act} - E_{ohm} = E_{rev} - \left(1 + \frac{1}{\alpha}\right)\frac{RT}{nF}\ln\left(\frac{j_L}{j_L - j}\right) - \frac{RT}{n\alpha F}\ln\left(\frac{j + j_{leak}}{j_0}\right) - j\frac{t_{mem}}{\sigma_{mem}} \quad (14)$$

*2.4. Finite-Time Thermodynamic Performance Analysis of HT-PEMFC*

All analyses are based on the following assumptions:
1. With the HT-PEMFC system operating in a quasi-steady state, provided that the operating temperature and operating pressure are continuously changing, it is assumed that the operating pressure and operating temperature are constant at a fixed time;
2. The enthalpy of hydrogen entering the HT-PEMFC determines the maximum working capacity of the HT-PEMFC;
3. The exergy [33] mainly contains chemical exergy $\varepsilon_{chem}$ and physical exergy $\varepsilon_{phy}$, the kinetic and potential exergy of the hydrogen are neglected

$$\varepsilon = \varepsilon_{chem} + \varepsilon_{phy} \quad (15)$$

4. The energy required for compressing reactants is ignored.
- Output power density [34];

The output power density of the HT-PEMFC can be expressed as follows:

$$P = jU \quad (16)$$

- Output efficiency [35];

For any energy conversion device, the thermal efficiency is the energy output divided by the total energy input. Therefore, the output efficiency of HT-PEMFC can be shown as Equation (16):

$$\eta = \frac{P}{-\Delta H} \quad (17)$$

where $\Delta \dot{H}$ is the total energy absorbed from hydrogen and oxygen.

$$\Delta \dot{H} = -\frac{jA\Delta h}{nF} \quad (18)$$

where $\Delta h$ is the change of molar enthalpy.

- Exergy efficiency [36,37];

Exergy is an indicator used to evaluate energy quality. Exergy efficiency of HT-PEMFC, $\varphi$, is defined as the utilization degree of exergy, can be represented by:

$$\varphi = \frac{PA}{\dot{\varepsilon}_{in}} \quad (19)$$

$$\dot{\varepsilon}_{in} = \frac{jA}{nF}\left(\varepsilon_{H_2} + 0.5\varepsilon_{O_2}\right) \quad (20)$$

where $\dot{\varepsilon}_{in}$ is the total input exergy rate of $H_2$ and $O_2$; $A$ is the electrode effective surface area; $\varepsilon_{H_2}$ and $\varepsilon_{O_2}$ are the standard chemical exergy of $H_2$ and $O_2$.

- Ecological coefficient of performance [14,23];

In addition to energetic and exergetic analyses, the energy conversion performance of HT-PEMFC can also be analyzed by ecological standards. Angulo [38] proposed the ecological criterion function $E = P - T_L\dot{\delta}$ based on the heat engine, where $P$ is output power, $T_L\dot{\delta}$ reflects the power dissipation of the engine. The objective function not only optimizes the output power, but also takes the power dissipation into account, which makes the objective conform to the principle of long-term ecology. On this basis, Ust [39,40] presented a new ecological objective function, known as the ecological coefficient of performance (ECOP), which is the ratio of output power to power dissipation. Compared with the previous ecological objective function, the relationship between output power and power dissipation is improved. Its expression is as follows:

$$ECOP = \frac{PA}{T_0\dot{\delta}} \quad (21)$$

$$\dot{\delta} = \frac{-\Delta \dot{H} - P}{T} \quad (22)$$

### 2.5. Finite-Time Thermodynamic Optimization of HT-PEMFC

Since the output power and output efficiency of HT-PEMFC cannot reach the maximum at the same time, in order to minimize the power consumption of HT-PEMFC, the maximum output power is taken as the optimization objective. The power density of HT-PEMFC first increases and then decreases with the continuous increase of current density. Therefore, within a certain minimum range of current density, the power density of HT-PEMFC is bound to reach the maximum.

The power density of HT-PEMFC is related to current density $j$, operating temperature $T$, operating pressure $p$, relative humidity $RH$ and phosphoric acid doping level $DL$, which can be written as

$$P = f(j, T, p, RH, DL) \quad (23)$$

When $p$, $RH$ and $DL$ are constant, the power density is only related to current density and operating temperature, which can be presented as

$$P_1 = h(j, T) \quad (24)$$

When the operating temperature of HT-PEMFC is $T_1$, the power is only related to current density $j$, and the maximum power density is obtained, and the corresponding current density is $j_1$. Therefore, the maximum output power density at the operating temperature $T_1$ can be shown as

$$P^*_{1,max} = \max g(j) \quad (25)$$

By analogy, when the operating temperature of HT-PEMFC is $T_n$, the maximum power density is $P^*_{n,max}$, and the corresponding current density is $j_n$. Thus, the relationship between the maximum output power density $P^*_{n,max}$ of HT-PEMFC and operating temperature $T_n$ is deduced. In addition, similar methods can be used to study the influence of operating pressure, doping level and relative humidity on the maximum output power density of HT-PEMFC. The relationship between these parameters and the maximum output power density can also be acquired, which can further boost the power density of HT-PEMFC.

### 2.6. Output Efficiency, Exergy Efficiency and ECOP Based on Maximum Output Power Density

During the operation of HT-PEMFC, the output efficiency $\eta$ decreases with the increase of current density $j$. Therefore, the output efficiency reaches the maximum value in the low current density region, but the power density is low. In order to ensure the lowest power dissipation of HT-PEMFC, the current density $j_n$, corresponding to the maximum power density, is obtained according to the optimization criterion of the maximum power density. Thus, the output efficiency of HT-PEMFC corresponding to $j_n$ is the most efficient. In order to further improve the performance of HT-PEMFC, the effects of operating temperature, operating pressure, relative humidity and doping level on the output efficiency of HT-PEMFC are studied, and the relationship between these parameters and output efficiency is obtained.

As can be seen from the curve of exergy efficiency $\varphi$ and ecological coefficient of performance coefficient $ECOP$ of HT-PEMFC, with the continuous work of HT-PEMFC, $\varphi$ and $ECOP$ decrease with the increase of current density. Therefore, the method of obtaining the relationship between $\varphi$ and $ECOP$ and different parameters is similar to that of obtaining the relationship between output efficiency and different parameters.

### 2.7. Comparsion of Optimization Analysis of Different Objective Functions

Different objective functions include the maximum output power density $\overline{P}$, the optimal output efficiency $\overline{\eta}$ at the maximum output power density, the optimal exergy efficiency $\overline{\varphi}$ at the maximum output power density and ecological coefficient of performance $\overline{ECOP}$ at the maximum output power density, where $\overline{P}, \overline{\eta}, \overline{\varphi}$ and $\overline{ECOP}$ are dimensionless functions [41]. The dimensionless maximum output power density of HT-PEMFC can be expressed as follows:

$$\overline{P} = \frac{P_{max}}{P_{2,max}} \quad (26)$$

where $P_{max}$ is the maximum output power density of HT-PEMFC at different operating temperature, and $P_{2,max}$ is the maximum output power density of HT-PEMFC at the operating pressure $p = 2\ atm$. The dimensionless method of output efficiency, exergy efficiency and ecological coefficient of performance corresponding to the maximum power density is similar to that of the dimensionless maximum power density.

## 3. Results and Discussion
### 3.1. Model Verification

Figure 2 compares the predicted model voltage and the experimental data [42] of HT-PEMFC at 423 K and 448 K (p = 1 atm; DL = 5.6; RH = 0.38%), the experimental data are in good agreement with the predicted data. Figure 3 shows the relationship between reversible potential ($E_{rev}$), concentration overpotential ($E_{con}$), activation overpotential ($E_{act}$), ohmic overpotential ($E_{ohm}$), and output voltage (U) with current density. The reversible potential is a constant independent of current density. The three kinds of over-potential increase with the increase of current density, the concentration overpotential increases exponentially, the activation overpotential grows logarithmically, and the ohmic overpotential grows less.

In the low current density region, the rapid decline of output voltage is mainly due to the rapid increase of activation overpotential. In the region of high current density, the output voltage drops rapidly, mainly because the concentration overpotential increases rapidly.

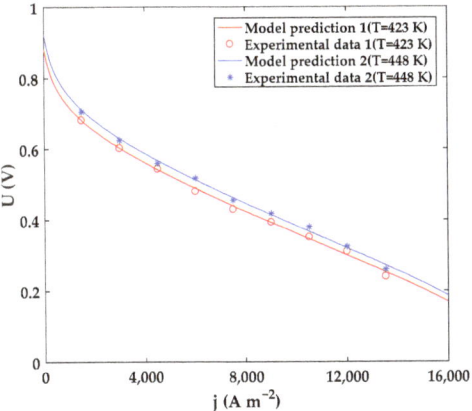

**Figure 2.** Comparisons of the predicted model voltage and experimental data.

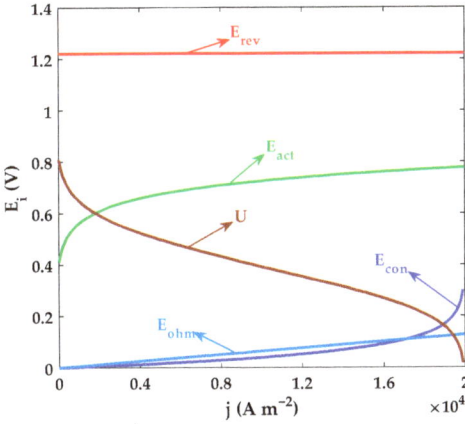

**Figure 3.** The relationship between reversible potential, concentration overpotential, activation overpotential, ohmic overpotential and current density.

### 3.2. Influences of the Operating Temperature

Figure 4a shows the variation of the dimensionless maximum power density of HT-PEMFC with operating temperature under different pressure. It can be seen that the maximum power density of the irreversible HT-PEMFC improves continuously with the increase of operating temperature. This is mainly because, as the operating temperature rises, the exchange current density grows, so the activation overpotential decreases. At the same time, the increase of operating temperature will enhance the proton conductivity, which will reduce the ohmic overpotential of HT-PEMFC. Therefore, the power loss produced by ohmic overpotential and activation overpotential will be cut down. Therefore, as the growth of operating temperature, the maximum power density of HT-PEMFC will increase constantly. When the operating pressure is 1 atm and the operating temperature is 403 K, the corresponding maximum power density is 3071.58 W $m^{-2}$. When the operating temperature rises up to 473 K, the corresponding maximum power density is 5291.60 W $m^{-2}$.

This indicates that the maximum power density of HT-PEMFC increases by 72%, when the operating temperature of HT-PEMFC increases from 403 K to 473 K. The maximum power density increases by 70% and 73%, respectively, at 2 atm and 3 atm. The result shows that HT-PEMFC can significantly increase its maximum power density in a suitable operating temperature range.

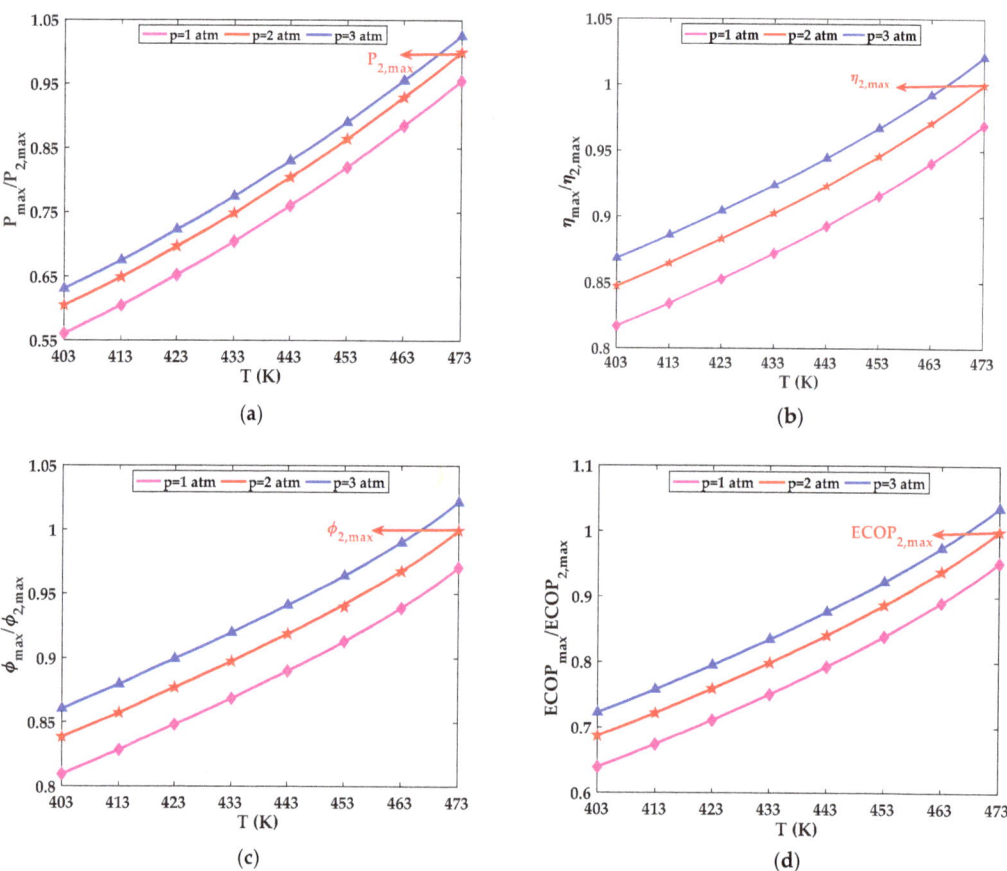

**Figure 4.** (a) $\overline{P}$ varying with operating temperature; (b) $\overline{\eta}$ varying with operating temperature; (c) $\overline{\phi}$ varying with operating temperature; (d) $\overline{ECOP}$ varying with operating temperature.

Figure 4b–d reflect the $\overline{\eta}$, $\overline{\phi}$ and $\overline{ECOP}$ corresponding to the maximum power density of HT-PEMFC. It is obvious that the increase of operating temperature can improve the output efficiency, exergy efficiency and ECOP of the irreversible process of HT-PEMFC. When the operating pressure is 1 atm and the operating temperature is 403 K, the corresponding output efficiency is 22.3%, exergy efficiency is 25.93% and ECOP is 37.23%. When the operating temperature is 473 K, the corresponding output efficiency is 26.46%, exergy efficiency is 31.3%, and ECOP is 55.97%. This shows that when the operating temperature of HT-PEMFC increased from 130 °C to 200 °C, its output efficiency, exergy efficiency and ECOP increased by 19, 21 and 50%, respectively. The increase of HT-PEMFC temperature accelerates the passage rate of protons and increases the conductivity of the proton exchange membrane, so the power generation of HT-PEMFC boosts, as well as the output efficiency and exergy efficiency. As the power consumption lessens, the entropy generated decreases and the output power increases, so the ratio of output power to power consumption increases, that is, the ecological coefficient of performance ECOP improves. However, though raising the operating temperature can improve the performance of HT-PEMFC, it can also cause many problems, such as high cost, poor stability, and long start-up time.

### 3.3. Influences of the Operating Pressure

Figure 5a shows that $\overline{P}$ of HT-PEMFC changes with operating pressure at different operating temperatures. Obviously, with the increase of operating pressure, the maximum power density of irreversible HT-PEMFC is continuously increasing. Owing that as the exchange current density rises with the rise of the operating pressure, the activation overpotential will decrease and the reversible potential will boost. Therefore, with the increase of operating pressure, the irreversibility of HT-PEMFC decreases and the maximum power density of HT-PEMFC increases continuously. When the operating temperature is 453 K and the operating pressure is 1 atm, the corresponding maximum power density is 4515.13 W m$^{-2}$. When the operating pressure rises up to 3 atm, the corresponding maximum power density is 4919.07 W m$^{-2}$. From the numerical point of view, when the operating temperature is 453 K and the operating temperature of HT-PEMFC increases from 1 atm to 3 atm, the maximum power density of HT-PEMFC only increases by 9%. This shows that operating pressure has little influence on HT-PEMFC.

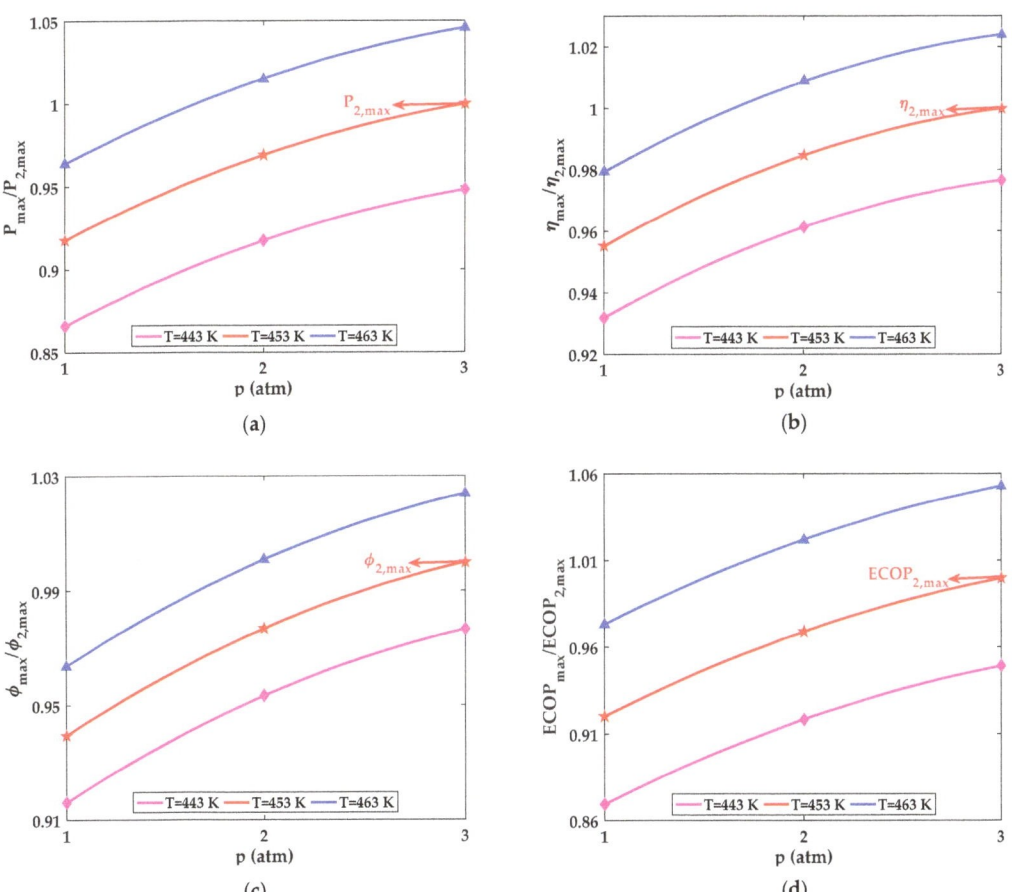

**Figure 5.** (a) $\overline{P}$ varying with operating pressure; (b) $\overline{\eta}$ varying with operating pressure; (c) $\overline{\phi}$ varying with operating pressure; (d) $\overline{ECOP}$ varying with operating pressure.

It can be seen from Figure 5b–d that $\bar{\eta}$, $\bar{\phi}$ and $\overline{ECOP}$ correspond to the maximum power density of HT-PEMFC. The increase of operating pressure can slightly improve the output efficiency, exergy efficiency and ECOP of the irreversible process of HT-PEMFC. When the operating temperature of HT-PEMFC increased from 1 atm to 3 atm, its output efficiency, exergy efficiency and ECOP increased by 5, 6 and 9%, respectively. According to the numerical analysis, the increase of operating pressure does not improve the performance of HT-PEMFC as significantly as the increase of operating temperature. In addition, increasing the operating pressure consumes extra power to compress the reactants in the inlet, resulting in higher costs.

### 3.4. Influences of the Doping Level

As shown in Figure 6, with the increase of current density, ohmic overpotential will improve. When doping level rises, the proton conductivity of HT-PEMFC increases, which reduces the ohmic overpotential. According to the relationship between $\bar{P}$ and DL in Figure 7a, it can be observed that the maximum power density raises endlessly with the increase of doping level. This is mainly because, as the rise of doping level, the ohmic overpotential decreases and the reversible potential improves. Hence, if the doping level of HT-PEMFC is raised appropriately, the maximum power density will become larger. When the doping level is 2, the corresponding maximum power density is 1868.52 W m$^{-2}$. When the doping level is 10, the corresponding maximum power is 4694.53 W m$^{-2}$. It is clear that when the doping level increases from 2 to 10 and relative humidity is 3.8%, the maximum power density of HT-PEMFC increases by 150%. This indicates that DL has a significant effect on the performance of HT-PEMFC.

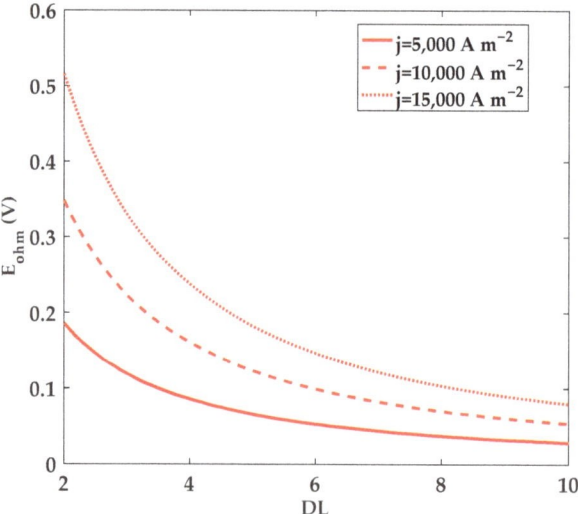

**Figure 6.** The relationship between ohmic overpotential and doping level at different current density (T = 453 K, p = 1 atm, RH = 3.8%).

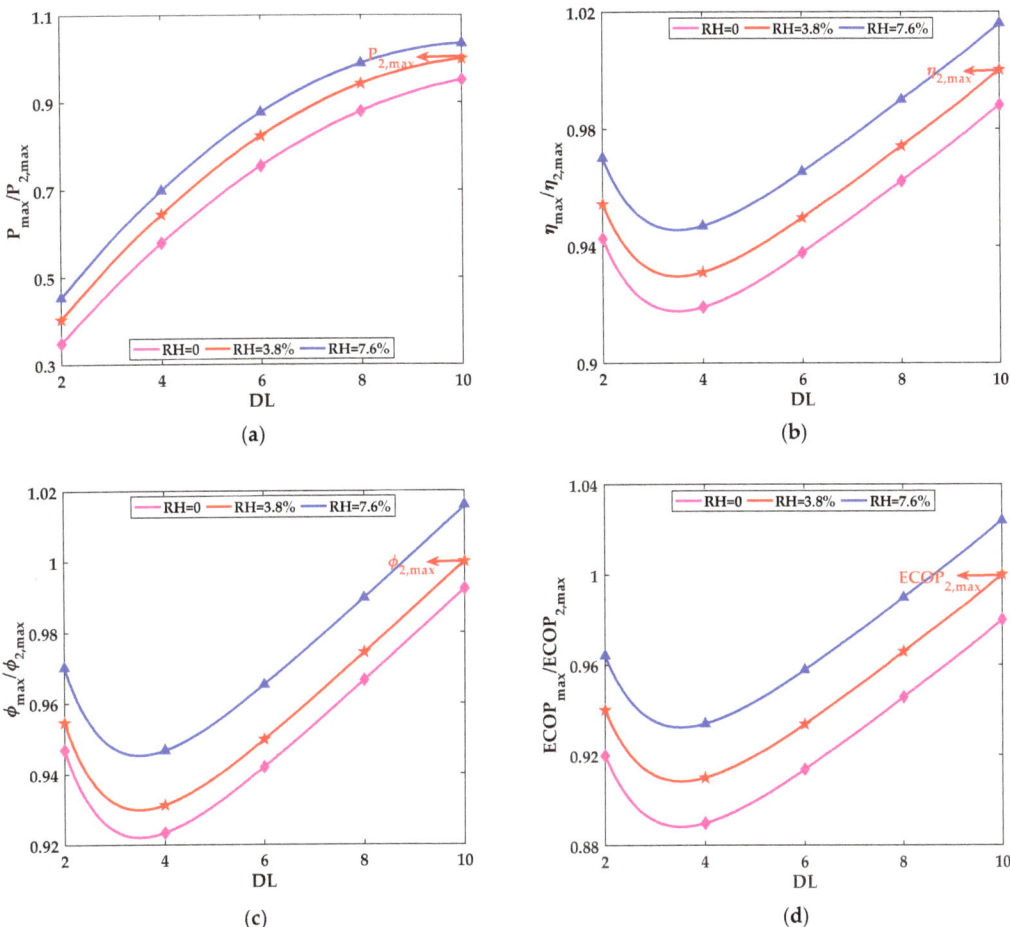

**Figure 7.** (a) $\overline{P}$ varying with doping level; (b) $\overline{\eta}$ varying with doping level; (c) $\overline{\phi}$ varying with doping level; (d) $\overline{ECOP}$ varying with doping level.

Figure 7b–d show the $\overline{\eta}$, $\overline{\phi}$ and $\overline{ECOP}$ corresponding to the maximum power density of HT-PEMFC varying with doping level at different relative humidity. It can be seen that when the doping level increases, the output efficiency, exergy efficiency and ECOP of HT-PEMFC all showed a trend of decreasing at first and then increasing. This is mainly because the current density of the maximum is different under different doping levels. When the doping level is small, the current density is low and the gap of current density of the maximum power density at different doping levels is bigger, thus causing ohmic potential increases with the increase of DL. When the doping level is high, the current density corresponding to the maximum power density is in the region of high current density, and the difference of the current density of the maximum power density is small at different doping level, so the ohmic overpotential decreases with the increase of DL, leading to the increase of reversible potential. When the relative humidity of HT-PEMFC is 3.8% and the doping level increases from 2 to 10, its output efficiency, exergy efficiency and ECOP increases by 5, 5 and 6%, respectively. This indicates that doping level significantly improved the maximum power density, but has little impact on output efficiency, exergy efficiency and ECOP.

## 3.5. Influences of the Relative Humidity

Figure 8a shows the change of $\overline{P}$ with relative humidity at different temperatures. It can be seen that with the increase of relative humidity, the maximum power density of irreversible HT-PEMFC rises. The main reason is that with the increase of relative humidity, the conductivity of the proton exchange membrane at high temperature will boost, resulting in the decrease of ohmic overpotential. When RH is 0, the corresponding maximum power density is 4064.78 W m$^{-2}$. When RH rises up to 7.6%, the corresponding maximum power density is 4168.1 W m$^{-2}$. This indicates that the maximum power density of HT-PEMFC increases by only 2% when the operating temperature is 453 K and the relative humidity increases from 0 to 7.6%. As shown in Figure 9, although relative humidity can improve proton conductivity, increasing relative humidity has less effect on the ohmic overpotential than the doping level. Therefore, relative humidity has little effect on the maximum power density of HT-PEMFC.

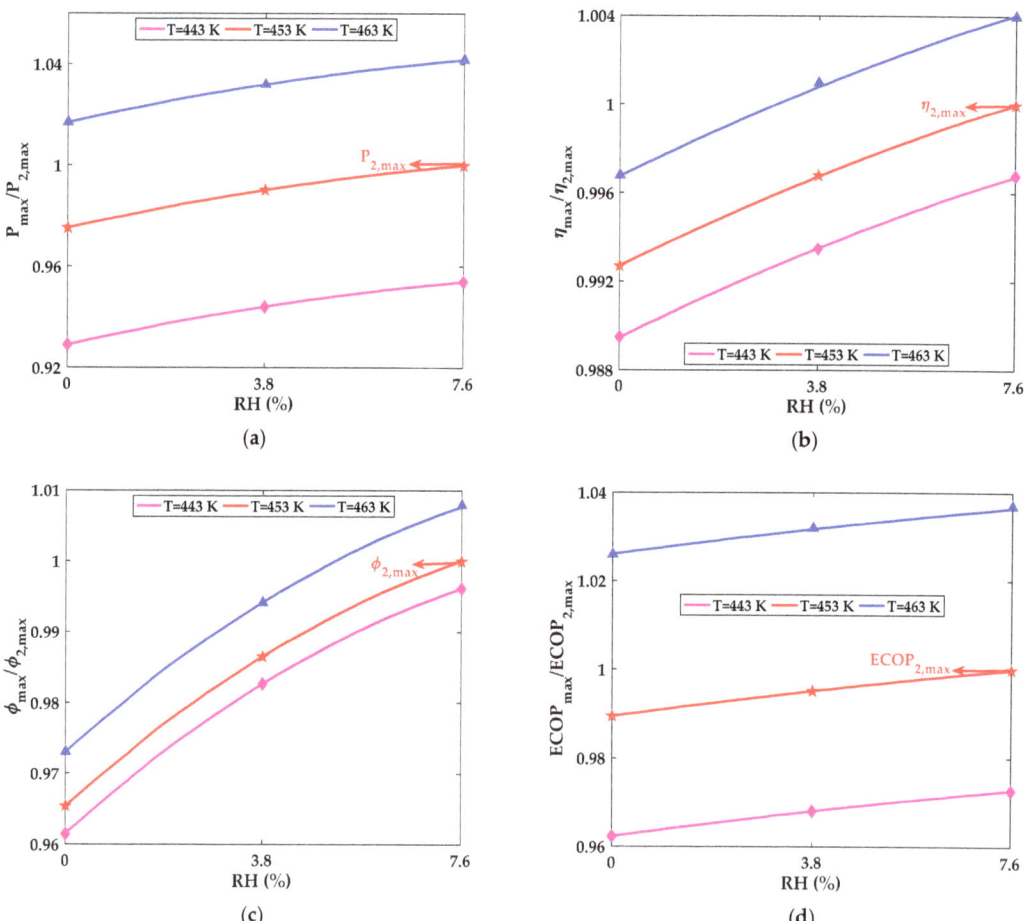

**Figure 8.** (a) $\overline{P}$ varying with relative humidity; (b) $\overline{\eta}$ varying with relative humidity; (c) $\overline{\phi}$ varying with relative humidity; (d) $\overline{ECOP}$ varying with relative humidity.

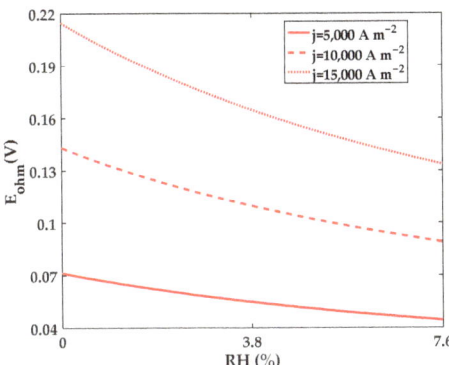

**Figure 9.** The relationship between ohmic overpotential and relative humidity at different current density (T = 453 K, p = 1 atm, DL = 8).

Figure 8b–d reflect $\bar{\eta}$, $\bar{\phi}$ and $\overline{ECOP}$ corresponding to the maximum power density of HT-PEMFC varying with relative humidity at different temperatures. It can be seen from the figures that the three indexes all show a monotonically increasing trend. When the relative humidity is 0 and the operating temperature is 453 K, the corresponding output efficiency is 24.5%, exergy efficiency was 25.1%, and ECOP is 48.16%. When the relative humidity was 7.6%, the corresponding output efficiency was 24.68%, exergy efficiency was 26%, and ECOP was 48.667%. It reflects that when the relative humidity of HT-PEMFC increased from 0 to 7.6%, its output efficiency, exergy efficiency and ECOP increased by 0.7, 4 and 1%, respectively. Compared with the consequences of doping levels, the numerical results of relative humidity are significantly lower. Therefore, relative humidity has little effect on the performance improvement of HT-PEMFC.

## 4. Conclusions

In this paper, the irreversibility caused by polarization and leakage current is considered, and the finite-time thermodynamic model of HT-PEMFC is established. The influence of operating temperature, operating pressure, doping level of phosphoric acid and relative humidity on the maximum output power density is studied. In addition, according to the maximum power density criterion, the influence of different parameters on output efficiency, exergy efficiency and ecological coefficient of performance is obtained. Among them, ECOP compromises the relationship between power and efficiency performance of HT-PEMFC.

Through numerical analysis and calculation, when the operating temperature increases from 403 K to 473 K, the maximum output power density increases by 72%, the output efficiency rises by 19%, the exergy efficiency rises by 21%, and the ecological performance coefficient boosts by 50%. When the doping level of phosphoric acid increases from 2 to 10, the maximum power density increases by 150%. However, the operating pressure and relative humidity have little influence on the maximum power density and the output efficiency, exergy efficiency and ECOP.

In the future, the extended irreversible thermodynamics [43,44] that expands the scope of classical irreversible thermodynamics into a new field could be considered to analyze the HT-PEMFC system.

**Author Contributions:** All of the authors contributed to publishing this article. The collection of materials and summarization of this article was done by D.L. and Y.L. The simulation and analysis were done by Z.L., S.L. and M.Z. The conceptual ideas, methodology and guidance for the research were provided by Z.M. and B.X. All authors have read and agreed to the published version of the manuscript.

**Funding:** We gratefully acknowledge the financial support of the National Natural Science Foundation of China (No. 51176069) and the Scientific Research Foundation of Nanjing Forestry University (No. GXL2018004).

**Institutional Review Board Statement:** Not applicable.

**Informed Consent Statement:** Not applicable.

**Data Availability Statement:** Not applicable.

**Conflicts of Interest:** The authors declare no conflict of interest.

## References

1. Haider, R.; Wen, Y.; Ma, Z.-F.; Wilkinson, D.P.; Zhang, L.; Yuan, X.; Song, S.; Zhang, J. High temperature proton exchange membrane fuel cells: Progress in advanced materials and key technologies. *Chem. Soc. Rev.* **2021**, *50*, 1138–1187. [CrossRef] [PubMed]
2. Devrim, Y.; Arıca, E.D. Multi-walled carbon nanotubes decorated by platinum catalyst for high temperature PEM fuel cell. *Int. J. Hydrog. Energy* **2019**, *44*, 18951–18966. [CrossRef]
3. Alpaydin, G.U.; Devrim, Y.; Colpan, C.O. Performance of an HT-PEMFC having a catalyst with graphene and multiwalled carbon nanotube support. *Int. J. Energy Res.* **2019**, *43*, 3578–3589. [CrossRef]
4. Nalbant, Y.; Colpan, C.O.; Devrim, Y. Energy and exergy performance assessments of a high temperature-proton exchange membrane fuel cell based integrated cogeneration system. *Int. J. Hydrog. Energy* **2020**, *45*, 3584–3594. [CrossRef]
5. Reddy, E.H.; Jayanti, S. Thermal management strategies for a 1 kWe stack of a high temperature proton exchange membrane fuel cell. *Appl. Therm. Eng.* **2012**, *48*, 465–475. [CrossRef]
6. Lee, D.; Lim, J.W.; Gil Lee, D. Cathode/anode integrated composite bipolar plate for high-temperature PEMFC. *Compos. Struct.* **2017**, *167*, 144–151. [CrossRef]
7. Oono, Y.; Sounai, A.; Hori, M. Influence of the phosphoric acid-doping level in a polybenzimidazole membrane on the cell performance of high-temperature proton exchange membrane fuel cells. *J. Power Sources* **2009**, *189*, 943–949. [CrossRef]
8. Pinar, F.J.; Cañizares, P.; Rodrigo, M.A.; Úbeda, D.; Lobato, J. Long-term testing of a high-temperature proton exchange membrane fuel cell short stack operated with improved polybenzimidazole-based composite membranes. *J. Power Sources* **2015**, *274*, 177–185. [CrossRef]
9. Li, Q.; Rudbeck, H.; Chromik, A.; Jensen, J.; Pan, C.; Steenberg, T.; Calverley, M.; Bjerrum, N.J.; Kerres, J. Properties, degradation and high temperature fuel cell test of different types of PBI and PBI blend membranes. *J. Membr. Sci.* **2010**, *347*, 260–270. [CrossRef]
10. Muthuraja, P.; Prakash, S.; Shanmugam, V.; Radhakrsihnan, S.; Manisankar, P. Novel perovskite structured calcium titanate-PBI composite membranes for high-temperature PEM fuel cells: Synthesis and characterizations. *Int. J. Hydrog. Energy* **2018**, *43*, 4763–4772. [CrossRef]
11. Miansari, M.; Sedighi, K.; Amidpour, M.; Alizadeh, E. Experimental and thermodynamic approach on proton exchange membrane fuel cell performance. *J. Power Sources* **2009**, *190*, 356–361. [CrossRef]
12. Ozen, D.N.; Timurkutluk, B.; Altinisik, K. Effects of operation temperature and reactant gas humidity levels on performance of PEM fuel cells. *Renew. Sustain. Energy Rev.* **2016**, *59*, 1298–1306. [CrossRef]
13. Esfeh, H.K.; Hamid, M.K.A. Temperature Effect on Proton Exchange Membrane Fuel Cell Performance Part II: Parametric Study. *Energy Procedia* **2014**, *61*, 2617–2620. [CrossRef]
14. Li, C.; Liu, Y.; Xu, B.; Ma, Z. Finite Time Thermodynamic Optimization of an Irreversible Proton Exchange Membrane Fuel Cell for Vehicle Use. *Processes* **2019**, *7*, 419. [CrossRef]
15. Li, C.J.; Liu, Y.; Ma, Z.S. Thermodynamic Analysis of the Performance of an Irreversible PEMFC. *Defect Diffus. Forum* **2018**, *388*, 350–360. [CrossRef]
16. Wei, F.F.; Huang, Y.W. Performance Characteristics of an Irreversible Proton Exchange Membrane (PEM) Fuel Cell. *J. Donghua Univ.* **2012**, *29*, 393–398.
17. Ay, M.; Midilli, A.; Dincer, I. Exergetic performance analysis of a PEM fuel cell. *Int. J. Energy Res.* **2005**, *30*, 307–321. [CrossRef]
18. Xu, B.; Chen, Y.; Ma, Z.S. Exergetic sustainability indicators of a polymer electrolyte membrane fuel cell at variable operating conditions. *Arch. Thermodyn.* **2021**, *42*, 183–204.
19. Guo, Y.; Guo, X.; Zhang, H.; Hou, S. Energetic, exergetic and ecological analyses of a high-temperature proton exchange membrane fuel cell based on a phosphoric-acid-doped polybenzimidazole membrane. *Sustain. Energy Technol. Assess.* **2020**, *38*, 100671. [CrossRef]
20. Haghighi, M.; Sharifhassan, F. Exergy analysis and optimization of a high temperature proton exchange membrane fuel cell using genetic algorithm. *Case Stud. Therm. Eng.* **2016**, *8*, 207–217. [CrossRef]
21. Ishihara, A.; Mitsushima, S.; Kamiya, N.; Ota, K.-I. Exergy analysis of polymer electrolyte fuel cell systems using methanol. *J. Power Sources* **2004**, *126*, 34–40. [CrossRef]
22. Xie, D.; Wang, Z.; Jin, L.; Zhang, Y. Energy and exergy analysis of a fuel cell based micro combined heat and power cogeneration system. *Energy Build.* **2012**, *50*, 266–272. [CrossRef]

23. Li, C.; Xu, B.; Ma, Z. Ecological Performance of an Irreversible Proton Exchange Membrane Fuel Cell. *Sci. Adv. Mater.* **2020**, *12*, 1225–1235. [CrossRef]
24. Barati, S.; Ghazi, M.M.; Khoshandam, B. Study of effective parameters for the polarization characterization of PEMFCs sensitivity analysis and numerical simulation. *Korean J. Chem. Eng.* **2018**, *36*, 146–156. [CrossRef]
25. Lu, X.; Li, B.; Guo, L.; Wang, P.; Yousefi, N. Exergy analysis of a polymer fuel cell and identification of its optimum operating conditions using improved Farmland Fertility Optimization. *Energy* **2021**, *216*, 119264. [CrossRef]
26. Xia, L.; Zhang, C.; Hu, M.; Jiang, S.; Chin, C.S.; Gao, Z.; Liao, Q. Investigation of parameter effects on the performance of high-temperature PEM fuel cell. *Int. J. Hydrog. Energy* **2018**, *43*, 23441–23449. [CrossRef]
27. Lin, D.; Han, Y.H.; Khodaei, H. Application of the meta-heuristics for optimizing exergy of a HT-PEMFC. *Int. J. Energ. Res.* **2020**, *44*, 3749–3761. [CrossRef]
28. Guo, X.; Zhang, H.; Zhao, J.; Wang, F.; Wang, J.; Miao, H.; Yuan, J. Performance evaluation of an integrated high-temperature proton exchange membrane fuel cell and absorption cycle system for power and heating/cooling cogeneration. *Energy Convers. Manag.* **2019**, *181*, 292–301. [CrossRef]
29. Lee, W.-Y.; Kim, M.; Sohn, Y.-J.; Kim, S.-G. Power optimization of a combined power system consisting of a high-temperature polymer electrolyte fuel cell and an organic Rankine cycle system. *Energy* **2016**, *113*, 1062–1070. [CrossRef]
30. Cheddie, D.; Munroe, N. Analytical correlations for intermediate temperature PEM fuel cells. *J. Power Sources* **2006**, *160*, 299–304. [CrossRef]
31. Olapade, P.O.; Meyers, J.P.; Borup, R.L.; Mukundan, R. Parametric Study of the Morphological Proprieties of HT-PEMFC Components for Effective Membrane Hydration. *J. Electrochem. Soc.* **2011**, *158*, B639–B649. [CrossRef]
32. Mamaghani, A.H.; Najafi, B.; Casalegno, A.; Rinaldi, F. Optimization of an HT-PEM fuel cell based residential micro combined heat and power system: A multi-objective approach. *J. Clean. Prod.* **2018**, *180*, 126–138. [CrossRef]
33. Al-Sulaiman, F.A.; Dincer, I.; Hamdullahpur, F. Exergy analysis of an integrated solid oxide fuel cell and organic Rankine cycle for cooling, heating and power production. *J. Power Sources* **2010**, *195*, 2346–2354. [CrossRef]
34. Wu, Z.; Zhu, P.; Yao, J.; Tan, P.; Xu, H.; Chen, B.; Yang, F.; Zhang, Z.; Ni, M. Thermo-economic modeling and analysis of an NG-fueled SOFC-WGS-TSA-PEMFC hybrid energy conversion system for stationary electricity power generation. *Energy* **2020**, *192*, 116613. [CrossRef]
35. Zhang, X.; Cai, L.; Liao, T.; Zhou, Y.; Zhao, Y.; Chen, J. Exploiting the waste heat from an alkaline fuel cell via electrochemical cycles. *Energy* **2018**, *142*, 983–990. [CrossRef]
36. Cohce, M.; Dincer, I.; Rosen, M. Energy and exergy analyses of a biomass-based hydrogen production system. *Bioresour. Technol.* **2011**, *102*, 8466–8474. [CrossRef]
37. Nguyen, H.Q.; Aris, A.M.; Shabani, B. PEM fuel cell heat recovery for preheating inlet air in standalone solar-hydrogen systems for telecommunication applications: An exergy analysis. *Int. J. Hydrog. Energy* **2016**, *41*, 2987–3003. [CrossRef]
38. Angulo-Brown, F. An ecological optimization criterion for finite-time heat engines. *J. Appl. Phys.* **1991**, *69*, 7465–7469. [CrossRef]
39. Ust, Y.; Sahin, B.; Sogut, O.S. Performance analysis and optimization of an irreversible dual-cycle based on an ecological coefficient of performance criterion. *Appl. Energy* **2005**, *82*, 23–39. [CrossRef]
40. Ust, Y.; Sahin, B.; Kodal, A. Performance analysis of an irreversible Brayton heat engine based on ecological coefficient of performance criterion. *Int. J. Therm. Sci.* **2006**, *45*, 94–101. [CrossRef]
41. E, Q.; Wu, F.; Chen, L.-G.; Qiu, Y.-N. Thermodynamic optimization for a quantum thermoacoustic refrigeration micro-cycle. *J. Cent. South. Univ.* **2020**, *27*, 2754–2762. [CrossRef]
42. Sousa, T.; Mamlouk, M.; Scott, K. An isothermal model of a laboratory intermediate temperature fuel cell using PBI doped phosphoric acid membranes. *Chem. Eng. Sci.* **2010**, *65*, 2513–2530. [CrossRef]
43. Chen, K.C.; Yeh, C.S. Extended Irreversible Thermodynamics Approach to Magnetorheological Fluids. *J. Non-Equilib. Thermodyn.* **2001**, *26*, 355–372. [CrossRef]
44. Versaci, M.; Palumbo, A. Magnetorheological Fluids: Qualitative comparison between a mixture model in the Extended Irreversible Thermodynamics framework and an Herschel–Bulkley experimental elastoviscoplastic model. *Int. J. Non-Linear Mech.* **2020**, *118*, 103288. [CrossRef]

Article

# SARS-CoV-2 Spread Forecast Dynamic Model Validation through Digital Twin Approach, Catalonia Case Study

Pau Fonseca i Casas [1,*], Joan Garcia i Subirana [1], Víctor García i Carrasco [1] and Xavier Pi i Palomés [2]

[1] Department of Statistics and Operations Research, Universitat Politècnica de Catalunya, 08034 Barcelona, Spain; joan.garcia-subirana@upc.edu (J.G.i.S.); victor.garcia.carrasco@upc.edu (V.G.i.C.)
[2] Open University of Catalonia, Computer Science, Multimedia and Telecommunications Studies, 08860 Barcelona, Spain; xpi@enginyers.net
* Correspondence: pau@fib.upc.edu

**Citation:** Fonseca i Casas, P.; Garcia i Subirana, J.; García i Carrasco, V.; Pi i Palomés, X. SARS-CoV-2 Spread Forecast Dynamic Model Validation through Digital Twin Approach, Catalonia Case Study. *Mathematics* **2021**, *9*, 1660. https://doi.org/10.3390/math9141660

Academic Editors: Maria Luminița and Catalin I. Pruncu

Received: 5 June 2021
Accepted: 12 July 2021
Published: 14 July 2021

**Publisher's Note:** MDPI stays neutral with regard to jurisdictional claims in published maps and institutional affiliations.

**Copyright:** © 2021 by the authors. Licensee MDPI, Basel, Switzerland. This article is an open access article distributed under the terms and conditions of the Creative Commons Attribution (CC BY) license (https://creativecommons.org/licenses/by/4.0/).

**Abstract:** The spread of the SARS-CoV-2 modeling is a challenging problem because of its complex nature and lack of information regarding certain aspects. In this paper, we explore a Digital Twin approach to model the pandemic situation in Catalonia. The Digital Twin is composed of three different dynamic models used to perform the validations by a Model Comparison approach. We detail how we use this approach to obtain knowledge regarding the effects of the nonpharmaceutical interventions and the problems we faced during the modeling process. We use Specification and Description Language (SDL) to represent the compartmental forecasting model for the SARS-CoV-2. Its graphical notation simplifies the different specialists' understanding of the model hypotheses, which must be validated continuously following a Solution Validation approach. This model allows the successful forecasting of different scenarios for Catalonia. We present some formalization details, discuss the validation process and present some results obtained from the validation model discussion, which becomes a digital twin of the pandemic in Catalonia.

**Keywords:** SARS-CoV-2; COVID-19; SEIRD (Susceptible, Exposed, Infected and Recovered and Death); SDL; Catalonia

## 1. Introduction

Computer simulations to forecast the infections, based on the Susceptible, Exposed, Infected and Recovered and Death (SEIRD) models are common since its preliminary inception in 1930 [1] These models can be used to represent a pandemic situation and to forecast the new cases due to SARS-CoV-2. This paper aims to use an extended SEIRD model to predict the pandemic situation in Catalonia.

In the current context of the pandemic situation caused by the spread of SARS-CoV-2, the use of mathematical models to forecast the its trend becomes a valuable tool [2]. Some of these models are focused on the analysis of the theoretical spread of the virus, understanding the dynamic behavior of the particles, as an example, using a cellular automaton to capture the spatial relations [3] or to understand the importance of the airborne route that dominates exposure during close contact [4]. Several models have been centered on the understanding of the evolution of the trend of the new cases [5,6], modeling different scenarios [7], and understanding through the evolution of the model the effects of different non-pharmaceutical interventions (NPIs) that can be applied to the population [8,9]. The stay at the home recommendation is analyzed in [10], and a plethora of different NPIs are analyzed in [11,12] to understand their effectiveness. Not only SEIRD-like models are used to predict the behavior of pandemics. Other approaches also exist, like the use of time series [13,14]. Furthermore, the models are not only used to represent the spread of the virus. In fact, they are also used for other aspects, like the effects of the pandemic on the economy [15–17]. Almost all these models are focused on a specific area because they need to be validated using a specific dataset (country, region, etc.).

They need a good description of the trend of the evolution of new cases. From this trend, one can obtain other variables of interest, like the hospitalization rates, etc. Furthermore, understanding the effects of the NPIs over the trend, one can infer the effectiveness of the NPIs over the population. We want to model new cases trend in Catalonia, a region in Europe located in the Occidental Mediterranean Sea. To model SARS-CoV-2 spread in Catalonia, we must estimate parameters that detail the virus propagation, like effective reproductive number ($R_t$). To measure the transmission potential of a disease, we use the basic reproduction number $R_0$; this value represents the average number of secondary infections produced by a single infection in a population. The average number of secondary cases will be lower than the basic reproduction number because not all the population will be susceptible; this will be represented by the effective reproductive number ($R_t$). If $R_t > 1$, the number of cases will increase, on $R = 1$, remains stable, and with $R < 1$, the number of cases will decrease. We can calculate $R_t$ as $R_t = R_0 P_s$, being $P_s$, the fraction of susceptible population. We also must consider the complexity to cope with the large amount of information generated by all the scientists working on SARS-CoV-2 along with the project evolution, information that must be discussed while building the model. The analysis of this phenomenon needs experts from different areas collaborating in the study using different viewpoints. Moreover, when experts make a model, they can introduce errors due to a wrong understanding of the system behavior. Also, the model implementation can introduce errors [18].

This proposal improves collaboration between experts through a formal model defined using Specification and Description Language (SDL). The formalization provides a common language for modeling and a concrete mechanism to generate correct simulations, an approach that has shown to be helpful in health care [19]. Our approach allows testing different containment measures for forecasting the spread of SARS-CoV-2 in Catalonia. However, the unknowns regarding the system analyzed imply carefully analyzing all the assumptions we use on the model. We use a Model Comparison Validation and Verification technique to solve this issue, which consists of using different models to improve error detection. The first model is a System Dynamics (SD) SEIRD model [20]. This model implements an initial parametrization to perform a preliminary analysis of the system and represents an overview of the system behavior.

We develop a second model, a SEIRD Python-coded model, that is a refinement of the first SD model. This second model aims to obtain the parametrization details used later in the last SDL model. The third model, on SDL, allows us to extend the model semantics and structure, allowing the model representation over a cellular automaton (CA) [21], providing results not only for all Catalonia, but also for different geographical zones. The tree models act as a digital twin of the pandemic situation in Catalonia, enabling the model-based discussion. A digital twin is a digital reproduction of a system, potentially at any time (present, past, and future), driven by a simulation or a set of simulation models. The approach we use in this research is to define a digital twin of the pandemic situation of Catalonia since the model assumptions will be updated coherently to the system modifications and validated continuously using the actual data we obtain from the system.

*A Digital Twin Approach*

Validation of the solution implies the use of real-time data to assess the model. The framework used in this work corresponds to the notion of digital twin (DT), introduced in 2002 by Michael Grieves [22] and now formalized by the Digital Twin Consortium (www.digitaltwinconsortium.org accessed on 14 July 2021). We define a digital twin as a virtual reproduction of a system based on simulations, real-time and historical data that allows representing, understanding, and predicting scenarios of the past, present, and future, with verified and validated models, and synchronized at a specified frequency and fidelity with the system. Kritzinger [23] identifies in the literature the extensive use of the digital model and digital shadow notions as the two constituent parts of the digital twin.

The Digital Model comprises requirements, specifications, and theoretical models, and it is possible to create asset simulators. On the other hand, the digital shadow (DS) contains models based on data captured from the actual world, via observation or by automatic measurements using sensors, in an Internet of Things context.

Stark [24] from the Fraunhofer Institute, introduced the concept of the digital master (DM), defined as the set of digital models used in the digital twin. The formula DT = DM + DS expresses the relation between DM, DS, and DT. Digital twins can be the basis of digitalization. We can apply the Reference Architecture Industry 4.0 (RAMI 4.0), formalized as the IEC/PAS 63088 standard, defining a graphical notation for digitalized components (I4.0 Components). It introduces the administration shell concept, which is metaphorically a digital shell that can cover any asset, converting it into an I4.0 Component. This approach permits representing any physical or virtual asset as a software agent, which can send and receive messages with an arbitrary amount of data. Furthermore, an administration shell can specify a list of incoming messages and a list of out-coming messages in the same way as UML components. We can represent the notion of digital twin using internally four main agent components (DM, DS, simulators, and simulators' traces), see Figure 1.

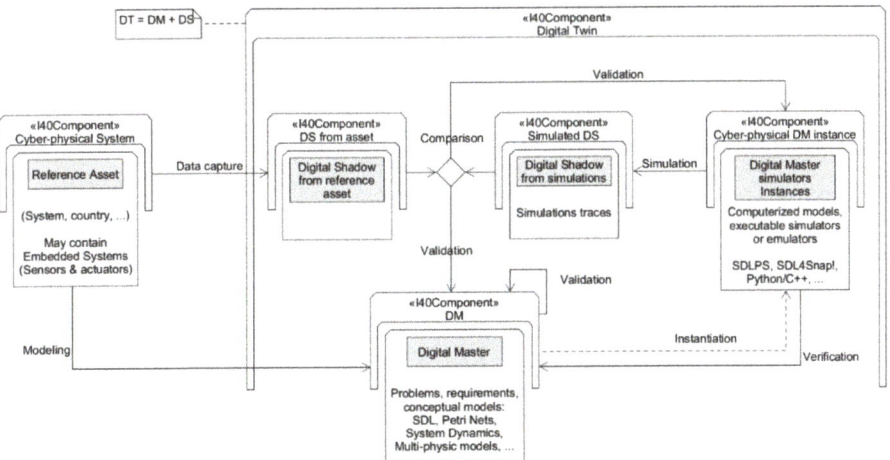

**Figure 1.** Digital Twin and the V&V Loop.

The leftmost I4.0 Component, as shown in Figure 1, corresponds to the reference asset, and in our case, is the geographic area of Catalonia and its population. Using statistical tools, the data collected over time (the digital shadow) is the basis for building data-based models. In our case, we have used datasets from different countries to compare the results.

A daily process helps in the calibration of the simulation model using the data taken from the datasets. Hence, we can obtain an evolutive simulator that produces traces over time, which defines a virtual digital shadow. Therefore, the continuous comparison between the real asset's digital shadow with the virtual simulator-generated digital shadow determines how the model must be calibrated or modified through the so called verification and validation loop (V&V Loop).

Because it is needed to perform a validation of the model continuously, the use of a standard formal language (SDL) that has a flowchart-like graphical representation, that is unambiguous (consistent) and complete, and has features that connect it to the IoT paradigm [25], simplifies this process. Essentially. SDL becomes the common language used to understand the model assumptions. We have ultimately developed a dashboard based on the Digital Twin Administration Shell, which is available at http://pand.sdlps.com (accessed on 14 July 2021).

## 2. Materials and Methods

As we mention, we develop three models to establish a validation through a model comparison approach. The first model is a classical SEIRD model and allows a fast conceptualization of the main model characteristics and features; this model is mainly used to test the preliminary assumptions and draw the model's main boundaries. The second model, coded with Python, allows executing an optimization algorithm to find the points on the curve where the trend changes and calculate the transmission rate, a key parameter in our models. This model will accurately represent the trend of the pandemic situation in Catalonia, but that does not consider the real cases nor the confinement options, details we add on the SDL model. SDL and Python models must have similar behavior when comparing the variable Infective Detected (in the SDL model) and Infective (in the Python model). Both variables will also compare with the new cases time series obtained from the local authorities database [26]. In the following sections, we detail these three models and the interaction that exists between them.

### 2.1. System Dynamics Model

A classical SEIRD model is the basis of the system dynamics (SD) model, as depicted in Figure 2. We use this model to obtain a boundary that expresses the trend. These boundaries will be helpful to perform a validation based on the model comparisons approach [27] since the shapes and the trends of the other models will be similar to those obtained on this model. If not, we must reanalyze the models to detect the errors.

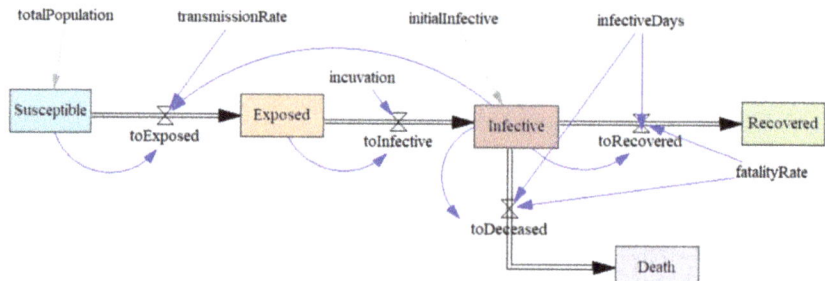

**Figure 2.** SD models implemented in Vensim PLE.

Using this model, one can test the different assumptions faster, analyze the shape of the resulting curves and understand how the effects of the model modification, for example, the addition of new compartments, will affect the other models. We opt to keep this model as simple as possible, but it helps us define the boundaries of the forecast, the general trend, and the direction of the movement. Notice, however, that this model does not intend to make a forecast. It only serves as a basis for the discussions and becomes a tool for fast prototyping, in the initial stages of the model's development. Essentially, it is used in the implementation stages to be able to detect errors in both coding or model definition through the model comparison approach.

The equations that drive the evolution of the model (Equations (1)–(5)) presented below represent the population variation of each compartment due to the flows between them, as shown in Figure 2, assuming that the total population remains constant (Equation (6)):

$$S' = -\frac{\beta S I}{N} \qquad (1)$$

$$E' = \frac{\beta S I}{N} - \alpha E \qquad (2)$$

$$I' = \alpha E - \gamma I \qquad (3)$$

$$R' = \gamma(1-\mu)I \tag{4}$$

$$D' = \gamma\mu I \tag{5}$$

$$N = S + E + I + R + D \tag{6}$$

The $\alpha$ (latency rate), $\gamma$ (recovery rate) and $\mu$ (mortality rate) parameters will be considered constant, based on the measurements made by other studies, and $N$ will be the total population, assuming that there is no pre-existing level of immunization to SARS-Cov2 and that therefore the whole population is initially susceptible ($S_0 = N$).

The mean incubation period ($\frac{1}{\alpha}$) and the mean infectious period ($\frac{1}{\gamma}$) come from the inverse of the above parameters. The fatality rate ($\mu$) gives us the fraction of sick people who do not recover. The transmission rate ($\beta$) is the product of the contact rate per day and infectivity to make more detailed modeling. Finally, to calibrate $\beta$ values, we will code a SEIRD model in Python.

*2.2. Python Model*

Python model runs an optimization model representing the different turning points that change the $\beta$ we use on the SDL model. We use a simulated annealing algorithm called "dual annealing" included in the SciPy libraries to proceed with the multiparametric optimization by least-squares minimization [28]. In addition, we will also be able to estimate a value for the effective ($Rt$) and the basic reproduction number ($R_0$) from the containment factor ($\rho$), the transmission rate ($\beta$), and the recovery rate ($\gamma$) with the Equation (7). The basic reproduction number, $R_0$, see [29], represents the average number of secondary cases that result from the introduction of a single infectious case in a susceptible population during the infectiousness period. The effective reproduction number, $Rt$, is the same concept but after applying the containment measures:

$$R_t = \rho R_0 = \frac{\rho \beta}{\gamma} \tag{7}$$

$\beta_i$ varies according to a confinement coefficient $\rho_i$ so that effective transmission rate can be calculated for each segment of the curve:

$$\beta_i = \rho_i \beta \tag{8}$$

As input data, we have used the 7-day cumulative incidence and the number of turning points; using the Generalized Simulated Annealing algorithm also contained in the SciPy package [30] to optimize the values of the confinement coefficients ($\rho$) at the same time as the date on which they start to be applied. Thus, by establishing several turning points, we can reproduce different regimes of the epidemic curve depending on these non-pharmaceutical actions.

To fit the position of all the changing points and the transmission rate of each regime and ensure that the algorithm converges, we will proceed in a stepwise approach. So, we will consolidate the search boundaries of the previous fitted values to find the next ones and so on iteratively.

We selected this technique after trying several and finding that it worked well for us. The model allows randomly worsening the intermediate solutions to find a final solution closer to the global optimum. However, as the number of change points has increased, the difficulty of adjustment has also increased rapidly. Therefore, we end up approaching the problem in parts, so it would be interesting to improve the speed of the solution to look for other heuristics that could converge with a single execution and more efficiently. However, these alternatives are not needed at this point of the research since the time to obtain the solution is low, about one hour to do the whole process.

The main goal of the Python model is to estimate the changing points, as shown in Figure 3 that presents the different regimes on the curve. The aim is to define a model that behaves the same that the data we will obtain from the authorities (digital shadow). The assumption is that these points should correspond to the effect of the different government interventions such as confinements, reopening, distribution of masks, among others.

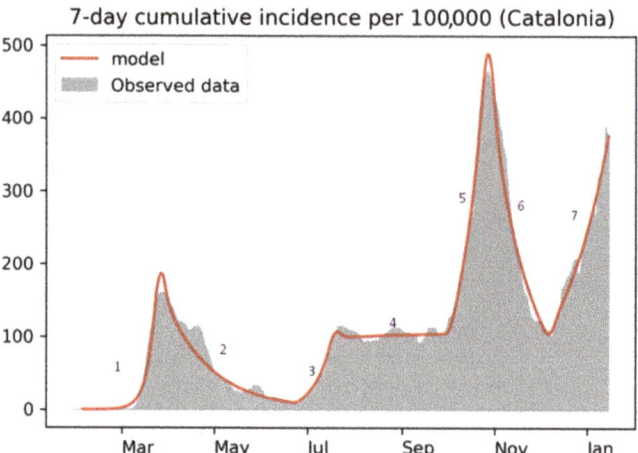

**Figure 3.** Fitting the changing points to detect regime shifts in the cumulative incidence curve (7 regimes in this picture).

*2.3. The SDL Model*

Specification and Description Language (SDL) is a graphical object-oriented language with unambiguous formal semantics. The International Telecommunication Union standardized it. SDL uses four hierarchized building blocks: (i) SYSTEM, (ii) BLOCK, (iii) PROCESS, and (iv) PROCEDURES. Regarding the notation we use to describe the SDL diagrams, every time we refer to an SDL element, we write it in CAPS, while the name of the elements will be in *Italic*. The SYSTEM and BLOCK diagrams represent the model's structure, using a hierarchical decomposition.

On the other hand, PROCESS and PROCEDURES define the model's behavior. BLOCK and PROCESS are AGENTS that establish the communication sending SIGNALS through CHANNELS. SIGNALS act as a trigger, generating the execution of a set of actions in a PROCESS. All SIGNALS, sorted by time and priority in the input queue of every input channel, own a delay parameter that represents the time.

For those not used with SDL in the following Appendix A, we will detail the main elements of the language.

The SDL model is used to test the assumptions, and from them, effects of the NPIs. We define different models that incorporate these modifications. Table 1 shows the different models we develop in the frame of the project. All models are defined using SDL, but the modeling technique differs. 1.X models are the result of SD models, while 2.X models use cellular automaton (CA) structures, mainly to define the propagation in the different Health regions (a Health Region is an administrative division in Catalonia). 3.X models, not detailed here, are based on multi-agent simulation models (MAS) and are defined on SDL.

Figure 4 shows the 2.5 SDL model *BLayer* BLOCK diagram. Squares (BLOCKS in SDL semantics) represent the different compartments. The semantics of the compartments is the same as in the SD or Python models, representing the *Susceptible* people and the *Exposed* people and so on. On the SDL model, we add some refinement with more compartments to better detail the population's behavior, shown in grey. The main improvements of the SDL model are: (i) BLOCK *BConfinement*, representing the confined population excluded from becoming exposed. (ii) BLOCK *BInfectiveDetected*, representing the infected detected. We assume that over time the detection improves. (iii) BLOCK *BContentionActions* that controls the actions taken by the authorities to prevent SARS-CoV-2 spread. (iv) CA defines each cell as a model that details the spread of the SARS-CoV-2 over a Catalonia Health Region, Table 2.

**Table 1.** Models defined on the frame of the digital twin Catalonia's pandemics modeling process.

| Model Number | Description | Valid (at the Time of Writing This Paper) |
|---|---|---|
| 1.9 | The initial model contains the initial growth and the total lockdown | No, the total lockdown was open. |
| 2.5 | Optimistical return to normality (schools and work). | No |
| 2.6 | Increase online learning and teleworking. | No |
| 2.7 | Pessimistic return to normality. | No |
| 2.8 | More NPIs application. | No |
| 2.9 | Readjusted the effect of the holidays and January restraints added. Adding the effects of the vaccination on the population. | Yes |

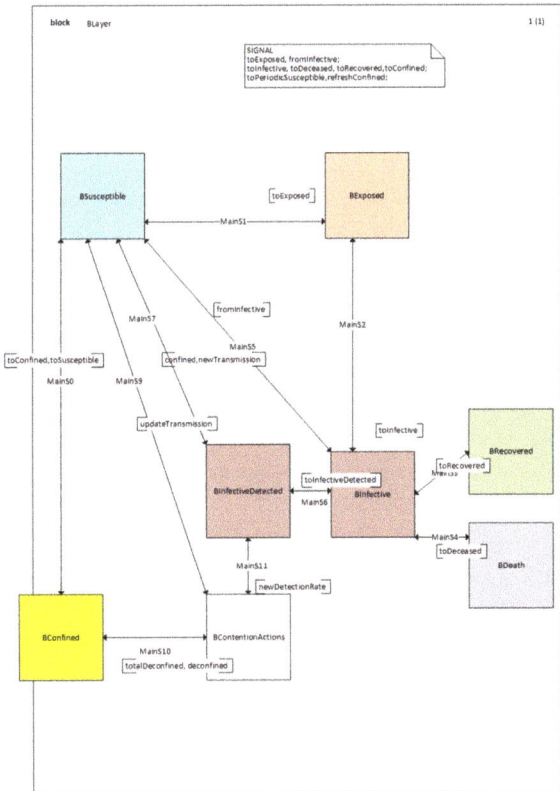

**Figure 4.** SDL 2.5 model. Each color on the diagram represents a Compartment on the original SEIRD model (using the same colors). The exception is for the *BContentionActions* BLOCK that represents the different NPIs applied. Notice that *BInfectiveDetected* and *BInfective* refer to the single Infective compartment on the initial SEIRD model because on the SDL 2.5 version model, we can distinguish between real and detected cases. Notice that the 1(1) in the upper corner defines the number of pages that detail this diagram and the current page number.

Table 2. Catalonia Health regions.

| Id. | Code | Description | Population |
|---|---|---|---|
| 6100 | LL | Lleida | 362,850 |
| 6200 | CT | Camp de Tarragona | 607,999 |
| 6300 | TE | Terres de l'Ebre | 176,817 |
| 6400 | GR | Girona | 861,753 |
| 6700 | CC | Catalunya Central | 526,959 |
| 7100 | AA | Vall d'Aran | 67,277 |
| 7801 | BS | Barcelona Sud | 1,370,709 |
| 7802 | BN | Barcelona Nord | 1,986,032 |
| 7803 | BC | Barcelona Ciutat | 1,693,449 |
| All | CAT | Catalunya | 7,653,845 |

The use of the Health regions for validation allows us to see if the different curves for the different Health regions follow the data correctly. We keep the parameters that define the dynamics of the pandemic fixed and modify just the specificities of each health region, like the time of the first case or the population. A CA secondary layer details the number of persons included in each cell. Secondary layers represent static information needed for the model that remains unchanged during the model execution, like the initial total population for each health region.

As we mention, Model 2.5 is not the last model we develop; in SDL, model 2.9 represents the last refinement and contains other BLOCKS that allow us to describe the vaccination process accurately. To represent this vaccination process, we use the digital shadow of the system to introduce the vaccination rate on the model. Also, we can forecast the trend of this vaccination process to forecast the trend of the pandemic evolution. We can access the SDL model conceptualization at https://doi.org/10.6084/m9.figshare.13153100 (accessed on 14 July 2021).

## 3. Results

In this section, we present the process we follow to perform the continuous calibration of the models. We consider this process also a result since, through this discussion, we will be able to detect when and why a model is no longer valid, learning regarding the validity of the model assumptions we use.

### 3.1. Models Coding and Calibration

With the diagrams that compose the SDL model, one can implement it automatically with software that understands SDL, in our case using SDLPS (https://sdlps.com/ accessed on 14 July 2021). We define the same time discretization used in the SD model, using $\Delta_t = 0.1$ days as the model time steep, following an activity scanning approach. Since the SDL model is the one that we will modify continuously, this automatic implementation will simplify the verification process (we can assume that the model implementation is correct).

To validate the model's accuracy, we compare the daily new cases forecast with the dataset that contains the daily new cases in Catalonia. This dataset is available through the Open Data service in Catalonia (http://governobert.gencat.cat/en/dades_obertes/ accessed on 14 July 2021).

Since the model becomes a digital twin of the pandemic in Catalonia, it must include the NPIs applied to reduce the expansion of SARS-CoV-2. As we mention, these NPIs are part of the SDL model through events detailed on *BContentionActions*, defining a scenario parametrization.

### 3.2. Second Wave Calibration

To propose $\beta$ values for a potential second outbreak, we use patterns of pandemic evolution in countries that have started earlier the school year, such as South Korea (August 25) or Israel (September 1), as shown in Figure 5.

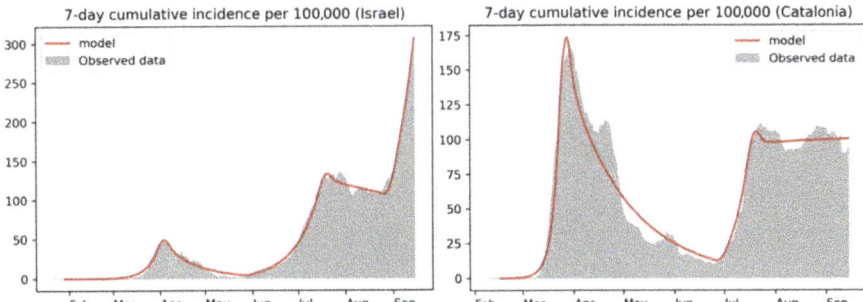

**Figure 5.** SEIRD python-coded model results for Israel after the September 1 outbreak (top left) and for Catalonia before schools opening (September 14) (right). See the similarity in the patterns before the outbreak.

To estimate the $\beta$ in those countries, we use our Python model. We have used the 7-day cumulative incidence to obtain the turning points representing the NPIs effects. The regimes are the first outbreak in spring, followed by a hard lockdown, a stabilization in summer, and a second outbreak after the academic year restart. Table 3 represents the parametrization for the SDL model version 2.5 that forecast the second wave. Several questions yet exist, like the percent of detection that will be estimated using the percent of asymptomatic [31].

**Table 3.** Parameters for model 2.5 to be considered due to NPIs. % Det denotes real cases reporting level.

| Event | Date (2020) | $\beta$ | % Det | % Conf | NPIs |
|---|---|---|---|---|---|
| 1 | 29 January | 1.2 | 0.1 | 0% | First infected |
| 2 | 08 Febrary | 1.2 | 0.25 | 0% | Initial tests |
| 3 | 15 March | 0.6 | 0.45 | 35% | Confinement |
| 4 | 23 March | 0.24 | 0.45 | 35% | Air space closes |
| 5 | 13 April | 0.2 | 0.45 | 25% | Workers partial comeback |
| 6 | 20 April | 0.18 | 0.45 | 25% | Free masks |
| 7 | 25 May | 0.18 | 0.45 | 25% | Phase 1 for some regions |
| 8 | 18 June | 0.18 | 0.54 | 0% | Phase 3 for BCN |
| 9 | 24 June | 1.2 | 0.54 | 0% | National day |
| 10 | 25 June | 0.18 | 0.54 | 0% | Phase 3 for BCN |
| 11 | 02 July | 0.3 | 0.54 | 0% | New normality |
| 12 | 17 July | 0.21 | 0.54 | 0% | Summer plateau |
| 13 | 15 September | 0.24 | 0.7 | 0% | School returns |

To estimate the percent of confinement, we use public data that allows calculating this percent. On March 15, the total closure of the country represents a maximum of 35% of confinement, but the return of the industrial sector implies 10% on April 13, 2020 [32]. For the second wave, October 10, the university population has been confined, these consist about 3% of the population, but we must increase this value on October 25 because leisure closes. Half of the public workers go online, representing about 7% more of the population, which is about 10% of the population that remain at home until November 23. To estimate the percent of detection, we use the prevalence study [31,33–36] for the 2.9 model, while for the previous models, we use an estimation based on [37,38].

Model 2.5 begins on January 29. On March 15, the government confined the entire population, except essential sector workers. On March 23, air space closes, on April 13, the industrial sector workers returned to work, and on April 20, the government gave all population basic masks. During May and June until July, the restrictions gradually decimated where we enter the "new normal."

In Catalonia on September 14, we got $\beta \approx 0.16$. The time-series analysis shows that during the summer (that defines a sort of plateau), the value of $\beta$ is peaking about 0.21, which can be considered a reference value for some kind of citizen's behavior. The return to normality, with an increasing movement of citizens, can represent an inflection point. Different tests have been performed, concluding with a $\beta = 0.24$ for 2.5 model, is this the value obtained from the analysis done in the Israel case, using the Catalonia cumulative incidence (85) and Israel (119), and considering that $\beta = 0.21$ will be increased by the return to the normality and will become no smaller than the value of a total lockdown, see Table 4. We consider a 70% detection rate due to the increasing testing, being the 30% the asymptomatic rate we use [38].

**Table 4.** $\beta$ comparison. (*) In South Korea, the summer outbreak is so low that the model is not proper. (**) The forecast of a new outbreak with a $\beta$ close to Israel proved to be accurate. (?) Information is not yet available, or the situation did not occur at the time of analysis of model 2.5.

| Regime | Israel | S. Korea | Catalonia |
|---|---|---|---|
| First outbreak | 0.55 | 0.95 | 0.95 |
| Lockdown | 0.12 | 0.09 | 0.15 |
| Summer outbreak | 0.34 | (*) | 0.43 |
| Summer plateau | 0.20 | (*) | 0.20 |
| Reopening outbreak | 0.33 | 0.48 | 0.30 (**) |
| Partial lockdown | (?) | 0.12 | (?) |

Figure 6 shows the new infections forecast from the different scenarios that we ran on the SDL model. SDL attempts to forecast the maximum values (worst scenario).

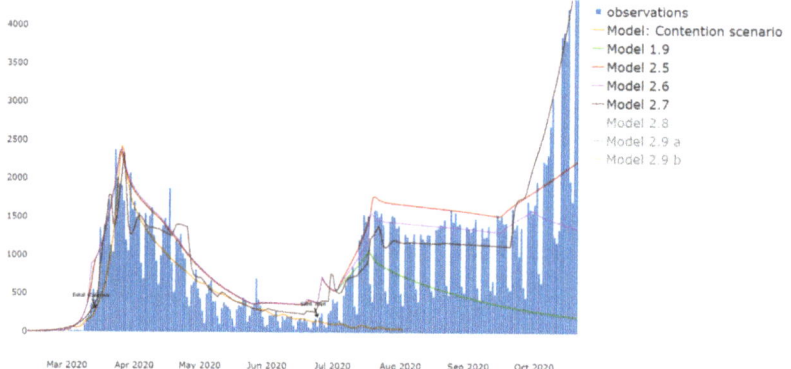

**Figure 6.** Forecast for the new infections (y-axis), the red line for the 2.5 SDL model (being this an optimistic scenario). We consider other scenarios in case 2.5 model hypotheses fail, model 2.6, and model 2.7.

### 3.3. Third-Wave Calibration

For the calibration of the third wave, we use a historical data comparison to obtain more insights regarding the behavior of the pandemic in Catalonia. We define a new model named 2.8 that includes some new improvements over the previous models.

As in the previous models, we must define the $\beta$ with the NPIs applied, selecting the same $\beta$ that in the previous wave due to the coincidence on the scenario (schools reopening after the Christmas holidays). We use the $\beta$ we apply on model 2.7 that represents model 2.5 with the adjusted $\beta$ due to the observed cases.

Model 2.8 represents a new wave, but that does not show a uniform growth like on the second wave. That depends on how the people confine due to the Christmas holidays.

We detect one issue with the prevision during the validation process that does not invalidate the growth but avoids obtaining a better fit. We suppose that the growth does not continue to point A (see Figure 7) because we do not consider increasing the transmission due to increase contacts during the Christmas holidays. We suppose that remains at the same level observed in the summer, but clearly because now the people interact in a closed environment, this affects the transmission.

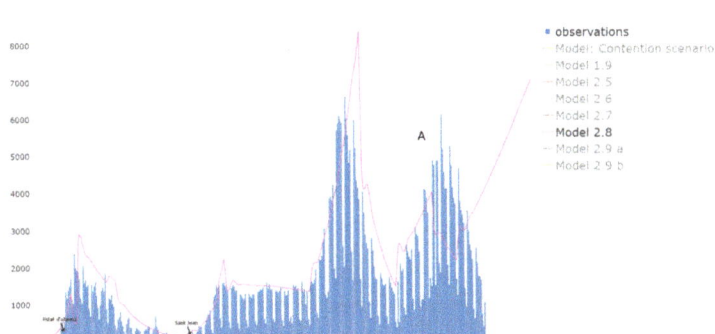

**Figure 7.** Model 2.8 with the forecast for the second wave for the new cases (y-axis). Notice the divergence between the trend of the model and point A, the highest value of the data. We are presenting a scenario considering that the Christmas holidays have no acute effects on the $\beta$.

If we compare with the data, we see that the curve has an inflection point when the NPIs that include increasing the restrictions in the leisure sector (restaurants, theaters, among others) are applied, obtaining a $\beta$ similar to the $\beta$ we see on the descending trend of the second wave. Table 5 shows the parametrization for the 2.8 model.

**Table 5.** Parameters used on the 2.8 simulation scenarios: [1] as a result of the validation process, the correct parameter for this $\beta$ is 0.3, being added on the 2.9 model. [2] the organization of the online courses needs a week to be fully implemented.

| Event | Date | $\beta$ | %Det | % Conf | Description Event |
|---|---|---|---|---|---|
| 1 | 01 December 2019 | - | - | - | Start of simulation |
| 2 | 01 December 2019 | 0.81 | 0 | 0 | - |
| 3 | 11 December 2019 | 0.81 | 0.11 | 0 | Pandemic Beginning |
| 4 | 15 March 2020 | 0.81 | 0.17 | 0 | Confinement |
| 5 | 15 March 2020 | 0.81 | 0.17 | 0.35 | Confinement |
| 6 | 15 March 2020 | 0.25 | 0.17 | 0.35 | Confinement |
| 7 | 13 April 2020 | 0.25 | 0.17 | 0.2 | Workers partial comeback |
| 8 | 20 April 2020 | 0.16 | 0.17 | 0.2 | Free Masks |
| 9 | 06 May 2020 | 0.16 | 0.18 | 0.2 | Phase 1 for some regions |
| 10 | 01 June 2020 | 0.16 | 0.25 | 0.2 | Phase 3 for some regions |
| 11 | 18 June 2020 | 0.465 | 0.25 | 0.2 | Phase 3 for BCN |
| 12 | 18 June 2020 | 0.465 | 0.25 | 0 | Phase 3 for BCN |
| 13 | 22 June 2020 | 0.465 | 0.6 | 0 | New normality |
| 14 | 16 July 2020 | 0.21 | 0.6 | 0 | Summer plateau |
| 15 | 15 September 2020 | 0.34 | 0.6 | 0 | School returns |
| 16 | 20 October 2020 | 0.34 | 0.6 | 0.03 | University online [2] |
| 17 | 25 October 2020 | 0.34 | 0.6 | 0.1 | Movement and restaurants restrictions |
| 18 | 25 October 2020 | 0.15 | 0.6 | 0.1 | Movement and restaurants restrictions |
| 19 | 23 November 2020 | 0.3 | 0.6 | 0.1 | Reopening restaurants |
| 20 | 23 November 2020 | 0.3 | 0.6 | 0.03 | Reopening restaurants |
| 21 | 23 December 2020 | 0.21 (0.3) [1] | 0.6 | 0.03 | Holidays |
| 22 | 11 January 2021 | 0.3 | 0.6 | 0.03 | Schools Returns |

New NPIs have been applied to the system, implying the modification of the $\beta$ parameter, suggesting the need to revise the model hypotheses and the execution of the model following the continuous validation and verification loop.

## 4. Discussion

Applying a digital twin approach and focusing on a continuous validation of the model assumptions enables testing the possible causality effects for the NPIs applied to the population. The forecasts provided by the model seem accurate and allows us to understand the future evolution of the system. It will be a crucial aspect to size the resources needed at each stage of the pandemic evolution.

We calibrated the model to show always a pessimistic view of the situation. Notice that the forecast always tries to be over the observation's maximum values. It will help decision-makers understand the worst scenario and prepare consistently for it.

We can access the dashboard to see the Key Performance Indicators (KPI's) and model forecast, validated continuously against the actual data in a solution validation approach, becoming a digital twin of the pandemic. Figure 8 shows the dashboards for the model 1.9 prevision on September 16, previous to the publication of the 2.5 model.

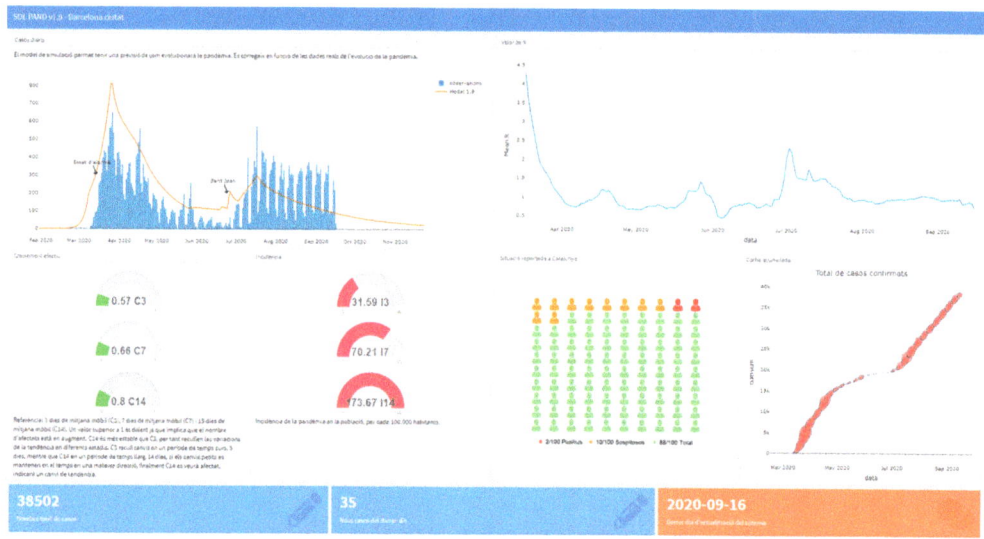

**Figure 8.** KPIs for Barcelona and its health area CA cells Model 1.9.

When the model invalidates, a divergence between the digital shadows of the system (Catalonia SARS-CoV-2 new cases obtained from Socrata Open Data Source) and the master's digital shadows (simulation traces) appears. Then, one must reevaluate the hypotheses and recalibrate the model, providing valuable information to understand the situation. It will happen continuously since new NPIs will be applied to control the spread of the virus. An interesting example of NPIs is when the government provides free masks to the population. This action seems to positively affect the population, which helps increase the confidence in the mask and change the population's mindset.

By analyzing the different models, we can discuss the possibilities and the effects of the NPIs before their application on the system. As an example, model 2.5, with a $\beta = 2.4$, shows a scenario that is not optimistic but that presents a situation that implies an increase in the number of cases. Considering the model comparison approach and using the data we obtain from other countries, we see that the $\beta$s are highest than the proposed ones. At this point, the model suggests that if the growth happens, this will become faster, implying that a discussion based on models must include the needed analysis with different scenarios.

Another example is on model 2.8, where the situation presents the Christmas holidays, using for the analysis data that comes from our observations in Catalonia. We detect that the assumption that Christmas does not affect the $\beta$ is not valid. The restoration reopening seems to impact the increase of the number of cases in the population. It has been detected and corrected on the 2.9 model using an accurate value for the $\beta$ at this point. Notice that the current valid model, 2.9, correctly predicts the current trend of the pandemic in Catalonia, suggesting that the end of the cases will be possible at the end of the summer if the current conditions do not change (NPIs and no a variant with increased transmissibility appears). Figure 9 shows the forecast for the current trend of the pandemic situation in Catalonia using the current model.

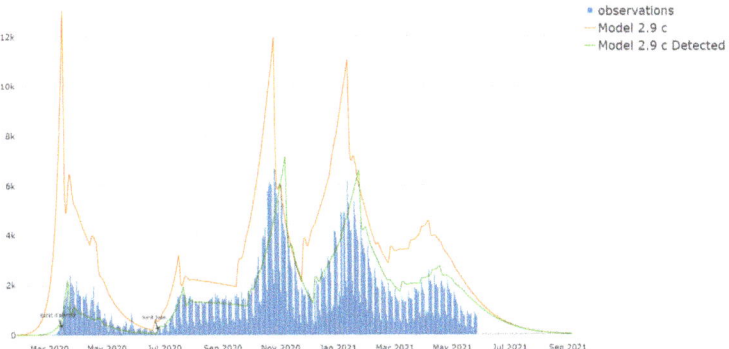

**Figure 9.** Model 2.9 now forecasts for new cases (*y*-axis) the end of the pandemic situation in Catalonia if the NPIs remain similar to the current (at 05 June 2021) and no new variants appear.

This forecast again needs a continuous validation process but shows a positive effect of the vaccines for the disease contention. It suggests that the number of people who receive the vaccines will not have to reach 70% to achieve the desired herd immunity to control the spread; this time (05 June 2021), the number of people with complete vaccination is below 24%. We will confirm this scenario if the NPIs remain similar to the current or no new variants with the highest $\beta$ appears. During the preparation of this document model, 2.9c was invalidated. The Delta variant spread and the reduction of some NPIs implied an increase in the number of cases in Catalonia. We therefore define a new scenario (2.9d) applying a $\beta$ in our SDL model based on [39] and considering the reduction of the NPIs due to the reopening of the leisure [40] to be able to validate our forecast again. At this point, accurate monitoring of the epidemic using models will become an excellent tool to understand the implications and the effects of the NPIs if the situation worsens. Figure 10 shows the different models developed until now. Notice that every time an effective NPI is applied to the population, the model invalidates.

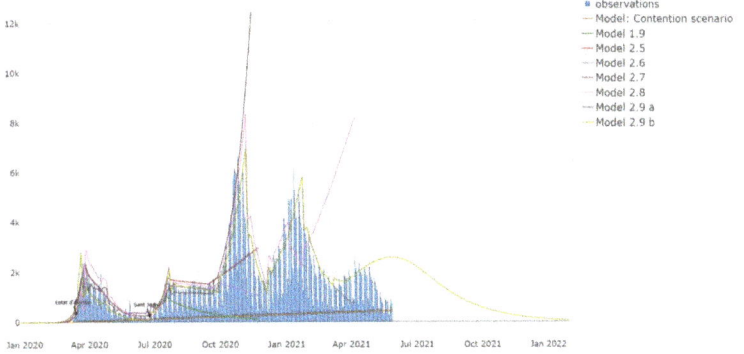

**Figure 10.** The different models that we develop due to the continuous validation process, new cases (*y*-axis).

## 5. Conclusions

The current last model at the time of writing this paper, scenario 2.9c, needs accurate monitoring due to the inclusion of the vaccination effect in the model. An increase in the number of vaccines delivered every week to the population implies a substantial modification on the number of new cases detected later. Also, an eventual modification on the factors (NPIs, infectivity, among others) implies a necessary change of the model parameters to fit again with the system, following the shadows comparison of the digital twin approach.

The proposed methodology, defining a digital twin of the pandemic, must serve precisely to facilitate this monitoring, providing early warnings if the trend of the new cases differs from the trend of the forecasted model (shadows comparison). Because of the modification applied to the system, mainly the NPIs, to contain the virus spread, the representation of the model assumptions in a graphical language simplifies the immediate modification of the model. It makes easier the agreement between the different specialists involved in the project. This involvement of the specialists in the digital twin maintenance and continuous validation must serve to increase the model's credibility, providing a way to achieve accreditation in a continuous verification and validation loop.

In the context of Industry 4.0, the model-based discussion becomes a central element to understand the causality of complex systems. The continuous validation of simulation models, which includes a clear representation of the causal relations, is part of the proposed methodology. We cannot make decisions without understanding the possible causality rules. Although we do not fully understand that the things happen as detailed in the model, the discussion based on these causal relations helps the decision-makers understand these rules, and more interestingly, when the assumptions and the hypotheses assumed on the model are not valid and must be reconsidered.

**Author Contributions:** Conceptualization, P.F.i.C., X.P.i.P. and J.G.i.S.; methodology, P.F.i.C.; software, V.G.i.C. and J.G.i.S.; validation, V.G.i.C., P.F.i.C. and J.G.i.S.; formal analysis, P.F.i.C. and V.G.i.C.; investigation, P.F.i.C., X.P.i.P. and V.G.i.C.; resources, P.F.i.C.; data curation, V.G.i.C., P.F.i.C. and J.G.i.S.; writing—original draft preparation, P.F.i.C., X.P.i.P. and V.G.i.C.; writing—review and editing, P.F.i.C., X.P.i.P. and V.G.i.C.; visualization, P.F.i.C. and V.G.i.C.; supervision, P.F.i.C.; project administration, P.F.i.C.; funding acquisition, P.F.i.C. All authors have read and agreed to the published version of the manuscript.

**Funding:** This research was partially funded by CCD, grant 2020-L015.

**Institutional Review Board Statement:** Not applicable.

**Informed Consent Statement:** Not applicable.

**Data Availability Statement:** The SDL 2.5 model conceptualization can be reviewed at https://doi.org/10.6084/m9.figshare.13153100 (accessed on 14 July 2021). and the model can be downloaded at https://figshare.com/articles/dataset/SDL-PAND_model_2_5/14910627 (accessed on 14 July 2021). The system's dataset can be obtained at https://analisi.transparenciacatalunya.cat/ca/Salut/Registre-de-casos-de-COVID-19-realitzats-a-Catalun/xuwf-dxjd (accessed on 14 July 2021). Other SDL models can be accessed and downloaded at https://sdlps.com/Experiments (accessed on 14 July 2021), the results of the different scenarios can be accessed at http://pand.sdlps.com (accessed on 14 July 2021).

**Acknowledgments:** Thanks to the help of inLab FIB in the administration of the project.

**Conflicts of Interest:** The authors declare no conflict of interest.

## Appendix A. SDL

Specification and Description Language (SDL) is an object-oriented, formal language. The International Telecommunication Union—Telecommunication Standardization Sector (ITU–T) defines it as a standard language. The document that defines the language structure is Recommendation Z.100. Originally the language was designed to specify event-driven, real-time, complex interactive applications that involve many concurrent activities. SDL bases the communication between the different model elements on the use of discrete signals [41,42].

The conceptualization of a simulation model needs to define the following components: (i) Structure: system, blocks, processes, and the hierarchy of processes, (ii) Communication: signals, including the parameters and channels that the signals need to travel from element to element, (iii) Behavior: defined through the processes. (iv) Data: based on Abstract Data Types (ADT), and (v) Inheritances: describe the relationships between and the specializations of the model elements. At least the first two components, the structure, and the behavior, must be described in a formal language to be able to do a correct conceptualization of a simulation model.

The language has four levels: (i) system, (ii) blocks, (iii) processes, and (iv) procedures.

A model always starts with the definition of what we want to represent. Using an approach based on a formal language, like SDL, we will start with a simple box that explains what the system to be modeled will be. This system, represented by the SYSTEM diagram, shows the uppermost level of the model's structure. It clearly shows the elements that we consider in the analysis. All the other elements not present inside the SYSTEM diagram will become the ENVIRONMENT, and we express the communications between them using the CHANNELS. Following the SDL terminology, SYSTEM is an AGENT. Other AGENTS can appear in an SDL diagram, the BLOCKS, and the PROCESS. These other elements appear in lower levels of the model. The CHANNELS can be unidirectional or bidirectional and are using PORTS to connect with the BLOCKS or PROCESS. The PORTS guarantee the independence of the different AGENTS we will use in our models. It guarantees modularity, a crucial aspect of conceptual models since it allows us to reuse the different model elements. Also, modularity allows performing an incremental Validation and Verification process. It means that we do not need to Validate the whole model and focus on a subset of the model, a set of the model's AGENTS. Because of modularity, an AGENT only knows what it sends and receives events using a specific PORT or CHANNEL.

To define a simulation model conceptualization, we need to detail the model structure and behavior. SYSTEM diagrams and BLOCK diagrams represent the model structure. It is a hierarchical decomposition of the different model elements; some good examples are available in [42]. The behavior will be described on the PROCESS diagrams and below, on the PROCEDURES, although these elements are not AGENTS and only represent some pieces of code. The pieces of information that travels on the CHANNELS, from AGENT to AGENT, are the SIGNALS. When a PROCESS receives a specific SIGNAL, it acts as a trigger, starting to execute a set of actions in no time. To represent the time, all the SIGNALS own a parameter, delay that represents when this SIGNAL can be processed. Every PROCESS has an input queue for every input CHANNEL, containing the received SIGNALS ordered by time and priority.

Being SDL a graphical language, a PROCESS uses different graphical elements to represent the model's behavior. In the following lines, we describe some of the more essential elements synthetically to simplify understanding the SDL model definition.

**Start.** This element defines the initial condition for a PROCESS diagram.

**State.** The PROCESS must always start in a STATE, and this owns a name.

**Input.** The PROCESS starts an execution when an INPUT receives the SIGNAL for this INPUT. All the STATES can own several different INPUTS to work with the different SIGNALS one can receive.

**Create.** This element allows the creation of an AGENT.

**Task.** To interpret a piece of code, we can use the TASK element. In our approach, we can use C on this element.

**Output.** To send a SIGNAL, we must use the OUTPUT element. We can also add parameters to the SIGNALS and describe the destination if ambiguity about the signal destination exists. We can direct the communication specifying destinations using a PROCESS identifier (PId), an identifier that must own all the PROCESS. Also, we can send using the sentence via path. We can use four PId expressions: (i) **self**, an agent's own identity; (ii) **parent**, the agent that created the agent (Null for initial agents); (iii) **offspring**, the most recent agent created by the agent; (iv) **sender**, the agent that sent the last signal input (null before any signal received). Also, we can use {CUR_CELLS} and {ALL_CELL} to send the information to a specific cell of the CA.

**Decision**. ◇ To define a bifurcation, a decision point, we can use the DECISION.

Finally, mention that on the last level of the SDL language (PROCEDURE diagrams), we can describe pieces of code, also graphically, making the language complete. We can use these pieces of code in the PROCESS with the PROCEDURE CALL ▭.

## References

1. Anderson, R. Discussion: The Kermack-McKendrick epidemic threshold theorem. *Bull. Math. Biol.* **1991**, *53*, 3–32. [CrossRef]
2. Wynants, L.; van Calster, B.; Collins, G.S.; Riley, R.D.; Heinze, G.; Schuit, E.; Bonten, M.M.J.; Dahly, D.L.; Damen, J.A.; Debray, T.P.A.; et al. Prediction models for diagnosis and prognosis of covid-19: Systematic review and critical appraisal. *BMJ* **2020**, *369*, m1328. [CrossRef]
3. Nava, A.; Papa, A.; Rossi, M.; Giuliano, D. Analytical and cellular automaton approach to a generalized SEIR model for infection spread in an open crowded space. *Phys. Rev. Res.* **2020**, *2*, 043379. [CrossRef]
4. Chen, W.; Zhang, N.; Wei, J.; Yen, H.-L.; Li, Y. Short-range airborne route dominates exposure of respiratory infection during close contact. *Build. Environ.* **2020**, *176*, 106859. [CrossRef]
5. Fonseca i Casas, P.; García i Carrasco, V.; Garcia i Subirana, J. SEIRD COVID-19 Formal Characterization and Model Comparison Validation. *Appl. Sci.* **2020**, *10*, 5162. [CrossRef]
6. Ndaïrou, F.; Area, I.; Nieto, J.J.; Silva, C.J.; Torres, D.F. Fractional model of COVID-19 applied to Galicia, Spain and Portugal. *Chaos Solitons Fractals* **2021**, *144*, 110652. [CrossRef] [PubMed]
7. Ogden, N.H.; Fazil, A.; Arino, J.; Berthiaume, P.; Fisman, D.N.; Greer, A.L.; Ludwig, A.; Ng, V.; Tuite, A.R.; Turgeon, P.; et al. Modelling scenarios of the epidemic of COVID-19 in Canada. *Can. Commun. Dis. Rep.* **2020**, *46*, 198–204. [CrossRef]
8. Faniran, T.; Bakare, E.; Potucek, R.; Ayoola, E. Global and Sensitivity Analyses of Unconcerned COVID-19 Cases in Nigeria: A Mathematical Modeling Approach. *WSEAS Trans. Math.* **2021**, *20*, 218–234. [CrossRef]
9. Arenas, A.; Cota, W.; Gómez-Gardeñes, J.; Gómez, S.; Granell, C.; Matamalas, J.T.; Soriano-Paños, D.; Steinegger, B. Modeling the Spatiotemporal Epidemic Spreading of COVID-19 and the Impact of Mobility and Social Distancing Interventions. *Phys. Rev. X* **2020**, *10*, 041055. [CrossRef]
10. Lymperopoulos, I.N. SIR-Neurodynamical epidemic modeling of infection patterns in social networks. *Expert Syst. Appl.* **2021**, *165*, 113970. [CrossRef]
11. Li, Y.; Campbell, H.; Kulkarni, D.; Harpur, A.; Nundy, M.; Wang, X.; Nair, H. The temporal association of introducing and lifting non-pharmaceutical interventions with the time-varying reproduction number (R) of SARS-CoV-2: A modelling study across 131 countries. *Lancet Infect. Dis.* **2021**, *21*, 193–202. [CrossRef]
12. Flaxman, S.; Mishra, S.; Gandy, A.; Unwin, H.J.T.; Mellan, T.A.; Coupland, H.; Whittaker, C.; Zhu, H.; Berah, T.; Eaton, J.W.; et al. Estimating the effects of non-pharmaceutical interventions on COVID-19 in Europe. *Nature* **2020**, *584*, 257–261. [CrossRef] [PubMed]
13. Ding, Y.; Huang, R.; Shao, N. Time Series Forecasting of US COVID-19 Transmission. *Altern. Ther. Health Med.* **2021**, *27*, 4–11. [PubMed]
14. Maleki, M.; Mahmoudi, M.R.; Wraith, D.; Pho, K.-H. Time series modelling to forecast the confirmed and recovered cases of COVID-19. *Travel Med. Infect. Dis.* **2020**, *37*, 101742. [CrossRef]
15. Vasiljeva, M.; Neskorodieva, I.; Ponkratov, V.; Kuznetsov, N.; Ivlev, V.; Ivleva, M.; Maramygin, M.; Zekiy, A. A Predictive Model for Assessing the Impact of the COVID-19 Pandemic on the Economies of Some Eastern European Countries. *J. Open Innov. Technol. Mark. Complex.* **2020**, *6*, 92. [CrossRef]
16. Deb, P.; Furceri, D.; Ostry, J.; Tawk, N. The Effect of Containment Measures on the COVID-19 Pandemic. *IMF Work. Pap.* **2020**, *20*, 166. [CrossRef]
17. Chen, S.; Prettner, K.; Kuhn, M.; Bloom, D.E. The economic burden of COVID-19 in the United States: Estimates and projections under an infection-based herd immunity approach. *J. Econ. Ageing* **2021**, *20*, 100328. [CrossRef]
18. Sargent, R.G. Verification and Validation of Simulation Models. In Proceedings of the 2007 Winter Simulation Conference, Washington, WA, USA, 9–12 December 2007; pp. 124–137.
19. Leiva, J.; Fonseca i Casas, P.; Ocaña, J. Modeling anesthesia and pavilion surgical units in a Chilean hospital with Specification and Description Language. *Simulation* **2013**, *89*, 1020–1035. [CrossRef]
20. Vynnycky, E.; White, R. *An Introduction to Infectious Disease Modelling*; Oxford University Press: Oxford, UK, 2010; ISBN 978-0198565765.
21. Fonseca i Casas, P.; Garcia i Subirana, J.; Garcia i Carrasco, V.; Silva de Barcellos, J.L.; Roma, J.; Pi, X. SDL Cellular Automaton COVID-19 conceptualization. In *Proceedings of the 12th System Analysis and Modelling Conference on ZZZ*; ACM: New York, NY, USA, 2020; pp. 144–153.
22. Grieves, M.; Vickers, J. Digital Twin: Mitigating Unpredictable, Undesirable Emergent Behavior in Complex Systems. In *Transdisciplinary Perspectives on Complex Systems*; Springer: Berlin, Germany, 2017; pp. 85–113.

23. Kritzinger, W.; Karner, M.; Traar, G.; Henjes, J.; Sihn, W. Digital Twin in manufacturing: A categorical literature review and classification. *IFAC-PapersOnLine* **2018**, *51*, 1016–1022. [CrossRef]
24. Stark, R.; Kind, S.; Neumeyer, S. Innovations in digital modelling for next generation manufacturing system design. *CIRP Ann.* **2017**, *66*, 169–172. [CrossRef]
25. Sherratt, E.; Ober, I.; Gaudin, E.; Fonseca i Casas, P.; Kristoffersen, F. SDL—The IoT Language. In *Lecture Notes in Computer Science (Including Subseries Lecture Notes in Artificial Intelligence and Lecture Notes in Bioinformatics)*; Springer: Berlin/Heidelberg, Germany, 2015; Volume 9369, pp. 27–41. ISBN 9783319249117.
26. Generalitat de Catalunya. Registre de Casos de COVID-19 Realitzats a Catalunya. Segregació per Sexe i Edat. Available online: https://analisi.transparenciacatalunya.cat/en/Salut/Registre-de-casos-de-COVID-19-realitzats-a-Catalun/qwj8-xpvk (accessed on 21 February 2021).
27. Robinson, S.; Brooks, R.J. Independent Verification and Validation of an Industrial Simulation Model. *Simulation* **2009**, *86*, 405–416. [CrossRef]
28. SciPy. scipy.optimize.dual_annealing. Available online: https://docs.scipy.org/doc/scipy/reference/generated/scipy.optimize.dual_annealing.html (accessed on 21 May 2021).
29. Anastassopoulou, C.; Russo, L.; Tsakris, A.; Siettos, C. Data-based analysis, modelling and forecasting of the COVID-19 outbreak. *PLoS ONE* **2020**, *15*, e0230405. [CrossRef]
30. Xiang, Y.; Gubian, S.; Suomela, B.; Hoeng, J. Generalized Simulated Annealing for Global Optimization: The GenSA Package an Application to Non-Convex Optimization in Finance and Physics. *R J.* **2013**, *5*, 13–28. [CrossRef]
31. Pollán, M.; Pérez-Gómez, B.; Pastor-Barriuso, R.; Oteo, J.; Hernán, M.A.; Perez-Olmeda, M.; Sanmartín, J.L.; Fernández-García, A.; Cruz, I.; de Larrea, N.F.; et al. Prevalence of SARS-CoV-2 in Spain (ENE-COVID): A nationwide, population-based seroepidemiological study. *Lancet* **2020**, *396*, 535–544. [CrossRef]
32. Taylor, L. Study Highlights Costa del Sol's Malaga as One of the Best Quarantined Cities in Spain During Phase 0. Available online: https://www.euroweeklynews.com/2020/05/14/study-highlights-costa-del-sols-malaga-as-one-of-the-best-quarantined-cities-in-spain-during-phase-0/ (accessed on 12 July 2021).
33. Ministerio de Sanidad estudio Ene-Covid19: Primera Ronda. Available online: https://www.mscbs.gob.es/ciudadanos/ene-covid/docs/ESTUDIO_ENE-COVID19_PRIMERA_RONDA_INFORME_PRELIMINAR.pdf (accessed on 14 July 2021).
34. Ministerio de Sanidad Estudio Ene-Covid19: Segunda Ronda. Available online: https://www.mscbs.gob.es/ciudadanos/ene-covid/docs/ESTUDIO_ENE-COVID19_SEGUNDA_RONDA_INFORME_PRELIMINAR.pdf (accessed on 14 July 2021).
35. Ministerio de Sanidad Estudio Ene-Covid: Informe Final. Available online: https://www.mscbs.gob.es/ciudadanos/ene-covid/docs/ESTUDIO_ENE-COVID19_INFORME_FINAL.pdf (accessed on 14 July 2021).
36. Ministerio de Sanidad Estudio Ene-Covid: Cuarta Ronda. Available online: https://www.mscbs.gob.es/gabinetePrensa/notaPrensa/pdf/15.12151220163348113.pdf (accessed on 14 July 2021).
37. Jung, C.-Y.; Park, H.; Kim, D.W.; Choi, Y.J.; Kim, S.W.; Chang, T.I. Clinical Characteristics of Asymptomatic Patients with COVID-19: A Nationwide Cohort Study in South Korea. *Int. J. Infect. Dis.* **2020**, *99*, 266–268. [CrossRef] [PubMed]
38. Nishiura, H.; Kobayashi, T.; Miyama, T.; Suzuki, A.; Jung, S.-M.; Hayashi, K.; Kinoshita, R.; Yang, Y.; Yuan, B.; Akhmetzhanov, A.R.; et al. Estimation of the asymptomatic ratio of novel coronavirus infections (COVID-19). *Int. J. Infect. Dis.* **2020**, *94*, 154–155. [CrossRef] [PubMed]
39. Reed, R. SDL-2000 for New Millennium Systems. *Telektronikk* **2000**, *96*, 20–35.
40. Doldi, L. *SDL Illustrated—Visually Design Executable Models*, 1st ed.; TMSO Systems, Ed.; TMSO Systems: Old Main, PA, USA, 2001; ISBN 978-2951660007.
41. Campbell, F.; Archer, B.; Laurenson-Schafer, H.; Jinnai, Y.; Konings, F.; Batra, N.; Pavlin, B.; Vandemaele, K.; van Kerkhove, M.D.; Jombart, T.; et al. Increased transmissibility and global spread of SARS-CoV-2 variants of concern as at June 2021. *Eurosurveillance* **2021**, *26*, 2100509. [CrossRef]
42. Agencias Catalunya Levanta a Partir del Lunes la Prohibición de Celebrar Fiestas Mayores. Available online: https://www.lavanguardia.com/vida/20210604/7504929/catalunya-levanta-lunes-prohibicion-celebrar-fiestas-mayores-procicat-coronavirus-covid.html (accessed on 4 June 2021).

Article

# Power Spectral Density Analysis of Nanowire-Anchored Fluctuating Microbead Reveals a Double Lorentzian Distribution

Gregor Bánó [1], Jana Kubacková [2], Andrej Hovan [1], Alena Strejčková [3], Gergely T. Iványi [4], Gaszton Vizsnyiczai [5], Lóránd Kelemen [5], Gabriel Žoldák [6], Zoltán Tomori [2] and Denis Horvath [6,*]

[1] Department of Biophysics, Faculty of Science, P. J. Šafárik University, Jesenná 5, 041 54 Košice, Slovakia; gregor.bano@upjs.sk (G.B.); andrej.hovan@upjs.sk (A.H.)
[2] Institute of Experimental Physics SAS, Department of Biophysics, Watsonova 47, 040 01 Košice, Slovakia; kubackova@saske.sk (J.K.); tomori@saske.sk (Z.T.)
[3] Department of Chemistry, Biochemistry and Biophysics, University of Veterinary Medicine and Pharmacy, Komenského 73, 041 81 Košice, Slovakia; Alena.Strejckova@uvlf.sk
[4] Faculty of Science and Informatics, University of Szeged, Dugonics Square 13, 6720 Szeged, Hungary; itgergo@gmail.com
[5] Biological Research Centre, Institute of Biophysics, Eötvös Loránd Research Network (ELKH), Temesvári krt. 62, 6726 Szeged, Hungary; vizsnyiczai.gaszton@brc.hu (G.V.); kelemen.lorand@brc.hu (L.K.)
[6] Center for Interdisciplinary Biosciences, Technology and Innovation Park, P. J. Šafárik University, Jesenná 5, 041 54 Košice, Slovakia; gabriel.zoldak@upjs.sk
* Correspondence: denis.horvath@upjs.sk

**Citation:** Bánó, G.; Kubacková, J.; Hovan, A.; Strejčková, A.; Iványi, G.T.; Vizsnyiczai, G.; Kelemen, L.; Žoldák, G.; Tomori, Z.; Horvath, D. Power Spectral Density Analysis of Nanowire-Anchored Fluctuating Microbead Reveals a Double Lorentzian Distribution. *Mathematics* **2021**, *9*, 1748. https://doi.org/10.3390/math9151748

Academic Editors: Maria Luminita Scutaru and Junseok Kim

Received: 26 May 2021
Accepted: 21 July 2021
Published: 24 July 2021

**Publisher's Note:** MDPI stays neutral with regard to jurisdictional claims in published maps and institutional affiliations.

**Copyright:** © 2021 by the authors. Licensee MDPI, Basel, Switzerland. This article is an open access article distributed under the terms and conditions of the Creative Commons Attribution (CC BY) license (https://creativecommons.org/licenses/by/4.0/).

**Abstract:** In this work, we investigate the properties of a stochastic model, in which two coupled degrees of freedom are subordinated to viscous, elastic, and also additive random forces. Our model, which builds on previous progress in Brownian motion theory, is designed to describe water-immersed microparticles connected to a cantilever nanowire prepared by polymerization using two-photon direct laser writing (TPP-DLW). The model focuses on insights into nanowires exhibiting viscoelastic behavior, which defines the specific conditions of the microbead. The nanowire bending is described by a three-parameter linear model. The theoretical model is studied from the point of view of the power spectrum density of Brownian fluctuations. Our approach also focuses on the potential energy equipartition, which determines random forcing parametrization. Analytical calculations are provided that result in a double-Lorentzian power density spectrum with two corner frequencies. The proposed model explained our preliminary experimental findings as a result of the use of regression analysis. Furthermore, an a posteriori form of regression efficiency evaluation was designed and applied to three typical spectral regions. The agreement of respective moments obtained by integration of regressed dependences as well as by summing experimental data was confirmed.

**Keywords:** nanowire cantilever; stochastic model; double Lorentzian spectrum

## 1. Introduction

Many of the problems addressed by current nanosciences can be traced back to statistical mechanics and the concept of fluctuations. Fundamental problems constantly arise in nanosciences that go beyond conventional findings, complementing the emphasis and motivations of statistical mechanics. Related fields, now broadly referred to as stochastic processes, continue to pose a mathematical challenge.

Stochastic oscillations of anchored mechanical systems immersed in fluidic media or kept in vacuum have attracted significant attention in the past and are important in many ways today. The Brownian motion of a millimeter-sized mirror suspended from a torsion wire was utilized by Kappler back in 1931 to measure the Avogadro constant [1]. The thermal fluctuations of resonant micron-scale mechanical oscillators have been studied extensively, mostly in connection with AFM (atomic force microscopy) cantilevers and MEMS-based (micro- electromechanical) resonators [2–6]. Thermal fluctuations of glass nanofibers and silicon nitride cantilevers have been used to characterize and calibrate such systems for single-molecule force measurements [7,8]. Brownian motion has also been used at a smaller scale in TPM (tethered particle motion) experiments to investigate the properties of linear macromolecules such as DNA [9–11]. In contrast, thermal noise represents the main disturbing and limiting factor in experiments that rely on highly sensitive mechanical and opto-mechanical systems. Examples are inertial sensors [12,13] as well as recently proposed gravitational-wave and dark matter sensors [14–16].

Under certain conditions, system noises and their corresponding statistical quantities can become valuable, measurable features of technological devices and measurement instruments. When a particle immersed in a dissipative medium is simultaneously exposed to thermal noise, it reaches an equilibrium state with time, which provides a good possibility to measure statistical properties. Mechanical system fluctuations may also be related to intrinsic damping mechanisms such as internal friction, thermoelastic losses, or losses to the anchor system. Theoretical and experimental interest may then be directed toward elucidating the relationships between damping strength and noise. The well-known "fluctuation-dissipation theorem", which is also used in this work, addresses these general relationships.

The significance of the outputs of the monitored processes is clearly influenced by the equipment parameters and laboratory conditions. As a result, analytical methods differ. The power spectrum of thermal fluctuations can be derived for the case of damped harmonic oscillators [17], which is a good approximation for AFM cantilevers in liquids and gases [3]. The basic oscillator theory was modified by Saulson [17] to account for the thermal noise of mechanical systems, whose losses are dominated by processes occurring inside the material. High external dissipation conditions represent another extreme, which usually happens for low-stiffness micron-scale [7] or even smaller (molecular) systems [18]. When immersed in viscous liquids, these structures operate in a non-resonant, overdamped regime. Interestingly, the same overdamping conditions are present in optical tweezers experiments, where the fluctuation theory was elaborated thoroughly [19,20].

In this work, we investigate the thermal fluctuations of micron-scale viscoelastic mechanical systems submerged in water. In this particular case, as we show below, both the dissipation to the surrounding fluid and the intrinsic damping play an important role.

We are interested in the stochastic motion of microbeads attached to cantilevered photopolymer nanowires prepared by two-photon polymerization direct laser writing (TPP-DLW) (see Figure 1a,b) [21,22]. The nanowire thickness can be tuned during the fabrication process [23]. In the limiting case of thin nanowires, the stochastic thermal forces exerted on the microbeads cause clearly detectable Brownian motion behavior (see Figure 1c), specifically in the direction perpendicular to the nanowire axis, as depicted in Figure 1d). Moreover, as recently demonstrated, photopolymer nanowires possess viscoelastic material properties [24,25], which define the specific confinement forces investigated in the present work.

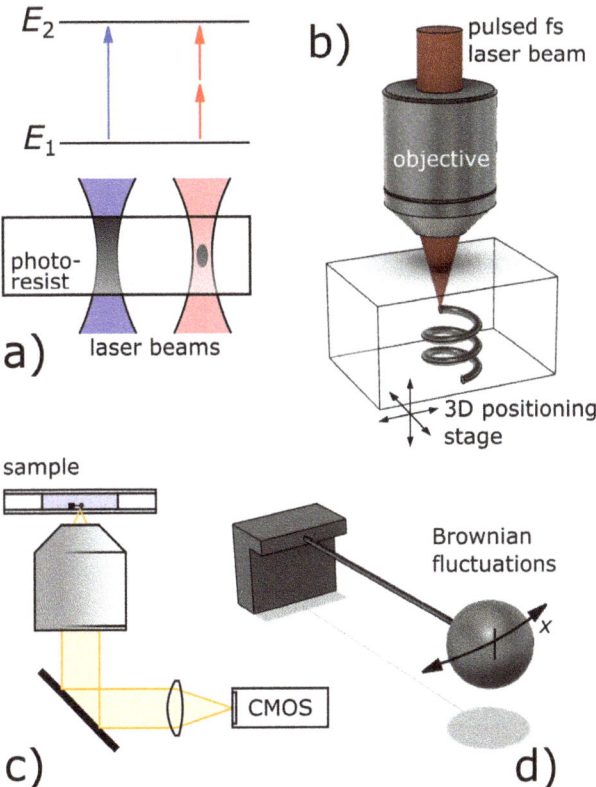

**Figure 1.** The main steps from sample preparation to measurement. (**a**) Light-sensitive (photoresist) material exposed to the laser beam. Single photon (left part, in blue) and two-photon (right part, in red) polymerization is depicted separately. (The exceptional spatial resolution can be reached by the two-photon process. The polymerized material is indicated in gray.) (**b**) The illumination used to produce three-dimensional design in the photoresist volume employing TPP-DLW. (**c**) Setup for motion detection and data production. The application of a CMOS image sensor, which provides encoded light information about position $x(t)$ that can be converted into digital data records by particle tracking algorithms. (**d**) A closer look at the mechanics of Brownian fluctuations under anchoring conditions. The studied fluctuations in the horizontal plane are indicated by the arrows.

Our present work is closely related to the preceding study focused on the bending recovery motion of photopolymer microbead-nanowire systems [25]. A three-parameter linear mechanical model of viscoelastic behavior has been found to provide a good explanation of the recovery time-dependence in this study. We aim to use the above theoretical description to include the thermal motion of the microstructure. This can be done analogously to other works. The original mechanistic model, which was first developed and validated in [25], can be generalized to reflect random forces. The problem can then be solved using the Fourier transform within the limits of the stochastic steady state in accordance with the experimental setup. The aim is to obtain the corresponding power spectrum and autocorrelation function of Brownian motions of the microstructure analytically. To summarize, our present approach promotes a practical and empirically supported transition from deterministic to stochastic frameworks.

## 2. Initial Considerations and Model Assumptions

Our main objective is to describe the stochastic motion of viscoelastic microstructures composed of cantilevered nanowires and spherical beads (see Figure 1) immersed in Newtonian liquids. The 18 µm long nanowire equipped with a 5 µm sphere (both made of Ormocomp) was prepared in a similar way as described in [25]. The previous work also provides all relevant experimental details. We assume that the nanowire bending is characterized by a 3-parameter linear mechanical model of viscoelastic behavior (see Figure 2). In the thin nanowire limit, the external viscous damping and the external thermal forces acting on the nanowire itself are neglected. In this approximation, the liquid surroundings interact only with the attached bead. We focus our attention on the nanowire bending oscillations in the horizontal plane perpendicular to the nanowire axis (see Figure 1d).

The equivalent mechanical model, which consists of ideal spring and dashpot elements (shown in Figure 2), contains two parts. The left branch (A) stands for the nanowire forces exerted on the microbead. The photopolymer viscoelastic properties are characterized by the two elastic terms $k_1$, $k_2$ and the viscoelastic damping coefficient $\delta$. The right branch (B) includes the damping of the surrounding medium, with $\gamma$ denoting the hydrodynamic resistance. The inertia of the particle and the displaced fluid are neglected. Therefore, the results obtained represent a low-frequency approximation [20].

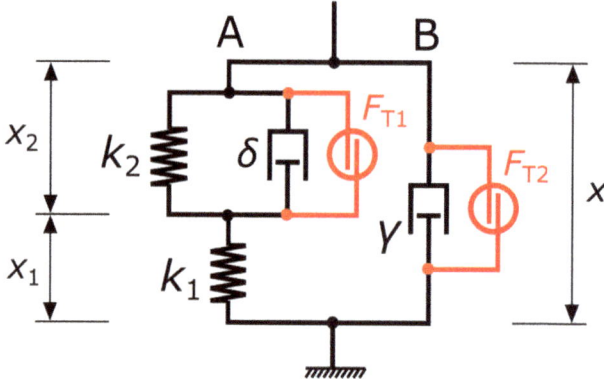

**Figure 2.** The schematic depiction of the linear mechanical model for the microbead motion. Arms A and B represent the nanowire and the hydrodynamic damping by the surrounding medium, respectively. The internal characteristics $x_1$, $x_2$ are related to the total observable parameter $x$.

The basic mechanical model (i.e., model depicted in Figure 2, which is free of random forces $F_{T1} = F_{T2} = 0$) is identical to the one given in [25]. This original form is amended here by adding stochastic forces to the system. In agreement with the fluctuation-dissipation theorem, uncorrelated Gaussian random forces ($F_{T1}$ and $F_{T2}$) are introduced in parallel with the dissipative elements $\delta$ and $\gamma$. Due to random excitation terms, the two displacement coordinates $x_1$ and $x_2$ fluctuate stochastically, which translates to the overall microbead displacement $x = x_1 + x_2$. Unlike $x_1$ and $x_2$, the value of x can be observed experimentally. The power spectral density and the autocorrelation function of the microbead stochastic oscillations are derived by solving the system of Langevin equations describing the proposed model.

*Research Design*

Technological advances in microprinting 3D polymer patterns can stimulate and motivate progress in the formulation of appropriate mathematical and physical models. In this paper, we present a research framework associated with advanced two-photon microfabrication that can be applied in both practical and theoretical directions. Linear relationships are used to characterize the viscous and elastic properties of micromechanical systems (microbead nanowire systems) operating under stochastic (thermal) dynamic conditions. To evaluate observable coordinate changes, digital data from a CMOS image sensor is processed. This data can be represented mathematically as a one-dimensional time series.

The present study is motivated by experimental results, which, after analyzing the power spectral densities of the corresponding time series, show a kind of double Lorentzian form. The double Lorentzian form of the spectrum appears to us as an attractive research problem but also as a feature that has not yet been explored under the experimental conditions described above.

We use a regression approach with an objective function containing position-dependent weights in the spectrum to compare the theoretical model with experiments and perform the best model parameter finding as well. In addition, our analysis focuses on a validation strategy based on comparisons of differently obtained spectral moments.

Our study reflects the assumption that statistical mechanics models can reveal efficient ways to parameterize optically fabricated systems that exhibit significant fluctuations due to their size.

### 3. Model

The model consisting of two first-order stochastic differential equations is obtained based on the following considerations: (i) the forces acting in the upper and lower part of branch A are equal, and (ii) the sum of branch A and branch B forces is zero. We study the stochastic system for two displacement coordinates

$$
\begin{aligned}
\delta \frac{dx_2}{dt} + k_2 x_2 - k_1 x_1 &= F_{T1}, \\
\gamma \left( \frac{dx_1}{dt} + \frac{dx_2}{dt} \right) + k_1 x_1 &= F_{T2}
\end{aligned}
\tag{1}
$$

formulated for the uncorrelated stationary Gaussian and white noise in time $t$ random forces $F_{T1}(t)$, $F_{T2}(t)$. In such a framework, a set of assumptions applies to the mean values

$$
\begin{aligned}
\langle F_{T1}(t) \rangle &= 0, \quad \langle F_{T2}(t) \rangle = 0, \\
\langle F_{T1}(t) F_{T2}(t+t') \rangle &= 0 \quad \text{for all } t, t'; \\
\langle F_{T1}(t) F_{T1}(t+t') \rangle &= C_{FT1} \hat{\delta}(t'), \\
\langle F_{T2}(t) F_{T2}(t+t') \rangle &= C_{FT2} \hat{\delta}(t')
\end{aligned}
\tag{2}
$$

written by means of the Dirac delta function $\hat{\delta}(.)$ (Here, the label $\hat{\delta}$ is selected to distinguish it from the parameter $\delta$). The details and the physical rationale regarding the new parameter pair $C_{FT1}$, $C_{FT2}$ will be provided later in Section 4.3. To indicate the mean value in the space of repetitive random variants, we use the symbol $\langle \ldots \rangle$, which is identified with the physical literature. Important here is the mathematical note that random forces are the derivatives of the corresponding Wiener processes.

## 4. Results

### 4.1. Solution of the Stochastic Problem

Using a linear transformation involving multiplication by the terms $1/\delta$, $1/\gamma$, as well as subtraction to remove the combination of derivatives on the left-hand side, we obtained a more standard form of the stochastic differential equation

$$\left[\frac{d}{dt} + k_1\left(\frac{1}{\gamma} + \frac{1}{\delta}\right)\right]x_1 - \frac{k_2}{\delta}x_2 = F_{S1},$$
$$\left(\frac{d}{dt} + \frac{k_2}{\delta}\right)x_2 - \frac{k_1}{\delta}x_1 = F_{S2}, \tag{3}$$

where the respective dissipative terms ($\sim dx_1/dt$, $\sim dx_2/dt$) are counterbalanced by the auxiliary random forces $F_{S1}$, $F_{S2}$. There are no more mixed derivatives of variables in one equation, which results in a qualitative change from white to colored noise. The properties of $F_{S1}$, $F_{S2}$ can be represented by the linear relations

$$F_{S1}(t) = \frac{F_{T2}(t)}{\gamma} - \frac{F_{T1}(t)}{\delta}, \qquad F_{S2}(t) = \frac{F_{T1}(t)}{\delta}. \tag{4}$$

For computational purposes, the system of Equation (3) is converted to the Fourier domain in a standard way. Fourier images (coefficients), which we begin to denote by a tilde become functions of the angular frequency $\omega$. Despite the fact that $\ldots_{(\omega)}$ or $\ldots(\omega)$ symbols implying frequency dependence may be redundant in the case of white noise, it emphasizes the dependence on $\omega$ for general reasons in other situations. It is also worth noting that the complex conjugate's asterisk label appears after the transition to the Fourier representation. Assuming that cross-correlations of Fourier images $\tilde{F}_{T1}$, $\tilde{F}_{T2}$ vanish as a result of Equations (2) and (4), for the relations of the first and the second-order moments, we have

$$\langle \tilde{F}_{S1}\rangle_{(\omega)} = \langle \tilde{F}_{S2}\rangle_{(\omega)} = 0,$$
$$\langle \tilde{F}^*_{S1}\tilde{F}_{S1}\rangle_{(\omega)} = \frac{1}{\gamma^2}\langle \tilde{F}^*_{T2}\tilde{F}_{T2}\rangle_{(\omega)} + \frac{1}{\delta^2}\langle \tilde{F}^*_{T1}\tilde{F}_{T1}\rangle_{(\omega)},$$
$$\langle \tilde{F}^*_{S2}\tilde{F}_{S2}\rangle_{(\omega)} = \frac{1}{\delta^2}\langle \tilde{F}^*_{T1}\tilde{F}_{T1}\rangle_{(\omega)}, \tag{5}$$
$$\langle \tilde{F}^*_{S1}\tilde{F}_{S2}\rangle_{(\omega)} = \langle \tilde{F}^*_{S2}\tilde{F}_{S1}\rangle_{(\omega)} = -\frac{1}{\delta^2}\langle \tilde{F}^*_{T1}\tilde{F}_{T1}\rangle_{(\omega)}.$$

Of course, the properties of the above averages are sufficient to determine multivariate Gaussian random force statistics. More precisely, the consequences of the Gaussian process from the postulates for $F_{T1,2}$ towards the statements for $F_{S1,2}$ can be easily justified.

According to the Equation (3), the respective coefficients $\tilde{x}_1(\omega)$, $\tilde{x}_2(\omega)$, $\tilde{F}_{S1}(\omega)$, $\tilde{F}_{S2}(\omega)$ are present in

$$\mathbf{G}(\omega)\begin{pmatrix}\tilde{x}_1(\omega)\\ \tilde{x}_2(\omega)\end{pmatrix} = \begin{pmatrix}\tilde{F}_{S1}(\omega)\\ \tilde{F}_{S2}(\omega)\end{pmatrix}, \tag{6}$$

where

$$\mathbf{G}(\omega) = \begin{bmatrix} i\omega + k_1\left(\frac{1}{\gamma} + \frac{1}{\delta}\right) & -\frac{k_2}{\delta} \\ -\frac{k_1}{\delta} & i\omega + \frac{k_2}{\delta} \end{bmatrix}. \tag{7}$$

Note that here $\mathbf{G}(\omega)$ is the label of newly introduced $\omega$-dependent matrix. The linearity of the problem implies that the solution

$$\begin{pmatrix}\tilde{x}_1(\omega)\\ \tilde{x}_2(\omega)\end{pmatrix} = \mathbf{G}^{-1}(\omega)\begin{pmatrix}\tilde{F}_{S1}(\omega)\\ \tilde{F}_{S2}(\omega)\end{pmatrix} \tag{8}$$

can be expressed in the terms of the inverse matrix $\mathbf{G}^{-1}(\omega)$. The following elements of the matrix represent a solution of the linear response type

$$(\mathbf{G}^{-1})_{11}(\omega) = \tfrac{1}{\det(\mathbf{G}(\omega))}\left(i\omega + \tfrac{k_2}{\delta}\right), \quad (\mathbf{G}^{-1})_{12}(\omega) = \tfrac{1}{\det(\mathbf{G}(\omega))}\tfrac{k_2}{\delta},$$
$$(\mathbf{G}^{-1})_{21}(\omega) = \tfrac{1}{\det(\mathbf{G}(\omega))}\tfrac{k_1}{\delta}, \quad (\mathbf{G}^{-1})_{22}(\omega) = \tfrac{1}{\det(\mathbf{G}(\omega))}\left[i\omega + k_1\left(\tfrac{1}{\gamma} + \tfrac{1}{\delta}\right)\right]. \tag{9}$$

We see that the formulas contain $\det(\mathbf{G}) = \mathcal{G}_R + i\mathcal{G}_I$, in the form $1/\det(\mathbf{G}) = (\mathcal{G}_R - i\mathcal{G}_I)/(\mathcal{G}_R^2 + \mathcal{G}_I^2)$ with the auxiliary real-valued components $\mathcal{G}_R(\omega)$ and $\mathcal{G}_I(\omega)$. We also state that

$$\mathcal{G}_R(\omega) = -\omega^2 + \frac{k_1 k_2}{\gamma \delta}, \quad \mathcal{G}_I(\omega) = \omega\left[\frac{k_2}{\delta} + k_1\left(\frac{1}{\gamma} + \frac{1}{\delta}\right)\right]. \tag{10}$$

Next, we will use

$$(\det(\mathbf{G}))^* \det(\mathbf{G}) = \mathcal{G}_R^2 + \mathcal{G}_I^2 \tag{11}$$

often reflected in the results.

### 4.2. Statistical Averages, Responses to Random Perturbations

This section is about the change to mean values, which are important for the measurement process, interpretation, and data processing. Only the statistics of the sum $\tilde{x}_1(\omega) + \tilde{x}_2(\omega)$, not isolated $\tilde{x}_1(\omega)$, $\tilde{x}_2(\omega)$ is observable in the experiment and allows comparison with the model. Thus, for many aspects of the study, only the behavior of $\tilde{x}_1(\omega) + \tilde{x}_2(\omega)$ needs to be used to determine experimentally relevant correlations. To understand the statistics of $x$, we focus on the Fourier spectrum of autocorrelation function

$$\begin{aligned}C_{xx}(\omega) &\equiv \langle\,(\tilde{x}_1^*(\omega) + \tilde{x}_2^*(\omega))(\tilde{x}_1(\omega) + \tilde{x}_2(\omega))\,\rangle \\ &= \langle\tilde{x}_1^*\tilde{x}_1\rangle_{(\omega)} + \langle\tilde{x}_2^*\tilde{x}_2\rangle_{(\omega)} + \langle\tilde{x}_1^*\tilde{x}_2\rangle_{(\omega)} + \langle\tilde{x}_2^*\tilde{x}_1\rangle_{(\omega)}.\end{aligned} \tag{12}$$

It is of course convenient to divide it into four independent terms. In the following, these are treated independently by means of Equation (8). Partial results (so far without an emphasis on the $\omega$ dependence) are

$$\begin{aligned}\langle\tilde{x}_1^*\tilde{x}_1\rangle =&(\mathbf{G}^{-1})_{11}^*(\mathbf{G}^{-1})_{11}\langle\tilde{F}_{S1}^*\tilde{F}_{S1}\rangle + (\mathbf{G}^{-1})_{11}^*(\mathbf{G}^{-1})_{12}\langle\tilde{F}_{S1}^*\tilde{F}_{S2}\rangle \\ &+(\mathbf{G}^{-1})_{12}^*(\mathbf{G}^{-1})_{11}\langle\tilde{F}_{S2}^*\tilde{F}_{S1}\rangle + (\mathbf{G}^{-1})_{12}^*(\mathbf{G}^{-1})_{12}\langle\tilde{F}_{S2}^*\tilde{F}_{S2}\rangle, \\[4pt] \langle\tilde{x}_2^*\tilde{x}_2\rangle =&(\mathbf{G}^{-1})_{21}^*(\mathbf{G}^{-1})_{21}\langle\tilde{F}_{S1}^*\tilde{F}_{S1}\rangle + (\mathbf{G}^{-1})_{21}^*(\mathbf{G}^{-1})_{22}\langle\tilde{F}_{S1}^*\tilde{F}_{S2}\rangle \\ &+(\mathbf{G}^{-1})_{22}^*(\mathbf{G}^{-1})_{21}\langle\tilde{F}_{S2}^*\tilde{F}_{S1}\rangle + (\mathbf{G}^{-1})_{22}^*(\mathbf{G}^{-1})_{22}\langle\tilde{F}_{S2}^*\tilde{F}_{S2}\rangle, \\[4pt] \langle\tilde{x}_1^*\tilde{x}_2\rangle =&(\mathbf{G}^{-1})_{11}^*(\mathbf{G}^{-1})_{21}\langle\tilde{F}_{S1}^*\tilde{F}_{S1}\rangle + (\mathbf{G}^{-1})_{11}^*(\mathbf{G}^{-1})_{22}\langle\tilde{F}_{S1}^*\tilde{F}_{S2}\rangle \\ &+(\mathbf{G}^{-1})_{12}^*(\mathbf{G}^{-1})_{21}\langle\tilde{F}_{S2}^*\tilde{F}_{S1}\rangle + (\mathbf{G}^{-1})_{12}^*(\mathbf{G}^{-1})_{22}\langle\tilde{F}_{S2}^*\tilde{F}_{S2}\rangle, \\[4pt] \langle\tilde{x}_2^*\tilde{x}_1\rangle =&(\mathbf{G}^{-1})_{21}^*(\mathbf{G}^{-1})_{11}\langle\tilde{F}_{S1}^*\tilde{F}_{S1}\rangle + (\mathbf{G}^{-1})_{21}^*(\mathbf{G}^{-1})_{12}\langle\tilde{F}_{S1}^*\tilde{F}_{S2}\rangle \\ &+(\mathbf{G}^{-1})_{22}^*(\mathbf{G}^{-1})_{11}\langle\tilde{F}_{S2}^*\tilde{F}_{S1}\rangle + (\mathbf{G}^{-1})_{22}^*(\mathbf{G}^{-1})_{12}\langle\tilde{F}_{S2}^*\tilde{F}_{S2}\rangle.\end{aligned} \tag{13}$$

We can achieve a clearer relationship by including the correlations between the initially imposed random forces $\tilde{F}_{T1}$ and $\tilde{F}_{T2}$ from Equation (5). The pairwise correlations take the form

$$\langle \tilde{x}_s^* \tilde{x}_m \rangle = \left( \frac{\langle \tilde{F}_{T2}^* \tilde{F}_{T2} \rangle}{\gamma^2} + \frac{\langle \tilde{F}_{T1}^* \tilde{F}_{T1} \rangle}{\delta^2} \right) \bar{\bar{g}}_{1;s,m} + \frac{\langle \tilde{F}_{T1}^* \tilde{F}_{T1} \rangle}{\delta^2} \bar{\bar{g}}_{2;s,m} \qquad (14)$$

with $s \in \{1,2\}$; $m \in \{1,2\}$, which define the following eight coefficients

$$\begin{aligned}
\bar{\bar{g}}_{1;1,1} &= \hat{g}_{11,11}, & \bar{\bar{g}}_{2;1,1} &= \hat{g}_{12,12} - \hat{g}_{11,12}^{\text{sym}}, \\
\bar{\bar{g}}_{1;2,2} &= \hat{g}_{21,21}, & \bar{\bar{g}}_{2;2,2} &= \hat{g}_{22,22} - \hat{g}_{21,22}^{\text{sym}}, \\
\bar{\bar{g}}_{1;1,2} &= \hat{g}_{11,21}, & \bar{\bar{g}}_{2;1,2} &= \hat{g}_{12,22} - \hat{g}_{11,22} - \hat{g}_{12,21}, \\
\bar{\bar{g}}_{1;2,1} &= \hat{g}_{21,11}, & \bar{\bar{g}}_{2;2,1} &= \hat{g}_{22,12} - \hat{g}_{21,12} - \hat{g}_{22,11}.
\end{aligned} \qquad (15)$$

For the relations above, we use a notation that also includes the auxiliary symbols $\hat{g}_{ij,kl}$ and $\hat{g}_{ij,kl}^{\text{sym}}$. They are interrelated to the combinations $(\mathbf{G}^{-1})_{ij}^* (\mathbf{G}^{-1})_{ij}$ of the prior $\mathbf{G}^{-1}$ terms

$$\hat{g}_{ij,kl} = (\mathbf{G}^{-1})_{ij}^* (\mathbf{G}^{-1})_{kl}, \qquad \hat{g}_{ij,kl}^{\text{sym}} = \hat{g}_{ij,kl} + \hat{g}_{kl,ij}. \qquad (16)$$

Obviously, the emphasis on the symmetry $\hat{g}_{ij,kl}^{\text{sym}} = \hat{g}_{kl,ij}^{\text{sym}}$ will help us to handle the complex numbers. Furthermore, we recognize that the identical pairs of indices provide that $\hat{g}_{ij,ij} = (1/2)\hat{g}_{ij,ij}^{\text{sym}}$. The advantage of the auxiliary notation by means of $\hat{g}_{...}$ is that we obtain $C_{xx}$ from Equation (12) in the compact form

$$\begin{aligned}
C_{xx} &= \frac{\langle \tilde{F}_{T2}^* \tilde{F}_{T2} \rangle}{\gamma^2} \left( \hat{g}_{11,11} + \hat{g}_{21,21} + \hat{g}_{11,21}^{\text{sym}} \right) \\
&+ \frac{\langle \tilde{F}_{T1}^* \tilde{F}_{T1} \rangle}{\delta^2} \left( \hat{g}_{11,11} + \hat{g}_{12,12} + \hat{g}_{21,21} + \hat{g}_{22,22} \right. \\
&+ \left. \hat{g}_{11,21}^{\text{sym}} + \hat{g}_{12,22}^{\text{sym}} - \hat{g}_{11,12}^{\text{sym}} - \hat{g}_{21,22}^{\text{sym}} - \hat{g}_{11,22}^{\text{sym}} - \hat{g}_{12,21}^{\text{sym}} \right).
\end{aligned} \qquad (17)$$

We continue the calculation to reveal the terms introduced by Equation (16)

$$\begin{aligned}
\hat{g}_{11,11} &= \frac{1}{\mathcal{G}_R^2 + \mathcal{G}_I^2} \left[ \omega^2 + \left(\frac{k_2}{\delta}\right)^2 \right], & \hat{g}_{21,21} &= \frac{1}{\mathcal{G}_R^2 + \mathcal{G}_I^2} \left(\frac{k_1}{\delta}\right)^2, \\
\hat{g}_{22,22} &= \frac{1}{\mathcal{G}_R^2 + \mathcal{G}_I^2} \left[ \omega^2 + k_1^2 \left(\frac{1}{\gamma} + \frac{1}{\delta}\right)^2 \right], & \hat{g}_{12,12} &= \frac{1}{\mathcal{G}_R^2 + \mathcal{G}_I^2} \left(\frac{k_2}{\delta}\right)^2,
\end{aligned} \qquad (18)$$

$$\begin{aligned}
\hat{g}_{11,21}^{\text{sym}} &= \frac{1}{\mathcal{G}_R^2 + \mathcal{G}_I^2} \left(\frac{2k_1 k_2}{\delta^2}\right), & \hat{g}_{12,22}^{\text{sym}} &= \frac{1}{\mathcal{G}_R^2 + \mathcal{G}_I^2} \left(\frac{2k_2 k_1}{\delta}\right)\left(\frac{1}{\gamma} + \frac{1}{\delta}\right), \\
\hat{g}_{11,12}^{\text{sym}} &= \frac{1}{\mathcal{G}_R^2 + \mathcal{G}_I^2} \left(\frac{2k_2^2}{\delta^2}\right), & \hat{g}_{21,22}^{\text{sym}} &= \frac{1}{\mathcal{G}_R^2 + \mathcal{G}_I^2} \left(\frac{2k_1^2}{\delta}\right)\left(\frac{1}{\gamma} + \frac{1}{\delta}\right), \\
\hat{g}_{11,22}^{\text{sym}} &= \frac{1}{\mathcal{G}_R^2 + \mathcal{G}_I^2} \left[ 2\omega^2 + \frac{2k_1 k_2}{\delta}\left(\frac{1}{\gamma} + \frac{1}{\delta}\right) \right], & \hat{g}_{12,21}^{\text{sym}} &= \frac{1}{\mathcal{G}_R^2 + \mathcal{G}_I^2} \left(\frac{2k_1 k_2}{\delta^2}\right).
\end{aligned} \qquad (19)$$

Note that we used $\mathcal{G}_R$ and $\mathcal{G}_I$ from Equation (10) to express the result. After substituting these elements into Equation (17), we come to the relation

$$C_{xx}(\omega) = \frac{1}{\gamma^2 (\mathcal{G}_R^2 + \mathcal{G}_I^2)} \left\{ \langle \tilde{F}_{T2}^* \tilde{F}_{T2} \rangle \left[ \omega^2 + \left(\frac{k_1 + k_2}{\delta}\right)^2 \right] + \langle \tilde{F}_{T1}^* \tilde{F}_{T1} \rangle \left(\frac{k_1}{\delta}\right)^2 \right\}, \qquad (20)$$

which is important to derive measurable results. We will apply a similar procedure later to determine the pairwise correlations to prove the validity of the equipartition theorem.

### 4.3. Towards Fusing of Theory and Experiment

Suppose that there are two finite formal limits relevant for the obtaining of the power spectral density in the form

$$C_{\text{FT1}} = \lim_{T_{\text{msr}} \to \infty} \frac{\langle \tilde{F}^*_{\text{T1}} \tilde{F}_{\text{T1}} \rangle}{T_{\text{msr}}}, \quad C_{\text{FT2}} = \lim_{T_{\text{msr}} \to \infty} \frac{\langle \tilde{F}^*_{\text{T2}} \tilde{F}_{\text{T2}} \rangle}{T_{\text{msr}}}. \quad (21)$$

The formula is understood as a postulate, which introduces the duration of the measurement time $T_{\text{msr}}$ [19] into a part of the procedure at the formal level. The correlations in the Fourier domain can be formally taken as infinite for frequency $f = \frac{\omega}{2\pi} \gg 1/T_{\text{msr}}$. The formal nature of the limits given by Equation (21) makes it evident that considerations are not fully compatible with the Fourier framework because the measurement is dependent on assumptions about the large time ($T_{\text{msr}}$) of the measurement.

The occurrence of $C_{\text{FT1}}$ and $C_{\text{FT2}}$ later in Equation (23) can be interpreted as the contribution of the power spectral densities of two random force variants given by the Wiener–Khinchin theorem

$$\lim_{T_{\text{msr}} \to \infty} \frac{\langle \tilde{F}^*_{\text{T}j} \tilde{F}_{\text{T}j} \rangle}{T_{\text{msr}}} = \int_{-\infty}^{\infty} dt'\, e^{-2\pi i f t'} \langle F_{\text{T}j}(t) F_{\text{T}j}(t+t') \rangle = C_{\text{FT}j} \quad (22)$$

considered for $j = 1, 2$ alternatives (see Equation (2), where the correlation function is defined and integrated). If we extend the application of the formal limit by dividing Equation (20) with $T_{\text{msr}}$, we obtain the power spectrum density in the form

$$S_{xx}(f) = \lim_{T_{\text{msr}} \to \infty} \frac{C_{xx}(f)}{T_{\text{msr}}}$$
$$= \frac{1}{\gamma^2 (\mathcal{G}_R^2 + \mathcal{G}_I^2)_{\omega = 2\pi f}} \left\{ C_{\text{FT2}} \left[ 4\pi^2 f^2 + \left( \frac{k_1 + k_2}{\delta} \right)^2 \right] + C_{\text{FT1}} \left( \frac{k_1}{\delta} \right)^2 \right\}. \quad (23)$$

In this way, the physical meaning of the coefficients $C_{\text{FT}\,1,2}$ is revealed. They can also be represented in an independent way by expressing their relation to the absolute temperature

$$C_{\text{FT1}} = 2k_B T \delta, \quad C_{\text{FT2}} = 2k_B T \gamma. \quad (24)$$

However, this construct also provides information about the dissipative mechanisms. This is built with the idea that fluctuations from random forces are dissipated by the mechanisms represented by the parameters $\delta, \gamma$. As provided below in Section 4.5, the mean potential energy for the respective degrees of freedom can be compared to determine the equilibrium level of energy flow controlled by $C_{\text{FT1}}$ and $C_{\text{FT2}}$. The Equation (24) given above is essentially the case of the general *fluctuation–dissipation theorem* introducing the natural heat unit $k_B T$.

The fluctuation–dissipation theorem is a statistical thermodynamics statement that explains how fluctuations in a detailed balanced system determine its response to applied disturbances. According to this theorem, two opposing mechanisms are responsible for creating a detailed equilibrium in mechanical systems. On the one hand, there are the consequences of the dynamics of a microsphere attached to a nanowire that is damped by the surrounding fluid. Contributions from the internal damping mechanisms of the nanowire also fall into the same category. Even with this damping combination, the mechanical energy is converted into heat. On the other hand, the presence of damping is necessarily accompanied by fluctuations born in the viscous environment. In the case of the surrounding liquid, these fluctuations result in typical random Brownian collisions of liquid molecules with the microbead. In a standard way, the process is interpreted so that on microscopic scales, heat can be converted back into the mechanical energy of the microbead. The internal damping inside the nanowire acts likewise. Summarizing the above statements, we arrive at a specific form of the fluctuation-dissipation theorem, which states that a constant dissipation flux keeps the mean mechanical energy input invariant, while ensuring the production of new fluctuations.

As a result, let us emphasize an important point: Equation (23) can be modified to account for the temperature effect. With the intention of linking theory with experiment, we attain the expression

$$S_{xx}(f) = \frac{2k_B T}{\gamma} \frac{4\pi^2 f^2 + K_A}{(4\pi^2 f^2 - K_B)^2 + 4\pi^2 f^2 K_C}. \tag{25}$$

The asymptotic, high-frequency consequence of this general result is

$$S_{xx}(f) \simeq \frac{k_B T}{2\pi^2 \gamma f^2}, \quad f \gg \frac{1}{2\pi} \max\left\{\sqrt{K_A}, \sqrt{K_B}, \sqrt{K_C}\right\}. \tag{26}$$

At this stage, we benefit from the choice of auxiliary parameters $K_A, K_B, K_C$. Returning to material details is possible using transformations

$$K_A = \left(\frac{k_1 + k_2}{\delta}\right)^2 + \frac{\gamma}{\delta}\left(\frac{k_1}{\gamma}\right)^2, \quad K_B = \frac{k_1 k_2}{\gamma \delta}, \quad K_C = \left[\frac{k_2}{\delta} + k_1\left(\frac{1}{\gamma} + \frac{1}{\delta}\right)\right]^2. \tag{27}$$

These auxiliary parameters are positive for a given model specification that operates exclusively with positive $k_1, k_2, \delta, \gamma$. However, there is also another, more sophisticated level of interpretation. It is interesting and also productive to assume that the result can be written as a sum of two weighted Lorentzian functions

$$\frac{4\pi^2 f^2 + K_A}{(4\pi^2 f^2 - K_B)^2 + 4\pi^2 f^2 K_C} \stackrel{!}{=} \frac{1}{\Gamma_2 - \Gamma_1}\left(\frac{K_A - \Gamma_1}{\Gamma_1 + 4\pi^2 f^2} + \frac{\Gamma_2 - K_A}{\Gamma_2 + 4\pi^2 f^2}\right). \tag{28}$$

Here $\Gamma_{1,2}$ play the role of free parameters, which incorporate information coming from previously introduced $K_A, K_B, K_C$. The change to $\Gamma_{1,2}$ should be considered as an intermediate step along with other consequences. The key consequence is double Lorentzian form

$$S_{xx}(f) = \frac{k_B T}{2\pi^2 \gamma (f_{C2}^2 - f_{C1}^2)}\left(\frac{\frac{K_A}{4\pi^2} - f_{C1}^2}{f_{C1}^2 + f^2} + \frac{f_{C2}^2 - \frac{K_A}{4\pi^2}}{f_{C2}^2 + f^2}\right). \tag{29}$$

It is based on the assumption that there exist some relations between $\Gamma_{1,2}$ and the corner frequencies $f_{C1,2}$. When Equations (28) and (29) are combined, we obtained

$$f_{C1,2}^2 = \frac{\Gamma_{1,2}}{4\pi^2} = \frac{1}{4\pi^2}\left[\frac{K_C}{2} - K_B \mp \frac{1}{2}\sqrt{K_C(K_C - 4K_B)}\right]. \tag{30}$$

In the above solution, we use the consensus that the plus sign corresponds to $f_{C2}$. The constraints that allow for such a solution are as follows:

$$\frac{K_C}{2} - K_B \geq \frac{1}{2}\sqrt{K_C(K_C - 4K_B)}, \quad K_C \geq 0, \quad K_C \geq 4K_B. \tag{31}$$

If we consider the transformation to physical parameters in the sense of Equation (27) to analyze the satisfaction of the above constraints, we obtain

$$\frac{K_C}{2} - K_B = \frac{k_1^2}{2}\left(\frac{1}{\gamma} + \frac{1}{\delta}\right)^2 + \frac{k_1 k_2}{\delta^2} + \frac{k_2^2}{2\delta^2} \geq 0, \tag{32}$$

$$K_C - 4K_B = \left(\frac{k_2}{\delta} - \frac{k_1}{\gamma}\right)^2 + \left(\frac{k_1}{\delta}\right)^2 + \frac{2k_1}{\delta}\left(\frac{k_2}{\delta} + \frac{k_1}{\gamma}\right) \geq 0. \tag{33}$$

Using Equation (27), we confirm that $K_C \geq 0$. Moreover, the trivial $K_B^2 \geq 0$ implies $(K_C/2) - K_B \geq (1/2)\sqrt{K_C(K_C - 4K_B)}$. Therefore, there is no obvious contradiction with the fact that $\Gamma_1, \Gamma_2$ correspond to $f_{C1}^2, f_{C2}^2$. It is also notable that the inverse transformations $(f_{C1}, f_{C2}) \to (\Gamma_1, \Gamma_2) \to (K_B, K_C)$ become

$$K_B = \sqrt{\Gamma_1 \Gamma_2} = 4\pi^2 f_{C1} f_{C2},$$
$$K_C = 2\left(\sqrt{\Gamma_1 \Gamma_2} + \frac{\Gamma_1 + \Gamma_2}{2}\right) = 4\pi^2 (f_{C1} + f_{C2})^2. \tag{34}$$

The result is intriguing in terms of revealing the central tendency in $K_B(\Gamma_1, \Gamma_2)$ and $K_C(\Gamma_1, \Gamma_2)$ as representatives of the pair $\Gamma_1, \Gamma_2$.

### 4.4. Autocorrelation Function

The findings presented above can be augmented by using direct time representation. According to the well-known *Wiener–Khinchin* relation, we have the consequence for the autocorrelation function in the form

$$R_{xx}(t) = \int_{-\infty}^{\infty} df\, S_{xx}(f) \exp(2\pi i f t). \tag{35}$$

As a result, for Equation (29) as a specific version of $S_{xx}(f)$, we obtain a two-exponential autocorrelation function

$$R_{xx}(t) = \mathcal{R}_0 \left( \mathcal{R}_1 e^{-2\pi f_{C1}|t|} + \mathcal{R}_2 e^{-2\pi f_{C2}|t|} \right), \tag{36}$$

where

$$\mathcal{R}_0 = \frac{k_B T}{2\pi \gamma (f_{C2}^2 - f_{C1}^2)}, \quad \mathcal{R}_1 = \frac{\frac{K_A}{4\pi^2} - f_{C1}^2}{f_{C1}}, \quad \mathcal{R}_2 = \frac{f_{C2}^2 - \frac{K_A}{4\pi^2}}{f_{C2}}. \tag{37}$$

It is worth noting that corner frequency parameters have a significant impact on autocorrelation decrease over time. At a first glance, we can see the essential property here where a pair of frequencies in the Lorentz form corresponds to a pair of damping terms with the typical decay times proportional to $1/f_{C1}$ and $1/f_{C2}$. It should also be noted that, assuming that the physical parameters are constant, the temperature is directly manifested only in the amplitude $\mathcal{R}_0$. The calculations above were performed with the help of a well-known auxiliary relation

$$\int_{-\infty}^{\infty} df\, \frac{\exp(2\pi i f t)}{f^2 + f_C^2} = \frac{\pi}{f_C} \exp(-2\pi |t| f_C) \tag{38}$$

with some auxiliary parameter $f_C$.

### 4.5. Sharing of Elastic Energy; Rationale for Choosing $C_{FT1}, C_{FT2}$

According to the principle of energy equipartition, average energy is evenly distributed among the various degrees of freedom of ergodic systems. As shown here, the implications of this principle are valuable tools for calculating the amplitudes $(C_{FT1}, C_{FT2})$ of a pair of random forces. The equipartition principle can be applied to the mean elastic energies. We start by writing the energy for the Fourier modes corresponding to $\omega$. It is worth noting that since the inertial term is considered negligible, the zero limit of the kinetic energy has no effect on the equipartition issues.

Using the integration techniques already discussed, we continue to utilize the formal limit approach ($T_{\text{mrs}} \to \infty$) for the integration of the spectrum and averaging over the respective potential energy fluctuations as follows

$$U_{P1} = \frac{k_1}{2} \int_0^\infty d\omega \, \langle \tilde{x}_1^* \tilde{x}_1 \rangle_{(\omega)} = I_{P11} C_{FT1} + I_{P12} C_{FT2},$$
$$U_{P2} = \frac{k_2}{2} \int_0^\infty d\omega \, \langle \tilde{x}_2^* \tilde{x}_2 \rangle_{(\omega)} = I_{P21} C_{FT1} + I_{P22} C_{FT2}. \quad (39)$$

The formulas below can be applied to complete the integration

$$I_{P11} = \frac{k_1}{2\delta^2} I_{E2}, \qquad I_{P12} = \frac{k_1}{2\gamma^2}\left[I_{E2} + \left(\frac{k_2}{\delta}\right)^2 I_{E0}\right],$$
$$I_{P21} = \frac{k_2}{2\delta^2}\left[I_{E2} + \left(\frac{k_1}{\gamma}\right)^2 I_{E0}\right], \qquad I_{P22} = \frac{k_2 k_1^2}{2\gamma^2 \delta^2} I_{E0}. \quad (40)$$

These four coefficients include two spectral integrals

$$I_{E0} = \int_0^\infty \frac{d\omega}{\pi} \frac{1}{\mathcal{G}_R^2(\omega) + \mathcal{G}_I^2(\omega)}, \qquad I_{E2} = \int_0^\infty \frac{d\omega}{\pi} \frac{\omega^2}{\mathcal{G}_R^2(\omega) + \mathcal{G}_I^2(\omega)}. \quad (41)$$

Going back to a spectral decomposition using a pair of Lorentzian forms (see Equation (29)) in combination with Equations (30) and (34) gives the following result

$$I_{E0} = \frac{1}{2(\Gamma_2 - \Gamma_1)}\left(\frac{1}{\sqrt{\Gamma_1}} - \frac{1}{\sqrt{\Gamma_2}}\right) = \frac{1}{2K_B \sqrt{K_C}},$$
$$I_{E2} = \frac{1}{2(\Gamma_2 - \Gamma_1)}\left(\sqrt{\Gamma_2} - \sqrt{\Gamma_1}\right) = \frac{1}{\sqrt{K_C}}. \quad (42)$$

As a consequence, the following relationship $I_{E2} = I_{E0} k_1 k_2 / (\gamma \delta)$ can be used in the mean potential energies listed below

$$U_{P1} = \frac{k_1 k_2}{2\gamma \delta}\left[\left(\frac{k_1}{\delta}\right)\frac{C_{FT1}}{\delta} + \left(\frac{k_1}{\gamma} + \frac{k_2}{\delta}\right)\frac{C_{FT2}}{\gamma}\right] I_{E0},$$
$$U_{P2} = \frac{k_1 k_2}{2\gamma \delta}\left[\left(\frac{k_1}{\gamma} + \frac{k_2}{\delta}\right)\frac{C_{FT1}}{\delta} + \left(\frac{k_1}{\gamma}\right)\frac{C_{FT2}}{\gamma}\right] I_{E0}. \quad (43)$$

Finally, in accordance with Equation (24), we have the confirmation of the equipartition in the form

$$U_{P1} = U_{P2} = K_B \sqrt{K_C} I_{E0} k_B T = \frac{1}{2} k_B T. \quad (44)$$

*4.6. The Spectrum Moments*

In this subsection, we discuss the usefulness of introducing power spectral density integrals in cases where the frequency domain over which we integrate is divided into non-intersecting intervals. Frequency integration is motivated by the fact that providing excessive detail for spectrum characterization may be unnecessary in certain contexts. The second reason is that aggregation of data helps to suppress statistical errors. The third reason is the possibility of comparing only a few moments with the moments estimated by direct data processing.

Naturally, the analytical form of the model moments simplifies further processing. In our case, the specificity of the moments corresponding to Lorentzian and related spectral forms supports the overall validation process. Let the *regression-related* (rr) moments obtained by analytical integration be referred to as $\mathcal{M}_{Sxx}^{rr}(f_L, f_H)$. This notation is used to mean that integration has occurred within the range between the lowest $f_L$ and the highest $f_H$ frequencies. Then

$$\begin{aligned}\mathcal{M}^{rr}_{Sxx}(f_L, f_H) &= \int_{f_L}^{f_H} df\, S_{xx}(f) \\ &= \frac{\mathcal{R}_0}{\pi}\left[\mathcal{R}_1\left(\arctan\left(\frac{f_H}{f_{C1}}\right) - \arctan\left(\frac{f_L}{f_{C1}}\right)\right)\right. \\ &\quad \left. + \mathcal{R}_2\left(\arctan\left(\frac{f_H}{f_{C2}}\right) - \arctan\left(\frac{f_L}{f_{C2}}\right)\right)\right].\end{aligned} \qquad (45)$$

Because the interval length may diverge, we decided to use non-normalized moments. Recall that $\mathcal{R}_1$ and $\mathcal{R}_2$ are the two respective amplitudes of the exponentials corresponding to the autocorrelation function (see Equation (37)). On this basis, using $f_{C1}, f_{C2}$ as natural boundaries, we can define the system of three specific *regression-related* spectral moments

$$\begin{aligned}\mathcal{M}^{rr}_{Sxx}(0, f_{C1}) &= \frac{\mathcal{R}_0}{\pi}\left[\mathcal{R}_1 \frac{\pi}{4} + \mathcal{R}_2 \arctan\left(\frac{f_{C1}}{f_{C2}}\right)\right], \\ \mathcal{M}^{rr}_{Sxx}(f_{C1}, f_{C2}) &= \frac{\mathcal{R}_0}{\pi}\left[\mathcal{R}_1 \arctan\left(\frac{f_{C2}}{f_{C1}}\right) - \mathcal{R}_2 \arctan\left(\frac{f_{C1}}{f_{C2}}\right) + (\mathcal{R}_2 - \mathcal{R}_1)\frac{\pi}{4}\right], \\ \mathcal{M}^{rr}_{Sxx}(f_{C2}, \infty) &= \frac{\mathcal{R}_0}{\pi}\left[(2\mathcal{R}_1 + \mathcal{R}_2)\frac{\pi}{4} - \mathcal{R}_1 \arctan\left(\frac{f_{C2}}{f_{C1}}\right)\right]\end{aligned} \qquad (46)$$

with the total sum $(1/2)\mathcal{R}_0 (\mathcal{R}_1 + \mathcal{R}_2)$. Other suitable boundary options are, of course, possible, such as those that do not depend on regression results but instead emerge entirely from generalized averaging procedures of the experimental spectrum.

### 4.7. Experimental Results and Their Regression

After the successful implementation of the experiment, we obtained data representing the observed dynamics $x(t)$, which we have then transformed into corresponding Fourier images. The aim was to obtain an experimental power spectrum density $\{S^{ex}_{xx,j}\}_{j=1}^{N^{ex}}$ for the system of $N^{ex}$ frequencies $\{f_j\}_{j=1}^{N^{ex}}$ (see Figure 3). Some of the evaluations have been performed according to the work of [19]. Preprocessing with grouping of the adjacent experimental spectral points is a necessary methodological peculiarity. Frequency and spectrum groupings with eight points over the frequency decade were introduced. The effectiveness with which the representative grouping frequencies were allocated was evaluated. Naturally, the grouping process affects not only the locations of representative frequencies, but also the statistics of spectral points, potentially increasing the regression's feasibility. The optimization of parametric combinations is made possible by data knowledge. Let us formally encapsulate the unknown model parameters in a single symbol Par, resulting in the parameterized form of the double Lorentzian model $S_{xx}(f_j, \text{Par})$ (see Equation (29)). In addition to identifying the optimum, we will focus on estimating errors for various components of Par.

The problem-specific emphasis is on the asymptotic behavior of the spectrum. Despite the fact that the density of the power spectrum decreases as $\sim f^{-2}$, the high frequency domain must be properly included in the regression due to its physical significance. Hence, a weighted regression of the squares of $S_{xx}(f_j, \text{Par}) - S^{ex}_{xx,j}$ deviations has been implemented. The preference can be defined as the minimization of the objective function

$$\text{Obj\_F}(\text{Par}) \equiv \sum_{j=1}^{N^{ex}} \left[\frac{S_{xx}(f_j, \text{Par}) - S^{ex}_{xx,j}}{S^{ex}_{xx,j}}\right]^2. \qquad (47)$$

Here, the parameters and their combinations appear to be formally merged into the vector

$$\text{Par} \equiv \left( f_{C1}, f_{C2}, \frac{K_A}{4\pi^2}, \frac{k_B T}{2\pi^2 \gamma} \right). \tag{48}$$

This is subject to optimization. We used the standard global function optimizer, which was built on the concept of the [26] work with the implementation (scypy.optimize.curve-_fit(...)) to the SciPy library [27]. The regression corresponding to Obj_F provides the corner frequencies

$$f_{C1} = (0.443 \pm 0.093)\,(\text{Hz}), \qquad f_{C2} = (4.82 \pm 0.23)\,(\text{Hz}). \tag{49}$$

Along with them

$$\frac{K_A}{4\pi^2} = (0.47 \pm 0.19)\,(\text{Hz}^2). \tag{50}$$

Finally, there is also fixed corresponding parametric combination

$$\frac{k_B T}{2\pi^2 \gamma} = (3.53 \pm 0.13) \times 10^{-15}\,(\text{m}^2\,\text{Hz}), \tag{51}$$

which represents the constant factor in $S_{xx}(f, \text{Par})$ as defined by Equation (29). The regression outcomes are depicted in Figure 3.

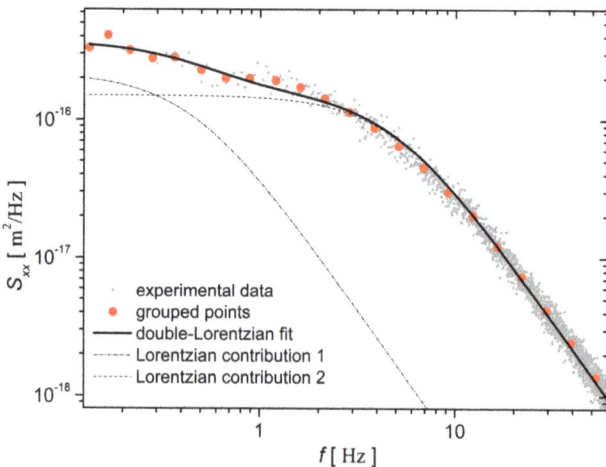

**Figure 3.** Power spectral density of microsphere fluctuations. The solid line belongs to the model according to Equation (29). The fit was set after optimization of Obj_F(Par). Two Lorentzian contributions spanning the entire spectrum are also shown.

Now there is a standard way to find out the autocorrelation function (see Equation (37)) via the respective parameters

$$\frac{\mathcal{R}_0}{\pi} = \frac{\left(\frac{k_B T}{2\pi^2 \gamma}\right)}{f_{C2}^2 - f_{C1}^2} = 1.532 \times 10^{-16}\,(\text{m}^2/\text{Hz}), \tag{52}$$

$$\mathcal{R}_1 = 0.623\,(\text{Hz}), \qquad \mathcal{R}_2 = 4.723\,(\text{Hz}).$$

A posteriori evaluation methodology following the regression results provides an implication for the values of the *regression-related* spectral moments

$$\begin{aligned}
\mathcal{M}^{rr}_{Sxx}(0, f_{C1}) &= 1.414 \times 10^{-16} \, (\text{m}^2), \\
\mathcal{M}^{rr}_{Sxx}(f_{C1}, f_{C2}) &= 5.682 \times 10^{-16} \, (\text{m}^2), \\
\mathcal{M}^{rr}_{Sxx}(f_{C2}, \infty) &= 5.772 \times 10^{-16} \, (\text{m}^2).
\end{aligned} \quad (53)$$

We see that the sum of the moments $1.286 \times 10^{-15}$ (m$^2$) equals $(\mathcal{R}_0/2)(\mathcal{R}_1 + \mathcal{R}_2)$, as predicted.

Adjusting the integration boundaries can be important for the design of some alternative test moments. The premise of the adjustment is that these variants should be more closely linked to the measurement process, conditioned by the need to avoid spectral distortions known as "aliasing" and "motion blur". The effects occur due to too superficial and insufficient sampling of the signal $x(t)$ captured by the camera. This means that the calculations must focus on bands with frequencies less than $f_{upp}$, which in our case was set to around a quarter of the Nyquist frequency. The lower limit value $f_{low}$ prevents the use of extremely low frequencies. Respecting the lower limit suppresses distortions caused by the apparatus background noise. Numerically, the boundaries we introduce are $f_{low} = 0.1333$ [Hz] and $f_{upp} = 59.97$ [Hz]. The following three moments

$$\begin{aligned}
\mathcal{M}^{ex}_{Sxx}(f_{low}, f_{C1}) &= 0.924 \times 10^{-16} \, (\text{m}^2), \\
\mathcal{M}^{ex}_{Sxx}(f_{C1}, f_{C2}) &= 5.687 \times 10^{-16} \, (\text{m}^2), \\
\mathcal{M}^{ex}_{Sxx}(f_{C2}, f_{upp}) &= 4.952 \times 10^{-16} \, (\text{m}^2)
\end{aligned} \quad (54)$$

were created to express the properties of the experimental data set, which was achieved by partly reducing the impact of the regression results. Here we see that Simpson's integration quadrature based on uniform data sampling (without grouping) also provides us with variants of spectral moments. However, even when using numerical integration, we must be careful if we subsequently perform comparisons and interpretations. The reason is that certain integrals approximated by a suitable summation can become dependent on the previous regression only by their integration boundaries when these are linked to regression parameters ($f_{C1}$ and $f_{C2}$). Independence from regression can be achieved using descriptive spectrum characteristics (analogous to descriptive statistics). This means using characteristic frequencies in the role of integration boundaries. Then the results of the calculation are generalized spectral averages. The fact that we do not present more moment variants here is mainly related to the focus of this work.

When comparing the Equations (53) and (54), we see that only the central moments for the $[f_{C1}, f_{C2}]$ band are close enough to each other, which means that $\mathcal{M}^{rr}_{Sxx}(0, f_{C1})$ and $\mathcal{M}^{rr}_{Sxx}(f_{C2}, \infty)$ are not sufficient approximations of $\mathcal{M}^{ex}_{Sxx}(f_{low}, f_{C1})$ and $\mathcal{M}^{ex}_{Sxx}(f_{C2}, f_{upp})$, respectively. Results show that in the case of regression-related moments, there is only a slight and negligible rise in moments compared to the use of Simpson's rule

$$\begin{aligned}
\sim 1.2\% \text{ increase}: & \\
\mathcal{M}^{rr}_{Sxx}(0, f_{C1}) - \mathcal{M}^{rr}_{Sxx}(0, f_{low}) &= 0.935 \times 10^{-16} (\text{m}^2) \\
&\gtrsim \mathcal{M}^{ex}_{Sxx}(f_{low}, f_{C1}),
\end{aligned} \quad (55)$$

$$\begin{aligned}
\sim 4.6\% \text{ increase}: & \\
\mathcal{M}^{rr}_{Sxx}(f_{C2}, \infty) - \mathcal{M}^{rr}_{Sxx}(f_{upp}, \infty) &= 5.184 \times 10^{-16} (\text{m}^2) \\
&\gtrsim \mathcal{M}^{ex}_{Sxx}(f_{C2}, f_{upp}).
\end{aligned}$$

## 5. Discussion

Two kinds of spectral moments ($\mathcal{M}^{rr}_{Sxx}(.)$ and $\mathcal{M}^{ex}_{SSx}(.)$) were designed, calculated and compared for the experimental input data given. We have shown by analysis that subsequent regression-related and the numerical integration outputs can be globally or locally compared and evaluated. This would have an effect on the choice of the overall regression process or model, thereby affecting at least one of them.

The studied model is based on the assumption that a pair of different corner frequencies is needed to describe the spectrum. Consider a situation that, under certain parametric conditions, only a small gap between $f_{C1}$ and $f_{C2}$ is observable. Alternatively, one of the two spectrum amplitudes before the terms $1/(f_{C1}^2 + f^2)$, $1/(f_{C2}^2 + f^2)$ may be negligible (see Equation (29)). All these cases lead to a special limit of the single Lorentz power spectral density.

It should be noted that we did not study beyond the level of a few phenomenological parameters provided in Par on purpose. This is related to the reasons for which the correct determination of the four constants $k_1, k_2, \gamma, \delta$ is part of a wider methodological issue that demands the use of several independent observations. The nanowire material properties, rather than the model's less universal parameters, will likely be the focus of our future research.

## 6. Conclusions

Nanowires prepared by TPP-DLW from Ormocomp possess viscoelastic material properties. This viscoelasticity determines the mechanical behavior of microstructures comprising such nanowires.

The bending recovery motion of cantilevered nanowire systems equipped with a microbead at the free end and immersed in Newtonian liquids was studied previously [25]. The same microstructures, in isolation, exhibit significant thermal fluctuations. In this work, the previous mechanical model was extended with stochastic forces to explain the power spectral density and autocorrelation function of the microstructure thermal fluctuations. In principle, the Brownian fluctuations of cantilevered nanowires can be utilized for micronscale viscosity measurements. Our results pave the way for a quantitative analysis of such micro-viscometer systems.

The calculation of the correlation functions of the characteristic coordinate was carried out in the frequency domain by introducing symmetric forms for the corresponding coefficients. The implications of the calculation concern the steady state, which can be alternatively characterized by the spectral power density. An interesting aspect of our approach is that the weighted regression results for the spectra have been validated using the spectral moment system.

Theoretical considerations confirmed the double-Lorentzian power density spectrum and the doubly-exponential autocorrelation function. As we have shown by a special implementation of weighted regression, reasonable agreement with our experimental observations was obtained for the spectrum. Simultaneous weighted regression of multiple functions, including autocorrelation, is reserved for further empirically oriented work.

It is concluded that the source of thermal fluctuations is related both to energy dissipation inside the photopolymer material and the surrounding liquid. In this partial dissipation problem, we have shown that the equipartition theorem allows us to correctly parameterize random forces. Furthermore, we believe that the stochastic mechanical model we present has the potential to be used for further analysis and prediction of characteristics for similar nanowire-based systems.

**Author Contributions:** Methodology, J.K. and Z.T. and A.S. and G.T.I. and G.V. and L.K.; conceptualization, D.H. and G.B.; data curation, G.B. and J.K and A.H. and Z.T.; writing—original draft, D.H. and G.B.; writing—review and editing, G.B. and D.H. and G.V. and L.K. and G.Ž.; funding acquisition, G.B. and G.Ž. and Z.T. and D.H. and L.K.; visualization G.B. and A.H. All authors have read and agreed to the published version of the manuscript.

**Funding:** This work was supported by the Slovak Research and Development Agency (grant APVV-18-0285), the Slovak Ministry of Education (grants VEGA 2/0094/21, KEGA No. 012 UVLF – 4/2018), the Operational Program Integrated Infrastructure, funded by the ERDF (Project: OPENMED, ITMS2014+: 313011V455), the joint project of Slovak and Hungarian Academies of Sciences (NKM-

88/2019), and by the GINOP-2.3.2-15-2016-00001 and the GINOP-2.3.3-15-2016-00040 programs. This project also received funding from the European Union's Horizon 2020 research and innovation program under grant agreement No. 654148 Laserlab-Europe.

**Institutional Review Board Statement:** Not applicable.

**Informed Consent Statement:** Not applicable.

**Data Availability Statement:** Not applicable.

**Conflicts of Interest:** The authors declare no conflict of interest.

## References

1. Kappler, E. Versuche zur Messung der Avogadro-Loschmidtschen Zahl aus der Brownschen Bewegung einer Drehwaage. *Ann. Der Phys.* **1931**, *403*, 233–256. [CrossRef]
2. Sader, J.E. Frequency response of cantilever beams immersed in viscous fluids with applications to the atomic force microscope. *J. Appl. Phys.* **1998**, *84*. [CrossRef]
3. Boskovic, S.; Chon, J.W.M.; Mulvaney, P.; Sader, J.E. Rheological measurements using microcantilevers. *J. Rheol.* **2002**, *46*, 891–899. [CrossRef]
4. Kara, V.; Sohn, Y.I.; Atikian, H.; Yakhot, V.; Lončar, M.; Ekinci, K.L. Nanofluidics of Single-Crystal Diamond Nanomechanical Resonators. *Nano Lett.* **2015**, *15*, 8070–8076. [CrossRef]
5. Miller, J.M.L.; Ansari, A.; Heinz, D.B.; Chen, Y.H.; Flader, I.B.; Shin, D.D.; Villanueva, L.G.; Kenny, T.W. Effective quality factor tuning mechanisms in micromechanical resonators. *Appl. Phys. Rev.* **2018**, *5*, 041307. [CrossRef]
6. Paul, M.R.; Clark, M.T.; Cross, M.C. The stochastic dynamics of micron and nanoscale elastic cantilevers in fluid: Fluctuations from dissipation. *Nanotechnology* **2006**, *17*, 4502. [CrossRef]
7. Meyhöfer, E.; Howard, J. The force generated by a single kinesin molecule against an elastic load. *Proc. Natl. Acad. Sci. USA* **1995**, *92*, 574–578. [CrossRef]
8. Viani, M.B.; Schäffer, T.E.; Chand, A.; Rief, M.; Gaub, H.E.; Hansma, P.K. Small cantilevers for force spectroscopy of single molecules. *J. Appl. Phys.* **1999**, *86*, 2258–2262. [CrossRef]
9. Fan, H.F.; Ma, C.H.; Jayaram, M. Single-Molecule Tethered Particle Motion: Stepwise Analyses of Site-Specific DNA Recombination. *Micromachines* **2018**, *9*, 216. [CrossRef]
10. Kovari, D.T.; Yan, Y.; Finzi, L.; Dunlap, D. Tethered Particle Motion: An Easy Technique for Probing DNA Topology and Interactions with Transcription Factors. *Methods Mol. Biol.* **2018**, *1665*, 317–340. [CrossRef]
11. Manghi, M.; Destainville, N.; Brunet, A. Statistical physics and mesoscopic modeling to interpret tethered particle motion experiments. *Methods* **2019**, *169*, 57–68. [CrossRef]
12. Huang, Y.; Flores, F.; Jaime, G.; Li, Y.; Wang, W.; Wang, D.; Goldberg, N.; Zheng, J.; Yu, M.; Lu, M.; et al. A Chip-Scale Oscillation-Mode Optomechanical Inertial Sensor Near the Thermodynamical Limits. *Laser Photonics Rev.* **2020**, *14*, 1800329. [CrossRef]
13. Hines, A.; Richardson, L.; Wisniewski, H.; Guzman, F. Optomechanical inertial sensors. *Appl. Opt.* **2020**, *59*, G167–G174. [CrossRef]
14. Geraci, A.; Bradley, C.; Gao, D.; Weinstein, J.; Derevianko, A. Searching for Ultralight Dark Matter with Optical Cavities. *Phys. Rev. Lett.* **2019**, *123*, 031304. [CrossRef] [PubMed]
15. Catano-Lopez, S.B.; Santiago-Condori, J.G.; Edamatsu, K.; Matsumoto, N. High-Q Milligram-Scale Monolithic Pendulum for Quantum-Limited Gravity Measurements. *Phys. Rev. Lett.* **2020**, *124*, 221102. [CrossRef]
16. Sharifi, S.; Banadaki, Y.; Cullen, T.; Veronis, G.; Dowling, J.; Corbitt, T. Design of microresonators to minimize thermal noise below the standard quantum limit. *Rev. Sci. Instruments* **2020**, *91*, 054504. [CrossRef] [PubMed]
17. Saulson, P.R. Thermal noise in mechanical experiments. *Phys. Rev. D* **1990**, *42*, 2437–2445. [CrossRef]
18. Howard, J. *Mechanics of Motor Proteins and the Cytoskeleton*; Sinauer Associates: Sunderland, MA, USA, 2001; Volume 55. [CrossRef]
19. Berg-Sorensen, K.; Flyvbjerg, H. Power spectrum analysis for optical tweezers. *Rev. Sci. Instruments* **2004**, *75*, 594–612. [CrossRef]
20. Lukić, B.; Jeney, S.; Sviben, Ž.; Kulik, A.J.; Florin, E.L.; Forró, L. Motion of a colloidal particle in an optical trap. *Phys. Rev. E* **2007**, *76*, 011112. [CrossRef] [PubMed]
21. Malinauskas, M.; Farsari, M.; Piskarskas, A.; Juodkazis, S. Ultrafast laser nanostructuring of photopolymers: A decade of advances. *Phys. Rep. Rev. Sect. Phys. Lett.* **2013**, *533*, 1–31. [CrossRef]
22. LaFratta, C.N.; Fourkas, J.T.; Baldacchini, T.; Farrer, R.A. Multiphoton fabrication. *Angew. Chem. Int. Ed.* **2007**, *46*, 6238–6258. [CrossRef]
23. Nakanishi, S.; Shoji, S.; Kawata, S.; Sun, H.B. Giant elasticity of photopolymer nanowires. *Appl. Phys. Lett.* **2007**, *91*, 063112. [CrossRef]
24. Cayll, D.R.; Ladner, I.S.; Cho, J.H.; Saha, S.K.; Cullinan, M.A. A MEMS dynamic mechanical analyzer for in situ viscoelastic characterization of 3D printed nanostructures. *J. Micromech. Microeng. Struct. Devices Syst.* **2020**, *30*, 075008. [CrossRef]
25. Kubacková, J.; Iváni, G.T.; Kažiková, V.; Strejčková, A.; Hovan, A.; Žoldák, G.; Vizsnyiczai, G.; Kelemen, L.; Tomori, Z.; Bánó, G. Bending dynamics of viscoelastic photopolymer nanowires. *Appl. Phys. Lett.* **2020**, *117*, 013701. [CrossRef]

26. Burden, R.; Faires, J. *Numerical Analysis*, 4th ed.; The Prindle, Weber and Schmidt Series in Mathematics; PWS-Kent Publishing Company: Boston, FL, USA, 1989.
27. Virtanen, P.; Gommers, R.; Oliphant, T.E.; Haberland, M.; Reddy, T.; Cournapeau, D.; Burovski, E.; Peterson, P.; Weckesser, W.; Bright, J.; et al. SciPy 1.0: Fundamental Algorithms for Scientific Computing in Python. *Nat. Methods* **2020**, *17*, 261–272. [CrossRef] [PubMed]

Article

# Thermodynamic Optimization of a High Temperature Proton Exchange Membrane Fuel Cell for Fuel Cell Vehicle Applications

Bing Xu, Dongxu Li, Zheshu Ma *, Meng Zheng and Yanju Li

College of Automobile and Traffic Engineering, Nanjing Forestry University, Nanjing 210037, China; xb18260078388@163.com (B.X.); Ldx961203@163.com (D.L.); mengzai19950929@163.com (M.Z.); njfulyj@163.com (Y.L.)
* Correspondence: mazheshu@njfu.edu.cn; Tel.: +86-137-7665-9269

**Abstract:** In this paper, a finite time thermodynamic model of high temperature proton exchange membrane fuel cell (HT-PEMFC) is established, in which the irreversible losses of polarization and leakage current during the cell operation are considered. The influences of operating temperature, membrane thickness, phosphoric acid doping level, hydrogen and oxygen intake pressure on the maximum output power density $P_{max}$ and the maximum output efficiency $\eta_{max}$ are studied. As the temperature rises, $P_{max}$ and $\eta_{max}$ will increase. The decrease of membrane thickness will increase $P_{max}$, but has little influence on the $\eta_{max}$. The increase of phosphoric acid doping level can increase $P_{max}$, but it has little effect on the $\eta_{max}$. With the increase of hydrogen and oxygen intake pressure, $P_{max}$ and $\eta_{max}$ will be improved. This article also obtains the optimization relationship between power density and thermodynamic efficiency, and the optimization range interval of HT-PEMFC which will provide guidance for applicable use of HT-PEMFCs.

**Keywords:** HT-PEMFC; irreversibility; finite time thermodynamic optimization; power density; thermodynamic efficiency

## 1. Introduction

With the decreasing of oil resources and the worsening of the natural environment, most countries have developed new/renewable energy systems and are trying to change their existing energy structure [1–3]. The proton exchange membrane fuel cell (PEMFC) has had considerable attention paid to it due to its excellent performance, including high energy conversion efficiency, low operating temperature, short start-up time, high power density and small size, etc. As a reliable power source, it has been widely used in the field of traffic engineering.

A conventional low-temperature proton exchange membrane fuel cell (LT-PEMFC) operates at 40–80 °C. Liquid water produced by the reaction affects conductivity and gas transmission, which makes the water management and gas management more complicated. In addition, the Nafion membrane at low temperature has low tolerance to CO and S. The high-temperature proton exchange membrane fuel cell (HT-PEMFC) equipped with phosphoric acid-doped polybenzimidazole membrane (PA/PBI) can increase the operating temperature to over the water boiling point (100–200 °C) and maintain high proton conductivity under high temperature operating conditions. Therefore, the corresponding water management system can be simplified greatly and the reaction rate of the cathode and anode can be improved, with a cell efficiency that is higher than LT-PEMFC [4].

In terms of modeling HT-PEMFCs, Cheddie et al. [5] established the output voltage and power density model of HT-PEMFC based on PBI membrane which considered the irreversibilities of activation polarization and ohmic polarization. The results found that the voltage loss was caused by activation polarization and ohmic polarization, and membrane conductivity and catalyst performance affected the polarization loss greatly. Hu et al. [6] presented a two-dimensional model of HT-PEMFC with PA/PBI membrane, in which the polarization loss on the cathode side was mainly considered. The mathematical model of the cathode exchange current density was established by linear sweep voltammetry (LSV) and the ohmic polarization of the cathode is estimated by electrochemical impedance spectroscopy (EIS). Numerical simulation results were in good agreement with the experimental results. Scott et al. [7,8] developed the output voltage and power model of HT-PEMFC and investigated the influence of operating temperature and operating pressure on open-circuit voltage, exchange current density and diffusion coefficient. Kim et al. [9,10] studied the influence of operating conditions on performance degradation of HT-PEMFC. The results showed that the doping level and current density had a significant effect on the durability of HT-PEMFC.

The above mathematical models consider the influence of polarization losses on cell output and degradation, but the influence of irreversibility of leakage current on cell operation is neglected. In addition, in the application of a fuel cell engine, not only the amount but also the quality of energy should be considered, so the efficiency model should be added into the model to study the overall optimization performance.

The fundamental purpose of finite time thermodynamics is to seek ultimate performance of thermodynamic processes and systems with the goal of reducing irreversibility in finite time or under the constraints of finite size [11–16]. In terms of thermodynamic optimization research of fuel cells, Watowich et al. [17] applied the optimal control theory to determine the limit of the fuel cell operation process, and the current path and optimal terminal state of the constrained cell in a limited time, so as to provide maximum output power, maximum efficiency and maximum profit. Li et al. [18] utilized finite time thermodynamics in HT-PEMFC performance analysis to investigate the effects of kinds of parameters. Although leakage current was considered, concentration potential was not contained in the reversible potential. Sieniutycz et al. [19,20] established the steady-state model of PEMFC based on finite time thermodynamics. The influence of design and operation parameters on the performance of fuel cells was analyzed, and the power limit was predicted from the perspective of thermodynamic optimization. Li et al. [21] conducted ecological analysis on LT-PEMFC and derived ecological coefficient of performance. In addition to the studies of HT-PEMFC model, lots of development was devoted to materials [22–25] and component degradation issues [26].

Firstly, aiming at the lack of irreversible factors, this paper established a thermodynamic model that considered various polarizations and leakage current. Secondly, the influences of operating temperature on $P_{max}$ and thermodynamic efficiency $\eta_{max}$ were studied. The influences of operating temperature, membrane thickness, phosphoric acid doping level, hydrogen and oxygen intake pressure on the optimal performance are discussed. Finally, the optimal relationship between power density and thermodynamic efficiency is analyzed, and the optimal interval of power density and thermodynamic efficiency is obtained.

## 2. Thermodynamic Model
### 2.1. Internal Processes of HT-PEMFC

As shown in Figure 1, HT-PEMFC converts chemical energy into electrical energy through electrochemical reaction of hydrogen and oxygen.

The mass transfer mechanism of HT-PEMFC is much different from LT-PEMFC. LT-PEMFC basically uses the Nafion membrane and the proton transport carrier sulfonic acid functional group can separate hydrogen ions and form hydronium ions with water molecules under humidified conditions; while in HT-PEMFCs, phosphoric acid replaces the humidified water and chemical reaction and mass transfer are based on a so-called Grotthuss mechanism. The chemical reaction and mass transfer within the anode, the cathode and the membrane can be expressed as Equations (1)–(3):

$$\text{Anode}: H_2PO_4^- + H^+ = H_3PO_4 \tag{1}$$

$$\text{Membrane}: H_3PO_4 + PBI = H_2PO_4^- + PBI \cdot H^+ \tag{2}$$

$$\text{Cathode}: \text{PBI} \cdot \text{H}^+ = \text{PBI} + \text{H}^+ \quad (3)$$

When supplying hydrogen to the anode and oxygen to the cathode, hydrogen atoms are separated into hydrogen ions and electrons under the action of anodic catalyst. Hydrogen ions pass through proton exchange membrane and electrons flow to the cathode through the external circuit load; hydrogen ions combine with oxygen atoms and electrons at the cathode to form water molecules at relative higher temperatures over water boiling point. Therefore, water molecules are discharged in the gas phase avoiding water management systems like LT-PEMFCs. The total electrochemical reaction of HT-PEMFC can be formulated as Equations (4)–(6):

$$\text{Anode reaction}: H_2 \to 2H^+ + 2e^- \quad (4)$$

$$\text{Cathodic reaction}: 2H^+ + \frac{1}{2}O_2 + 2e^- \to H_2O(gas) + heat \quad (5)$$

$$\text{Total reaction}: H_2 + \frac{1}{2}O_2 \to H_2O(gas) + heat + electricity \quad (6)$$

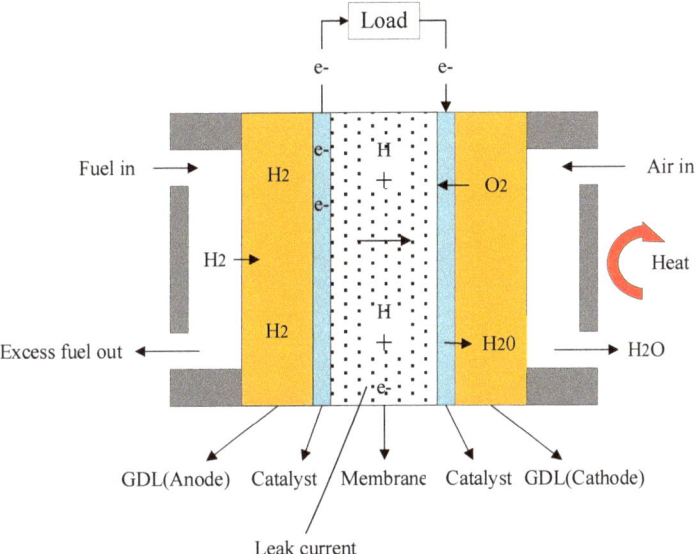

**Figure 1.** Working principle of HT–PEMFC.

*2.2. Reversible Output Voltage of HT-PEMFC*

For HT-PEMFC, the reversible output voltage can be written as Equation (7),

$$E_r = E_r^0 + \frac{\Delta S}{nF}(T - T_0) + \frac{RT}{nF} \ln(\frac{p_{H_2} p_{O_2}^{0.5}}{p_w}) \quad (7)$$

where, $E_r$ is obtained under isothermal conditions; $E_r^0$ [18] is the reference standard voltage at ambient temperature and pressure (298.15 K, 1 atm), and its value is 1.185 V; $\Delta S$ is the change of standard molar entropy, $T$ is operating temperature of the HT-PEMFC, $T_0$ is the ambient temperature, $R$ is the gas constant, $p_{H_2}$ is the intake pressure of hydrogen, $p_{O_2}$ is the intake pressure of oxygen, and $p_w$ is the pressure of discharging water vapor.

The entropy change is related to the operating temperature, Equation (8) [27]:

$$\frac{\Delta S}{n} = -18.449 - 0.01283(T) \quad (8)$$

## 2.3. Irreversible Loss of HT-PEMFC

### 2.3.1. Polarization Phenomenon

For a single electrode, the electrode potential with no current passing through is the equilibrium potential ($E_r$), and the electrode potential with current passing through is $E_{cell}$. In general, $E_{cell} < E_r$, the absolute value of electrode potential difference ($|E_r - E_{cell}|$) is overpotential.

The activation overpotential can be written as Equation (9) [5,28]:

$$E_{act} = \frac{RT}{2\alpha F} \ln\left(\frac{I + I_{leak}}{I_0}\right) \tag{9}$$

where, $I$ is current density, $I_0$ is exchange current density, $I_{leak}$ is leakage current density, $\alpha$ is transfer coefficient, $E_{act}$ is activation overpotential. Exchange current density can be expressed as Equation (10) [7]:

$$\ln(I_0) = 2.2266 \times \frac{1000}{T} - 0.4959 \tag{10}$$

Ohmic overpotential can be calculated as the following Equation (11) [27]:

$$E_{ohm} = I\left(\frac{l_m}{K_m} + \frac{2l_d}{\sigma_d^{eff}}\right) \tag{11}$$

where, $l_m$ is membrane thickness, $K_m$ is proton conductivity in the membrane phase, $l_d$ is thickness of diffusion layer, $\sigma_d^{eff}$ is electron conductivity. The proton conductivity can be presented as Equation (12) [8]:

$$K_m = \frac{100}{T} \exp\left[8.0219 - \left(\frac{2605.6 - 70.1X}{T}\right)\right] \tag{12}$$

where, $X$ is the doping level of phosphoric acid of the proton exchange membrane.

If the reactant gas or oxidant is not supplied in time, the electrode surface cannot maintain the reactant concentration, and concentration polarization will occur. In HT-PEMFC, the value of the concentration overpotential is already included in the reversible potential.

### 2.3.2. Leakage Current

Theoretically, the electrolyte is an ionic conductor and has no electron transport. However, in actual operation, some hydrogen and electrons will diffuse from the anode to the cathode through the electrolyte, and a small number of electrons will flow outward through the proton membrane. Such leakage current includes internal current and cross current [29]. Therefore, the total current density generated by the fuel cell is equal to the sum of the output current density and leakage current density,

$$I_{gross} = I + I_{leak} \tag{13}$$

where $I_{gross}$ represents the total current density generated on the fuel cell electrode, $I$ is the working current density that can be measured through the external load, and $I_{leak}$ represents the leakage current density.

Haji [30] found that the leakage current increased with the rise of operating temperature and concluded that leakage current density and operating temperature met the functional relationship:

$$\ln I_{leak} = \left(-2342.9\frac{1}{T} + 9.0877\right) \times \ln 10 \tag{14}$$

## 2.4. Irreversible Output Voltage

The irreversible output voltage of HT-PEMFC can be expressed as Equation (15),

$$\begin{aligned} E_{cell} &= E_r - E_{act} - E_{ohm} - E_{con} \\ &= 1.185 - (1.91 \times 10^{-4} + 1.33 \times 10^{-7}T)(T - 298.15) + 4.13 \\ &\times 10^{-5} T \ln \frac{P_{H_2} P_{O_2}^{0.5}}{0.0243} - 1.72 \times 10^{-5} T \ln \frac{I + 88458.17 exp\left(\frac{-2342.9}{T}\right)}{3.95 \times 10^{-6} T^3 - 0.00424 T^2 + 1.523 - 183} \\ &+ I \left( \frac{l_m T}{304696.11 exp\left(\frac{70.1X - 2605.6}{T}\right)} \right) \end{aligned} \quad (15)$$

## 2.5. Power Density and Efficiency of HT-PEMFC

The power density [31] of HT-PEMFC can be expressed as Equation (16),

$$\begin{aligned} P = E_{cell} \cdot I &= (E_r - E_{act} - E_{ohm} - E_{con}) \cdot I \\ &= [1.185 - (1.91 \times 10^{-4} + 1.33 \times 10^{-7}T)(T - 298.15) \\ &+ 4.13 \\ &\times 10^{-5} T \ln \frac{P_{H_2} P_{O_2}^{0.5}}{0.0243} \\ &- 1.72 \times 10^{-5} T \ln \frac{I + 88458.17 exp\left(\frac{-2342.9}{T}\right)}{3.95 \times 10^{-6} T^3 - 0.00424 T^2 + 1.523 - 183} \\ &+ I \left( \frac{l_m T}{304696.11 exp\left(\frac{70.1X - 2605.6}{T}\right)} \right) ] I \times 10^{-3} \end{aligned} \quad (16)$$

The total energy absorbed from hydrogen and oxygen is enthalpy of reaction [32],

$$\Delta H = \sum_k \left| \frac{d_e n_k}{d_t} \right| h_k(T) - \sum_j \left| \frac{d_e n_j}{d_t} \right| h_j(T) = \frac{I}{nF} \Delta h(T) \quad (17)$$

where $\Delta h(T)$ is molar enthalpy at temperature $T$, $j$ is component of the reactants, $k$ is component of the products in the reaction.

For the energy conversion device, the basic definition of thermodynamic efficiency is the ratio of actual useful work to the total energy input [33]. Therefore, thermodynamic efficiency of HT-PEMFC can be expressed as Equation (18),

$$\begin{aligned} \eta = -\frac{P}{\Delta} H &= -\frac{(E_r - E_{act} - E_{ohm}) \cdot I}{\Delta H} \\ &= -192970 \times [1.185 - (1.91 \times 10^{-4} + 1.33 \times 10^{-7}T)(T - 298.15) \\ &+ 4.13 \\ &\times 10^{-5} T \ln \frac{P_{H_2} P_{O_2}^{0.5}}{0.0243} \\ &- 1.72 \\ &\times 10^{-5} T \ln \frac{I + 88458.17 exp\left(\frac{-2342.9}{T}\right)}{3.95 \times 10^{-6} T^3 - 0.00424 T^2 + 1.523 - 183} \\ &+ I \left( \frac{l_m T}{304696.11 exp\left(\frac{70.1X - 2605.6}{T}\right)} \right) ] I \times 10^{-3} / \cdot h(T) \end{aligned} \quad (18)$$

## 2.6. Thermodynamic Optimization

For HT-PEMFC, $P$ is related to $I$, $T$, $l_m$, $p_{H_2}$, $p_{O_2}$ and $X$, so $P$ can be expressed as Equation (19) [34–37]:

$$P = f(I, T, l_m, p_{H_2}, p_{O_2}, X) \quad (19)$$

When the membrane thickness ($l_m$), hydrogen intake pressure ($p_{H_2}$), oxygen intake pressure ($p_{O_2}$) and membrane acid doping level ($X$) are determined, the power density ($P$) is only related to current density ($I$) and cell operating temperature ($T$), so $P$ can be expressed as Equation (20):

$$P = g(I, T) \quad (20)$$

In Equation (22), when $T = T_1$, the power density ($P$) is only related to the current density ($I$), the maximum power density $P_{max}$ ($P_1$) is obtained; when $T = T_2$, the power density ($P$) is only related to the current density ($I$), the maximum power density $P_{max}$ ($P_2$) is obtained. By analogy, when $T = T_n$, the maximum power density $P_{max}$ ($P_n$) can be obtained. And the relationship between $P_{max}$ and $T$ can be acquired.

Similarly, when operating temperature $T$, film thickness $l_m$, oxygen intake pressure $p_{O_2}$, and film acid doping level $X$ are determined, the corresponding curve between $P_{max}$ and $p_{H_2}$ can be obtained.

When the operating temperature $T$, film thickness $l_m$, hydrogen intake pressure $p_{H_2}$ and film acid doping level $X$ are determined, the corresponding curve between $P_{max}$ and $p_{O_2}$ can be obtained.

Similarly, thermodynamic efficiency ($\eta$) is related to current density ($I$), cell operating temperature ($T$), membrane thickness ($l_m$), hydrogen intake pressure ($p_{H_2}$), oxygen intake pressure ($p_{O_2}$) and membrane acid doping level ($X$), so it can be expressed as formula (21):

$$\eta = h(I, T, l_m, p_{H_2}, p_{O_2}, X) \quad (21)$$

When the membrane thickness ($l_m$), hydrogen intake pressure ($p_{H_2}$), oxygen intake pressure ($p_{O_2}$) and membrane acid doping level ($X$) are determined, the thermodynamic efficiency ($\eta$) is only related to current density ($I$) and cell operating temperature ($T$), so $\eta$ can be expressed as Equation (22):

$$\eta = j(I, T) \quad (22)$$

In Equation (22), when $T = T_1$, the thermodynamic efficiency ($\eta$) is only related to the current density ($I$), the maximum thermodynamic efficiency $\eta_{max}$ ($\eta_1$) is received; when $T = T_2$, the thermodynamic efficiency ($\eta$) is only related to the current density ($I$), the maximum thermodynamic efficiency $\eta_{max}$ ($\eta_2$) is received. By analogy, when $T = T_n$, the maximum thermodynamic efficiency $\eta_{max}$ ($\eta_n$) can be obtained. And the relationship between $\eta_{max}$ and $T$ can be acquired.

Similarly, when operating temperature $T$, film thickness $l_m$, oxygen intake pressure $p_{O_2}$, and film acid doping level $X$ are determined, the corresponding curve between $\eta_{max}$ and $p_{H_2}$ can be obtained.

When the operating temperature $T$, film thickness $l_m$, hydrogen intake pressure $p_{H_2}$ and film acid doping level $X$ are determined, the corresponding curve between $\eta_{max}$ and $p_{O_2}$ can be obtained.

## 3. Results and Discussion

The relevant parameters in the HT-PEMFC model are shown in Table 1.

**Table 1.** Relevant data of HT-PEMFC.

| Parameter | Value |
|---|---|
| current density, $I$ (A m$^{-2}$) | 0–20,000 [27] |
| operating temperature, $T$ (K) | 373–473 [20] |
| intake pressure, $p_{H_2}, p_{H_2}$ (atm) | 1–3 [27] |
| thickness of membrane, $l_m$ (μm) | 20, 60, 100 [27] |
| doping level, $X$ | 2, 6, 10 [27] |
| electronic number, $n$ | 2 |
| faraday constant, $F$ (C mol$^{-1}$) | 96,485 |
| ambient temperature, $T_0$ (K) | 298.15 |
| transfer coefficient, $\alpha$ | 0.25 [20] |
| diffusion layer thickness, $l_d$ (m) | $2.6 \times 10^{-4}$ [20] |
| electron conductivity, $\sigma_d^{eff}$ (S m$^{-1}$) | 53 [20] |

As shown in Table 1, some parameters referenced in this paper are from the literature [27,38]. The choice of any type of fuel cell as the vehicle power is mainly based on the application scenario of the vehicle and the demand of the power plant. Power density, dynamic response, output efficiency, durability and life, fuel form, emission and other aspects are selected as key indicators for passenger vehicles. PEMFC is superior to other fuel cells in terms of efficiency, dynamic response, durability and life, fuel form and emission.

## 3.1. Model Validation

Figure 2 compares model Formula (15) prediction and experimental data at 398 K and 448 K ($l_m = 20$ μm; $X = 6$; $p_{H_2} = 1$ atm; $p_{O_2} = 1$ atm). The experimental studies in the literature [8] and model studies in the literature [27,38] are both based on HT-PEMFC equipped with PA/PBI membrane and have the same specifications. The results show that the curve predicted by the cell output voltage model has a good agreement with the experimental data. When leakage current is not considered in the model, the output voltage is significantly higher than that predicted by the model, especially in the low current density region. This is mainly because leakage current mainly affects the activation polarization, as shown in Figure 3, the activation polarization potential changes most significantly in the low current density region.

Figure 3 shows the variation curve of reversible voltage, polarized overpotential and irreversible output voltage versus current density of HT-PEMFC. It can be seen that the reversible voltage is a constant independent of current density. The activation overpotential and ohm overpotential increase with the increase of current density, and the activation overpotential changes greatly in the area of low current density. The irreversible output voltage decreases as the current density increases.

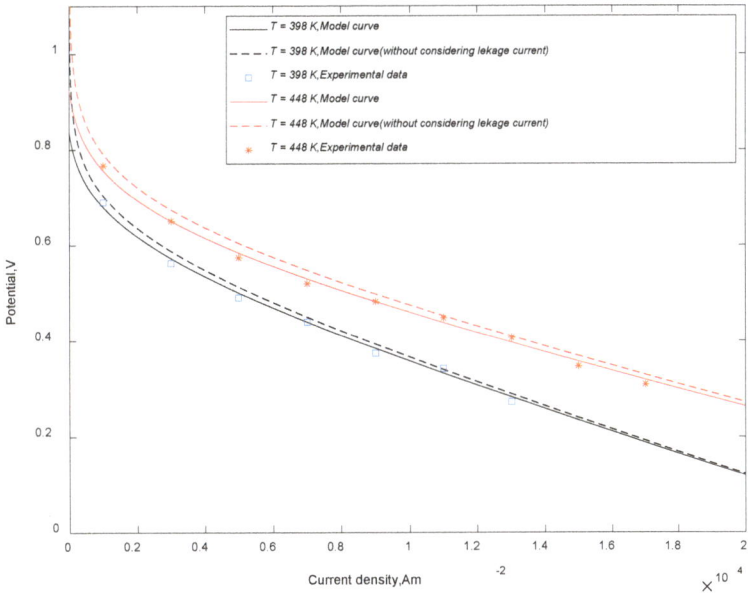

**Figure 2.** Comparison of model curve and experimental data.

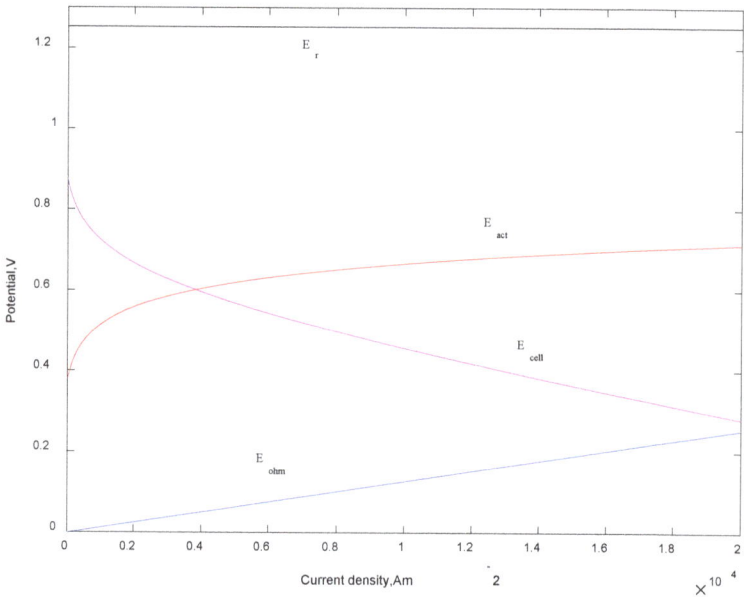

**Figure 3.** Reversible voltage, polarized overpotential and irreversible output voltage of HT-PEMFC.

*3.2. Maximum Output Performance at a Given Temperature*

Figure 4 reflects the influence of operating temperature on $P_{max}$ and $\eta_{max}$ of HT-PEMFC. As can be seen from the figures, $P_{max}$ and $\eta_{max}$ raise with the increase of temperature. From the perspective of electrochemical kinetics, the increase of operating temperature enhances the proton conductivity of the membrane and reduces the electrochemical polarization of the HT-PEMFC. Moreover, the increase of operating temperature can improve the exchange current density and reduce the activation overpotential, so $P_{max}$ and $\eta_{max}$ will be boosted.

Figure 4a shows the impact of operating temperature on $P_{max}$ and $\eta_{max}$ of HT-PEMFC under different $l_m$. It can be seen that with the decrease of proton membrane thickness, $P_{max}$ and $\eta_{max}$ will increase. If the thickness of proton membrane is reduced, the barrier of ions passing through proton membrane and the ohmic overpotential decrease. Therefore, $P_{max}$ and $\eta_{max}$ will be improved. When the temperature is 433 K, $P_{max}$ increases by 50% and the $\eta_{max}$ raises by 1.8% under the change of membrane thickness, which indicates that the membrane thickness has little influence on $\eta_{max}$. The reduction of membrane thickness can improve the performance, but fuel penetration, short circuit and other problems always limit the thickness.

Figure 4b reveals the effect of operating temperature on $P_{max}$ and $\eta_{max}$ of HT-PEMFC under different $X$. It is obvious that with the increase of the acid doping level of the membrane, $P_{max}$ and $\eta_{max}$ will improve, and the variation range of the low temperature zone is greater than that of the high temperature zone. The doping level directly affects the proton conductivity of the membrane. The increase of doping level enhances the proton conductivity of the membrane, thus reducing ohmic overpotential. Therefore, $P_{max}$ and $\eta_{max}$ will be improved. When the temperature is 373 K, $P_{max}$ increases by 84% under the change of doping level; while when the temperature is 473 K, $P_{max}$ improves by 52% under the change of doping level. When the temperature is 433 K, the $\eta_{max}$ raises by 0.8% under the change of acid doping level. This indicates that doping level has little effect on $\eta_{max}$.

Figure 4c displays the influence of operating temperature on $P_{max}$ and $\eta_{max}$ of HT-PEMFC under different $p_{H_2}$. Obviously, with the increase of hydrogen intake pressure, $P_{max}$ and $\eta_{max}$ will improve. The increase in pressure, on the one hand, increases the diffusion rate of the gas and improves the mass transfer of the reaction gas, which increases the reversible electromotive force of the HT-PEMFC; on the other hand, it increases the gas concentration and reduces the effect of concentration polarization on the reversible electromotive force. At the temperature of 433 K, $P_{max}$ boosts by 12.5% and the $\eta_{max}$ enhanced by 4.5% under the influence of hydrogen intake pressure.

Figure 4d shows the effect of operating temperature on $P_{max}$ and $\eta_{max}$ of HT-PEMFC under different $p_{O_2}$. It can be seen that with the increase of oxygen intake pressure, $P_{max}$ and $\eta_{max}$ will increase. At the temperature of 433 K, $P_{max}$ and the $\eta_{max}$ raises by 6.7% and 2.9% respectively under the change of oxygen intake pressure.

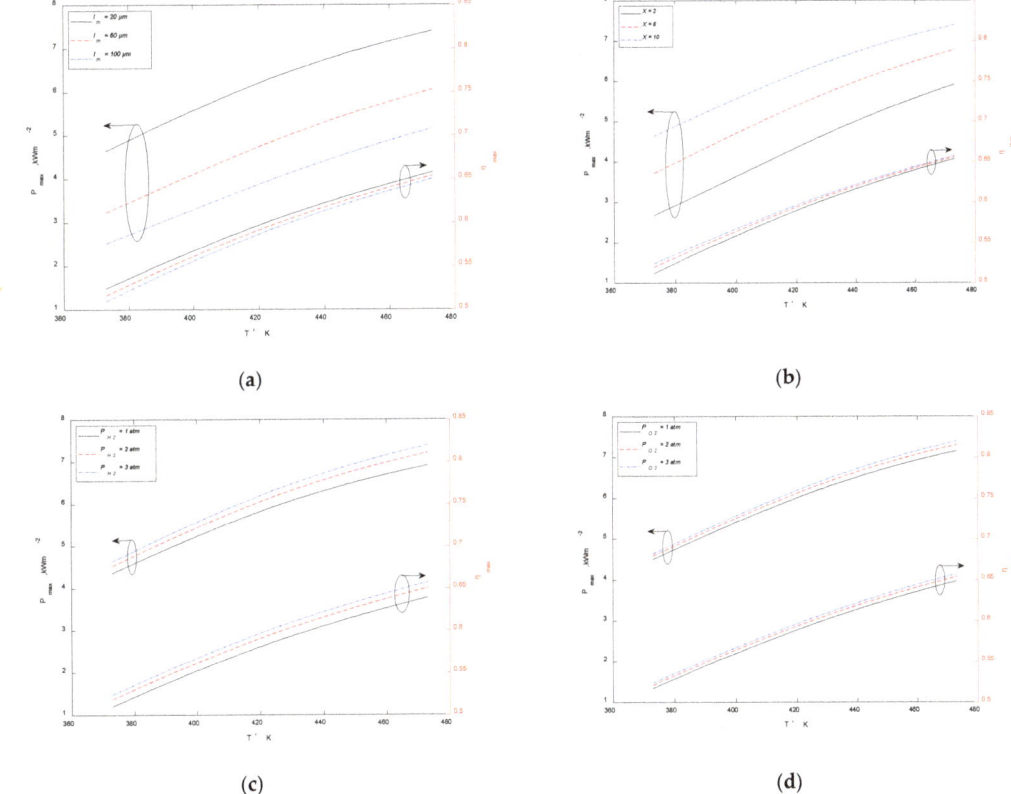

**Figure 4.** Effect of operating temperature on $P_{max}$ and $\eta_{max}$ of HT-PEMFC under different parameters. (**a**) Different $l_m$; (**b**) different $X$; (**c**) different $p_{H_2}$; (**d**) different $p_{O_2}$.

### 3.3. Maximum Output Performance at a Given $p_{H_2}$

Figure 5 reflects the effect of hydrogen inlet pressure on $P_{max}$ and $\eta_{max}$ of HT-PEMFC with. It can be seen from several figures that $P_{max}$ and $\eta_{max}$ increase with the raise of $p_{H_2}$, but from a numerical point of view the increase is not large. The rise of pressure not only increases the diffusivity of the bipolar gas, but also improves the concentration of the bipolar gas and boosts the mass transfer of the reaction gas, which will strength the mass transfer of the reaction gas and reduce the influence of concentration polarization on the reversible potential.

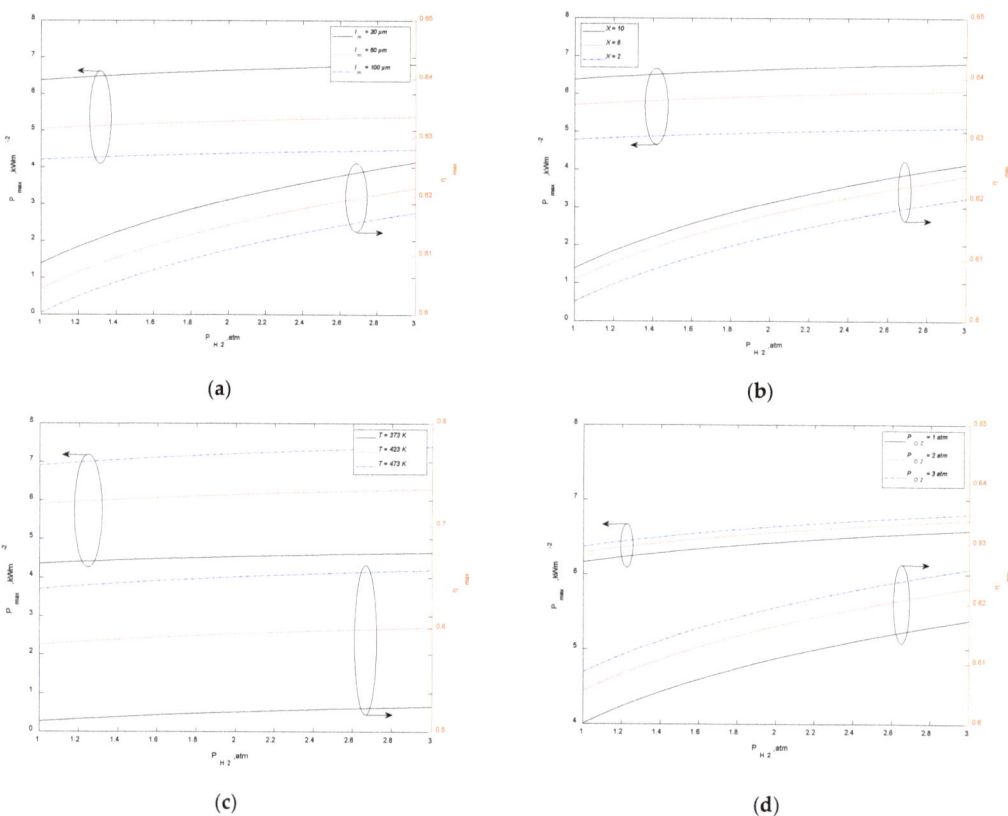

**Figure 5.** Effect of operating temperature on $P_{max}$ and $\eta_{max}$ of HT-PEMFC under different parameters. (**a**) Different $l_m$; (**b**) different X; (**c**) different T; (**d**) different $p_{O_2}$.

Figure 5a shows the impact of $p_{H_2}$ on $P_{max}$ and $\eta_{max}$ of HT-PEMFC under different $l_m$. It can be seen that as $l_m$ decreases, $P_{max}$ and $\eta_{max}$ will increase. The thinner the proton membrane is, the smaller the barrier for ions to pass through the proton membrane, that is, the ohmic overpotential decreases.

Figure 5b reveals the influence of $p_{H_2}$ in $P_{max}$ and $\eta_{max}$ of HT-PEMFC under different X. It is obvious that as the doping level of phosphoric acid increases, $P_{max}$ and $\eta_{max}$ will increase. The improvement of doping level increases the proton conductivity of the membrane.

Figure 5c displays the impact of $p_{H_2}$ on $P_{max}$ and $\eta_{max}$ of HT-PEMFC under different T. It is obvious that as the temperature rises, $P_{max}$ and $\eta_{max}$ will increase, and the increase is relatively large. When the operating temperature increases, the proton conductivity of the membrane improves and electrochemical polarization decreases; at the same time, the increase of operating temperature can boost the exchange current density, which reduces the activation overpotential.

Figure 5d describes the effect of $p_{H_2}$ on $P_{max}$ and $\eta_{max}$ of HT-PEMFC under different $p_{O_2}$. As can be seen from Figure 5d, with the increase of $p_{O_2}$, $P_{max}$ and $\eta_{max}$ will strength. Obviously, $p_{O_2}$ has little effect on the maximum output of the cell.

## 3.4. Maximum Output Performance at a Given $p_{O_2}$

The influence of oxygen intake pressure on $P_{max}$ and $\eta_{max}$ of HT-PEMFC is shown in Figure 6. It can be seen from several figures that $P_{max}$ and $\eta_{max}$ raise with the increase of $p_{O_2}$, but from a numerical point of view the improvement is not large. The increase in pressure improves the mass transfer of the reaction gas and reduces the influence of concentration polarization on the reversible potential.

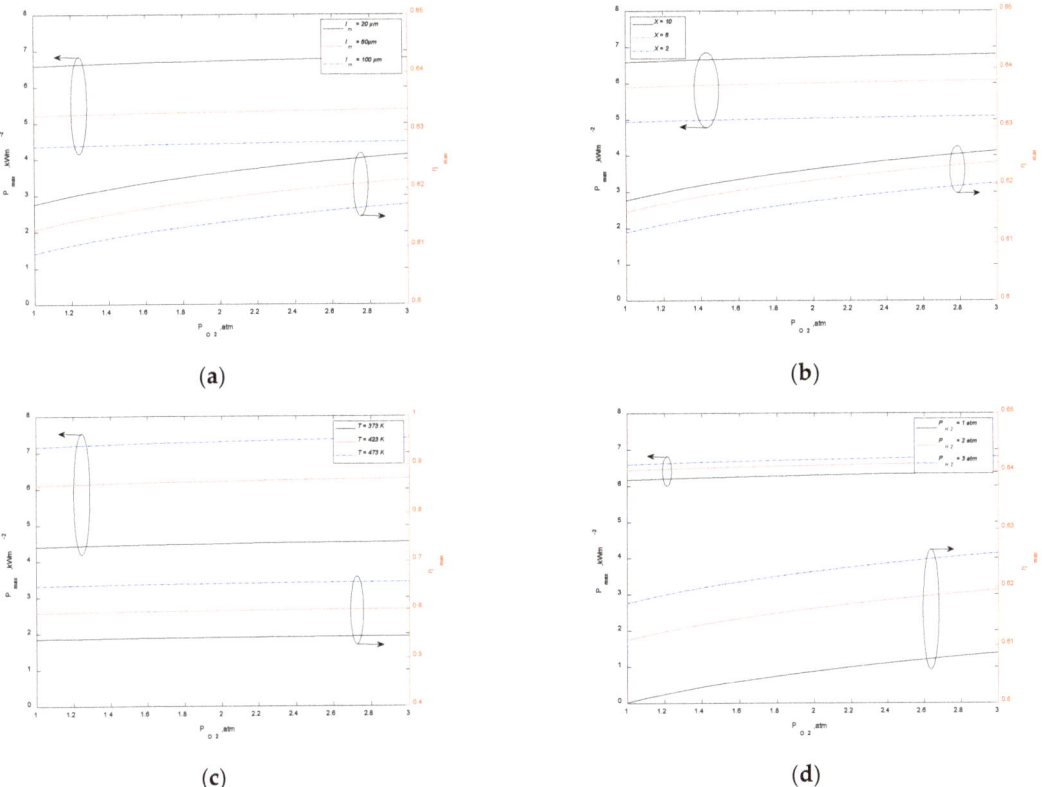

**Figure 6.** Effect of operating temperature on $P_{max}$ and $\eta_{max}$ of HT-PEMFC under different parameters. (**a**) Different $l_m$; (**b**) different X; (**c**) different T; (**d**) different $p_{H_2}$.

Figure 6a reflects the effect of $p_{O_2}$ on $P_{max}$ and $\eta_{max}$ of HT-PEMFC under different $l_m$. It can be seen that as the thickness of the proton film decreases, $P_{max}$ and $\eta_{max}$ will increase.

As shown in Figure 6b, $p_{O_2}$ has a significant effect on $P_{max}$ and $\eta_{max}$ of HT-PEMFC under different X. It is obvious that as the doping level of phosphoric acid raises, $P_{max}$ and $\eta_{max}$ will improve.

Figure 6c shows the impact of $p_{O_2}$ on $P_{max}$ and $\eta_{max}$ of HT-PEMFC under different T. Obviously, as the temperature increases, $P_{max}$ and $\eta_{max}$ will increase and the increase is relatively large.

Figure 6d reveals the effect of $p_{O_2}$ on $P_{max}$ and $\eta_{max}$ of HT-PEMFC under different $p_{H_2}$. It is obvious that with the increase of $p_{H_2}$, $P_{max}$ and $\eta_{max}$ will enhance. It is clear that $p_{H_2}$ has little effect on the maximum output power $P_{max}$.

## 3.5. Maximum Output Performance at a Given $p_{O_2}$

In the application of fuel cell vehicle, not only the quantity but also the quality of energy should be considered. Figure 7 shows the relation curve between power density and thermodynamic efficiency under operating temperature $T$ (453 K), hydrogen and oxygen intake pressure $p_{O_2}$, $p_{O_2}$ (3 atm), membrane thickness $l_m$ (20 μm) and membrane acid doping level $X$ (10). In order to improve the calculation accuracy, $P/P_{max}$ is chosen to transform the engineering problem into a mathematical problem. The curve is the willow leaf curve going back to the origin. As shown in Figure 7, when $P = P_B$, $\eta = \eta_{max}$ and when $\eta = \eta_A$, $P = P_{max}$. Thus, the optimal region of HT-PEMFC can be obtained.

$$P_B \leq P \leq P_{max}, \eta_A \leq \eta \leq \eta_{max}$$

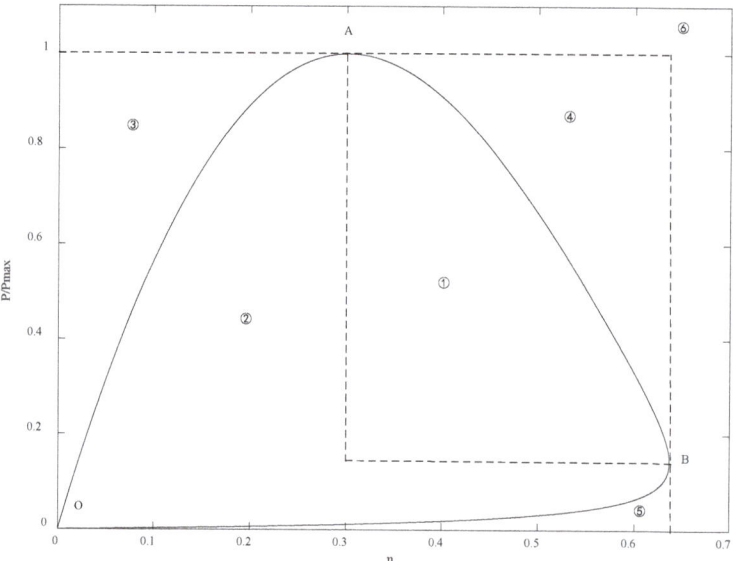

**Figure 7.** Optimization relationship between $P$ and $\eta$.

Curve OABO is an optimization curve derived from the power and efficiency model. When the operating point of the cell is located on the curve AB, its performance reaches the best. When the running point is located in region ①, it has better performance; when the running point is located in the region ②, it has the worst performance. Regions ③④⑤ are unstable region, because these three regions are outside OABO curve; there exists no operation points in the region ⑥ because $P > P_{max}$, $\eta > \eta_{max}$.

## 4. Discussion

In this paper, a finite time thermodynamic model of HT-PEMFC is established, which takes the irreversibility caused by polarization and leakage current into account. The influences of operating temperature, proton membrane thickness, proton membrane phosphoric acid doping level, hydrogen intake pressure and oxygen intake pressure on $P_{max}$ and $\eta_{max}$ at a given temperature are studied. The results show that $P_{max}$ and $\eta_{max}$ both increase with the increase of temperature. When the operating temperature is 433 K, with the decrease of proton membrane thickness, $P_{max}$ improves greatly, but the decrease of membrane thickness has little effect on $\eta_{max}$. As the doping level of proton membrane phosphoric acid increases, $P_{max}$ increases by 84% at the temperature of 373 K, and by 52% at the temperature of 473 K. However, the increase of phosphate doping level has little influence on $\eta_{max}$. The increase of hydrogen intake pressure and oxygen intake pressure will increase $P_{max}$ and

$\eta_{max}$. The optimal relationship between power density and thermodynamic efficiency of HT-PEMFC is also studied. The optimal interval of power density and thermodynamic efficiency is $P_B \leq P \leq P_{max}$, $\eta_A \leq \eta \leq \eta_{max}$.

**Author Contributions:** Conceptualization, Z.M.; formal analysis, B.X.; investigation, B.X. and D.L.; project administration, Z.M.; resources, M.Z.; software, D.L.; supervision, Z.M.; validation, M.Z.; visualization, Y.L.; writing—original draft, B.X.; writing—review & editing, Z.M. All authors have read and agreed to the published version of the manuscript.

**Funding:** We gratefully acknowledge the financial support of the National Natural Science Foundation of China (No. 51176069) and Scientific Research Foundation of Nanjing Forestry University (No. GXL2018004).

**Institutional Review Board Statement:** Not applicable.

**Informed Consent Statement:** Not applicable.

**Data Availability Statement:** Not applicable.

**Conflicts of Interest:** The authors declare no conflict of interest.

## References

1. Liu, S.Q.; Jia, L.M. Review on sustainable development of forest-based biodiesel. *J. Nanjing For. Univ. Nat. Sci. Ed.* **2021**, *44*, 216–224.
2. Yao, J.J.; Feng, X.Q.; Xiao, H.; Zheng, Y.; Zhang, C.L. Improvement effects of different solid waste and their disposal by products on saline-alkali soil in Huanghua Port. *J. Nanjing For. Univ. Nat. Sci. Ed.* **2021**, *45*, 45–52. [CrossRef]
3. Zhang, Z.G. Researches on green features and category architecture of green strategies of renewable-resource-based enterprises: A case study of forestry enterprise. *J. Nanjing For. Univ. Nat. Sci. Ed.* **2020**, *44*, 1–8. [CrossRef]
4. Sun, Y.F. High Temperature Proton Exchange Membrane Technology Improvement Research. *Appl. Energy Technol.* **2018**, *6*, 50–52.
5. Cheddie, D.; Munroe, N. Analytical correlations for intermediate temperature PEM fuel cells. *J. Power Sources* **2006**, *160*, 299–304. [CrossRef]
6. Hu, J.W.; Zhang, H.M.; Hu, J.; Zhai, Y.F.; Yi, B.L. Two dimensional modeling study of PBI/H3PO4 high temperature PEMFCs based on electrochemical methods. *J. Power Sources* **2006**, *160*, 1026–1034. [CrossRef]
7. Cheddie, D.F.; Munroe, N.D.H. A two-phase model of an intermediate temperature PEM fuel cell. *Int. J. Hydrogen Energy* **2007**, *32*, 832–841. [CrossRef]
8. Scott, K.; Pilditch, S.; Mamlouk, M. Modelling and experimental validation of a high temperature polymer electrolyte fuel cell. *J. Appl. Electrochem.* **2007**, *37*, 1245–1259. [CrossRef]
9. Kang, T.; Kim, M.; Kim, J.; Sohn, Y.J. Numerical modeling of the degradation rate for membrane electrode assemblies in high temperature proton exchange membrane fuel cells and analyzing operational effects of the degradation. *Int. J. Hydrogen Energy* **2015**, *40*, 5444–5455. [CrossRef]
10. Kim, M.; Kang, T.; Kim, J.; Sohn, Y.J. One-dimensional modeling and analysis for performance degradation of high temperature proton exchange membrane fuel cell using PA doped PBI membrane. *Solid State Ionics* **2014**, *262*, 319–323. [CrossRef]
11. Cheddie, D.; Munroe, N. Mathematical model of a PEMFC using a PBI membrane. *Energy Convers. Manag.* **2006**, *47*, 1490–1504. [CrossRef]
12. Qin, W.X.; Qin, X.Y.; Chen, L.G. Finite time thermodynamics optimization of an irreversible KCS-34 cycle coupled to variable temperature heat reservoirs. *Energy Conserv.* **2018**, *37*, 69–74.
13. He, S.; Lin, L.Y.; Wu, Z.X.; Chen, Z.M. Application of Finite Element Analysis in Properties Test of Finger-jointed Lumber. *J. Bioresour. Bioprod.* **2020**, *5*, 124. [CrossRef]
14. Zhao, X.Y.; Huang, Y.J.; Fu, H.Y.; Wang, Y.L.; Wang, Z. Deflection test and modal analysis of lightweight timber floors. *J. Bioresour. Bioprod.* **2021**, *6*, 266–278. [CrossRef]
15. Yang, J.; Zhang, Y.C.; Zhou, L.; Zhang, F.S.; Jing, Y.; Huang, M.Z.; Liu, H.B. Quality-related monitoring of papermaking wastewater treatment processes using dynamic multiblock partial least squares. *J. Bioresour. Bioprod.* **2021**. [CrossRef]
16. Hu, T.P.; Yu, Z.; Guo, L.; Xu, C.Y. Thermodynamic self-consistent dynamic model of wood dust explosion. *J. For. Eng.* **2019**, *4*, 29–34.
17. Stanley, J.; Watowich, R.; Stephen, B. Optimal current paths for model electrochemical systems. *J. Phys. Chem. B* **1986**, *90*, 4624–4631.
18. Li, D.; Li, S.; Ma, Z.; Xu, B.; Lu, Z.; Li, Y.; Zheng, M. Ecological Performance Optimization of a High Temperature Proton Exchange Membrane Fuel Cell. *Mathematics* **2021**, *9*, 1332. [CrossRef]
19. Sieniutycz, S. Thermodynamics of Power Production in Fuel Cells. *Chem. Process. Eng.-Inz.* **2010**, *31*, 81–105.
20. Sieniutycz, S.; Poswiata, A. Thermodynamic aspects of power production in thermal, chemical and electrochemical systems. *Energy* **2012**, *45*, 62–70. [CrossRef]

21. Li, C.J.; Liu, Y.; Xu, B.; Ma, Z.S. Finite Time Thermodynamic Optimization of an Irreversible Proton Exchange Membrane Fuel Cell for Vehicle Use. *Processes* **2019**, *7*, 419. [CrossRef]
22. Muthuraja, P.; Prakash, S.; Shanmugam, V.M.; Radhakr sihnan, S.; Manisankar, P. Novel perovskite structured calcium titanate-PBI composite membranes for high-temperature PEM fuel cells: Synthesis and characterizations. *Int. J. Hydrogen Energy* **2018**, *43*, 4763–4772. [CrossRef]
23. Miansari, M.; Sedighi, K.; Amidpour, M.; Alizadeh, E.; Miansari, M. Experimental and thermodynamic approach on proton exchange membrane fuel cell performance. *J. Power Sources* **2009**, *190*, 356–361. [CrossRef]
24. Peng, X.R.; Zhang, Z.K.; Zhao, L.Y. Analysis of Raman spectroscopy and XPS of plasma modified polypropylene decorative film. *J. For. Eng.* **2020**, *5*, 45–51.
25. Yu, P.J.; Zhang, W.; Chen, M.Z.; Zhou, X.Y. Plasma-treated thermoplastic resin film as adhesive for preparing environmentally-friendly plywood. *J. For. Eng.* **2020**, *5*, 41–47.
26. Lobato, J.; Rodrigo, M.A.; Linares, J.J.; Scott, K. Effect of the catalytic ink preparation method on the performance of high temperature polymer electrolyte membrane fuel cells. *J. Power Sources* **2006**, *157*, 284–292. [CrossRef]
27. Guo, X.R.; Zhang, H.C.; Zhao, J.P.; Wang, F.; Wang, J.T.; Miao, H.; Yuan, J.L. Performance evaluation of an integrated high-temperature proton exchange membrane fuel cell and absorption cycle system for power and heating/cooling cogeneration. *Energy Convers. Manag.* **2019**, *181*, 292–301. [CrossRef]
28. Chan, S.H.; Khor, K.A.; Xia, Z.T. A complete polarization model of a solid oxide fuel cell and its sensitivity to the change of cell component thickness. *J. Power Sources* **2001**, *93*, 130–140. [CrossRef]
29. Andreadis, G.M.; Podias, A.K.M.; Tsiakaras, P.E. The parasitic current on the direct ethanol PEM fuel cell operation. *J. Power Sources* **2008**, *181*, 214–227. [CrossRef]
30. Haji, S. Analytical modeling of PEM fuel cell i-V curve. *Renew. Energy* **2011**, *36*, 451–458. [CrossRef]
31. Wu, Z.; Zhu, P.F.; Yao, J.; Tan, P.; Xu, H.R.; Chen, B.; Yang, F.S.; Zhang, Z.X.; Ni, M. Thermo-economic modeling and analysis of an NG-fueled SOFC-WGS-TSA-PEMFC hybrid energy conversion system for stationary electricity power generation. *Energy* **2020**, *192*, 116613. [CrossRef]
32. Zhao, Y.R.; Ou, C.J.; Chen, J.C. A new analytical approach to model and evaluate the performance of a class of irreversible fuel cells. *Int. J. Hydrogen Energy* **2008**, *33*, 4161–4170. [CrossRef]
33. Zhang, X.; Cai, L.; Liao, T.J.; Zhou, Y.H.; Zhao, Y.R.; Chen, J.C. Exploiting the waste heat from an alkaline fuel cell via electrochemical cycles. *Energy* **2018**, *142*, 983–990. [CrossRef]
34. Zhou, S.J.; Wang, P.; Zhang, M.; Chen, S.Z.; Xu, W.; Zhu, L.T.; He, X.Q.; Gong, S.R. Effects of atmospheric acid deposition on root physiological characteristics of *Pinus massoniana* seedlings. *J. Nanjing For. Univ. Nat. Sci. Ed.* **2021**, *44*, 111–118. [CrossRef]
35. Xiong, G.K.; Li, Y.Q.; Xiong, Y.Q.; Duan, A.G.; Cao, D.C.; Sun, J.J.; Nie, L.Y.; Sheng, W.T. Effects of low stand density afforestation on the growth, stem-form and timber assortment structure of *Cunninghamia lanceolata* plantations. *J. Nanjing For. Univ. Nat. Sci. Ed.* **2021**, *45*, 165–173. [CrossRef]
36. Ji, X.L.; Yang, P. The exploration of the slope displacement with vegetation protection under different rainfall intensity. *J. For. Eng.* **2020**, *5*, 152–156.
37. Li, X.S.; Deng, T.T.; Wang, M.H.; Ju, S.; Li, X.C.; Li, M. Linear positioning algorithm improvement of wood acoustic emission source based on wavelet and signal correlation analysis methods. *J. For. Eng.* **2020**, *5*, 138–143.
38. Guo, Y.H.; Guo, X.R.; Zhang, H.C.; Hou, S.J. Energetic, exergetic and ecological analyses of a high-temperature proton exchange membrane fuel cell based on a phosphoric-acid-doped polybenzimidazole membrane. *Sustain. Energy Technol.* **2020**, *38*, 100671. [CrossRef]

Article

# Thermal Scaling of Transient Heat Transfer in a Round Cladded Rod with Modern Dimensional Analysis

Botond-Pál Gálfi [1], Ioan Száva [2], Daniela Șova [2,*] and Sorin Vlase [2,3,*]

1. Autolive Romania, Brașov, Bucegi, Str. 8, 500053 Brașov, Romania; janoska@clicknet.ro
2. Department of Mechanical Engineering, Transilvania University of Brașov, B-dul Eroilor 20, 500036 Brașov, Romania; eet@unitbv.ro
3. Romanian Academy of Technical Sciences, Bulevardul Dacia 26, 030167 Bucharest, Romania
* Correspondence: sova.d@unitbv.ro (D.Ș.); svlase@unitbv.ro (S.V.); Tel.: +40-722-643-020 (S.V.)

**Abstract:** Heat transfer analysis can be studied efficiently with the help of so-called modern dimensional analysis (MDA), which offers a uniform and easy approach, without requiring in-depth knowledge of the phenomenon by only taking into account variables that may have some influence. After a brief presentation of the advantages of this method (MDA), the authors applied it to the study of heat transfer in straight bars of solid circular section, protected but not thermally protected with layers of intumescent paints. Two cases (two sets of independent variables) were considered, which could be easily tracked by experimental measurements. The main advantages of the model law obtained are presented, being characterized by flexibility, accuracy, and simplicity. Additionally, this law and the MDA approach allow us to obtain much more advantageous models from an experimental point of view, with the geometric analogy of the model with the prototype not being a necessary condition. To the best knowledge of the present authors there are no studies reporting the application of the MDA method as it was used in this paper to heat transfer.

**Keywords:** geometric analogy; similarity theory; dimensional analysis; model law; heat transfer; straight bar

## 1. Introduction

### 1.1. General Considerations

The idea of dimensional analysis and its practical application dates from the end of the 18th century. The introduction of fundamental units allowed for the creation of some theoretical bases for the application of dimensional analysis in the verification of the correctness of some obtained formulas.

The method of dimensional analysis was conceived and developed in the last century by mathematicians and engineers in order to facilitate experimental investigations of complex structures, as well as difficult to reproduce phenomena, through the easier study of their small-scale models.

This method involves attaching a model (usually scaled down) to the actual structure, called a prototype. The experimental and theoretical study will be carried out/performed on the model, and the results obtained will be transferred to the prototype based on the rigorous application of the model law, specific to dimensional analysis.

The law of the model consists of a finite and well-determined number of dimensionless variables, established by Buckingham's theorem, which have as a starting point precisely the set of variables that intervene in the description of the respective physical phenomenon.

In the classical version (classical dimensional analysis—CDA), obtaining the model law, involves following one of the following paths:

- by the direct application of Buckingham's theorem, presented in detail in the papers mentioned in the paper;

- by applying the method of partial differential equations on the fundamental differential relations, which describe the phenomenon, when the initial variables are transformed into dimensionless quantities (through a normalization process) and by their appropriate grouping the desired dimensionless groups will result;
- identification of the complete form, but also the simplest of the equation (equations) that describe the phenomenon, which we will transform into dimensionless forms, from which the desired dimensionless groups will be identified.

These ways of obtaining the desired dimensionless groups, which in fact constitute the law of the model, represent quite a difficult and at the same time arbitrary method, which also presuppose the thorough knowledge of the pursued phenomenon.

Compared to these, the method called modern dimensional analysis (MDA) offers a unique and simple way to obtain the model law, requiring only the consideration of all variables that could have an influence on the phenomenon, which is a clear advantage to the MDA. In this case, the complete set of dimensionless groups is obtained, and thus the complete version of the model law.

From this complete variant, based on the exclusion of some physical or dimensional variables irrelevant to the studied phenomenon, will result the model law, which most accurately describes the model–prototype correlation. Thus, based on a unique and simple approach, those correlations will be established, i.e., the model law, which ensures the transfer of the information obtained on the model to the prototype.

In this paper, the authors established that only the law of the model, as shown in paragraph 3.2 (of the variant I studied), can be applied to a concrete case.

A series of papers present the advantages of dimensional analysis [1,2] and the limitations of using this method [3,4]. The basic results in the application of this method have been obtained in recent decades [5–8]. The fundamentals of the method are consistently developed and used in applications [9–13].

From all the fields in which the method of dimensional analysis has been applied, we referred only to its application to heat transfer, which will be the subject of this article.

Some particular cases of heat transfer have been used in the literature. The complexity of a heat transfer problem is significantly reduced using the dimensional analysis method and transforming the problem in a scale-free form. For example, this method is used to study the dimensionless groups in irradiated particle-laden turbulence [14]. For such systems it is concluded that two dimensionless groups are important in the system's thermal response.

An experimental study on the convection heat transfer coefficient and pressure drop values of $CO_2$ led to the use of the dimensional analysis technique to develop correlations between Nusselt numbers and pressure drops [15]. Other example of the dimensional analysis in the case of heat transfer are presented in the literature [16–20].

The complexity and nonlinearity of mechanical or thermal phenomena require a new approach regarding the correlation of experimental results with theoretical data, which requires the development of pertinent mathematical models [21]. The conventional analysis usually involves many trials and diagrams with measurement results.

*1.2. Dimensional Modelling, a Design Tool for Heat Transfer Analysis*

Starting from the geometric analogy, a first more efficient approach is given by the similarity theory [22,23], where alongside the prototype, the model—usually a small-scaled model—is defined. The governing equations applied to the prototype are obtained by means of the model's behavior [24,25]. The model must accurately reflect the behavior of the prototype. The similarity between prototype and model is structural or functional. The structural similarity highlights mainly the geometric similarity between prototype and model, while the functional similarity aims to find corresponding equations that describe both prototype and model. Additionally, geometric similarity supposes proportionality between length and angle equality for the prototype and model. Thus, homologous points, lines, surfaces, and volumes of the prototype and model can be defined. Functional similarity involves similar processes in both systems, prototype and model, that take place at similar times, i.e., the accomplishment of the similarity of all physical properties that govern the analyzed process. This kind of similarity can be kinematic or dynamic, and the phenomena occur so that, in homologous points, at homologous times, each dimension $\eta$ is characterized by a constant ratio between the values corresponding to the model and prototype, $S_\eta$. These dimensionless ratios, which are constant in time and space, are scale factors of the dimensions involved or similarity ratios. The scale factor $S_\eta$ is defined as the ratio between the value of the dimension corresponding to the model ($\eta_2$) and the prototype, respectively ($\eta_1$):

$$S_\eta = \frac{\eta_2}{\eta_1} \ [-], \tag{1}$$

The reverse of $S_\eta$ represents the coefficient of transition from the original to the model [21]. There are as many scale factors as dimensions describing the phenomenon. Practically, the mathematical solution of the complex equations that theoretically describe the actual phenomenon is replaced by correlations between dimensionless parameters, which are obtained from the fundamental relations of the phenomenon by a suitable grouping of dimensions, called similarity parameters, such as $Nu, Re, St, Pr$, etc. Therefore, the dimensions are replaced by the corresponding scale factors, multiplied by constants, and by an appropriate grouping, the similarity parameters are obtained, and correlations among them, such as $Nu = f(Re, Pr, Gr, \ldots)$, are also obtained. By means of experimental measurements, these correlations simplify the analysis performed and allow a reduction in the number of measurements in order to obtain important parameters of the phenomenon.

Among the basic theorems of similarity, two of them can be highlighted:

- for two similar phenomena the homologous dimensionless groups are the same;
- the conditions that are necessary and sufficient for two phenomena to present similarity are:
  - to be of the same nature;
  - to have the same determinant parameters of similarity;
  - to have the same initial and boundary conditions.

In the case of complex phenomena, the number of dimensionless parameter scales of involved variables and correlations increases very much and therefore the similarity theory must be replaced by a more efficient method that is the dimensional analysis [26]. The main aspects concerning the similarity theory and dimensional analysis are indicated in [27–30].

*1.3. Classical Dimensional Analysis (CDA)*

There is in this case a model that will be analyzed instead of the prototype, and as a result of the experiments carried out on the model, by means of dimensionless relations (dimensionless groups $\pi_j$), the behavior of the prototype can be predicted, obviously in conditions of similarity.

By using the $\pi_j$ groups, CDA simplifies very much the experimental investigations and the graphical representations, and the results have a high degree of abstraction and generality. The works [26,29] present in detail the main $\pi_j$ groups that describe thermal energy processes.

CDA is not a substitute for experimental measurements and does not have the purpose of explaining physical phenomena; it aims to simplify and optimize the design of experiments by grouping measurable parameters of a phenomenon in dimensionless groups, defined by Buckingham's $\pi$ theorem. Both model and prototype obey in their behavior the conditions set out in the $\pi_j$ group.

By using CDA, the $\pi_j$ groups can be set in one of the following ways:

- by direct application of Buckingham's $\pi$ theorem;
- by applying the method of partial differential equations to fundamental differential relations that describe the phenomenon; the initial variables are transformed into

dimensionless quantities and then, by their suitably grouping, the $\pi_j$ groups are obtained;
- by identifying the full form, but also the simplest equation(s) that describe the phenomenon, which will be transformed into dimensionless forms from which the desired $\pi_j$ groups will be selected.

According to [24,29] the Buckingham's $\pi$ theorem has the following statement: the required number of independent dimensionless groups formed by combining the variables of a phenomenon is equal to the total number of these quantities minus the number of primary units of measurement that is necessary to express the dimensional relations of the physical quantities.

Consider a process that can be described by a set of independent parameters $y_i$, $i = 1, 2, \ldots, n$ by means of the general relation:

$$f(y_1, y_2, y_3, \ldots, y_n) = 0, \tag{2}$$

For describing the $n$ quantities, $m$ primary units of measurement are required and thus, from Buckingham's theorem, $(n - m)$ independent $\pi_j$ dimensionless groups can be formed that are able to describe the considered process. They are in a similar relation:

$$F(\pi_1, \pi_2, \ldots, \pi_{n-m}) = 0, \tag{3}$$

The set of relations is given by:

$$\pi_j = F_j(\pi_1, \pi_2, \ldots, \pi_{n-m}), \ j = 1, 2, \ldots, (n-m), \tag{4}$$

The functional relationship among the $\pi_j$ groups is obtained from trials.
As mentioned in [21], CDA involves three steps, namely:
1. the selection of parameters and primary units that can most accurately describe the phenomenon;
2. the determination of $\pi_j$ groups by identifying the exponents of the independent variables;
3. the experimental determination of the functional relations among the $\pi_j$ groups.

Thus, the $\pi_j$ groups are defined as products of the representative quantities that are involved in describing the phenomenon having unknown exponents $(a, b, c, \ldots)$. From the condition that all the $\pi_j$ groups are dimensionless (the sum of the exponents of each primary dimension must be zero), a system of equations will be obtained where the unknowns are the exponents. It is a multiple indeterminate system, where convenient values are given from the beginning to the exponents of the primary units, while the rest of the unknown exponents are determined from the solution of the system. Finally, the total number of $\pi_j$ groups will be obtained.

Unfortunately, all approaches of the CDA show several shortcomings. That is why the original method described in [31,32], called modern dimensional analysis (MDA), is according to the authors, the most efficient and easy way to approach dimensional analysis.

*1.4. Objectives and Purpose of the Paper*

This paper represents a theoretical and experimental study on the implementation of modern dimensional analysis (MDA) in solving the problem of heat transfer, especially to the metal structures used in civil and industrial constructions, protected or unprotected with layers of intumescent paints. A fire protection, in addition to maintaining the flexibility of the original structure, leads both to maintaining the initial load-bearing capacity of the resistance structure for a longer time in case of fire and to increase the guaranteed time for evacuation of persons and property subjected to fire. Other recent studies concerning dimensional analysis are presented in [33–41].

In this article, the authors set out to achieve the following major objectives:

- Comparative analysis of methods that use the analysis of the phenomenon on models instead of prototypes, such as geometric analogy, theory of similarity, and classical dimensional analysis;
- Brief presentation of the MDA method and its net advantages in the study of the prototype-model correlation;
- Application of MDA to the study of heat transfer of straight metal bars of full circular section (but with the possibility of extending these results to rings of annular section) protected or unprotected by layers of intumescent paints;
- In this sense, the laws of the model are presented, which govern the heat transfer in these thermally protected or unprotected bars, the application of which leads to a significant simplification of the analysis of this complex and important phenomenon.

The aim of the manuscript is to apply modern dimensional analysis to the heat transfer in a circular bar. The heat transfer in the bar is transitory. The bar is placed in air; therefore, the boundary condition is convection. The heat transfer coefficients were considered among the other variables in applying *MDA*. As indicated in the manuscript, when using *MDA*, the relations of the model law are correlations among variables that are involved in the phenomenon, and they must not be compared with the physical relations that describe the phenomenon. In contrast with the classical dimensional analysis, MDA considers the variables that might influence the phenomenon, without requiring a thorough knowledge of the phenomenon and the governing relations. The relations of the model law can be extended to bars with tubular section and structures of bars with annular cross-section. This is also an advantage in using *MDA*. To the best knowledge of the authors, the heat transfer in a circular bars described by *MDA* has not been reported before in the literature.

## 2. Method of Analysis in Modern Dimensional Analysis (MDA)

In a physical relation there is a single dependent variable and a finite number of independent variables. The variables are denoted by $(H_1, H_2, H_3, \ldots)$, while their dimensions are denoted by $(h_1, h_2, h_3, \ldots)$. The derived dimensions are obtained from the combination of previously selected primary dimensions, such as $h_1^{r_1} \cdot h_2^{r_2} \cdot h_3^{r_3} \cdot \ldots \cdot h_n^{r_n}$ (where, $r_1, r_2, r_3 \ldots$ are the exponents of the primary dimensions, while $n$ is the number of the involved primary dimensions). A variable $H_j$ has the dimension $[H_j] = \varphi_j \cdot h_1^{r_{1j}} \cdot h_2^{r_{2j}} \cdot h_3^{r_{3j}} \cdots$, where $\varphi_j$ is a coefficient.

The author of works [31,32] indicates the following steps for analysis, which were presented in [33]:

- the dimensional matrix (*DM*) is defined; it consists of the exponents of all involved dimensions $h_i$ that describe all independent variables $H_k$ and the dependent one. In the case of four variables, among one is dependent (for instance $H_1$), the dimensional relations are:

$$H_1 = h_1^{\alpha_1} \cdot h_2^{\beta_1} \cdot h_3^{\gamma_1} \cdot h_4^{\delta_1} \; ; H_2 = h_1^{\alpha_2} \cdot h_2^{\beta_2} \cdot h_3^{\gamma_2} \cdot h_4^{\delta_2} \; ; H_3 = h_1^{\alpha_3} \cdot h_2^{\beta_3} \cdot h_3^{\gamma_3} \cdot h_4^{\delta_3} \; ; H_4 = h_1^{\alpha_4} \cdot h_2^{\beta_4} \cdot h_3^{\gamma_4} \cdot h_4^{\delta_4}. \quad (5)$$

The dimensional matrix contains the exponents of these dimensions and is indicated in rel. (6):

$$\begin{array}{c|cccc} & H_1 & H_2 & H_3 & H_4 \\ h_1 & \alpha_1 & \alpha_2 & \alpha_3 & \alpha_4 \\ h_2 & \beta_1 & \beta_2 & \beta_3 & \beta_4 \\ h_3 & \gamma_1 & \gamma_2 & \gamma_3 & \gamma_4 \\ h_4 & \delta_1 & \delta_2 & \delta_3 & \delta_4 \end{array} \quad (6)$$

Matrix *M*, associated with the dimensional matrix, is:

$$M = \begin{bmatrix} \alpha_1 & \alpha_2 & \alpha_3 & \alpha_4 \\ \beta_1 & \beta_2 & \beta_3 & \beta_4 \\ \gamma_1 & \gamma_2 & \gamma_3 & \gamma_4 \\ \delta_1 & \delta_2 & \delta_3 & \delta_4 \end{bmatrix}, \quad (7)$$

In the general case, there are $N_V$ total variables and $N_d$ primary dimensions that define both the dimensional matrix and the associated one, as a matrix consisting of $N_d$ lines and $N_V$ columns.

- it is to find the quadratic submatrix A, starting with the upper right elements of matrix M, which has the highest rank, r and which will also be the rank of the dimensional matrix $R_{DM} = r$. For this purpose, some rows (dimensions that cannot be selected arbitrarily, but will result from the model law) and columns (dependent variables) are eliminated from matrix M, and those independent variables are set that have the exponents of the dimensions included in matrix A. Matrix A must not be singular ($\det|A| \neq 0$), and the rows contain the exponents of the primary dimensions of the remaining independent variables. The model law can comprise one or more correlations among independent and dependent variables, as will later be indicated.
- the remaining rows of matrix M represent the reduced dimensional matrix $M_1$. They contain the primary dimensions (i.e., the dimensions that can be arbitrarily selected). The columns of matrix $M_1$, which are not included in matrix A, represent matrix B.
- the dimensional set is defined; it comprises the reduced dimensional matrix (B + A), matrix $C = -(A^{-1} \cdot B)^T$ and the unit matrix of order n, $D \equiv I_{nxn}$, as indicated by (8) and (9) [31,32,34].

| | | B | A |
|---|---|---|---|
| The rows correspond to the remaining primary dimensions $k = N_d$ after defining matrix A | 1.<br>2.<br>3.<br>4.<br>...<br>k. | | |
| The rows correspond to n columns (dependent variables) that had matrix B; the number of the rows is the same as that of the $\pi_j$ , resulting in dimensionless quantities | 1.<br>2.<br>3.<br>4.<br>...<br>...<br>n. | $D \equiv I_{nxn}$ | $C = -(A^{-1} \cdot B)^T$ |

$$D \equiv I_{nxn}, \quad (8)$$

It should be mentioned that matrix C is obtained from the relation:

$$C = -\left(A^{-1} \cdot B\right)^T, \quad (9)$$

Relation (9) is valid if the set of new variables contains only $\pi_j$ dimensionless quantitates and matrix D is a unit matrix.

- the rows $j = 1, 2, \ldots, n$ of matrixes D and C define all $\pi_j$ dimensionless quantitates. Thus, row j of the common matrix (D and C) contains the exponents that are involved in defining $\pi_j$, which is the product between a dependent variable (from matrix B, having the exponent 1) and all involved independent variables (from matrix A, having the exponents from the row j of matrix C). In order to find the model law, the expressions of all $\pi_j$ dimensionless variables are equal to one. In all products of matrix D there is only one dependent variable with exponent 1, while in those of matrix C there are all independent variables with the exponents obtained from relation (9).

As mentioned before, in the matrices A, B and C the exponents ($h_1, h_2, \ldots, h_m$) of the basic dimensions involved intervene, which helps us to describe the set of variables involved ($H_1, H_2, H_3, \ldots, H_n$), and in matrix D (which is a unit matrix) these unit values will also represent exponents of dependent variables.

The illustration of how to obtain the elements of the model law is given in Figure 1:

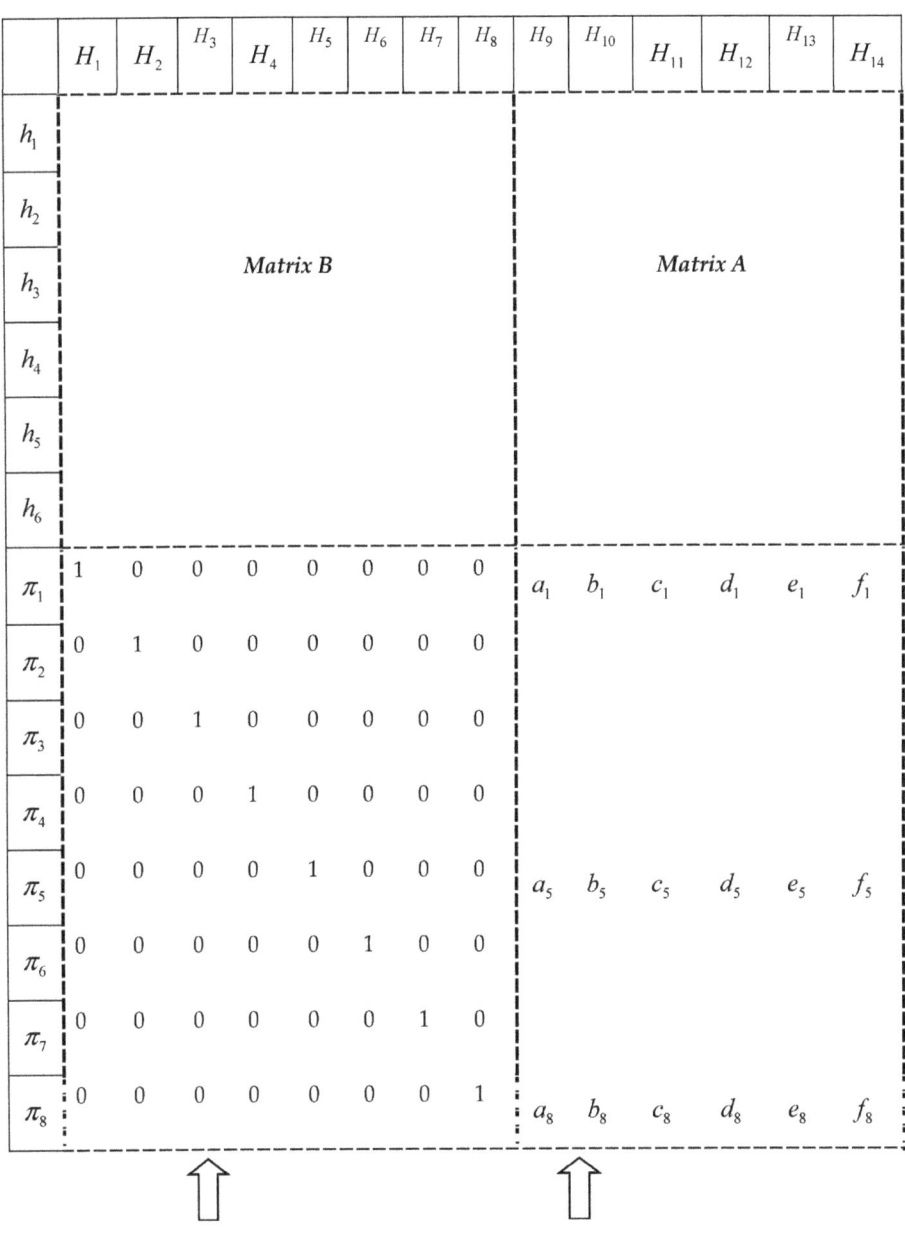

**Figure 1.** The illustration of how to obtain the elements of the model law.

If considering, for example, the dimensionless variable $\pi_5$, on its line there are the exponents of all involved independent variables ($H_9, \ldots, H_{14}$), the exponents of the independent variables ($a_5, \ldots, f_5$), as well as the exponent of the dependent variable

($H_5$), which is 1, being positioned on the main diagonal of matrix $D$. Consequently, $\pi_5$ can be written as:

$$\pi_5 = (H_5)^1 \cdot (H_9)^{a_5} \cdot (H_{10})^{b_5} \cdot (H_{11})^{c_5} \cdot (H_{12})^{d_5} \cdot (H_{13})^{e_5} \cdot (H_{14})^{f_5}, \qquad (10)$$

As shown before, relation (10) is equal to the unit, and from this equality the dependent variable is expressed (here being $H_5$), i.e.,

$$\pi_5 = (H_5)^1 \cdot (H_9)^{a_5} \cdot (H_{10})^{b_5} \cdot (H_{11})^{c_5} \cdot (H_{12})^{d_5} \cdot (H_{13})^{e_5} \cdot (H_{14})^{f_5} = 1 \Rightarrow$$
$$\Rightarrow H_5 = \frac{1}{(H_9)^{a_5} \cdot (H_{10})^{b_5} \cdot (H_{11})^{c_5} \cdot (H_{12})^{d_5} \cdot (H_{13})^{e_5} \cdot (H_{14})^{f_5}} \ . \qquad (11)$$

Then, the involved variables ($H_5$, $H_9$, ... , $H_{14}$) are replaced by the corresponding scale factors ($S_{H_n}$), and finally, the desired expression of the fifth element of the model law is obtained.

Obviously, some of the exponents involved being negative, the relationship obtained will be in the form of an ordinary fraction, where both the numerator and the denominator will have expressions of scale factors at certain powers.

Some observations can be formulated as:

- in this case, the model law will consist of eight elements, since eight dimensional variables resulted from the calculations ($\pi_1$, ... , $\pi_8$);
- at the same time, this law includes the complete set of dimensionless variables $\pi_k$ involved in the description of the analyzed physical phenomenon, and the way to obtain these dimensionless variables is the easiest and safest, which cannot be achieved with the rest of the methods mentioned above;
- for simplification, $\pi_j$ variables can be further grouped.
- Some conclusions can be drawn from the previous *MDA* analysis, namely:
- as compared to *CDA*, the relations of the Model obtained from *MDA* are correlations among variables that are involved in the phenomenon, which actually represent connections between the scale factors of the involved variables. They must not be compared with the physical relations that describe the phenomenon
- if opting for the case in which the set of new variables comprises only $\pi_j$ dimensionless variables and matrix $D$ is quadratic, but not a unit matrix, then matrix $C$ is calculated from relation (10) [31,32]:

$$C = -D \cdot \left(A^{-1} \cdot B\right)^T, \qquad (12)$$

the final expressions of the $\pi_j$ variables do not change;
- the order of introducing the dependent variables in matrix $B$ and independent variables in matrix $A$ and thus, their positioning in the reduced dimensional matrix (B-A) and dimensional set (B-A-D-C), respectively, does not influence the $\pi_j$ relations and model law;
- the new approach proposed by *MDA* has the following advantages [31,32]:
    ○ all parameters that might have an influence upon the phenomenon are considered (total variables of the dimensional set). More information in defining the relevant variables increases the degree of freedom in selecting the properties of the model, and thus a more reliable description of the prototype is possible. Later, based on a careful analysis, the variables that have an insignificant influence can be excluded.
    ○ the $\pi_j$ variables can be easily and unitarily determined, which is impossible if *CDA* or the theory of similarity are used. It means that the dimensional set defined by Equation (8) represents the complete set of $\pi_j$ dimensionless products of variables $H_m$, $m = N_V$;
    ○ the calculations required for the arbitrary grouping and analysis used by the two previously mentioned methods, in order to obtain the $\pi_j$ groups, are

eliminated. They require a thorough knowledge of the phenomenon, thus making *CDA* difficult and inaccessible to many researchers;
- in contrast, *MDA* considers the variables that might influence the phenomenon without requiring a thorough knowledge of the phenomenon and the governing relations;
- in order to determine the model law that consists of the constitutive expressions of the $\pi_j$ variables, each $\pi_j$ variable is equal to one and each variable $\eta$ is replaced by the corresponding scale factor $S_\eta$. From these expressions, the scale factors of the dependent variables are determined as function of the independent ones, thus obtaining the components of the model law.

## 3. Application of *MDA* to the Heat Transfer in a Circular Bar. Case Study

### 3.1. General Approach

A metallic (steel) bar with a circular section is considered, being related to the reference system $xGrt$ (Figure 2).

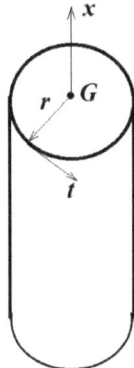

**Figure 2.** Bar with circular section.

Generally, the set of variables that govern the transient heat transfer in a bar with circular section that can be further analyzed in terms of dimensions are indicated in Table 1:

**Table 1.** The set of variables that govern the heat transient transfer in a beam with circular section.

| Name | Variable Symbol/Formula | Dimension |
|---|---|---|
| Heat * | $Q$ | $J = N \cdot m = \frac{kg \cdot m_x}{s^2} \cdot m_x = \frac{kg \cdot m_x^2}{s^2}$ |
| Heat rate | $\dot{Q} = \frac{dQ}{d\tau}$ | $W = \frac{J}{s} = \frac{kg \cdot m_x^2}{s^3}$ |
| Time | $\tau, \Delta\tau$ | $s$ |
| Density of material (steel, air, paint/insulating material) | $\rho$ | $\frac{kg}{m^3} = \frac{kg}{m_x \cdot m_r^2}$ |
| Constant-pressure specific heat of air | $c_p = \frac{1}{m} \cdot \frac{dQ}{dt}$ | $\frac{1}{kg} \cdot \frac{J}{^0C} = \frac{1}{kg} \cdot \frac{kg \cdot m_x^2}{s^2 \cdot ^0C} = \frac{m_x^2}{s^2 \cdot ^0C}$ ; |
| Specific heat capacity (steel, air) | $C = \frac{dQ}{dT}$ | $\frac{J}{^0C} = \frac{kg \cdot m_x^2}{s^2 \cdot ^0C}$ |

Table 1. Cont.

| Name | Variable Symbol/Formula | Dimension |
|---|---|---|
| Thermal conductivity(steel, paint), along directions | $\lambda_x$ (for steel) | $\frac{W}{m_x \cdot {}^0C} = \frac{J}{s} \cdot \frac{1}{m_x \cdot {}^0C} = \frac{1}{s} \cdot \frac{kg \cdot m_x^2}{s^2} \cdot \frac{1}{m_x \cdot {}^0C} = \frac{kg \cdot m_x}{s^3 \cdot {}^0C}$ |
| | $\lambda_r$ (for steel or paint coat) | $\frac{W}{m_r \cdot {}^0C} = \frac{J}{s} \cdot \frac{1}{m_r \cdot {}^0C} = \frac{1}{s} \cdot \frac{kg \cdot m_x^2}{s^2} \cdot \frac{1}{m_r \cdot {}^0C} = \frac{kg \cdot m_x^2}{s^3 \cdot m_r \cdot {}^0C}$ |
| Thermal diffusivity of air, along directions | $a_x = \frac{\lambda_x}{\rho \cdot c_p} = \frac{1}{\rho} \cdot \frac{1}{c_p} \cdot \lambda_x$ | $a_x = \frac{\lambda_x}{\rho \cdot c_p} = \frac{1}{\rho} \cdot \frac{1}{c_p} \cdot \lambda_x \left( \frac{m_x \cdot m_r^2}{kg} \cdot \frac{s^{2 \cdot 0}C}{m_x^2} \cdot \frac{kg \cdot m_x}{s^{3 \cdot 0}C} = \frac{m_r^2}{s} \right)$ |
| | $a_r = \frac{\lambda_r}{\rho \cdot c_p} = \frac{1}{\rho} \cdot \frac{1}{c_p} \cdot \lambda_r$ | $\frac{m_x \cdot m_r^2}{kg} \cdot \frac{s^{2 \cdot 0}C}{m_x^2} \cdot \frac{kg \cdot m_x^2}{s^{3 \cdot 0}C} = \frac{m_x \cdot m_r}{s}$ |
| Dynamic viscosity of air ** | $\eta = \tau_{0x} \cdot \frac{1}{\nabla w_0} = \frac{F_{0x}}{A} \cdot \frac{1}{\nabla w_0}$ | $\frac{kg}{s^2 \cdot m_t} \cdot \frac{1}{1/s} = \frac{kg}{s \cdot m_t}$ |
| | $\eta_r = \frac{\tau_{0r}}{\nabla w_0} = \frac{F_{0r}}{A} \cdot \frac{1}{\nabla w_0}$ | $\frac{kg}{s^2 \cdot m_r} \cdot \frac{1}{1/s} = \frac{kg}{s \cdot m_r}$ |
| Kinematic viscosity of air | $\nu_x = \frac{\eta_x}{\rho} = \frac{1}{\rho} \cdot \eta_x$ | $\frac{m_x \cdot m_r^2}{kg} \cdot \frac{kg}{s \cdot m_t} = \frac{m_x \cdot m_r^2}{s \cdot m_t}$ |
| | $\nu_y = \frac{\eta_r}{\rho} = \frac{1}{\rho} \cdot \eta_r$ | $\frac{m_x \cdot m_r^2}{kg} \cdot \frac{kg}{s \cdot m_r} = \frac{m_x \cdot m_r}{s}$ |
| Prandtl number of air, along directions | $\mathrm{Pr}_x = \frac{\nu_x}{a_x} = \nu_x \cdot \frac{1}{a_x}$ | $\frac{m_x \cdot m_r^2}{s \cdot m_t} \cdot \frac{s}{m_r^2} = \frac{m_x}{m_t}$ |
| | $\mathrm{Pr}_r = \frac{\nu_r}{a_r}$ *** | $\frac{m_x \cdot m_r}{s} \cdot \frac{s}{m_x \cdot m_r} = 1 = m_x^0 \cdot m_r^0 \cdot s^0$ |
| Convection heat transfer coefficient along directions | $\alpha_{nx}$ | $\frac{W}{m^{2 \cdot 0}C} = \frac{J}{s} \cdot \frac{1}{m^{2 \cdot 0}C} = \frac{kg \cdot m_x^2}{s^3} \cdot \frac{1}{m_r^{2 \cdot 0}C} = \frac{kg \cdot m_x^2}{s^3 \cdot m_r^{2 \cdot 0}C}$ |
| | $\alpha_{nr}$ (when the beam is protected (insulated) by a paint coat, then: $\alpha_{nf} = \alpha_{nr}$) | $\frac{W}{m^{2 \cdot 0}C} = \frac{J}{s} \cdot \frac{1}{m^{2 \cdot 0}C} = \frac{kg \cdot m_x^2}{s^3} \cdot \frac{1}{m_x \cdot m_t \cdot {}^0C} = \frac{kg \cdot m_x}{s^3 \cdot m_t \cdot {}^0C}$ |
| Thickness of the paint coat along the radial direction | $d_r = \delta_r$ | $m_r$ |
| Beam volume | $V$ | $m^3 = m_x \cdot m_r^2$ |
| Area of the beam cross section | $A_{tr}$ | $m_r^2$ |
| Lateral area | $A_{lat}$ | $m_x \cdot m_t$ |
| Beam dimensions | $L_x, L_r, L_t$ | $m_x, m_r, m_t$ |
| Shape factor of the cross-section | $\varsigma = \frac{A_{lat}}{V} = \frac{P}{A_{tr}}$; $P$ is the cross-section perimeter | $\frac{m_t}{m_r^2}$ |
| Gravitational acceleration | $g$ | $\frac{m}{s^2} = \frac{m_x}{s^2}$ |
| Temperature variation | $\Delta T(K)$ or $\Delta t\,(^\circ C)$ | $\Delta T(K)$ or $\Delta t\,(^\circ C)$ |
| Coefficient of volume expansion of steel or of fluid/air | $\beta$ | $\frac{1}{{}^0C}$ |
| Nusselt number, along directions | $Nu_x = \frac{\alpha_x \cdot l_x}{\lambda_{f,x}} = \alpha_x \cdot l_x \cdot \frac{1}{\lambda_{f,x}}$; $l_x$ ($m_x$)-characteristic length | $\frac{kg \cdot m_x^2}{s^3 \cdot m_r^{2 \cdot 0}C} \cdot m_x \cdot \frac{s^{3 \cdot 0}C}{kg \cdot m_x} = \frac{m_x^2}{m_r^2}$ |
| | $Nu_r = \frac{\alpha_r \cdot l_r}{\lambda_{f,r}} = \alpha_r \cdot l_r \cdot \frac{1}{\lambda_{f,r}}$; $l_r$ ($m_r$)-characteristic length | $\frac{kg \cdot m_x}{s^3 \cdot m_t \cdot {}^0C} \cdot m_r \cdot \frac{s^3 \cdot m_r \cdot {}^0C}{kg \cdot m_x^2} = \frac{m_r^2}{m_t \cdot m_x}$ |
| Reynolds number, along directions | $\mathrm{Re}_x = \frac{w_{0,x} \cdot l_x}{\nu_x} = w_{0,x} \cdot l_x \cdot \frac{1}{\nu_x}$; $w_0 \left(\frac{m}{s}\right)$ is the fluid velocity | $\frac{m_x}{s} \cdot m_x \cdot \frac{s \cdot m_t}{m_x \cdot m_r^2} = \frac{m_x \cdot m_t}{m_r^2}$ |

Table 1. Cont.

| Name | Variable | |
|---|---|---|
| | Symbol/Formula | Dimension |
| Péclet number, along directions | $Re_r = \frac{w_{0,r} \cdot l_r}{\nu_r} = w_{0,r} \cdot l_r \cdot \frac{1}{\nu_r}$ | $\frac{m_r}{s} \cdot m_r \cdot \frac{s}{m_x \cdot m_r} = \frac{m_r}{m_x}$ |
| | $Pe_x = Re_x \cdot Pr_x$ | $\frac{m_x \cdot m_t}{m_r^2} \cdot \frac{m_x}{m_t} = \frac{m_x^2}{m_r^2}$ |
| | $Pe_r = \frac{m_r}{m_x} Pe_r = Re_r \cdot Pr_r$ | $\frac{m_r}{m_x} \cdot 1 = \frac{m_r}{m_x}$ |
| Grashof number | $Gr_x = \frac{g \cdot \beta \cdot \Delta t \cdot l_x^3}{\nu_\zeta^2} = g \cdot \beta \cdot \Delta t \cdot l_x^3 \cdot \frac{1}{\nu_x^2}$ | $\frac{m_x}{s^2} \cdot \frac{1}{{}^0 C} \cdot {}^0 C \cdot m_x^3 \cdot \frac{s^2 \cdot m_t^2}{m_x^2 \cdot m_r^4} = \frac{m_x^2 \cdot m_t^2}{m_r^4}$ |
| Stanton number, along directions | $St_x = \frac{Nu_x}{Pe_x} = Nu_x \cdot \frac{1}{Pe_x}$ *** | $\frac{m_z^2}{m_z \cdot m_y} \cdot \frac{m_z \cdot m_y}{m_x^2} = 1$ |
| | $St_r = \frac{Nu_r}{Pe_r} = Nu_r \cdot \frac{1}{Pe_r}$ | $\frac{m_x^2}{m_x \cdot m_t} \cdot \frac{m_x}{m_r} = \frac{m_r}{m_t}$ |
| Fourier number, along directions | $Fo_x = \frac{a_x \cdot \tau}{l_x^2} = \frac{a_x \cdot \Delta \tau}{l_x^2} = a_x \cdot \Delta \tau \cdot \frac{1}{l_x^2}$ | $\frac{m_x^2}{s} \cdot s \cdot \frac{1}{m_x^2} = \frac{m_x^2}{m_x^2}$ |
| | $Fo_r = \frac{a_r \cdot \tau}{l_r^2} = \frac{a_r \cdot \Delta \tau}{l_r^2} = a_r \cdot \Delta \tau \cdot \frac{1}{l_r^2}$ | $\frac{m_x \cdot m_x}{s} \cdot s \cdot \frac{1}{m_r^2} = \frac{m_x}{m_r}$ |
| Biot number, along directions | $Bi_x = \frac{\alpha_x \cdot l_x}{\lambda_{s,x}} = \alpha_x \cdot l_x \cdot \frac{1}{\lambda_{s,x}}$ | $\frac{kg \cdot m_x^2}{s^3 \cdot m_r^2 \cdot {}^0 C} \cdot m_x \cdot \frac{s^3 \cdot {}^0 C}{kg \cdot m_x} = \frac{m_x^2}{m_r^2}$ |
| | $Bi_r = \frac{\alpha_r \cdot l_r}{\lambda_{s,r}} = \alpha_r \cdot l_r \cdot \frac{1}{\lambda_{s,r}}$ | $\frac{kg \cdot m_x}{s^3 \cdot m_t \cdot {}^0 C} \cdot m_r \cdot \frac{s^3 \cdot m_r \cdot {}^0 C}{kg \cdot m_x^2} = \frac{m_r^2}{m_t \cdot m_x}$ |

\* Heat is numerically equal to the dimension of work; the work is conventionally considered a product between a force having the direction along the bar, $F_x$ ($N_x = \frac{kg \cdot m_x}{s^2}$) and the displacement along the same direction $x$ ($m_x$). \*\* where the shear stress $\tau_0$ has one of the directions, $x$ or $r$, of the system $xGrt$, the applied force is $F_0$, while the surface $A$ where it occurs is in a plane that contains the direction of the shear stress; the velocity $w_0$ is normal to the plane where the shear stress is developed; $\nabla w_0$ represents its gradient. \*\*\* *this is not suitable for dimensional analysis* (Therefore, it cannot be used in the dimensional analysis).

Having the dimensions of the variables involved in the transient heat transfer, the MDA was applied as described by Szirtes in [31,32]. Additionally, for acquiring the simplest relations of the model law, according to [31,32], the dimensions were duplicated (in this case, the lengths were duplicated). This will contribute to the reduction in the number of $\pi_j$, $j = 1, \ldots, n$ dimensionless variables, once the dimensions of the variables involved increase. Thus, the reduced number of expressions of the Model Law will be obtained.

According to the principles mentioned in [31,32], the following two sets of independent variables were selected:

- for the first version (I): $[(Q, L_t, \Delta t, \tau, \lambda_{x\ steel}, \zeta]$;
- for the second version (II): $[\dot{Q}, L_t, \Delta t, \tau, \lambda_{x\ steel}, \zeta]$,

which are directly connected with the measurements that were performed and whose magnitude can be controlled during experiments carried out on the model.

These sets are included in matrix $A$; the other quantities, representing dependent variables, form matrix $B$.

It should be noted that the variables contained in matrix A are freely chosen, both for the prototype and for the model. The advantage of choosing these two sets of independent variables lies, inter alia, in the following:

- heating regimes can be chosen independently for prototype and model by:
  ○ accepting convenient and well-determined values for the amount of heat introduced into the system ($Q$ or $\dot{Q}$);
  ○ setting final temperatures compared to initial ones ($\Delta t$),
  ○ defining/accepting individual heating times ($\tau$) of the prototype and the model;
- length scales can also be chosen independently (expressed here as $L_t$, which can be extended to the rest of the dimensions, but it is not mandatory, because the rest of the dimensions are also included in matrix $B$, which represents a significant reserve for generalizing the model to the prototype);

- the factors $\varsigma$ (shape factor) of the cross sections can be chosen independently in the prototype and for the model, respectively;
- one can define the materials of the prototype and the model by $\lambda_x$, which do not necessarily have to be for both steel, which is also very important for the most favorable experiments (costs, manufacturing time, test times etc).

In the following, the obtained results for these two variants are analyzed.

### 3.2. First Case Study

Version I is based on the above-described protocol of the *MDA* and the following quantities were successively obtained:

- the dependent variables that define the heat transfer in the beam that is not coated with intumescent paint, based on experimental research: $\dot{Q}$, $A_{tr}$, $A_{lat}$, $r_{cyl}$, $L_x$, $L_r$
- the dependent variables that are useful for theoretical analyses:

$c_{p\ air}$, $C_{air}$, $C_{steel}$, $a_{x\ air}$, $a_{r\ air}$, $\rho_{air}$, $\rho_{steel}$, $\lambda_{r\ steel}$, $v_{x\ air}$, $v_{r\ air}$, $\alpha_{nx\ steel}$, $\alpha_{nr\ steel}$, $\eta_{x\ air}$, $\eta_{r\ air}$, $\beta_{air/steel}$

- the dependent variables that are useful for setting convection heat transfer correlations between dimensionless numbers (similarity criteria) $Crit01$, $Crit02$, $Crit03$, $Pr_x$, $Gr_{x\ air}$, $Fo_{x\ air}$, $Fo_{r\ air}$, $Re_{r\ air}$, $St_{r\ air}$ where the mentioned dimensionless numbers are:

$$Crit\ 01 = Re_r = Pe_r = \frac{m_x}{m_r};\ Crit\ 02 = Nu_x = Pe_x = Bi_x = \frac{m_x^2}{m_r^2};$$
$$Crit\ 03 = Nu_r = Bi_r = \frac{m_r^2}{m_x \cdot m_t},$$

- the properties of the paint layer: $\rho_{paint}$, $\lambda_{x\ paint}$, $\lambda_{r\ paint}$, $\alpha_{nr\ paint}$, $\delta_{r\ paint}$

The components of the reduced dimensional matrix $(B + A)$ are indicated in Tables 2–6, where, as mentioned before, these elements represent exactly the exponents of the dimensions involved in defining those variables.

**Table 2.** Matrix A, comprising independent variables.

| Dimensions | $Q$ | $L_t$ | $\Delta t$ | $\tau$ | $\lambda_{x\ steel}$ | $\zeta = P/A$ |
|---|---|---|---|---|---|---|
| $m_x$ | 2 | 0 | 0 | 0 | 1 | 0 |
| $m_r$ | 0 | 0 | 0 | 0 | 0 | −2 |
| $m_t$ | 0 | 1 | 0 | 0 | 0 | 1 |
| kg | 1 | 0 | 0 | 0 | 1 | 0 |
| s | −2 | 0 | 0 | 1 | −3 | 0 |
| °C | 0 | 0 | 1 | 0 | −1 | 0 |

**Table 3.** The quantities required by experiments (part of matrix B).

| Dimensions | $\dot{Q}$ | $A_{tr}$ | $A_{lat}$ | $r_{cyl}$ | $L_x$ | $L_r$ |
|---|---|---|---|---|---|---|
| $m_x$ | 2 | 0 | 1 | 0 | 1 | 0 |
| $m_r$ | 0 | 2 | 0 | 1 | 0 | 1 |
| $m_t$ | 0 | 0 | 1 | 0 | 0 | 0 |
| kg | 1 | 0 | 0 | 0 | 0 | 0 |
| s | −3 | 0 | 0 | 0 | 0 | 0 |
| °C | 0 | 0 | 0 | 0 | 0 | 0 |

Table 4. The quantities required by the theoretic analysis (part of matrix B).

| Dimensions | $c_{p\ air}$ | $C_{air}$ | $C_{steel}$ | $a_{x\ air}$ | $a_{r\ air}$ | $\rho_{air}$ | $\rho_{steel}$ | $\lambda_{r\ steel}$ | $v_{x\ air}$ | $v_{r\ air}$ | $\alpha_{nx\ steel}$ | $\alpha_{nr\ steel}$ | $\eta_{x\ air}$ | $\eta_{r\ air}$ | $\beta_{air/steel}$ |
|---|---|---|---|---|---|---|---|---|---|---|---|---|---|---|---|
| $m_x$ | 2 | 2 | 2 | 0 | 1 | −1 | −1 | 2 | 1 | 1 | 2 | 1 | 0 | 0 | 0 |
| $m_r$ | 0 | 0 | 0 | 2 | 1 | −2 | −2 | −1 | 2 | 1 | −2 | 0 | 0 | −1 | 0 |
| $m_t$ | 0 | 0 | 0 | 0 | 0 | 0 | 0 | 0 | −1 | 0 | 0 | −1 | −1 | 0 | 0 |
| kg | 0 | 1 | 1 | 0 | 0 | 1 | 1 | 1 | 0 | 0 | 1 | 1 | 1 | 1 | 0 |
| s | −2 | −2 | −2 | −1 | −1 | 0 | 0 | −3 | −1 | −1 | −3 | −3 | −1 | −1 | 0 |
| °C | −1 | −1 | −1 | 0 | 0 | 0 | 0 | −1 | 0 | 0 | −1 | −1 | 0 | 0 | −1 |

Table 5. The quantities required by the heat transfer correlations between dimensionless numbers (part of matrix B).

| Dimensions | Crit 01 | Crit 02 | Crit 03 | $Pr_{x\ air}$ | $Gr_{x\ air}$ | $Fo_{x\ air}$ | $Fo_{r\ air}$ | $Re_{x\ air}$ | $St_{r\ air}$ |
|---|---|---|---|---|---|---|---|---|---|
| $m_x$ | 1 | 2 | −1 | 0 | −1 | −2 | 1 | −1 | 0 |
| $m_r$ | −1 | −2 | 2 | 0 | 1 | 2 | 0 | 0 | 1 |
| $m_t$ | 0 | 0 | −1 | 0 | 0 | 0 | 0 | 1 | −1 |
| kg | 0 | 0 | 0 | 0 | 0 | 0 | 0 | 0 | 0 |
| s | 0 | 0 | 0 | 0 | 0 | 0 | −2 | 0 | 0 |
| °C | 0 | 0 | 0 | −1 | 0 | 0 | 0 | 0 | 0 |

Table 6. The properties of the intumescent paint (part of matrix B).

| | $\rho_{paint}$ | $\lambda_{x\ paint}$ | $\lambda_{r\ paint}$ | $\alpha_{nr\ paint}$ | $\delta_{r\ paint}$ |
|---|---|---|---|---|---|
| $m_x$ | −1 | 1 | 2 | 1 | 0 |
| $m_r$ | −2 | 0 | −1 | 0 | 1 |
| $m_t$ | 0 | 0 | 0 | −1 | 0 |
| kg | 1 | 1 | 1 | 1 | 0 |
| s | 0 | −3 | −3 | −3 | 0 |
| °C | 0 | −1 | −1 | −1 | 0 |

By performing the above-mentioned calculations, the elements of the Dimensional Set were finally obtained, from where all dimensionless $\pi_j$ expressions were extracted as corresponding lines of the Dimensional Set. In the following, this step-by-step procedure is presented just for the first expression of the model law (related to the dimensionless variable) and for the rest, only the final expressions of the model law are indicated. Thus, the following were obtained:

(a) From experiments on uncoated structures (prototype and model) the following expressions of the Model Law were obtained (that is, the final expressions in which the corresponding scale factors $S_\eta$ of the dependent variables were defined in function of the scale factors of the independent variables):

$$\pi_1 = \dot{Q} \cdot Q^{-1} \cdot L_t^0 \cdot \Delta t^0 \cdot \tau^1 \cdot \lambda_{x\ steel}^0 \cdot \varsigma^0 = \frac{\dot{Q} \cdot \tau}{Q} = 1 \Rightarrow \frac{S_{\dot{Q}} \cdot S_\tau}{S_Q} = 1 \Rightarrow S_{\dot{Q}} = \frac{S_Q}{S_\tau}, \quad (13)$$

$$\pi_2 : S_{A_{tr}} = \frac{S_{L_t}}{S_\varsigma} \quad (14)$$

$$\pi_3 : S_{A_{lat}} = \frac{S_Q \cdot S_{L_t}}{S_{\Delta t} \cdot S_\tau \cdot S_{\lambda_{x\ steel}}} \quad (15)$$

$$\pi_4 : S_{r_{cyl}} = \sqrt{\frac{S_{L_t}}{S_\varsigma}} \quad (16)$$

$$\pi_5 : S_{L_x} = \frac{S_Q}{S_{\Delta t} \cdot S_\tau \cdot S_{\lambda_x\, steel}} \tag{17}$$

$$\pi_6 : S_{L_r} = \sqrt{\frac{S_{L_t}}{S_\varsigma}} \tag{18}$$

(b) From experiments on coated structures (prototype and model) the set of previous expressions is completed with expressions specific to the coating paint, which are $(\pi_{31} \ldots \pi_{35})$. The following set of expressions of the Model Law is obtained $(\pi_1 \ldots \pi_6)$ and $(\pi_{31} \ldots \pi_{35})$.

$$\pi_1\, S_{\dot Q} = \frac{S_Q}{S_\tau}, \tag{19}$$

$$\pi_2 : S_{A_{tr}} = \frac{S_{L_t}}{S_\varsigma}, \tag{20}$$

$$\pi_3 : S_{A_{lat}} = \frac{S_Q \cdot S_{L_t}}{S_{\Delta t} \cdot S_\tau \cdot S_{\lambda_x\, steel}}, \tag{21}$$

$$\pi_4 : S_{r_{cyl}} = \sqrt{\frac{S_{L_t}}{S_\varsigma}}, \tag{22}$$

$$\pi_5 : S_{L_x} = \frac{S_Q}{S_{\Delta t} \cdot S_\tau \cdot S_{\lambda_x\, steel}}, \tag{23}$$

$$\pi_6 : S_{L_r} = \sqrt{\frac{S_{L_t}}{S_\varsigma}}, \tag{24}$$

$$\pi_{31} : S_{\rho_{paint}} = \frac{(S_{\Delta t})^3 \cdot (S_\tau)^5 \cdot (S_{\lambda_x\, steel})^3 \cdot S_\varsigma}{(S_Q)^2 \cdot S_{L_t}}, \tag{25}$$

$$\pi_{32} : S_{\lambda_x\, paint} = S_{\lambda_x\, steel}, \tag{26}$$

$$\pi_{33} : S_{\lambda_r\, paint} = \frac{S_Q}{S_{\Delta t} \cdot S_\tau} \cdot \sqrt{\frac{S_\varsigma}{S_{L_t}}}, \tag{27}$$

$$\pi_{34} : S_{\alpha_{nr}\, paint} = \frac{S_{\lambda_x\, steel}}{S_{L_t}}, \tag{28}$$

$$\pi_{35} : S_{\delta_r\, paint} = \sqrt{\frac{S_{L_t}}{S_\varsigma}}. \tag{29}$$

(c) For theoretical investigations of parameters dependence ($c_{p\ air}$, $C_{air}$, $C_{steel}$, $a_{x\ air}$, $a_{r\ air}$, $\rho_{air}$, $\rho_{steel}$, $\lambda_{r\ steel}$, $v_{x\ air}$, $v_{r\ air}$, $\alpha_{nx\ steel}$, $\alpha_{nr\ steel}$, $\eta_{x\ air}$, $\eta_{r\ air}$, $\beta_{air/steel}$) on the set of independent variables (of prototype and model), the following set of expressions will be used $(\pi_7 \ldots \pi_{21})$:

$$\pi_7 : S_{c_p\ air} = \frac{(S_Q)^2}{(S_{\Delta t})^3 \cdot (S_\tau)^4 \cdot (S_{\lambda_x\, steel})^2}, \tag{30}$$

$$\pi_8 : S_{C_{air}} = \frac{S_Q}{S_{\Delta t}}, \tag{31}$$

$$\pi_9 : S_{C_{steel}} = \frac{S_Q}{S_{\Delta t}}, \tag{32}$$

$$\pi_{10} : S_{a_{x\ air}} = \frac{S_{L_t}}{S_\tau \cdot S_\varsigma}, \tag{33}$$

$$\pi_{11}: S_{a_{r\ air}} = \frac{S_Q}{S_{\Delta t} \cdot (S_\tau)^2 \cdot S_{\lambda_{x\ steel}}}, \tag{34}$$

$$\pi_{12}: S_{\rho_{air}} = \frac{(S_{\Delta t})^3 \cdot (S_\tau)^5 \cdot (S_{\lambda_{x\ steel}})^3 \cdot S_\varsigma}{(S_Q)^2 \cdot S_{L_t}}, \tag{35}$$

$$\pi_{13}: S_{\rho_{steel}} = \frac{(S_{\Delta t})^3 \cdot (S_\tau)^5 \cdot (S_{\lambda_{x\ steel}})^3 \cdot S_\varsigma}{(S_Q)^2 \cdot S_{L_t}}, \tag{36}$$

$$\pi_{14}: S_{\lambda_{r\ steel}} = \frac{S_Q}{S_{\Delta t} \cdot S_\tau} \cdot \sqrt{\frac{S_\varsigma}{S_{L_t}}}, \tag{37}$$

$$\pi_{15}: S_{v_{x\ air}} = \frac{S_Q}{S_{\Delta t} \cdot (S_\tau)^2 \cdot S_{\lambda_{x\ steel}} \cdot S_\varsigma}, \tag{38}$$

$$\pi_{16}: S_{v_{r\ air}} = \frac{S_Q}{S_{\Delta t} \cdot (S_\tau)^2 \cdot S_{\lambda_{x\ steel}}}, \tag{39}$$

$$\pi_{17}: S_{\alpha_{nx\ steel}} = \frac{S_Q \cdot S_\varsigma}{S_{L_t} \cdot S_{\Delta t} \cdot S_\tau}, \tag{40}$$

$$\pi_{18}: S_{\alpha_{nr\ steel}} = \frac{S_{\lambda_{x\ steel}}}{S_{L_t}}, \tag{41}$$

$$\pi_{19}: S_{\eta_{x\ air}} = \frac{(S_{\Delta t})^2 \cdot (S_\tau)^3 \cdot (S_{\lambda_{x\ steel}})^2}{S_Q \cdot S_{L_t}}, \tag{42}$$

$$\pi_{20}: S_{\eta_{r\ air}} = \frac{(S_{\Delta t})^2 \cdot (S_\tau)^3 \cdot (S_{\lambda_{x\ steel}})^2}{S_Q} \cdot \sqrt{\frac{S_\varsigma}{S_{L_t}}}, \tag{43}$$

$$\pi_{21}: S_{\beta_{air/steel}} = \frac{1}{S_{\Delta t}}. \tag{44}$$

(d) For investigations of the dependence of the parameters on the set of independent variables and for setting of heat transfer correlations between dimensionless numbers based on the expressions of the model law (by combining them favorably), the next set of expressions ($\pi_{22}$ ... $\pi_{30}$) will be used:

$$\pi_{22}: S_{Crit\ 01} = \frac{S_{\Delta t} \cdot S_\tau \cdot S_{\lambda_{x\ steel}}}{S_Q} \cdot \sqrt{\frac{S_{L_t}}{S_\varsigma}}, \tag{45}$$

$$\pi_{23}: S_{Crit\ 02} = \frac{(S_Q)^2 \cdot S_\varsigma}{S_{L_t}(S_{\Delta t})^2 \cdot (S_\tau)^2 \cdot (S_{\lambda_{x\ steel}})^2}, \tag{46}$$

$$\pi_{24}: S_{Crit\ 03} = \frac{S_{\Delta t} \cdot S_\tau \cdot S_{\lambda_{x\ steel}}}{S_Q \cdot S_\varsigma}, \tag{47}$$

$$\pi_{25}: S_{Pr_{x\ air}} = \frac{S_Q}{S_{L_t} \cdot S_{\Delta t} \cdot S_\tau \cdot S_{\lambda_{x\ steel}}}, \tag{48}$$

$$\pi_{26}: S_{Gr_{x\ air}} = \frac{(S_Q)^2 \cdot (S_\varsigma)^2}{(s_{\Delta t})^2 \cdot (S_\tau)^2 \cdot (s_{\lambda_{x\ steel}})^2}, \tag{49}$$

$$\pi_{27}: S_{Fo_{x\ air}} = \frac{S_{L_t} \cdot (S_{\Delta t})^2 \cdot (S_\tau)^2 \cdot (S_{\lambda_{x\ steel}})^2}{(S_Q)^2 \cdot S_\varsigma}, \tag{50}$$

$$\pi_{28}: S_{Fo_{r\ air}} = \frac{S_Q}{S_{\Delta t} \cdot S_\tau \cdot S_{\lambda_{x\ steel}}} \cdot \sqrt{\frac{S_\varsigma}{S_{L_t}}}, \tag{51}$$

$$\pi_{29}: S_{Re_{x\ air}} = \frac{S_Q \cdot S_\varsigma}{S_{\Delta t} \cdot S_\tau \cdot S_{\lambda_{x\ steel}}}, \tag{52}$$

$$\pi_{30}: S_{St_{r\ air}} = \frac{1}{\sqrt{S_{L_t} \cdot S_\varsigma}}. \tag{53}$$

In order to show how the elements of the model law can be applied for correlating the prototype with the model, the following variables were selected:

- heat rate $\dot{Q}_1$;
- model length $L_{x\ 2}$;
- thickness of the paint layer used for the model $\delta_{r\ 2\ paint}$.

These variables are governed by relations (1), (17), and (35) of the model law.

As can be observed, $\dot{Q}_1$ is a quantity that refers to the prototype and cannot be measured, since experiments were carried out only on the model, while $L_{x\ 2}$ and $\delta_{r\ 2\ paint}$ are corresponding to the model and they can be determined only for the prototype; for the model they are obtained strictly from the elements of the model law.

Considering the set of independent variables, having the dimensions determined for both prototype and model, the scale factors ($S_Q$, $S_{L_t}$, $S_{\Delta t}$, $S_\tau$, $S_{\lambda_{steel}}$, $S_\varsigma$) are considered to be known, as well.

In order to obtain $\dot{Q}_1$, relation (1) is used, where the scale factor $S_{\dot{Q}}$ is the ratio between $\dot{Q}_2$ and $\dot{Q}_1$. Thus, the following is obtained:

$$\pi_1\ S_{\dot{Q}} = \frac{S_Q}{S_\tau} \Leftrightarrow \frac{\dot{Q}_2}{\dot{Q}_1} = \frac{S_Q}{S_\tau} \Rightarrow \dot{Q}_1 = \frac{S_\tau}{S_Q}\dot{Q}_2 \tag{54}$$

The model length $L_{x\ 2}$ is obtained from relation (17), as:

$$\pi_5: S_{L_x} = \frac{S_Q}{S_{\Delta t} \cdot S_\tau \cdot S_{\lambda_{x\ steel}}} \Leftrightarrow \frac{L_{x2}}{L_{x1}} = \frac{S_Q}{S_{\Delta t} \cdot S_\tau \cdot S_{\lambda_{x\ steel}}} \Rightarrow L_{x2} = \frac{S_Q}{S_{\Delta t} \cdot S_\tau \cdot S_{\lambda_{x\ steel}}} L_{x1} \tag{55}$$

The thickness of the paint layer that covers the model $\delta_{r\ 2\ paint}$ is acquired from relation (29):

$$\pi_{35}: S_{\delta_{r\ paint}} = \sqrt{\frac{S_{L_t}}{S_\varsigma}} \Leftrightarrow \frac{\delta_{r\ 2\ paint}}{\delta_{r\ 1\ paint}} = \sqrt{\frac{S_{L_t}}{S_\varsigma}} \Rightarrow \delta_{r\ 2\ paint} = \delta_{r\ 1\ paint} \cdot \sqrt{\frac{S_{L_t}}{S_\varsigma}}. \tag{56}$$

Considering the previous relations, some observations can be made:

(a) The dependent variable $\dot{Q}_1$, which has to be determined for the prototype, cannot be excluded from the dimensional set or the model law.
(b) The other dependent variables of the model (here $L_{x\ 2}$ and $\delta_{r\ 2\ paint}$) can be analyzed without so many restrictions, considering the set of independent variables, namely:

- if the scale factor is the same for all lengths, then $S_{L_t} = S_{L_x}$, and consequently the relation of the fifth element of the model law, $\pi_5$ can be neglected.
- if the thickness of the paint is the same for the prototype and model, then the relation of $\pi_{35}$ to the model law can be omitted.
- if it is aimed to conceive a more flexible model, then the model law allows us to consider different scales of the lengths along directions $(x, r, t)$ or different thicknesses of the paint layer, but strictly considering the elements of the model law.

As can be noticed, this is another major advantage of MDA, which cannot be obtained if the aforementioned methods are used.

### 3.3. Second Case Study

For the second significant version, II, where $Q$ was substituted by $\dot{Q}$, the following significant elements of the dimensional set were obtained, according to Tables 7–11:

**Table 7.** Matrix A, comprising independent variables.

| Dimensions | $\dot{Q}$ | $L_t$ | $\Delta t$ | $\tau$ | $\lambda_{x\ steel}$ | $\zeta = P/A$ |
|---|---|---|---|---|---|---|
| $m_x$ | 2 | 0 | 0 | 0 | 1 | 0 |
| $m_r$ | 0 | 0 | 0 | 0 | 0 | −2 |
| $m_t$ | 0 | 1 | 0 | 0 | 0 | 1 |
| kg | 1 | 0 | 0 | 0 | 1 | 0 |
| s | −3 | 0 | 0 | 1 | −3 | 0 |
| °C | 0 | 0 | 1 | 0 | −1 | 0 |

**Table 8.** The quantities required by experiments (part of matrix B).

| Dimensions | $Q$ | $A_{tr}$ | $A_{lat}$ | $r_{cyl}$ | $L_x$ | $L_r$ |
|---|---|---|---|---|---|---|
| $m_x$ | 2 | 0 | 1 | 0 | 1 | 0 |
| $m_r$ | 0 | 2 | 0 | 1 | 0 | 1 |
| $m_t$ | 0 | 0 | 1 | 0 | 0 | 0 |
| kg | 1 | 0 | 0 | 0 | 0 | 0 |
| s | −2 | 0 | 0 | 0 | 0 | 0 |
| °C | 0 | 0 | 0 | 0 | 0 | 0 |

**Table 9.** The quantities required by the theoretical analysis (part of matrix B).

| Dimensions | $c_{p\ air}$ | $C_{air}$ | $C_{steel}$ | $a_{x\ air}$ | $a_{r\ air}$ | $\rho_{air}$ | $\rho_{steel}$ | $\lambda_{r\ steel}$ | $v_{x\ air}$ | $v_{r\ air}$ | $\alpha_{nx\ steel}$ | $\alpha_{nr\ steel}$ | $\eta_{x\ air}$ | $\eta_{r\ air}$ | $\beta_{air/steel}$ |
|---|---|---|---|---|---|---|---|---|---|---|---|---|---|---|---|
| $m_x$ | 2 | 2 | 2 | 0 | 1 | −1 | −1 | 2 | 1 | 1 | 2 | 1 | 0 | 0 | 0 |
| $m_r$ | 0 | 0 | 0 | 2 | 1 | −2 | −2 | −1 | 2 | 1 | −2 | 0 | 0 | −1 | 0 |
| $m_t$ | 0 | 0 | 0 | 0 | 0 | 0 | 0 | 0 | −1 | 0 | 0 | −1 | −1 | 0 | 0 |
| kg | 0 | 1 | 1 | 0 | 0 | 1 | 1 | 1 | 0 | 0 | 1 | 1 | 1 | 1 | 0 |
| s | −2 | −2 | −2 | −1 | −1 | 0 | 0 | −3 | −1 | −1 | −3 | −3 | −1 | −1 | 0 |
| °C | −1 | −1 | −1 | 0 | 0 | 0 | 0 | −1 | 0 | 0 | −1 | −1 | 0 | 0 | −1 |

**Table 10.** The quantities required by the heat transfer correlations between dimensionless numbers (part of matrix B).

| Dimensions | $Crit\,01$ | $Crit\,02$ | $Crit\,03$ | $\mathbf{Pr}_{x\,air}$ | $\mathbf{Gr}_{x\,air}$ | $\mathbf{Fo}_{x\,air}$ | $\mathbf{Fo}_{r\,air}$ | $\mathbf{Re}_{x\,air}$ | $\mathbf{St}_{r\,air}$ |
|---|---|---|---|---|---|---|---|---|---|
| $m_x$ | 1 | 2 | −1 | 0 | −1 | −2 | 1 | −1 | 0 |
| $m_r$ | −1 | −2 | 2 | 0 | 1 | 2 | 0 | 0 | 1 |
| $m_t$ | 0 | 0 | −1 | 0 | 0 | 0 | 0 | 1 | −1 |
| kg | 0 | 0 | 0 | 0 | 0 | 0 | 0 | 0 | 0 |
| s | 0 | 0 | 0 | 0 | 0 | 0 | −2 | 0 | 0 |
| °C | 0 | 0 | 0 | −1 | 0 | 0 | 0 | 0 | 0 |

**Table 11.** The properties of the intumescent paint (part of matrix B).

|  | $\rho_{paint}$ | $\lambda_{x\,paint}$ | $\lambda_{r\,paint}$ | $\alpha_{nr\,paint}$ | $\delta_{r\,paint}$ |
|---|---|---|---|---|---|
| $m_x$ | −1 | 1 | 2 | 1 | 0 |
| $m_r$ | −2 | 0 | −1 | 0 | 1 |
| $m_t$ | 0 | 0 | 0 | −1 | 0 |
| kg | 1 | 1 | 1 | 1 | 0 |
| s | 0 | −3 | −3 | −3 | 0 |
| °C | 0 | −1 | −1 | −1 | 0 |

The corresponding elements of the model law are:

$$\pi_1 : S_Q = S_{\dot{Q}} \cdot S_\tau, \tag{57}$$

$$\pi_2 : S_{A_{tr}} = \frac{S_{L_t}}{S_\varsigma}, \tag{58}$$

$$\pi_3 : S_{A_{lat}} = \frac{S_{\dot{Q}} \cdot S_{L_t}}{S_{\Delta t} \cdot S_{\lambda_x\,steel}}, \tag{59}$$

$$\pi_4 : S_{r_{cyl}} = \sqrt{\frac{S_{L_t}}{S_\varsigma}}, \tag{60}$$

$$\pi_5 : S_{L_x} = \frac{S_{\dot{Q}}}{S_{\Delta t} \cdot S_{\lambda_x\,steel}}, \tag{61}$$

$$\pi_6 : S_{L_r} = \sqrt{\frac{S_{L_t}}{S_\varsigma}}, \tag{62}$$

$$\pi_7 : S_{c_{p\,air}} = \frac{\left(S_{\dot{Q}}\right)^2}{\left(S_{\Delta t}\right)^3 \cdot \left(S_\tau\right)^2 \cdot \left(S_{\lambda_x\,steel}\right)^2}, \tag{63}$$

$$\pi_8 : S_{C_{air}} = \frac{S_{\dot{Q}} \cdot S_\tau}{S_{\Delta t}}, \tag{64}$$

$$\pi_9 : S_{C_{steel}} = \frac{S_{\dot{Q}} \cdot S_\tau}{S_{\Delta t}}, \tag{65}$$

$$\pi_{10} : S_{a_{x\,air}} = \frac{S_{L_t}}{S_\tau \cdot S_\varsigma}, \tag{66}$$

$$\pi_{11} : S_{a_{r\,air}} = \frac{S_{\dot{Q}}}{S_{\Delta t} \cdot S_\tau \cdot S_{\lambda_x\,steel}} \cdot \sqrt{\frac{S_{L_t}}{S_\varsigma}}, \tag{67}$$

$$\pi_{12}: S_{\rho_{air}} = \frac{(S_{\Delta t})^3 \cdot (S_\tau)^3 \cdot (S_{\lambda_{x\ steel}})^3 \cdot S_\varsigma}{(S_{\dot{Q}})^2 \cdot S_{L_t}}, \qquad (68)$$

$$\pi_{13}: S_{\rho_{steel}} = \frac{(S_{\Delta t})^3 \cdot (S_\tau)^3 \cdot (S_{\lambda_{x\ steel}})^3 \cdot S_\varsigma}{(S_{\dot{Q}})^2 \cdot S_{L_t}}, \qquad (69)$$

$$\pi_{14}: S_{\lambda_{r\ steel}} = \frac{S_{\dot{Q}}}{S_{\Delta t}} \cdot \sqrt{\frac{S_\varsigma}{S_{L_t}}}, \qquad (70)$$

$$\pi_{15}: S_{v_{x\ air}} = \frac{S_{\dot{Q}}}{S_{\Delta t} \cdot S_\tau \cdot S_{\lambda_{x\ steel}} \cdot S_\varsigma}, \qquad (71)$$

$$\pi_{16}: S_{v_{r\ air}} = \frac{S_{\dot{Q}}}{S_{\Delta t} \cdot S_\tau \cdot S_{\lambda_{x\ steel}}} \cdot \sqrt{\frac{S_{L_t}}{S_\varsigma}}, \qquad (72)$$

$$\pi_{17}: S_{\alpha_{nx\ steel}} = \frac{S_{\dot{Q}} \cdot S_\varsigma}{S_{L_t} \cdot S_{\Delta t}}, \qquad (73)$$

$$\pi_{18}: S_{\alpha_{nr\ steel}} = \frac{S_{\lambda_{x\ steel}}}{S_{L_t}}, \qquad (74)$$

$$\pi_{19}: S_{\eta_{x\ air}} = \frac{(S_{\Delta t})^2 \cdot (S_\tau)^2 \cdot (S_{\lambda_{x\ steel}})^2}{S_{\dot{Q}} \cdot S_{L_t}}, \qquad (75)$$

$$\pi_{20}: S_{\eta_{r\ air}} = \frac{(S_{\Delta t})^2 \cdot (S_\tau)^2 \cdot (S_{\lambda_{x\ steel}})^2}{S_{\dot{Q}}} \cdot \sqrt{\frac{S_\varsigma}{S_{L_t}}}, \qquad (76)$$

$$\pi_{21}: S_{\beta_{air/steel}} = \frac{1}{S_{\Delta t}}. \qquad (77)$$

The mentioned dimensionless numbers have the same expressions:

$$Crit\ 01 = Re_r = Pe_r = \frac{m_x}{m_r}, \qquad (78)$$

$$Crit\ 02 = Nu_x = Pe_x = Bi_x = \frac{m_x^2}{m_r^2}, \qquad (79)$$

$$Crit\ 03 = Nu_r = Bi_r = \frac{m_r^2}{m_x \cdot m_t}, \qquad (80)$$

The elements of the model law are:

$$\pi_{22}: S_{Crit\ 01} = \frac{S_{\Delta t} \cdot S_{\lambda_{x\ steel}}}{S_{\dot{Q}}} \cdot \sqrt{\frac{S_{L_t}}{S_\varsigma}}, \qquad (81)$$

$$\pi_{23}: S_{Crit\ 02} = \frac{(S_{\dot{Q}})^2 \cdot S_\varsigma}{S_{L_t} \cdot (S_{\Delta t})^2 \cdot (S_{\lambda_{x\ steel}})^2}, \qquad (82)$$

$$\pi_{24}: S_{Crit\ 03} = \frac{S_{\Delta t} \cdot S_{\lambda_{x\ steel}}}{S_{\dot{Q}} \cdot S_\varsigma}, \qquad (83)$$

$$\pi_{25}: S_{Pr_{x\ air}} = \frac{S_{\dot{Q}}}{S_{L_t} \cdot S_{\Delta t} \cdot S_{\lambda_{x\ steel}}}, \qquad (84)$$

$$\pi_{26}: S_{Gr_x\ air} = \frac{\left(S_{\dot{Q}}\right)^2 \cdot (S_\varsigma)^2}{(S_{\Delta t})^2 \cdot \left(S_{\lambda_x\ steel}\right)^2}, \tag{85}$$

$$\pi_{27}: S_{Fo_x\ air} = \frac{S_{L_t} \cdot (S_{\Delta t})^2 \cdot \left(S_{\lambda_x\ steel}\right)^2}{\left(S_{\dot{Q}}\right)^2 \cdot S_\varsigma}, \tag{86}$$

$$\pi_{28}: S_{Fo_r\ air} = \frac{S_{\dot{Q}}}{S_{\Delta t} \cdot S_{\lambda_x\ steel}} \cdot \sqrt{\frac{S_\varsigma}{S_{L_t}}}, \tag{87}$$

$$\pi_{29}: S_{Re_x\ air} = \frac{S_{\dot{Q}} \cdot S_\varsigma}{S_{\Delta t} \cdot S_{\lambda_x\ steel}}, \tag{88}$$

$$\pi_{30}: S_{St_r\ air} = \frac{1}{\sqrt{S_{L_t} \cdot S_\varsigma}}, \tag{89}$$

The elements of the model law are:

$$\pi_{31}: S_{\rho_{paint}} = \frac{(S_{\Delta t})^3 \cdot (S_\tau)^3 \cdot \left(S_{\lambda_x\ steel}\right)^3 \cdot S_\varsigma}{\left(S_{\dot{Q}}\right)^2 \cdot S_{L_t}}, \tag{90}$$

$$\pi_{32}: S_{\lambda_x\ paint} = S_{\lambda_x\ steel}, \tag{91}$$

$$\pi_{33}: S_{\lambda_r\ paint} = \frac{S_{\dot{Q}}}{S_{\Delta t}} \cdot \sqrt{\frac{S_\varsigma}{S_{L_t}}}, \tag{92}$$

$$\pi_{34}: S_{\alpha_{nr}\ paint} = \frac{S_{\lambda_x\ steel}}{S_{L_t}}, \tag{93}$$

$$\pi_{35}: S_{\delta_r\ paint} = \sqrt{\frac{S_{L_t}}{S_\varsigma}}. \tag{94}$$

## 4. Discussion and Conclusions

The relations deduced in the paper for the case of the straight bar of the full circular section can be applied without problems to the tubular (ring) bars, both to the resistance structures formed/constituted by them, as well as the reticular structures used in the roofs of industrial halls, gyms, etc.

In these cases, of the structures made of straight bar elements, on the prototype and on the model, the homologous points (and sections) will be identified, with the help of which the thermal stresses on the model will be transferred to the prototype using of the model law.

It is clear that the internationally recognized work and achievements of Sedov [23], as well as other notable scientists [1–5,8,13,22,25–28,30], are not disputed in any way by the authors of this paper. However, a number of difficulties need to be highlighted in addressing the issue of dimensional analysis by them and other illustrious authors compared to the methodology developed by Szirtes, the author of the works [31,32] namely:

- the direct analysis of the differential relations that describe the phenomenon, in order to establish the dimensionless groups, does not always allow the unitary establishment of the complete set of these dimensionless groups;
- also, the classical methodology (CDA) is usually cumbersome and non-unitary, allowing different researchers to obtain different sets of dimensionless variables;
- in order to obtain these dimensionless groups, the authors of different works use, based on the application of Buckingham's theorem, either the normalization of the terms of the differential relations related to the phenomenon describing the phenomenon, or a rather arbitrary and unambiguous combination of variables involved in describing

the phenomenon of the main measure (dimensions), which takes place in each author according to his own logic, so it is a non-unitary approach to the phenomenon. Thus, based on these approaches, different sets of dimensionless variables may result, which may even represent combinations of those deduced by other authors [36,38–47].

- the classical methodology, i.e., CDA, presupposes from the very beginning a deep knowledge of the phenomenon and of the differential relations that govern the phenomenon, which for an ordinary researcher represents an impediment;
- the classical methodology, including those presented in the papers [22,23,25,27–30], does not explicitly allow highlighting from the very beginning of the set of independent variables or dependent variables, but applies a hard-to-follow (and often unexplained) logic of how these two sets were chosen;
- the involvement from the very beginning, in approaching with the help of the dimensional analysis of the phenomenon, of some very complicated differential relations whose analysis will eventually lead to the establishment of these dimensionless groups, discourages the vast majority of researchers/engineers from using a safe, unified, and simple way to approach the problem, as will happen with MDA;

On the contrary, the methodology, called MDA, developed by Szirtes [31,32], represents a unified approach, easy and particularly accessible to any engineer, without requiring deep/grounded knowledge of the phenomenon, but only reviewing all parameters/variables that could have any influence on it.

Here, they are defined, in a unitary and unambiguous way, on the basis of a clear and particularly accessible protocol/procedure:

- the set of main dimensions;
- the main variables (i.e., the independent ones), i.e., those that can be chosen a priori for both the prototype and the model;
- the dependent variables, i.e., those that can be chosen a priori only for the prototype, and for the model will result exclusively only through the rigorous application of the model law;
- the variables sought for the prototype, which cannot be obtained by direct measurements of the prototype, but only on the basis of the results of experimental investigations performed on the model and by the rigorous application of the model law;
- the complete set of dimensionless variables, without the existence of ambiguous variants, is unitary;
- here the independent variables of the dependent ones are clearly delimited from the very beginning, based on rigorous mathematical criteria, as well as on some practical criteria regarding the quantities that deserve and that can be determined/controlled by experimental measurements.

In the works [36–40,42,46,47] the classical approach is applied to determining the exponents, which will define the dimensionless groups. Thus, they are used either for the normalization of the known differential relations or the evaluation of the main dimensions and later the establishment of some combinations of the variables in order to obtain dimensionless groups.

In the paper [41], the dimensionless groups are arbitrarily defined, based on a combination, according to their own logic.

The only paper in which approaches closer to MDA were found is paper [35], where the determination of exponents was based on the methodology presented in [43], but does not specify how to choose independent or dependent variables, which is a deficiency of the methodology presented in [43] by Langhaar. In contrast, in Szirtes's work, i.e., in [31,32], each time, these independent variables are rigorously chosen, taking into account how an experiment of the model can be conducted more easily, allowing the model to be designed as favorably as possible for the experiments.

The author of the paper [44] uses the choice of independent and dependent variables but applies the standard methodology for determining exponents by solving the system of linear equations, which describes the phenomenon.

The main advantage of MDA in setting the content of these groups of variables is that the elimination of some variables from this whole set does not influence the ones that remain. In other words, the expressions of a certain set will not be influenced if some of the dependent variables are considered or not.

Accordingly, if the whole set of the variables specific to the beam coated with intumescent paint was conceived, representing 35 expressions that define the model law, a certain number of dependent variables can be neglected without affecting the rest of the expressions.

In the above-described protocols, the general cases are indicated, from which several particular cases can be obtained.

Moreover, if for the prototype and model, a certain variable has identical values, then they can be ignored due to the fact that their scale factor became $S_\eta = 1$ and consequently one will resolve useful particular cases similarly with the following:

- if both prototype and model are made of the same material (here: steel), then one has $S_{\alpha_{nx}\,steel} = S_{\alpha_{nr}\,steel} = S_{\lambda_x\,steel} = S_{\lambda_r\,steel} = S_{\rho_{steel}} = S_{C_{steel}} = S_{\beta_{steel}} = 1$;
- if environmental conditions for experiments are the same (the experiments are performed in the same environments) then: $S_{c_p\,air} = S_{C_{air}} = S_{a_x\,air} = S_{a_r\,air} = S_{\rho_{air}} = S_{\eta_x\,air} = S_{\eta_r\,air} = S_{v_x\,air} = S_{v_r\,air} = S_{\beta_{air}} = 1$;
- if the coating materials are identical for both prototype and model, then $S_{\rho_{paint}} = S_{\lambda_x\,paint} = S_{\lambda_r\,paint} = S_{\alpha_{nr}\,paint} = 1$, i.e., the expression corresponding to the dimensionnless variables $\pi_{31}, \ldots, \pi_{35}$ are eliminated, maintaining only the last one, $\pi_{35}$;
- if the same scales for lengths are adopted, other simplifications of the expressions of the model law will be obtained

It is also important to mention that, using the MDA, the model can be differently conceived from the prototype (another material, another coat of paint, etc.), which reveals once again the incontestable advantages of the method proposed in [30,31] as compared to the classical dimensional analysis;

Another conclusion is that for tubular sections, where the thickness of the tube is $\delta_r$, the expression of the model law corresponding to length $L_r$, which is identical to $r_{cyl}$, can be applied to the thickness of the tube too. Therefore, the model law is valid also for tubular sections if the same scale is adopted as for $L_r$ and $r_{cyl}$.

To the best knowledge of the present authors there are no studies reporting the application of the MDA method to the heat transfer in circular bars.

**Author Contributions:** Conceptualization, B.-P.G., I.S. and D.Ș.; methodology, B.-P.G., I.S. and D.Ș.; software, B.-P.G.; validation, B.-P.G., I.S., D.Ș. and S.V.; formal analysis, B.-P.G., I.S. and D.Ș.; investigation, B.-P.G., I.S. and D.Ș.; resources, B.-P.G., I.S., D.Ș. and S.V.; data curation, B.-P.G., I.S. and D.Ș.; writing—original draft preparation, I.S.; writing—review and editing, B.-P.G., I.S., D.Ș. and S.V.; visualization, B.-P.G., I.S., D.Ș. and S.V.; supervision, B.-P.G., I.S., D.Ș. and S.V.; project administration, I.S.; funding acquisition, B.-P.G., I.S., D.Ș. and S.V. All authors have read and agreed to the published version of the manuscript.

**Funding:** This research received no external funding. The APC was funded by the Transilvania University of Brasov.

**Institutional Review Board Statement:** Not applicable.

**Informed Consent Statement:** Not applicable.

**Data Availability Statement:** Not applicable.

**Conflicts of Interest:** The authors declare no conflict of interest.

## References

1. Schnittger, J.R. Dimensional Analysis in Design. *J. Vib. Accoustic Stress Reliab. Des. Trans. ASME* **1988**, *110*, 401–407. [CrossRef]
2. Carinena, J.F.; Santander, M. Dimensional Analysis. *Adv. Electron. Electron Phys.* **1988**, *72*, 181–258.
3. Canagaratna, S.G. Is dimensional analysis the best we have to offer. *J. Chem. Educ.* **1993**, *70*, 40–43. [CrossRef]
4. Bhaskar, R.; Nigam, A. Qualitative Physics using Dimensional Analysis. *Artif. Intell.* **1990**, *45*, 73–111. [CrossRef]
5. Romberg, G. Contribution to Dimensional Analysis. *Ingineiur. Arch.* **1985**, *55*, 401–412. [CrossRef]
6. Coyle, R.G.; Ballicolay, B. Concepts and Software for Dimensional Analysis in Modeling. *IEEE Trans. Syst. Man Cybern.* **1984**, *14*, 478–487. [CrossRef]
7. Barr, D.I.H. Consolidation of Basics of Dimensional Analysis. *J. Eng. Mech. ASCE* **1984**, *110*, 1357–1376. [CrossRef]
8. Remillard, W.J. Applying Dimensional Analysis. *Am. J. Phys.* **1983**, *51*, 137–140. [CrossRef]
9. Martins, R.D.A. The Origin of Dimensional Analysis. *J. Frankl. Inst.* **1981**, *311*, 331–337. [CrossRef]
10. Gibbings, J.C. A Logic of Dimensional Analysis. *J. Physiscs A Math. Gen.* **1982**, *15*, 1991–2002. [CrossRef]
11. Szekeres, P. Mathematical Foundations of Dimensional Analysis and the Question of Fundamental Units. *Int. J. Theor. Phys.* **1978**, *17*, 957–974. [CrossRef]
12. Carlson, D.E. Some New Results in Dimensional Analysis. *Arch. Ration. Mech. Anal.* **1978**, *68*, 191–210. [CrossRef]
13. Gibbings, J.C. Dimensional Analysis. *J. Phys. A Math. Gen.* **1980**, *13*, 75–89. [CrossRef]
14. Jofre, L.; del Rosario, Z.R.; Iaccarino, G. Data-driven dimensional analysis of heat transfer in irradiated particle-laden turbulent flow. *Int. J. Multiph. Flow* **2020**, *125*, 103198. [CrossRef]
15. Alshqirate, A.A.Z.S.; Tarawneh, M.; Hammad, M. Dimensional Analysis and Empirical Correlations for Heat Transfer and Pressure Drop in Condensation and Evaporation Processes of Flow Inside Micropipes: Case Study with Carbon Dioxide ($CO_2$). *J. Braz. Soc. Mech. Sci. Eng.* **2012**, *34*, 89–96.
16. Levac, M.L.J.; Soliman, H.M.; Ormiston, S.J. Three-dimensional analysis of fluid flow and heat transfer in single- and two-layered micro-channel heat sinks. *Heat Mass Transf.* **2011**, *47*, 1375–1383. [CrossRef]
17. Nakla, M. On fluid-to-fluid modeling of film boiling heat transfer using dimensional analysis. *Int. J. Multiph. Flow* **2011**, *37*, 229–234. [CrossRef]
18. Illan, F.; Viedma, A. Experimental study on pressure drop and heat transfer in pipelines for brine based ice slurry Part II: Dimensional analysis and rheological Model. *Int. J. Refrig. Rev. Int. Froid* **2009**, *32*, 1024–1031. [CrossRef]
19. Nezhad, A.H.; Shamsoddini, R. Numerical Three-Dimensional Analysis of the Mechanism of Flow and Heat Transfer in a Vortex Tube. *Therm. Sci.* **2009**, *13*, 183–196. [CrossRef]
20. Asgari, O.; Saidi, M. Three-dimensional analysis of fluid flow and heat transfer in the microchannel heat sink using additive-correction multigrid technique. In Proceedings of the Micro/Nanoscale Heat Transfer International Conference, PTS A and B. 1st ASME Micro/Nanoscale Heat Transfer International Conference, Tainan, Taiwan, 6–9 January 2008; pp. 679–689.
21. Carabogdan, G.I. *Methods of Analysis of Thermal Energy Processes and Systems*; Tehn: Bucharest, Romania, 1989.
22. Baker, W.E.; Westine, P.S.; Dodge, F.T. *Similarity Methods in Engineering Dynamics*; Elsevier: Amsterdam, The Netherlands, 1991.
23. Sedov, I.L. *Similarity and Dimensional Methods in Mechanics*; MIR Publisher: Moscow, Russia, 1982.
24. Șova, M.; Șova, D. *Thermotechnics, Vol.II*; Transilvania University Press: Brasov, Romania, 2001.
25. Zierep, J. *Similarity Laws and Modelling*; Marcel Dekker: New York, NY, USA, 1971.
26. Chen, W.K. Algebraic Theory of Dimensional Analysis. *J. Frankl. Inst.* **1971**, *292*, 403. [CrossRef]
27. Barenblatt, G.I. *Dimensional Analysis*; Gordon and Breach: New York, NY, USA, 1987.
28. Bridgeman, P.W. *Dimensional Analysis*; Reissued in Paperbound in 1963; Yale University Press: New Haven, CT, USA, 1922.
29. Buckingham, E. On Physically Similar Systems. *Phys. Rev.* **1914**, *4*, 345. [CrossRef]
30. Quintier, G.J. *Fundamentals of Fire Phenomena*; John Willey & Sons: Hoboken, NJ, USA, 2006.
31. Szirtes, T.H. The Fine Art of Modelling. *SPAR J. Eng. Technol.* **1992**, *1*, 37.
32. Szirtes, T.H. *Applied Dimensional Analysis and Modelling*; McGraw-Hill: Toronto, ON, Canada, 1998.
33. Trif, I.; Asztalos, Z.; Kiss, I.; Élesztős, P.; Száva, I.; Popa, G. Implementation of the Modern Dimensional Analysis in Engineering Problems; Basic Theoretical Layouts. *Ann. Fac. Eng. Hunedoara* **2019**, *17*, 73–76.
34. Száva, I.; Szirtes, T.H.; Dani, P. An Application of Dimensional Model Theory in The Determination of the Deformation of a Structure. *Eng. Mech.* **2006**, *13*, 31–39.
35. Allamsettya, S.; Mohapatro, S. Prediction of NO and $NO_2$ Concentrations in Ozone Injected Diesel Exhaust after NTP Treatment Using Dimensional Analysis. In Proceedings of the 10th International Conference on Applied Energy (ICAE2018), Hong Kong, China, 22–25 August 2018; pp. 4579–4585.
36. Phate, M.R.; Toney, S.B. Modeling and prediction of WEDM performance parameters for Al/SiCp MMC using dimensional analysis and artificial neural network. *Eng. Sci. Technol. Int. J.* **2019**, *22*, 468–476. [CrossRef]
37. Zhang, X.; Taira, H.; Liu, H. Error of Darcy's law for serpentine flow fields: Dimensional analysis. *J. Power Sources* **2019**, *412*, 391–397. [CrossRef]
38. He, Q.; Suorineni, F.T.; Ma, T.; Oh, J. Parametric study and dimensional analysis on prescribed hydraulic fractures in cave mining. *Tunn. Undergr. Space Technol.* **2018**, *78*, 47–63. [CrossRef]
39. Ashikhmin, V.N.; Kugaevskii, S.S. Dimensional Analysis in the Machining of Housing Components with Cast Holes. *Russ. Eng. Res.* **2013**, *33*, 509–513. [CrossRef]

40. Almeida, R.S.M.; Al-Qureshib, H.A.; Tushteva, K.; Rezwan, K. On the dimensional analysis for the creep rate prediction of ceramic fibers. *Ceram. Int.* **2018**, *44*, 15924–15928. [CrossRef]
41. Yao, S.; Yan, K.; Lu, S.; Xu, P. Prediction and application of energy absorption characteristics of thinwalled circular tubes based on dimensional analysis. *Thin-Walled Struct.* **2018**, *130*, 505–519. [CrossRef]
42. Ferro, V. Assessing flow resistaance law in vegetated channels by dimenisonal aalysis and self-similarity. *Flow Meas. Instrum.* **2019**, *69*, 101610. [CrossRef]
43. Langhaar, H.L. *Dimensional Analysis and Theory of Models*; John Wiley & Sons Ltd.: New York, NY, USA, 1951.
44. Kivade, S.B.; Murthy, C.S.N.; Vardhan, H. The use of Dimensional Analysis and Optimisation of Pneumatic Drilling Operations and Operating Parameters. *J. Inst. Eng. India Ser. D* **2012**, *93*, 31–36. [CrossRef]
45. Pankhurst, R.C. *Dimensional Analysis and Scale Factor*; Chapman & Hall Ltd.: London, UK, 1964.
46. Khan, M.A.; Shah, I.A.; Rizvi, Z.; Ahmad, J. A numerical study on the validation of thermal formulations towards the behaviours of RC beams. *Sci. Mater. Today Proc.* **2019**, *17*, 227–234. [CrossRef]
47. Yen, P.H.; Wang, J.C. Power generation and electrical charge desnity with temperature effect of alumina nanofluids using dimensional analysis. *Energy Convers. Manag.* **2019**, *186*, 546–555. [CrossRef]

Article

# Alternative Artificial Neural Network Structures for Turbulent Flow Velocity Field Prediction

Koldo Portal-Porras [1], Unai Fernandez-Gamiz [1,*], Ainara Ugarte-Anero [1], Ekaitz Zulueta [2] and Asier Zulueta [1]

[1] Nuclear Engineering and Fluid Mechanics Department, University of the Basque Country, UPV/EHU, Nieves Cano 12, Vitoria-Gasteiz, 01006 Araba, Spain; koldo.portal@ehu.eus (K.P.-P.); augarte060@ikasle.ehu.eus (A.U.-A.); azulueta@arrasate.es (A.Z.)

[2] System Engineering and Automation Control Department, University of the Basque Country, UPV/EHU, Nieves Cano 12, Vitoria-Gasteiz, 01006 Araba, Spain; ekaitz.zulueta@ehu.eus

* Correspondence: unai.fernandez@ehu.eus

**Citation:** Portal-Porras, K.; Fernandez-Gamiz, U.; Ugarte-Anero, A.; Zulueta, E.; Zulueta, A. Alternative Artificial Neural Network Structures for Turbulent Flow Velocity Field Prediction. *Mathematics* **2021**, *9*, 1939. https://doi.org/10.3390/math9161939

Academic Editors: Maria Luminița Scutaru and Efstratios Tzirtzilakis

Received: 15 June 2021
Accepted: 12 August 2021
Published: 14 August 2021

**Publisher's Note:** MDPI stays neutral with regard to jurisdictional claims in published maps and institutional affiliations.

**Copyright:** © 2021 by the authors. Licensee MDPI, Basel, Switzerland. This article is an open access article distributed under the terms and conditions of the Creative Commons Attribution (CC BY) license (https://creativecommons.org/licenses/by/4.0/).

**Abstract:** Turbulence in fluids has been a popular research topic for many years due to its influence on a wide range of applications. Computational Fluid Dynamics (CFD) tools are able to provide plenty of information about this phenomenon, but their computational cost often makes the use of these tools unfeasible. For that reason, in recent years, turbulence modelling using Artificial Neural Networks (ANNs) is becoming increasingly popular. These networks typically calculate directly the desired magnitude, having input information about the computational domain. In this paper, a Convolutional Neural Network (CNN) for predicting different magnitudes of turbulent flows around different geometries by approximating the equations of the Reynolds-Averaged Navier-Stokes (RANS)-based realizable $k$-$\varepsilon$ two-layer turbulence model is proposed. Using that CNN, alternative network structures are proposed to predict the velocity fields of a turbulent flow around different geometries on a rectangular channel, with a preliminary stage to predict pressure and vorticity fields before calculating the velocity fields, and the obtained results are compared with the ones obtained with the basic structure. The results demonstrate that the proposed structures clearly outperform the basic one, especially when the flow becomes uncertain. In addition, considering the results, the best network configuration is proposed. That network is tested with a domain with multiple geometries and a domain with a narrowing of the channel, which are domains with different conditions from the training ones, showing fairly accurate predictions.

**Keywords:** Deep Learning (DL); Computational Fluid Dynamics (CFD); Artificial Neural Network (ANN); Convolutional Neural Network (CNN); turbulent flow

## 1. Introduction

For many years, turbulence in fluids has been a popular research topic due to its impact on a wide variety of applications. Several experimental studies of turbulent flows have improved the understanding of turbulent behaviour and have been used to design more efficient systems. Even though these experiments have been very valuable, there are cases where experimentation is either too expensive or impractical. In these cases, CFD provides a more detailed insight into the physics of turbulent flows. Although CFD has the potential to predict accurately the behaviour of flows, decreasing the need for conducting experiments, it has two main disadvantages. The first disadvantage is the high computational cost of the simulations, which can be prohibitive in cases with very complex geometries or in cases where very accurate turbulence models, such as LES (Large Eddy Simulation) or DNS (Direct Numerical Simulation), are required. The second disadvantage is the influence of the user, especially in the generation of the mesh and the selection of the closure model. These problems, coupled with the growth of artificial intelligence, have led to an increasing number of studies using Deep Learning (DL) techniques applied to CFD, either as a complement to the simulations or to perform the simulations directly.

Numerous authors have used DL techniques to complement numerical simulations performed using CFD. Ray and Hesthaven [1] designed an ANN to detect cells where there is a discontinuity in the results. Liu et al. [2] established a method based on Deep Metric Learning (DML) to determine the optimal time-step value in non-stationary simulations. Bao et al. [3] applied a physically driven approach to improve the modelling and simulation capability of a coarse mesh, and Hanna et al. [4] designed a DL algorithm to predict and decrease the error of the results obtained on a coarse mesh. By using coarser meshes, substantial reductions in computational cost were obtained in both studies.

Several authors oriented DL methods to the analysed geometry. For example, Yan et al. [5], Zhang et al. [6], and Tao and Sun [7] improved the performance and efficiency of various geometries using DL techniques, and reached aerodynamic optimization.

However, the goal of the vast majority of studies using DL techniques applied to CFD is to obtain fluid characteristics. Guo et al. [8] applied a CNN to achieve slightly inaccurate but very fast predictions of stationary flow fields around solid objects. Ling et al. [9] used a Deep Neural Network (DNN) to model Reynolds stress tensors with Reynolds-Averaged Navier-Stokes (RANS) turbulence modelling, achieving a remarkable improvement of the results obtained in CFD simulations. Lee and You [10] predicted the shedding of non-stationary laminar vortices on a circular cylinder using a Generative Adversarial Network (GAN), focusing on explaining the learning potential of the solution of the Navier-Stokes equations. Liu et al. [11] and Deng et al. [12] designed impact and vortex detection methods, respectively, using CNN-based techniques.

Ribeiro et al. [13] and Kashefi et al. [14], using CNN architectures, achieved very accurate results for velocity and pressure fields of stationary fluids around simple shaped obstacles, with a computational cost three to five orders of magnitude lower than CFD simulations. In addition, in the study conducted by Kashefi et al. [14], different velocity and pressure fields were obtained with slight modifications of the geometry, which is essential for design optimization.

Among the previously mentioned studies, the study of Guo et al. [8] is the only one in which three-dimensional domains are analysed, the rest of them only analysing two-dimensional domains. Nowruzi et al. [15] analysed the behaviour of two airfoils in 2D and 3D using CFD and an ANN, and showed good agreements between the results obtained by both methods. Compared to two-dimensional systems, the main disadvantage of analysing three-dimensional systems using DL is the limited workspace [14]. For this reason, Mohan et al. [16] developed a DL-based infrastructure that performs a dimensional reduction of the geometry in order to analyse the flow characteristics susequently.

Although most studies are focused on laminar flows, there are several studies where turbulent flows are examined. Fang et al. [17] applied DL techniques for turbulent channel flow predictions, and Thuerey et al. [18] created a CNN to approximate the velocity and pressure fields of the RANS-based Spalart-Allmaras turbulence model on airfoils.

All the aforementioned studies have required prior CFD simulations to train the ANN. However, Sun et al. [19] designed a structured DNN architecture to approximate the solutions of the parametric Navier-Stokes equations. Instead of using data obtained from simulations, this DNN is trained by minimising only the error of the mass and momentum conservation laws of the flows, thus avoiding the computational expense of CFD simulations. Nonetheless, their study shows that data-driven ANNs are more accurate than this kind of network.

This paper aims to compare the basic network structure for velocity field prediction with alternative network structures, which include a previous stage to calculate pressure and vorticity fields, providing more information about the flow to the network. The remainder of the manuscript is divided as follows: Section 2 explains the methodology followed to conduct CFD simulations, designing the CNN and the different neural network structures and training them; Section 3 displays and compares the results obtained with the proposed different structures; and Section 4 shows an evaluation of the ability of the proposed neural network to make predictions under different conditions from the training ones.

## 2. Methodology

### 2.1. CFD Setup

Numerical simulations by means of CFD were conducted to obtain the required velocity, pressure, and vorticity fields for training, validating, and testing the studied CNNs. To perform these simulations, the Star-CCM+ [20] CFD commercial code was used.

The numerical domain consists of a two-dimensional 128 × 256 mm plate, with a geometry located on its geometrical center. The left and right sides of the plate are set as inlet and outlet, respectively, whereas top and bottom sides and the geometry contour are set as walls with no-slip conditions. A detailed view of the numerical domain is provided in Figure 1.

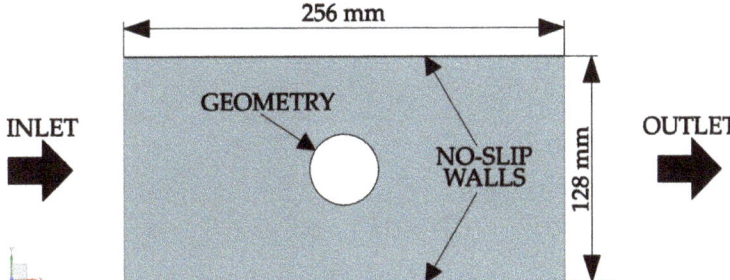

**Figure 1.** Numerical domain.

In order to collect enough samples for training the network, a total of 2065 simulations were performed. Each simulation was carried out with one of the geometries shown in Table 1. These geometries are based on the geometries of the study of Kashefi et al. [14], and they were generated by changing the size and orientation of eight basic geometries.

With the previously mentioned domain, an unstructured polygonal mesh of around 50,000 cells was generated. This mesh contains more cells near the boundaries in order to ensure good results on the most critical areas. An example of the used meshes can be shown in Figure 2.

To verify sufficient mesh resolution of the generated meshes, the General Richardson Extrapolation method [21] was performed, applied to the drag coefficient. For this study, the case of a circle with $a$ = 0.02 m is considered. This method consists of estimating the value of the analysed parameter when the cell quantity tends to infinite from a minimum of three meshes. Therefore, a coarse mesh (16,809 cells), a medium mesh (25,665 cells), and a fine mesh (43,963 cells) were considered. As summarized in Table 2, the convergence condition (R), which should be between 0 and 1 to ensure a monotonic convergence, is fulfilled, and the estimated values (RE) of the evaluated parameters are close to the ones obtained with the fine mesh. Therefore, the mesh is suitable for these simulations. In addition, the results were compared with the experimental ones of Roshko et al. [22] for Re = 6383, showing fairly similar values.

Regarding the fluid, incompressible turbulent unsteady air is considered. The density ($\rho$) of the fluid is equal to 1.18415 kg/m$^3$, and its dynamic viscosity ($\mu$) is equal to 1.85508·10$^{-5}$ Pa·s. These values are assumed to be constant. The velocity at the inlet ($u_\infty$) is set at 5 m/s, which means that the Reynolds number ($Re$) ranges between 6380 and 12,760, depending on the geometry and according to Expression (1).

$$Re = \frac{u_\infty \cdot L \cdot \rho}{\mu} \qquad (1)$$

where $L$ is the projection of the geometry on the direction of the flow.

Table 1. Tested geometries.

| Shape | Sketch | Orientation | Scale | Number of Data |
|---|---|---|---|---|
| Circle | | a | $a = 0.02$ m, $0.022$ m, $\ldots$, $0.04$ m | 11 |
| Equilateral triangle | | a $0°, 3°, \ldots, 177°$ | $a = 0.02$ m | 60 |
| Square | | a $0°, 3°, \ldots, 87°$ | $a = 0.02$ m | 30 |
| Equilateral pentagon | | a $0°, 3°, \ldots, 69°$ | $a = 0.02$ m | 24 |
| Equilateral hexagon | | a $0°, 3°, \ldots, 57°$ | $a = 0.02$ m | 20 |
| Ellipse | | $0°, 3°, \ldots, 177°$ | $a = 0.02$ m; $b/a = 1.1, 1.2, \ldots, 2$ | 600 |
| Rectangle | | $0°, 3°, \ldots, 177°$ | $a = 0.02$ m; $b/a = 1.1, 1.2, \ldots, 2$ | 600 |
| Triangle | | a $0°, 3°, \ldots, 357°$ | $a = 0.01$ m; $b/a = 1.5, 1.75$; $\gamma = 40°, 60°, 80°$ | 720 |

For turbulence modelling, the RANS-based realizable $k$-$\varepsilon$ two-layer [23] turbulence model is selected, since $k$-$\varepsilon$ models are the most common ones to obtain mean flow characteristics for turbulent flow conditions. RANS turbulence models provide closure relations for the RANS equations that govern the transport of the mean flow quantities. To obtain these equations, each flow variable is divided into a mean value and its fluctuating component, and then, the mean values are inserted into the Navier-Stokes equations, obtaining the mean mass and momentum transport Equations (2) and (3).

$$\nabla \cdot \overline{u} = 0 \qquad (2)$$

$$\frac{\partial}{\partial t}(\rho \overline{u}) + \nabla \cdot (\rho \overline{u} \otimes \overline{u}) = -\nabla \cdot \overline{p} I + \nabla \cdot (T + T_{RANS}) + f_b \qquad (3)$$

where $\overline{u}$ and $\overline{p}$ are the mean velocity and pressure, respectively; $I$ is the identity tensor; $T$ is the viscous stress tensor; and $f_b$ is the body force.

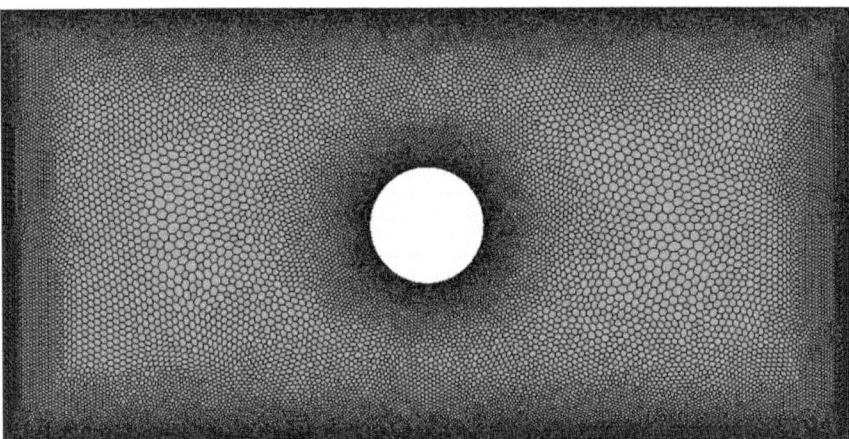

**Figure 2.** Mesh for a circle-shaped geometry.

**Table 2.** Mesh verification and comparison with experimental data for the case of a cylinder with $a = 0.02$ m.

| Mesh Resolution | | | Richardson Extrapolation | | | Experimental |
|---|---|---|---|---|---|---|
| Coarse | Medium | Fine | RE | p | R | |
| 0.796 | 0.835 | 0.858 | 0.907 | 0.681 | 0.566 | 0.91 |

Depending on the modelling of the stress tensor, there are different RANS model categories. The $k$-$\varepsilon$ model corresponds to the eddy viscosity models, which are based on the analogy between the molecular gradient-diffusion process and turbulent motion. This kind of models uses the turbulent dynamic eddy viscosity ($\mu_t$) to model the stress tensor as a function of mean flow quantities. In the present case, $T_{RANS}$ is modeled by means of the Boussinesq approximation (4).

$$T_{RANS} = 2\mu_t S - \frac{2}{3}(\mu_t \nabla \cdot \overline{u}) I \qquad (4)$$

where $S$ is the mean strain rate tensor defined by Equation (5).

$$S = \frac{1}{2}\left(\nabla \cdot \overline{u} + \nabla \cdot \overline{u}^T\right) \qquad (5)$$

The RANS-based $k$-$\varepsilon$ model is a two-equation model which consists of the model transportation equation for turbulent kinetic energy ($k$) (5), the model transportation equation for the dissipation rate ($\varepsilon$) (6) which is empirical, and the turbulent viscosity ($\mu_t$) specification (7).

$$\frac{\partial}{\partial t}(\rho k) + \nabla \cdot (\rho k \overline{u}) = \nabla \cdot \left[\left(\mu + \frac{\mu_t}{\sigma_k}\right)\nabla k\right] + P_k - \rho \varepsilon \qquad (6)$$

$$\frac{\partial}{\partial t}(\rho\varepsilon) + \nabla\cdot(\rho\varepsilon\bar{u}) = \nabla\cdot\left[\left(\mu + \frac{\mu_t}{\sigma_\varepsilon}\right)\nabla\varepsilon\right] + \frac{\varepsilon}{k}C_{\varepsilon 1}P_\varepsilon - \rho C_{\varepsilon 2}\frac{k}{k+\sqrt{v\varepsilon}}\left(\frac{\varepsilon^2}{k}\right) \qquad (7)$$

$$\mu_t = \rho C_\mu \frac{k^2}{\varepsilon} \qquad (8)$$

where $\sigma_k$, $\sigma_\varepsilon$, $C_{\varepsilon 1}$, $C_{\varepsilon 2}$, and $C_\mu$ are model coefficients; $P_k$ and $P_\varepsilon$ are production terms defined by (9) and (10), respectively; and $v$ is the kinematic viscosity ($v = \mu/\rho$).

$$P_k = G_k + G_b + \gamma_M \qquad (9)$$

$$P_\varepsilon = G_k + C_{\varepsilon 3}G_b \qquad (10)$$

where $C_{\varepsilon 3}$ is a model coefficient, whose value is 1 if $G_b \geq 1$ and 0 if $G_b < 0$; $G_k$ represents the turbulent production given by Equation (11); $G_b$ represents the buoyancy production given by Equation (12); and $\gamma_M$ represents the compressibility modification given by Equation (13).

$$G_k = \mu S^2 - \frac{2}{3}\rho k \nabla\cdot\bar{u} - \frac{2}{3}\mu_t(\nabla\cdot\bar{u})^2 \qquad (11)$$

$$G_b = -\frac{1}{\rho}\cdot\frac{\partial\rho}{\partial\bar{T}}\cdot\frac{\mu_t}{Pr_t}(\nabla\bar{T}\cdot g) \qquad (12)$$

$$\gamma_M = \frac{C_M k\varepsilon}{c^2} \qquad (13)$$

where $Pr_t$ is the turbulent Prandtl number; $\bar{T}$ is the mean temperature; $g$ is the gravitational vector; $C_M$ is a model coefficient equal to 2; and $c$ is the speed of sound.

Among the various forms of the k-ε model that are available, the realizable k-ε two-layer model is considered in the present study. This model combines the realizable k-ε model, which satisfies certain mathematical constraints on the normal stresses consistent with the physics of turbulence, and the two-layer approach, which allows the k-ε model to be applied in the viscous-affected layer. With the realizable k-ε two-layer model, the coefficients are identical to the ones of these two models separately, but the model gains the added flexibility of an all-$y^+$ wall treatment. For the selected turbulence model and the studied cases, the model coefficients are the following ones: $\sigma_k$ is equal to 1; $\sigma_\varepsilon$ is equal to 1.2; $C_{\varepsilon 1}$ is equal to 1.44; $C_{\varepsilon 2}$ is equal to 1.9; and $C_\mu$ is equal to 0.09.

Regarding the wall treatment, as mentioned before, the realizable k-ε two-layer model uses an all-$y^+$ wall treatment. This wall treatment emulates the low-$y^+$ wall treatment for fine meshes (near the boundaries), which resolves the viscous sublayer and needs little or no modelling to predict the flow across the wall boundary; the high-$y^+$ wall treatment for coarse meshes (far from the boundaries), which, instead of resolving the viscous sublayer, obtains the boundary conditions for the continuum equations.

For data generation, as non-stationary turbulence models are selected, the average values of the velocity, pressure, and vorticity fields are extracted. To obtain these average fields, 2 s of simulation are considered, once the flow is fully developed. The values of the fields are interpolated in order to fit the data into a 128 × 256 grid. Then, the procedure of Kashefi et al. [14] for data generation is followed.

First, the values are normalized following Expressions (14)–(17), to get dimensionless values.

$$u_x^* = \frac{u_x}{u_\infty} \qquad (14)$$

$$u_y^* = \frac{u_y}{u_\infty} \qquad (15)$$

$$p^* = \frac{p}{\rho\cdot u_\infty^2} \qquad (16)$$

$$\omega^* = \frac{\omega\cdot a}{u_\infty} \qquad (17)$$

where $u_x^*$, $u_y^*$, $p^*$, and $\omega^*$ are the dimensionless variables.

Finally, once the variables are dimensionless, they are scaled in a range of (0, 1), following Expression (18).

$$\Phi' = \frac{\Phi - \min(\Phi)}{\max(\Phi) - \min(\Phi)} \tag{18}$$

where $\Phi$ is replaced by each set of $u_x^*$, $u_y^*$, $p^*$, and $\omega^*$.

### 2.2. Convolutional Neural Network

#### 2.2.1. Domain Representation

In this study, the same layers used by Ribeiro et al. [13] are implemented to represent the domain. Therefore, numerical domain is represented by three different layers: the Flow Region Channel (FRC), the Signed Distance Function (SDF) of the geometry, and the SDF of the walls.

The FRC layer contains information about the boundary conditions of the domain. This layer consists of giving a number to each cell of the grid depending on the boundary condition assigned to that cell. In this case, the geometry is represented by a 0, the fluid by a 1, no-slip walls by a 2, the inlet by a 3, the outlet by a 4, and the outline of the geometry by a 5. A detailed example of a FRC layer is shown Figure 3.

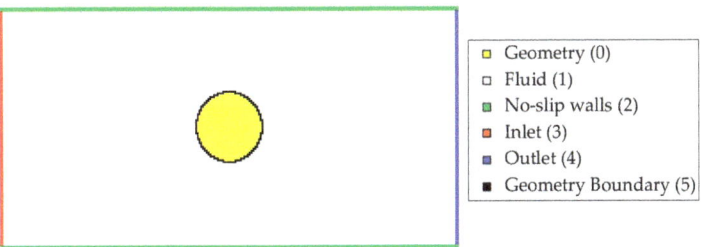

**Figure 3.** Flow Region Channel representation (not to scale).

The SDF layer represents the minimum distance between each cell and the outline of a specified contour. This function was proposed by Guo et al. [8], and as demonstrated in that study, it provides significantly smaller errors than the typical binary representation. In this study two different SDF layers are used, one for the geometry, shown in Figure 4a, and another one for the no-slip walls of the channel, shown in Figure 4b.

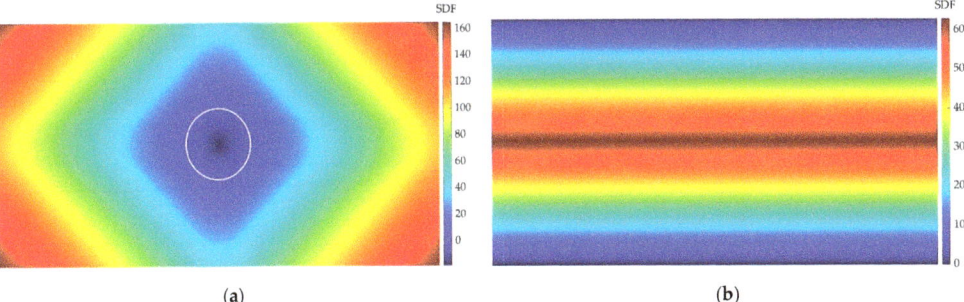

**Figure 4.** SDF layer examples. (**a**) SDF of the geometry (the outline of the geometry is represented in white); (**b**) SDF of the channel walls.

To create this layer, firstly the zero level set ($Z$) is created (19). This is the level where the analysed contour is located, and therefore, $SDF$ is equal to zero.

$$Z = \left\{ (X,Y) \in R^2 / SDF(x,y) = 0 \right\} \tag{19}$$

Then, the sign of $SDF$ is defined. $SDF(x,y) = 0$ if $(x,y)$ is on the geometry contour; $SDF(x,y) < 0$ if $(x,y)$ is inside the geometry; and $SDF(x,y) > 0$ if $(x,y)$ is outside the geometry.

Finally, the value of each cell is calculated following Expression (20).

$$SDF(x,y) = \min_{(X,Y) \in Z} |(x,y) - (X,Y)| \cdot sign \tag{20}$$

After generating all the input layers, they are scaled in a range of (0, 1), following the previously-mentioned Expression (18).

2.2.2. Neural Network Architecture

In the present paper, a CNN based on the previous works from Ribeiro et al. [13] and Thuerey et al. [18] is proposed. For this network, an U-Net architecture [24] is considered, which is a special case of an encoder-decoder network. In this case, the net consists of four encoder/decoder blocks. Each encoder block contains two convolutional layers. The first convolutional layer is followed by a ReLU (Rectifier Linear Unit) layer, and the second one by a ReLU layer and a Max Pooling layer. In the first two encoding blocks, the kernel size is equal to 5, and strided convolutions are performed on the first layer of the block, aiming to reduce the data size for the training step. In contrast, the kernel size of the las two encoding blocks is equal to 3. After each block, the number of filters is doubled. The decoding phase performs the reverse process. Encoder and decoder blocks are connected to each other by concatenation layers. A schematic view of the used CNN is provided in Figure 5. MATLAB 2021a [25] commercial code with its Deep Learning Toolbox [26] was used for designing and training the network.

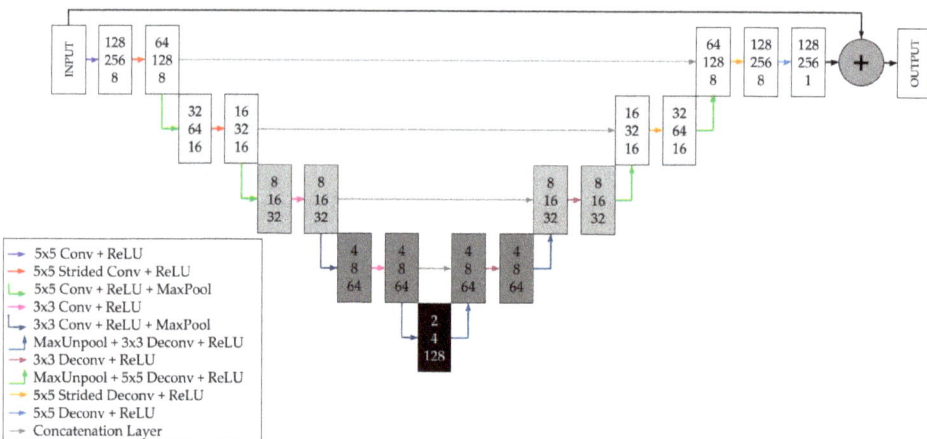

**Figure 5.** CNN architecture.

With regards to the network training process, the Adam [27] optimizer was used, with a learning rate of 0.001, a batch size of 64, and a weight decay of 0.005. The data were split into 60% training, 30% validation, and 10% test, and the validation was performed after each epoch. Three different configurations have been considered for the selection of the most appropriate data-splitting ratio: 60% training, 30% validation, and 10% test (the selected ratio); 70/20/10%; and 80/10/10%. Among these configurations, the selected one is the one that provides the best predictions of all the analysed magnitudes. Among these magnitudes, pressure is the most sensitive to the data-split, and vorticity the least sensitive, the differences between configurations being almost negligible for this magnitude. Considerable differences also appear in the velocity fields.

In order to validate the used net, the training Root-Mean Square Error (RMSE) curves obtained with the network used in the present study are compared with the ones obtained with the net of Ribeiro et al. [13], which was designed for laminar flow prediction on a channel. As this network was originally designed to predict velocity and pressure fields, only the training curves of these magnitudes are considered for this comparison. These curves can be shown in Figure 6.

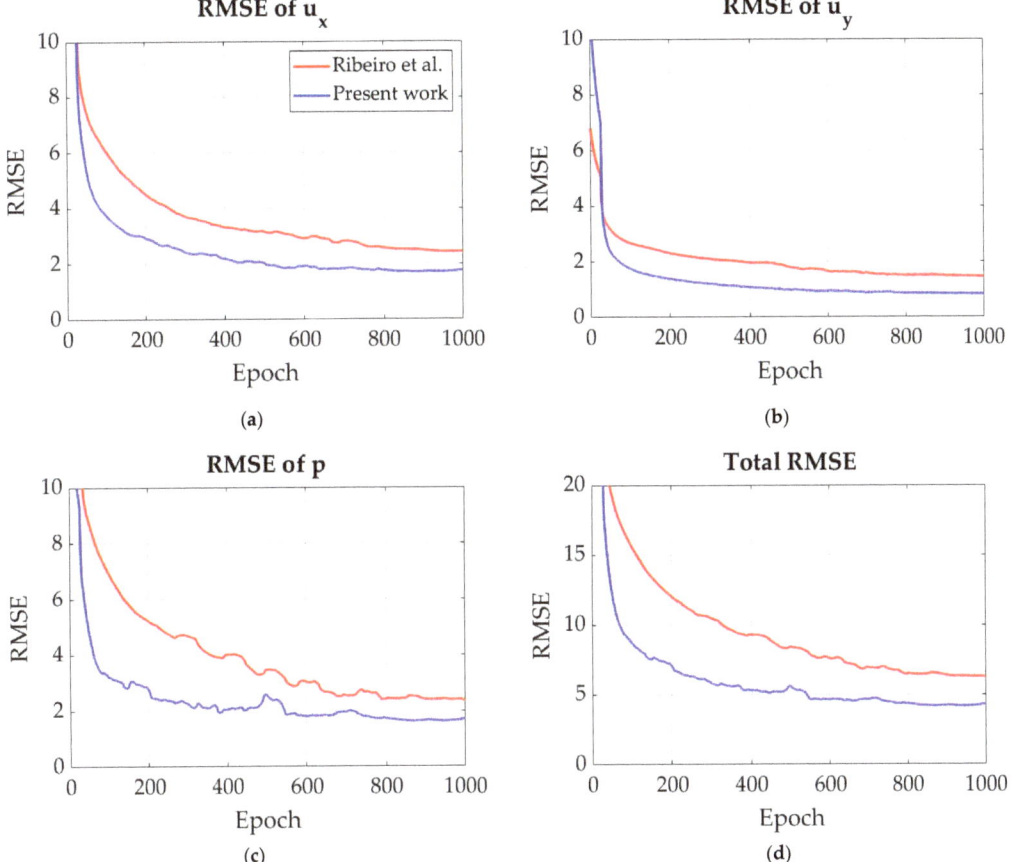

**Figure 6.** Comparison of the obtained training error curves with the ones obtained with the net of Ribeiro et al. [13] for 1000 epochs. (**a**) RMSE or $u_x$; (**b**) RMSE of $u_y$; (**c**) RMSE of $p$; (**d**) Total RMSE.

The curves show that the proposed network outperforms the baseline network for turbulent flow predictions. All the curves show a broadly similar trend. At the beginning of the training, up to approximately epoch 50, the error of the proposed network decreases significantly more than the error of the baseline network, except in the case of $u_y$. Thereafter, the difference between the two networks narrows, but when the results stabilize, the error of the proposed network is still lower. At the end of training, the proposed network has an error reduction of 28% in the case of $u_x$, 42% in $u_y$, and 30% in $p$.

2.2.3. Neural Network Configurations

In the present study, four different network structures are proposed in order to determine which provides the best predictions of the turbulent flow velocity fields. All structures have the domain characteristics as input and the velocity fields as output, but three of them have an additional intermediate stage to calculate the pressure and vorticity fields.

The basic structure (Figure 7a) is the most popular for predicting flow characteristics. This structure directly predicts the velocity fields. The pressure-based structure (Figure 7b) predicts the pressure field, and then the velocity fields considering the pressure field. The vorticity-based structure (Figure 7c) and pressure- and vorticity-based structure (Figure 7d) are equal to the pressure-based one, but these networks predict the vorticity field and both pressure and vorticity fields in the intermediate stage, respectively.

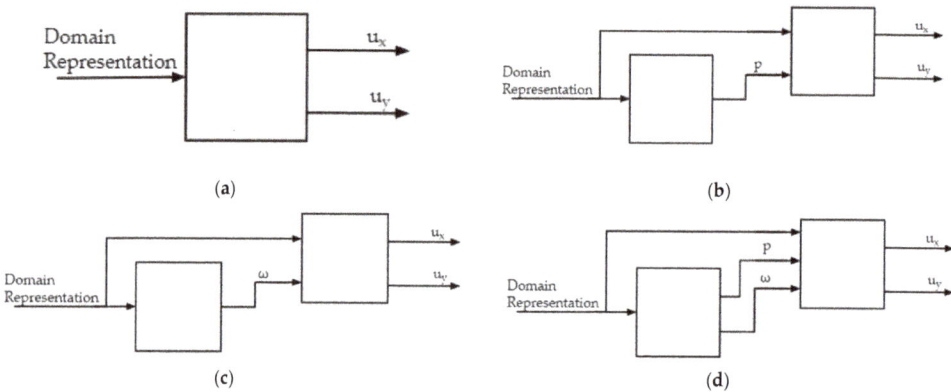

**Figure 7.** Studied ANN structures. (**a**) Basic structure; (**b**) Pressure-based structure; (**c**) Vorticity-based structure; (**d**) Pressure- and vorticity-based structure.

These magnitudes are selected for the intermediate stage as they are directly related to velocity. Regarding pressure, according to Bernoulli's principle, a decrease in the pressure occurs simultaneously with an increase in the velocity, and vice versa. Concerning vorticity, this magnitude determines the local rotation of the fluid, thus relating the velocity components.

## 3. Results and Discussion

Aiming to determine which of the previously mentioned ANN configurations provides the best predictions of turbulent velocity fields, a qualitative and quantitative comparison of the results obtained with all the structures is conducted. For these two studies, the results obtained with a test-set of 207 geometries (10% of all the samples) are considered.

### 3.1. Qualitative Study

The qualitative study is performed by comparing the predictions of the studied neural network structures with the results obtained by CFD simulations. To conduct this comparison, three different geometries, with different characteristics, are selected. The first geometry (Figure 8a) is an ellipse, and it is considered an easy-to-predict geometry because of its symmetry, its low Reynolds number and the fact of not having sharp corners. The second geometry (Figure 8b) is a triangle, and it is considered an aerodynamic geometry due to its low angle of attack and the small surface area on which the fluid directly impinges. The third geometry (Figure 8c) is also a triangle, and it is considered a non-aerodynamic geometry because of its high angle of attack and the big area where the fluid directly impacts.

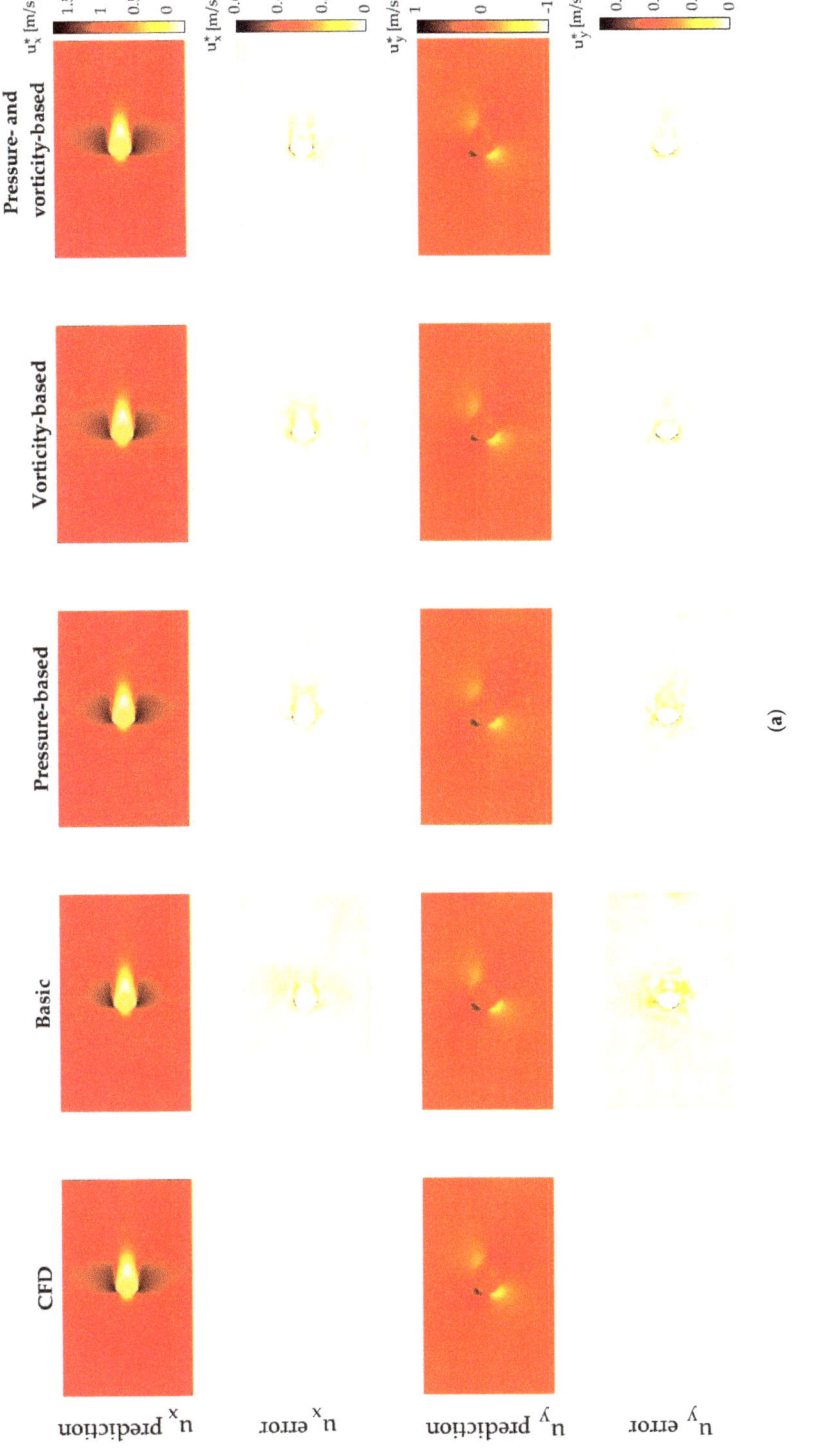

Figure 8. Cont.

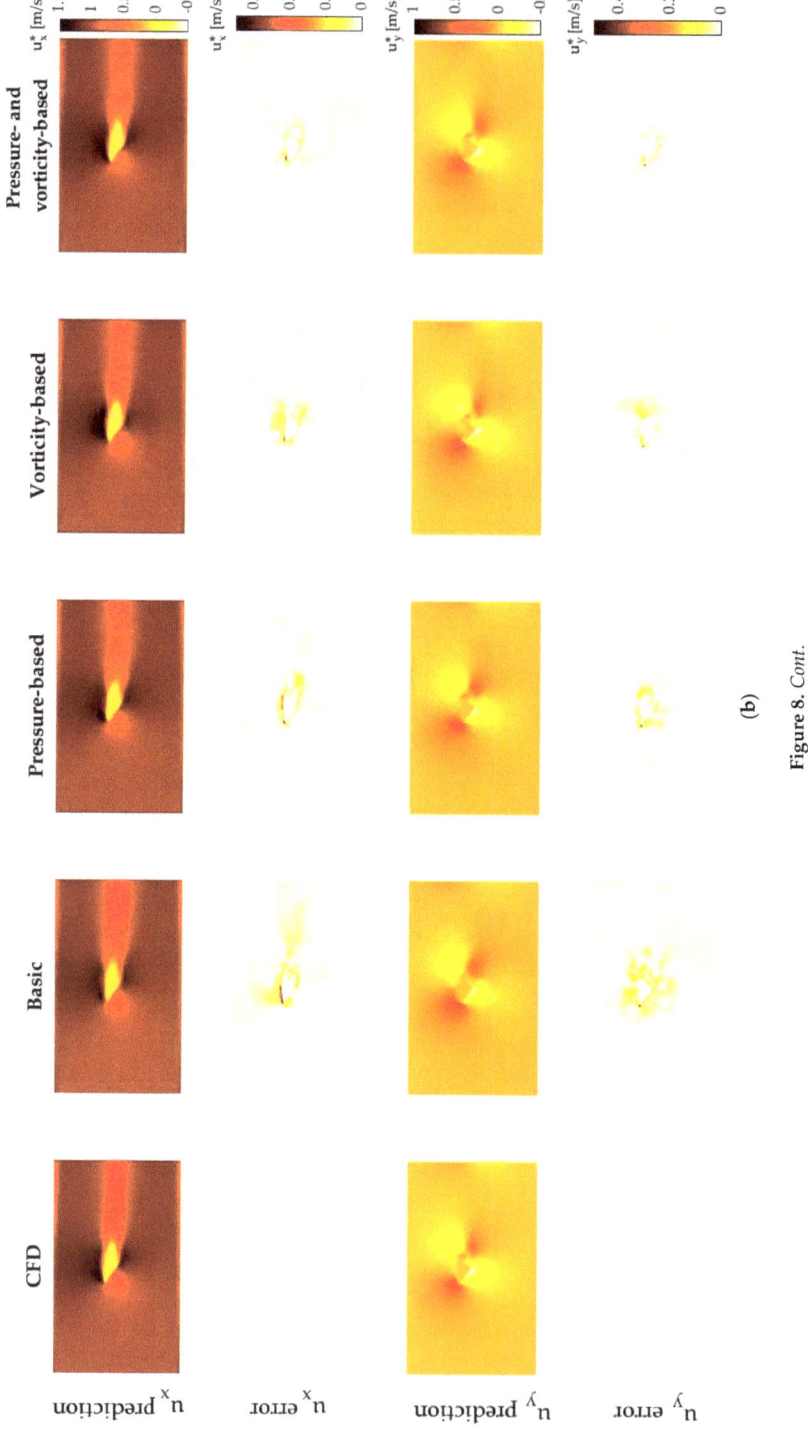

**Figure 8.** *Cont.*

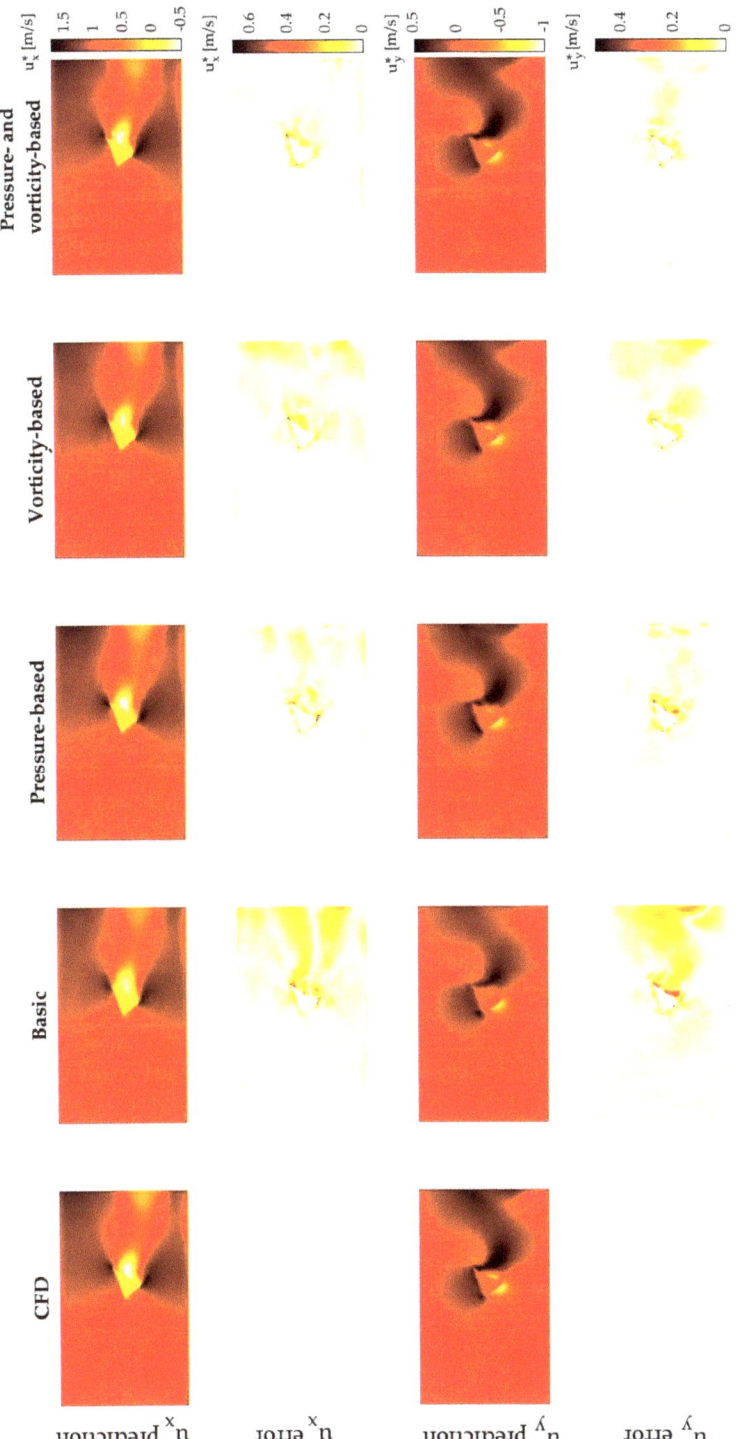

**Figure 8.** Qualitative comparison of the results obtained with the studied structures. (**a**) Easy-to-predict geometry; (**b**) Aerodynamic geometry; (**c**) Non-aerodynamic geometry.

The results show that, although some errors appear in the contour and in the wake behind the geometry, all the structures are able to predict the velocity fields of the easy-to-predict geometry and the aerodynamic geometry fairly precisely. For these two geometries, the basic structure is the one which provides the worst results, being the structure which shows the highest errors. Between the other three structures, no clear differences are visible, but the pressure- and vorticity-based structure seems to be the most accurate structure with very low errors across the predicted fields.

The predictions of the non-aerodynamic geometry show the larger differences between structures. In this case, the basic structure is clearly outperformed by the other ones, showing poor predictions on the boundary of the geometry and, most markedly, on the wake behind the geometry. This means that giving additional information about the flow to the network eases predictions when the flow behaviour is uncertain. The networks which include the prediction of the pressure show the best results on the wake behind the geometry, and the ones which include the prediction of the vorticity show the best results near the geometry. This is attributed to the fact that changes in the pressure field start to appear at a considerable distance from the geometry, but the largest changes in the vorticity field occur very close to the geometry, at its contour.

### 3.2. Quantitative Study

The quantitative study is focused on studying the error of the test-set. For a general overview of the error in the whole test-set, the mean error obtained with all the structures is analysed. Table 3 summarizes the mean error of the predicted velocity components.

**Table 3.** Mean absolute error of the velocity fields predicted by the tested structures.

| Structure | Mean Absolute $u_x$ Error (m/s) | Mean Absolute $u_y$ Error (m/s) |
|---|---|---|
| Basic | 0.1145 | 0.0851 |
| Pressure-based | 0.0619 | 0.0466 |
| Vorticity-based | 0.067 | 0.0331 |
| Pressure- and vorticity-based | 0.0636 | 0.0302 |

The mean error shows that all the proposed structures outperform the basic one. The proposed networks diminish the mean absolute $u_x$ error between 41.5% and 45.9%, the pressure-based structure being the most accurate structure. Regarding the mean absolute $u_y$ error, the proposed structures reduce the error between 45.2% and 64.5%, the pressure- and vorticity-based structure being the one which provides the best predictions.

In order to obtain more detailed information about the error distribution in the test-set, histograms with the absolute error of the velocity fields in both directions are made. These histograms can be shown in Figure 9.

The histograms of the absolute error confirm the results shown with the mean absolute errors. Although there is not a significant difference in the maximum $u_x$ error obtained with the tested structures, the predictions made with the basic structure have the highest errors in almost all ranges above 0.05 m/s. As for the rest of the structures, although they show quite similar error distributions, the pressure- and vorticity-based structure is the structure with the least errors up to 2 m/s, but above this value, this structure produces more errors than the other two structures. On the other hand, the pressure-based structure has more errors below 2 m/s, but fewer above this value. For that reason, the mean error of the pressure-based structure is the lower one.

The predictions of all structures show a very similar trend of $u_y$ error distribution, with the basic structure having the largest errors and the pressure- and vorticity-based structure having the smallest ones.

### 3.3. Performance Analysis

The main goal of calculating flow characteristics through neural networks is to reduce the high computational cost of CFD simulations. For this reason, the time required by each structure to make the predictions is considered a parameter of great relevance. Table 4 shows the time required to obtain the predictions with each studied method using a single core of an Intel Xeon 5420 CPU.

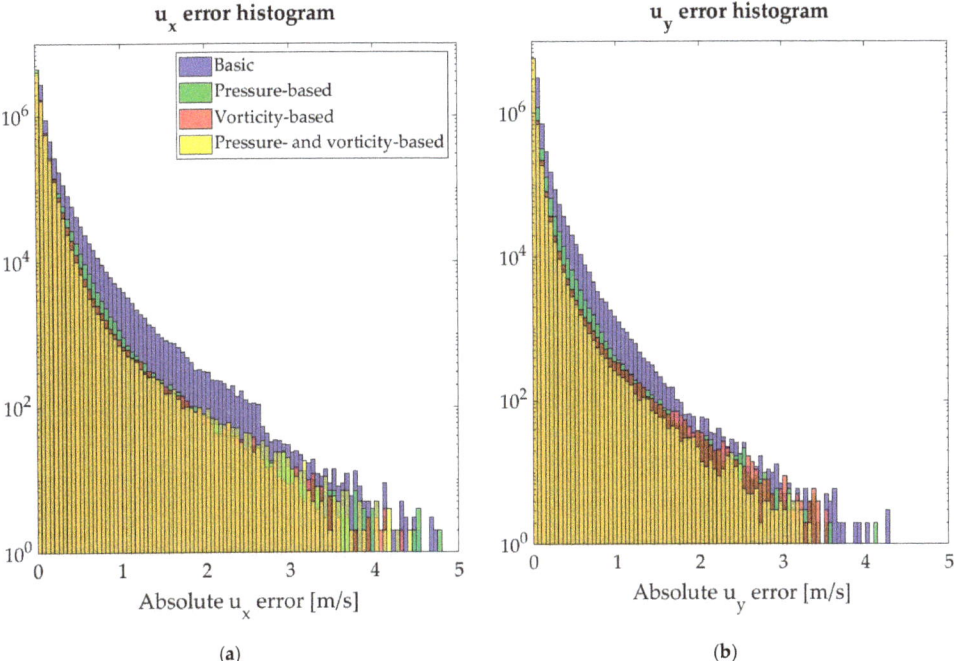

**Figure 9.** Absolute error distribution. (**a**) Absolute $u_x$ error; (**b**) Absolute $u_y$ error.

**Table 4.** Computational time required by each studied method to predict the velocity fields.

| Method | Time (s) | Speedup |
| --- | --- | --- |
| CFD | 23,053.2 | - |
| Basic | 21.4 | 1077.25 |
| Pressure-based | 27.35 | 842.9 |
| Vorticity-based | 24.01 | 960.15 |
| Pressure- and vorticity-based | 28.53 | 808.03 |

The computational times required by each method show that predicting velocity fields using CNN, in comparison with CFD, entails a reduction of around four orders of magnitude in terms of computational time. As expected, the simplest CNN structure (the basic structure) is the fastest one, and the most complex (the pressure- and vorticity-based structure) is the slowest one. Nonetheless, the differences between structures are negligible in comparison with the time required by CFD.

## 4. Network Testing

In order to evaluate the ability of the neural network proposed in this study to make predictions under different conditions from the training ones, two different domains are considered, one with two geometries and another one with a different channel.

To make the predictions of the velocity fields on these domains, the best neural network structure is considered. For predicting the $u_x$ velocity field, the pressure-based structure is considered, since it provides the minimum mean error and is the better one for predicting the wake behind the geometry, which is one of the most important parameters. To predict the $u_y$ velocity field, the pressure- and vorticity-based structure is considered, since it has been demonstrated to be the most accurate for this field. A schematic view of the structure of the used ANN is provided in Figure 10.

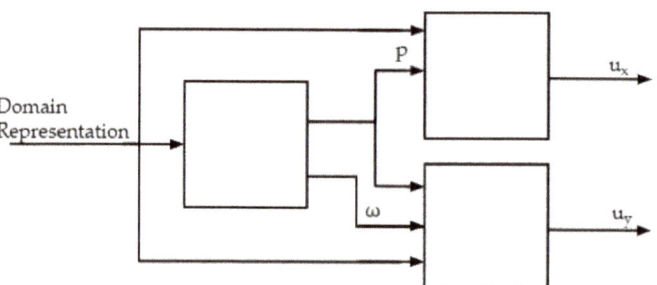

**Figure 10.** ANN structure to test the proposed network under different conditions.

*4.1. Multiple Objects*

The domain used in this case contains two geometries, particularly two circles. Therefore, the input layers for this case have some adaptations. The FRC layers contain two geometries and geometry boundaries, and the SDF layer of the geometry has two zero level sets. The SDF of the walls remains equal. A schematic view of the numerical domain and the input layers is provided in Figure 11. The predictions of the velocity fields performed by CFD and the proposed network are displayed in Figure 12.

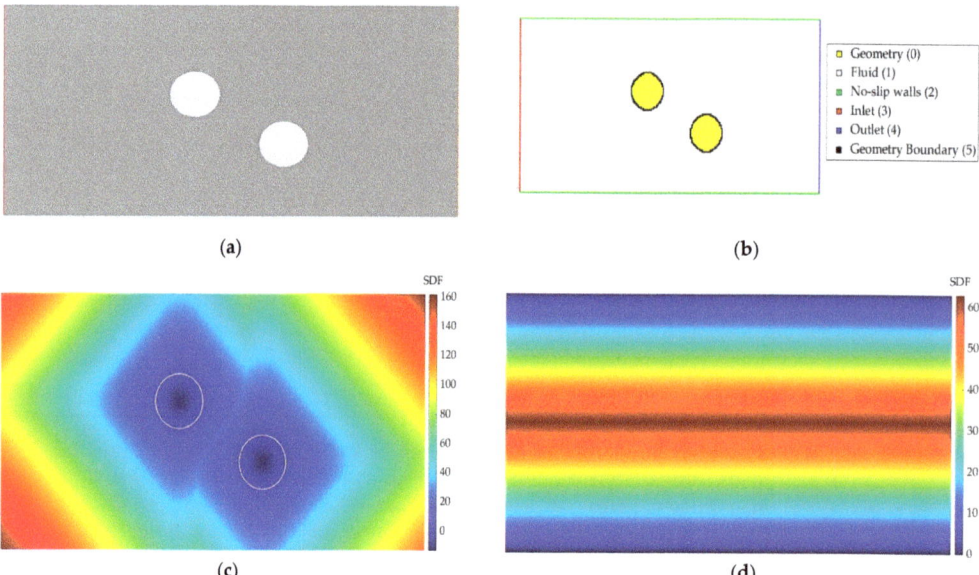

**Figure 11.** Information about the used domain with multiple geometry. (**a**) Numerical domain; (**b**) FRC layer; (**c**) SDF layer of the geometries (the contour of the geometries is drawn in white); (**d**) SDF layer of the walls.

The results show that the proposed network is able to predict the velocity fields of a domain with multiple geometries in a fairly reliable way, with tolerable errors. These results show two problematic areas. The first problematic area is the contour of the geometries. Whereas with a single geometry, these errors were already visible; in this case they are slightly higher. The second problematic area is the area between geometries. Although the errors are not very high, the neural network does not adequately predict the interaction between geometries, since the area with the most errors is the region where the wake of the first geometry impacts the second geometry.

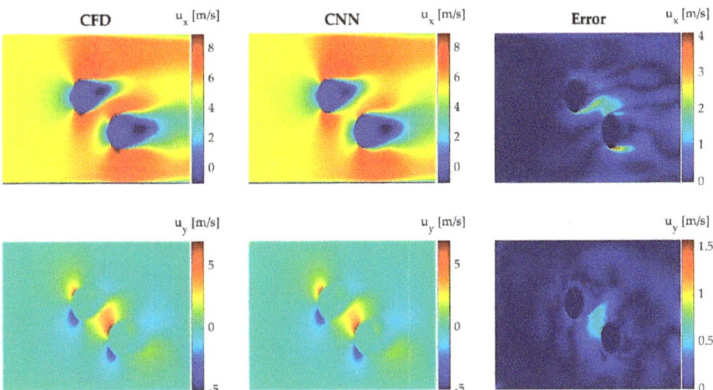

**Figure 12.** Comparison of the velocity fields of a domain with two geometries obtained with CFD and the proposed net.

### 4.2. Different Channel

This domain contains a single geometry, but the walls of the channel have a narrowing on their middle. Therefore, the input layers for this case also have to be adapted. In the FRC layer, although the outline of the walls remains with no-slip wall conditions, the inner part of the walls has been considered as geometry, since among the possible options it is considered to be the most appropriate one. The SDF layer of the geometry is equal to the training ones, but the SDF layer of the walls contains a zone of negative values. A schematic view of the numerical domain and the input layers is provided in Figure 13, and the predictions of the velocity fields performed by CFD and the proposed network are displayed in Figure 14.

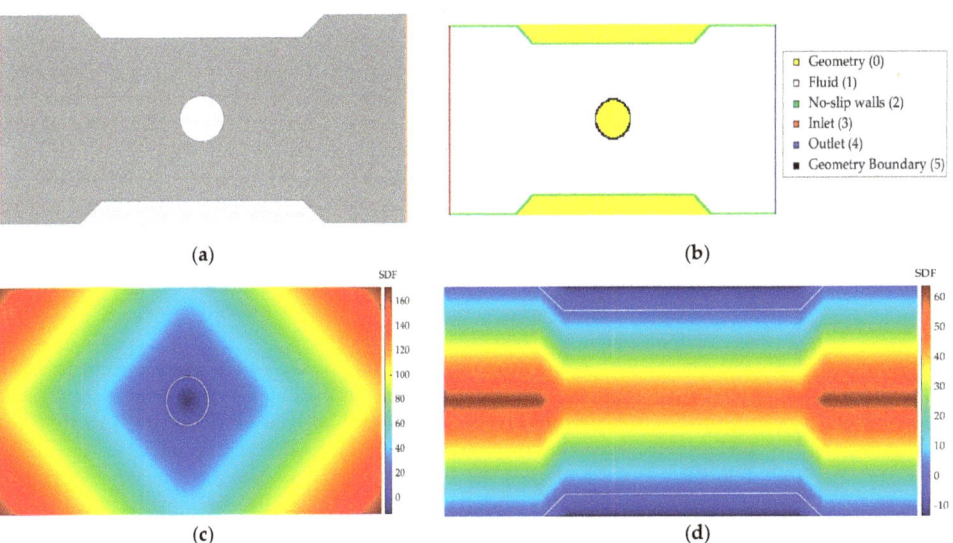

**Figure 13.** Information about the used domain with a different channel. (**a**) Numerical domain; (**b**) FRC layer; (**c**) SDF layer of the geometry (the contour of the geometry is drawn in white); (**d**) SDF layer of the walls (the contour of the walls is drawn in white).

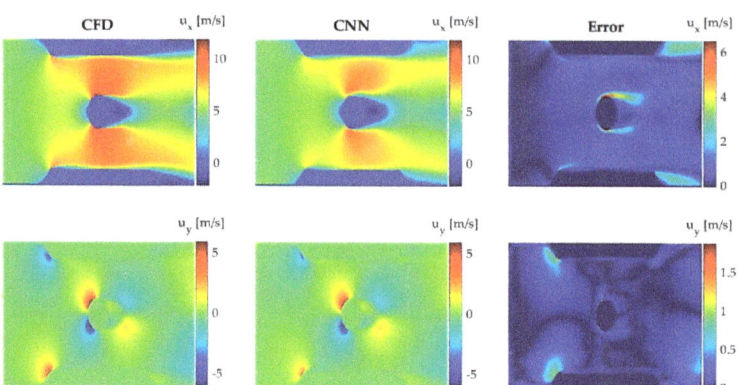

**Figure 14.** Comparison of the velocity fields of a domain with a different channel obtained with CFD and the proposed net.

There are three areas where considerable differences are observed. The first zone is the narrowing of the channel, where the neural network underpredicts the speed increase that occurs. The second, and most conflicting, area is the near-wall area. When the channel narrows, the network underpredicts the impact with the walls then (noticeable in $u_y$), and when the channel widens again, although a decrease in velocity is visible, the network is unable to predict the wake behind the walls (noticeable in $u_x$). The third region is the geometry contour. As with multiple geometries, when changing the channel, the errors in the geometry contour increase in comparison to the simple case.

## 5. Conclusions

In the present paper, a CNN for predicting different magnitudes of turbulent flows is proposed. With this CNN, alternative neural network structures for turbulent flow velocity field prediction are proposed. In contrast with the typical network structure, which directly calculated the velocity fields, the proposed structures perform a preliminary calculation of pressure and vorticity fields in order to obtain more information about the flow. Performing the predictions using the proposed networks instead of using CFD means a reduction of about four orders of magnitude in terms of computational time.

The results indicate that the proposed network structures outperform the basic structure, showing a decrease between 41.5% and 45.9% of mean absolute $u_x$ error and a decrease between 45.2% and 64.5% of mean absolute $u_y$ error. When the flow is simple, the results provided by the basic structure are correct, but the more uncertain the flow, the greater the differences between structures, with the structures with a preliminary calculation of pressure and vorticity fields being much more accurate than the basic one. These differences are more visible on the wake behind the geometry.

Finally, the best network structure is proposed considering the obtained results, and its ability to predict turbulent flow velocity fields in domains different from those used to train the network is evaluated. This network calculates the pressure and vorticity fields on the preliminary stage, and used the pressure field to calculate the horizontal component of the velocity field and both the pressure and vorticity fields to calculate the vertical component of the velocity field. To assess the network, two different domains are considered, one which has two geometries and another one which has a narrowing of the channel. The results show that the network is able to predict the velocity fields when the domain has two geometries, but when changing the channel, the errors are significant.

Therefore, this study demonstrates that obtaining fluid characteristics to predict the desired magnitude improves predictions substantially, and that this type of structure is applicable to different domains for the ones used for the training process.

**Author Contributions:** Conceptualization, K.P.-P.; methodology, K.P.-P. and A.U.-A.; software, E.Z. and A.Z.; validation, A.U.-A., E.Z. and U.F.-G.; formal analysis, K.P.-P. and A.U.-A.; investigation, K.P.-P. and A.U.-A.; resources, U.F.-G. and E.Z.; data curation, A.Z.; writing—original draft preparation, K.P.-P.; writing—review and editing, K.P.-P. and U.F.-G.; visualization, A.U.-A. and E.Z.; supervision, U.F.-G.; project administration, U.F.-G.; funding acquisition, U.F.-G. All authors have read and agreed to the published version of the manuscript.

**Funding:** The authors were supported by the government of the Basque Country through research grants ELKARTEK 21/10: BASQNET: Estudio de nuevas técnicas de inteligencia artificial basadas en Deep Learning dirigidas a la optimización de procesos industriales.

**Institutional Review Board Statement:** Not applicable.

**Informed Consent Statement:** Not applicable.

**Data Availability Statement:** The data presented in this study are available on request from the corresponding author.

**Acknowledgments:** The authors are grateful for the support provided by SGIker of UPV/EHU. This research was developed under the frame of the Joint Research Laboratory on Offshore Renewable Energy (JRL-ORE).

**Conflicts of Interest:** The authors declare no conflict of interest.

## Nomenclature

| | Definition |
|---|---|
| ANN | Artificial Neural Network |
| CFD | Computational Fluid Dynamics |
| CNN | Convolutional Neural Network |
| DL | Deep Learning |
| DML | Deep Metric Learning |
| DNN | Deep Neural Network |
| DNS | Direct Numerical Simulation |
| FRC | Flow Region Channel |
| GAN | Generative Adversarial Network |
| LES | Large Eddy Simulation |
| RANS | Reynolds-Averaged Navier-Stokes |
| ReLU | Rectifier Linear Unit |
| RMSE | Root-Mean-Square Error |
| SDF | Signed Distance Function |
| * | Dimensionless value |
| ' | Value ranged between [0, 1] |
| – | Mean value |
| $a$ | Characteristic length of the geometry |
| $b$ | Characteristic length of the geometry |
| $c$ | Speed of sound |
| $C_{\varepsilon 1}$ | Model coefficient |
| $C_{\varepsilon 2}$ | Model coefficient |
| $C_{\varepsilon 3}$ | Model coefficient |
| $C_\mu$ | Model coefficient |
| $C_M$ | Model coefficient |
| $\gamma$ | Characteristic angle of the geometry |
| $\gamma_M$ | Compressibility modification |
| $\varepsilon$ | Dissipation rate |
| $f_b$ | Resultant of body forces |
| $g$ | Gravitational vector |

| | |
|---|---|
| $G_b$ | Buoyancy production |
| $G_k$ | Turbulent production |
| $I$ | Identity tensor |
| $k$ | Turbulent kinetic energy |
| $L$ | Projection of the geometry on the direction of the flow |
| $\rho$ | Fluid density |
| $p$ | Pressure |
| $p$ | Order of accuracy (in General Richardson Extrapolation) |
| $P_k$ | Production term |
| $P_\varepsilon$ | Production term |
| $Pr_t$ | Turbulent Prandtl number |
| R | Convergence condition (in General Richardson Extrapolation) |
| Re | Reynolds number |
| RE | Estimated value (in General Richardson Extrapolation) |
| $\sigma_k$ | Model coefficient |
| $\sigma_\varepsilon$ | Model coefficient |
| $T$ | Viscous stress tensor |
| $\overline{T}$ | Mean temperature |
| $\mu$ | Fluid dynamic viscosity |
| $\mu_t$ | Turbulent viscosity |
| $u_x$ | Horizontal velocity |
| $u_y$ | Vertical velocity |
| $\omega$ | Vorticity |
| Z | Zero level set (in SDF) |

## References

1. Ray, D.; Hesthaven, J.S. An artificial neural network as a troubled-cell indicator. *J. Comput. Phys.* **2018**, *367*, 166–191. [CrossRef]
2. Liu, Y.; Lu, Y.; Wang, Y.; Sun, D.; Deng, L.; Wan, Y.; Wang, F. Key time steps selection for CFD data based on deep metric learning. *Comput. Fluids* **2019**, *195*, 104318. [CrossRef]
3. Bao, H.; Feng, J.; Dinh, N.; Zhang, H. Computationally efficient CFD prediction of bubbly flow using physics-guided deep learning. *Int. J. Multiph. Flow* **2020**, *131*, 103378. [CrossRef]
4. Hanna, B.N.; Dinh, N.T.; Youngblood, R.W.; Bolotnov, I.A. Coarse-Grid Computational Fluid Dynamic (CG-CFD) Error Prediction Using Machine Learning. *arXiv* **2017**, arXiv:171009105.
5. Yan, X.; Zhu, J.; Kuang, M.; Wang, X. Aerodynamic shape optimization using a novel optimizer based on machine learning techniques. *Aerosp. Sci. Technol.* **2019**, *86*, 826–835. [CrossRef]
6. Zhang, X.; Xie, F.; Ji, T.; Zhu, Z.; Zheng, Y. Multi-Fidelity deep neural network surrogate model for aerodynamic shape optimization. *Comput. Methods Appl. Mech. Eng.* **2021**, *373*, 113485. [CrossRef]
7. Tao, J.; Sun, G. Application of deep learning based multi-fidelity surrogate model to robust aerodynamic design optimization. *Aerosp. Sci. Technol.* **2019**, *92*, 722–737. [CrossRef]
8. Guo, X.; Li, W.; Iorio, F. Convolutional neural networks for steady flow approximation. In Proceedings of the 22nd ACM SIGKDD International Conference on Knowledge Discovery and Data Mining, San Francisco, CA, USA, 13 August 2016; pp. 481–490.
9. Ling, J.; Kurzawski, A.; Templeton, J. Reynolds averaged turbulence modelling using deep neural networks with embedded invariance. *J. Fluid Mech.* **2016**, *807*, 155–166. [CrossRef]
10. Lee, S.; You, D. Prediction of laminar vortex shedding over a cylinder using deep learning. *arXiv* **2017**, arXiv:1712.07854.
11. Liu, Y.; Lu, Y.; Wang, Y.; Sun, D.; Deng, L.; Wang, F.; Lei, Y. A CNN-based shock detection method in flow visualization. *Comput. Fluids* **2019**, *184*, 1–9. [CrossRef]
12. Deng, L.; Wang, Y.; Liu, Y.; Wang, F.; Li, S.; Liu, J. A CNN-based vortex identification method. *J. Vis.* **2019**, *22*, 65–78. [CrossRef]
13. Ribeiro, M.D.; Rehman, A.; Ahmed, S.; Dengel, A. DeepCFD: Efficient steady-state laminar flow approximation with deep convolutional neural networks. *arXiv* **2020**, arXiv:2004.08826.
14. Kashefi, A.; Rempe, D.; Guibas, L.J. A Point-cloud deep learning framework for prediction of fluid flow fields on irregular geometries. *arXiv* **2020**, arXiv:2010.09469.
15. Nowruzi, H.; Ghassemi, H.; Ghiasi, M. Performance predicting of 2D and 3D submerged hydrofoils using CFD and ANNs. *J. Mar. Sci. Technol.* **2017**, *22*, 710–733. [CrossRef]
16. Mohan, A.; Daniel, D.; Chertkov, M.; Livescu, D. Compressed convolutional LSTM: An efficient deep learning framework to model high fidelity 3D turbulence. *arXiv* **2019**, arXiv:1903.00033.
17. Fang, R.; Sondak, D.; Protopapas, P.; Succi, S. Deep learning for turbulent channel flow. *arXiv* **2018**, arXiv:1812.02241.
18. Thuerey, N.; Weißenow, K.; Prantl, L.; Hu, X. Deep learning methods for reynolds-averaged navier–stokes simulations of airfoil flows. *AIAA J.* **2020**, *58*, 25–36. [CrossRef]

19. Sun, L.; Gao, H.; Pan, S.; Wang, J.-X. Surrogate modeling for fluid flows based on physics-constrained deep learning without simulation data. *Comput. Methods Appl. Mech. Eng.* **2020**, *361*, 112732. [CrossRef]
20. STAR-CCM+ V2019.1. Available online: https://www.plm.automation.siemens.com/ (accessed on 2 June 2020).
21. Richardson, L.F.; Gaunt, J.A. The deferred approach to the limit. *Philos. Trans. R. Soc. Lond. Ser. A Contain. Pap. Math. Phys. Character* **1927**, *226*, 299–361. [CrossRef]
22. Roshko, A. *Vortex Shedding from Circular Cylinder at Low Reynolds Number*; Cambridge University Press: Cambridge, UK, 1954.
23. Shih, T.-H.; Liou, W.W.; Shabbir, A.; Yang, Z.; Zhu, J. A new $k$-$\varepsilon$ Eddy viscosity model for high Reynolds number turbulent flows: Model development and validation. *Computers Fluids* **1995**, *24*, 227–238. [CrossRef]
24. Ronneberger, O.; Fischer, P.; Brox, T. U-net: Convolutional networks for biomedical image segmentation. In Proceedings of the Medical Image Computing and Computer-Assisted Intervention—MICCAI 2015, Munich, Germany, 5–9 October 2015; Navab, N., Hornegger, J., Wells, W.M., Frangi, A.F., Eds.; Springer: Cham, Switzerland, 2015; pp. 234–241.
25. MATLAB. Available online: https://es.mathworks.com/products/matlab.html (accessed on 9 June 2021).
26. Deep Learning Toolbox. Available online: https://es.mathworks.com/products/deep-learning.html (accessed on 3 July 2021).
27. Kingma, D.P.; Ba, J. Adam: A Method for Stochastic Optimization. *arXiv* **2017**, arXiv:1412.6980.

Article

# Machine Learning Approach for Modeling and Control of a Commercial Heliocentris FC50 PEM Fuel Cell System

Mohamed Derbeli *, Cristian Napole * and Oscar Barambones *

System Engineering and Automation Department, Faculty of Engineering of Vitoria-Gasteiz, Basque Country University (UPV/EHU), 01006 Vitoria-Gasteiz, Spain
* Correspondence: mderbeli001@ikasle.ehu.eus (M.D.); cristianmario.napole@ehu.eus (C.N.); oscar.barambones@ehu.eus (O.B.)

**Citation:** Derbeli, M.; Napole, C.; Barambones, O. Machine Learning Approach for Modeling and Control of a Commercial Heliocentris FC50 PEM Fuel Cell System. *Mathematics* **2021**, *9*, 2068. https://doi.org/10.3390/math9172068

Academic Editor: António M. Lopes

Received: 12 July 2021
Accepted: 25 August 2021
Published: 26 August 2021

**Publisher's Note:** MDPI stays neutral with regard to jurisdictional claims in published maps and institutional affiliations.

**Copyright:** © 2021 by the authors. Licensee MDPI, Basel, Switzerland. This article is an open access article distributed under the terms and conditions of the Creative Commons Attribution (CC BY) license (https://creativecommons.org/licenses/by/4.0/).

**Abstract:** In recent years, machine learning (ML) has received growing attention and it has been used in a wide range of applications. However, the ML application in renewable energies systems such as fuel cells is still limited. In this paper, a prognostic framework based on artificial neural network (ANN) is designed to predict the performance of proton exchange membrane (PEM) fuel cell system, aiming to investigate the effect of temperature and humidity on the stack characteristics and on tracking control improvements. A large part of the experimental database for various operating conditions has been used in the training operation to achieve an accurate model. Extensive tests with various ANN parameters such as number of neurons, number of hidden layers, selection of training dataset, etc., are performed to obtain the best fit in terms of prediction accuracy. The effect of temperature and humidity based on the predicted model are investigated and compared to the ones obtained from real-time experiments. The control design based on the predicted model is performed to keep the stack operating point at an adequate power stage with high-performance tracking. Experimental results have demonstrated the effectiveness of the proposed model for performance improvements of PEM fuel cell system.

**Keywords:** machine learning; deep learning; artificial neural network; ANN; PEM fuel cell; modeling; control

## 1. Introduction

Fuel cells (FC) are conversion devices which transform hydrogen into electrical energy through an electro-chemical process [1]. These are a trending research topic since their efficiency (more than 40%) is higher than other renewable alternatives such as wind turbines ($\approx$25%) or photovoltaic systems ($\approx$6–20%) [2]. As a consequence, several industries used FC for their applications like aviation [3], automotive [4] and maritime [5]. According to Wang et al. [6], the cutting-edge FC technologies that are currently under focus are polymer electrolyte membrane fuel cells (PEMFCs), solid oxide fuel cells (SOFCs), phosphoric acid fuel cells (PAFCs) and molten carbonate fuel cells (MCFCs). Among the several available types, PEMFC stands out due to its high efficiency, power density and durability [7].

A PEMFC is frequently built with membrane electrode assembly (MEA) that holds an anode and a cathode that are isolated by a proton conductive membrane [8]. The continuous hydrogen supply goes into the anode electrode while the cathode receives oxygen. As a consequence, protons and electrons are generated because of an oxidation reaction; the electrolyte exchange membrane allows the path division of these particles. The electrons move to an external electric circuit whereas the protons join the oxygen to output water [9]. To achieve a suitable system design in terms of efficiency, several PEMFC mathematical models had been developed in recent years to understand the main phenomena that can alter the device performance.

According to Fang, Di and Ru, PEMFC models are divided into operational mechanism and experimental data ones [10]. In regards to the first mentioned category, based on the regime, these are divided into static and dynamic. Saadi et al. [11] studied three well known models used in static analysis such as the Amphlett et al. [12,13], Larminie and Dicks [14] and Chamberlin-Kim [15]. A simulation showed that the three approaches had different outcomes where the Amphlett produced the most accurate results with high complexity implementation as a downside (same disadvantage for Larmini-Dicks). Conversely, Chamberlin-Kim appeared to be the most simple one but with low precision. Contrarily, dynamic models are used in transient regimes where the double layer effect heads this condition. Often, this phenomenon is modelled as with an electrical capacitor that depends on the electrodes and individual stack features [16,17].

On the other hand, experimental methods comprise mechanisms such as fuzzy identification which has been carried by authors of [18]. In this study, the dehydration of a PEMFC was analysed through classification based on the knowledge from an operator over a FC. Results revealed that suitable nonlinearities like electrical features and uncertainties can be mirrored with linguistic rules, an essential feature of this tool [19]. However, one of the main disadvantages of fuzzy logic strategies is the computational requirement when features are increased and thus, rules are expanded [20]. Another different approach was used by authors of [21] where they employed support vector machine (SVM) based on data-driven for fault diagnosis in PEMFC. In spite of the high accuracy obtained, the disadvantages are related to dynamics that can happen in a short period of time such as switches that are unable to be shown by the proposed model. Additionally, in certain cases, SVM required high computational resources which is associated with the accuracy of the model to be trained [22].

Despite the disadvantages of the mentioned strategies, another approach is the usage of trend tools such as artificial neural networks (ANN). This algorithmic scheme is based on a biological approach of human brain neurons which have the capabilities of recognising, acquiring information, and self-adjusting according to past actions (this is also known as neuroplasticity) [23]. Recently, Nanadegani et al. [24] provided a PEMFC study based on ANNs to increase the output power with a multilayer perceptron (MLP). Therefore, in this study, an in-depth investigation was conducted with a commercial PEMFC to generate a spread variety of ANNs with the aim of finding a suitable configuration that can match the behaviour of the real PEMFC. After finding a proper ANN configuration, this was used as a plant for the calibration of neural control algorithm.

The controller used in this research is an artificial neural network proportional-integral-derivative (ANN-PID) controller to track a reference current. Differently to conventional PID controllers, ANN-PID can self-tune its own gains with an inner mechanism based on simple ANN linked with Hebbs learning rule [25]. In this research, the ANN-PID has been contrasted with a conventional PID tuned with appropriate gains gathered in previous experiments.

The structure of this paper has been arranged as follows. Section 2 covers further explanation about the ANN methodology applied and its design, the data collection procedure for the ANN training, the precision of the trained ANNs, the control design, and the metric used to show the accuracy of tracking in later tests. Section 3 includes a contrast between the real PEMFC curves and the ones obtained with the chosen ANN with details about the temperature and humidity effect; additionally, this section ends with the control results of the ANN-PID (tuned with an ANN) that was embedded in a dSpace platform and contrasted with a conventional PID. Finally, Section 4 provides a summary of the most significant outcomes obtained along this study as well as future viewpoints of research.

## 2. Materials and Methods

*2.1. Artificial Neural Networks (ANNs) Model*

2.1.1. Introduction to ANNs

ANNs have been considered as attractive and powerful tools to predict and approximate linear, nonlinear and even complex models, based only on input-output data mapping [26–28]. Actually, an ANN consists of input and output layers and at least one hidden interconnection layer. A general architecture of ANN with $N_1$ inputs, $N_2$ outputs and $L$ hidden layers is depicted in Figure 1.

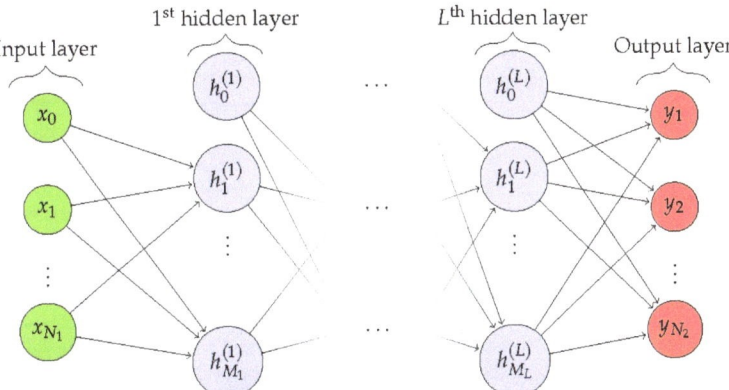

**Figure 1.** Network graph of $N_1$ input units, $N_2$ output units and $L$-layer perceptron where each hidden layer contains $M_j$ hidden units.

ANNs can manipulate information just like the human brain thanks to the computational features of their basic units (also called nodes or neurons) which take a set of inputs, multiply them by weighted values and put them through an activation function. The schematic structure of the $i$th hidden artificial neuron at the $j$th hidden layer can be depicted as Figure 2.

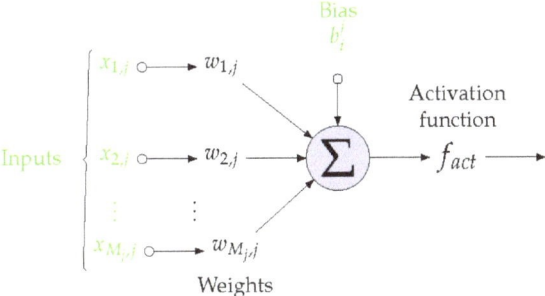

**Figure 2.** Structure of a single artificial neuron in a neural network.

There are several topologies of NNs in deep learning and they can be classified into two groups of algorithms. The first group contains the ones that were used for supervised deep learning problems such as fully-connected feed-forward algorithms (Multi-Layer Perceptron, Radial Basis Network, etc.), recurrent NNs algorithms (long short term memory, gated recurrent unit, gated feed-forward, etc.) and convolutions NNs algorithms (deep convolutional NNs, deep convolutional inverse graphics network, deconvolutional network, etc.). The second group contains the ones that were used for unsupervised deep learning problems such as restricted Boltzmann machine algorithms (deep belief network, deep Boltzmann machine, etc.) and ML auto-encoder algorithms (variational auto-encoder, denoising auto-encoder, sparse auto-encoder, etc.). However, since modelling the fuel cell is a supervised learning problem, different structures of feed-forward neural network perceptron (FFNNP) with back-propagation learning rule have been implemented in Matlab/Simulink$^R$ and Neureal Network Toolbox$^{TM}$ to predict the performance of a commercial fuel cell system (Heliocentris FC50).

2.1.2. Data Collection and Analysis

- **Data collection:** The first and the most important step in the supervised learning process is gathering the data. In other words, to carry out good training, vast amounts of real-world data (Big Data) is required since the more data we provide to the ML system, the faster the model can learn and improve. Besides, the collected dataset should be well distributed throughout the operation range so as to represent the behaviour of the fuel cell in each operating power point. To this end, a continuous triangular signal with a period of 15 s (7.5 s for each positive/negative slop) was built and supplied to the duty cycle of the boost converter so as to vary the stack current from the minimum to the maximum operating value. The selection of the period was made based on the characteristics of the fuel cell data acquisition software since it measures the data each 0.5 s. In other words, 15 samples in different operating current values will be measured fore each positive/negative slop. Figures 3 and 4 show, respectively, the Simulink blocks used to design the triangular signal and the generated signal. The maximum value of this signal (0.8) drives the fuel cell to operate at the highest current value [8–9A] where the minimum value (0.5) drives the fuel cell to operate at the lowest operating current [0.2–0.5A]. These values can be adjusted via the increase/decrease of the output load resistance value. We have avoided operating currents above 9A since the fuel cell used in this study (Heliocentris FC50) is occupied with a security system that turns off the fuel cell in case of higher currents/temperatures [29–31].

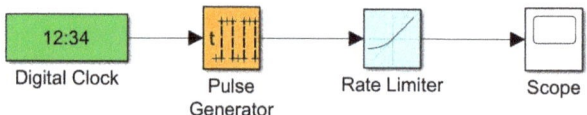

**Figure 3.** Triangular signal design.

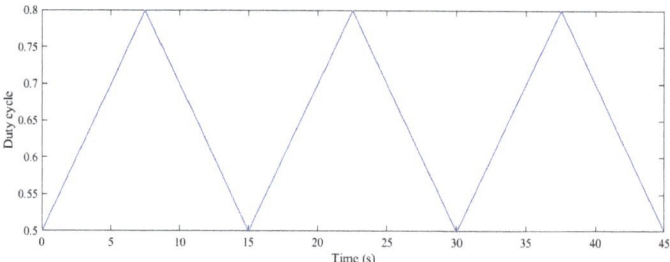

**Figure 4.** Triangular signal output.

To obtain data for different operating conditions, variations in temperature, humidity, hydrogen and airflow are required. It should be noted that the fuel cell contains an integrated control system that not only controls the supplied hydrogen but also provides an option to set the fans of the fuel cell at the automatic mode. By using the auto mode, the fans will automatically control the temperature, the humidity and the supplied airflow. However, to provide large degrees of freedom, the auto mode option of the fans was not considered. Therefore, a database containing 20,512 samples for different operating current, temperature and fan power were recorded and presented in Figure 5. This latter also shows the influence of the air flow on the fuel cell performance but the effect of temperature is still not well presented. Therefore, a 3D graph that clearly shows the effect of both temperature and air flow on the stack performance is presented in Figure 6. According to this latter, it is shown that at low air flow (fans power = 10%), by varying the temperature from 25 °C to 43 °C the stack performance improves in the beginning, then becomes almost constant and finally, it deteriorates for higher temperatures. At medium air flow (fans power = 50%), the stack performance improves with increasing temperature. However, for higher temperatures only slight improvements occur since the membrane requires an additional amount of water content. Regarding the last case at which the air flow is set at its maximum value (fans power = 100%), the stack performance improves largely with a temperature increase from $T = 25\,°C$ to over $40\,°C$. It is noticed that even for higher temperatures, the stack performance is still improving and this is due to the well humidification provided by the fans.

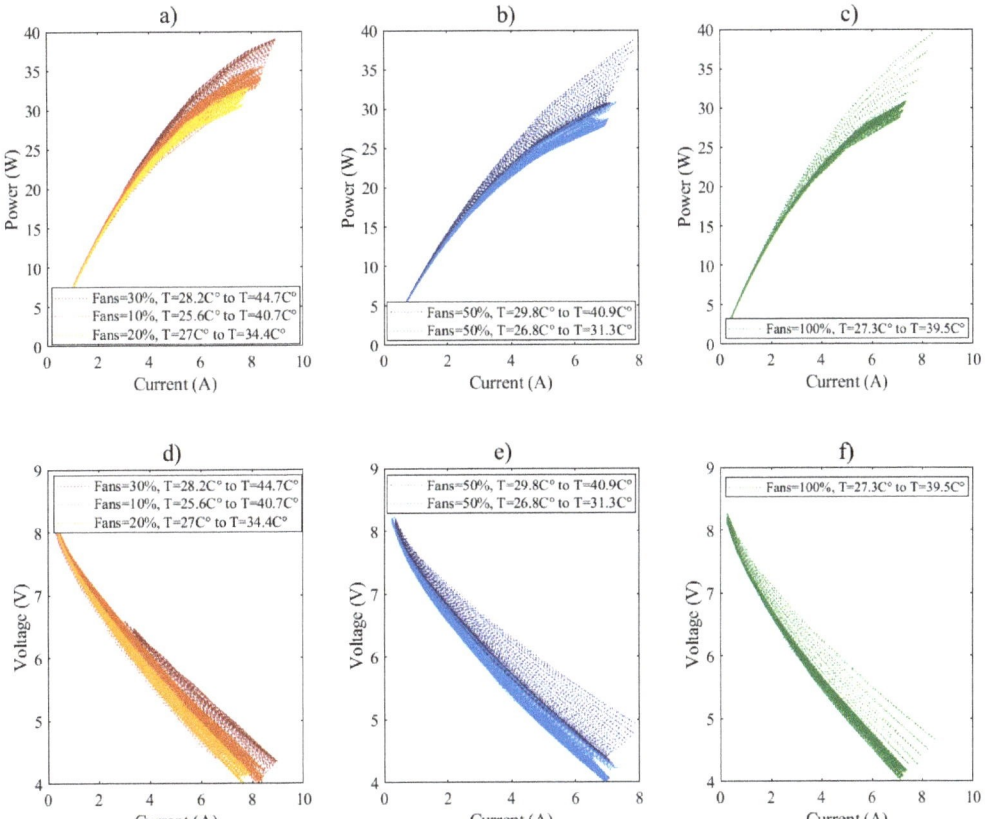

**Figure 5.** $I_{stack}$-$P_{stack}$ and $I_{stack}$-$V_{stack}$ measured data of Heliocentris FC50 fuel cell; (**a**,**d**): polarisation curves when Fans Power = [10%, 20%, 30%] and for different operating temperature; (**b**,**e**): polarisation curves when Fans Power = 50% and for different operating temperature; (**c**,**f**): polarisation curves when Fans Power = 100% and for different operating temperature.

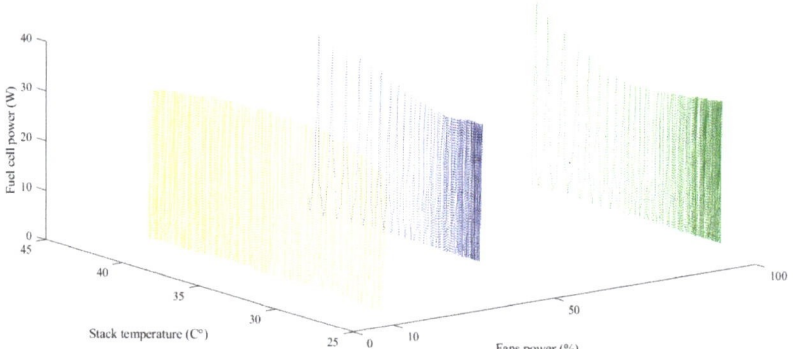

**Figure 6.** The Heliocentris FC50 stack power according to air flow and stack temperature.

- **Inputs and outputs selection:** Another factor that can improve the accuracy of the learned function is the selection of the inputs and outputs since the accuracy is strongly dependent on how the inputs are represented. The inputs should be entered as a feature vector that contains enough information to properly predict the output; but also, it should not be too large due to the dimensionality curse effect. In this study, the input variables are selected as: stack current $I_{stack}$ (A), stack temperature $T$ (°C) and fans power (%), to predict the stack voltage $V_{stack}$ (V).
- **Data division (training, validation and test):** When enough data is available, the next step is to split this data into three subsets which are training, validation and test. The training dataset needs to be fairly large and contains a variety of data in order to contain all the needed information. Many researchers have proposed a training set of 70%, 80% and 90% [32–35]; where the rest of data were divided between the validation and test. In this study, the recorded data was divided as the following: training = 14,358 data points (70% of whole data), validation = 3077 data points (15% of whole data) and test = 3077 data points (15% of whole data). The training subset is used to adjust the network via minimising its error. In other words, it is used for computing the gradient and updating the weights and biases of the NNs. The validation subset is used for measuring the network generalisation and to stop the training when the generalisation stops improving. In more detail, when the training begins to overfit the data, the validation error starts to rise. Therefore, the weights and biases of the network are saved at the minimum validation error point so as to balance the accuracy of the learned function versus overfitting. The test subset is used to evaluate the performance of learned function when applying a new set. Actually, the test subset has no influence on the determination of the learned function parameters, but it is a kind of 'final exam' to test the performance of each predicted function.

2.1.3. Designing the Network

Based on Figure 2, the output of the $h_i^{(j)}$ hidden layer unit can be calculated as Equation (1) [36].

$$h_i^{(j)} = f_{act}\left[\left(\sum_{i=0}^{M_j} w_{i,j} x_{i,j}\right) + b_i^j\right] \quad (1)$$

where, $j = [1,2,\ldots,L]$ refers to the $j$th hidden layer, $i = [1,2,\ldots,M_j]$ refers to the $i$th neuron in the hidden layer $j$, $M_j = [M_1, M_2, \ldots, M_L]$ refers to the number of neurons at each layer, $x \in \mathbb{R}^m$ are numerical inputs, $w \in \mathbb{R}^m$ are weights associated with the inputs, $b \in \mathbb{R}$ are biases. $f_{act}$ is the activation function which is used to introduce nonlinearity into the output of the artificial neuron. Actually, this is important since most of data in the real world is nonlinear and the neurons should learn these nonlinear representations. There are many activation functions that can be used in practice such as sigmoid, tanh, ReLu, etc. [37]. In this work a tansig function which is given in Equation (2) is used.

$$f_{act}(x) = \frac{2}{1+e^{-2x}} - 1 \qquad (2)$$

By using Equations (1) and (2), the $k^{th}$ output layer unit can be calculated as Equation (3).

$$y_k = f_{act}^{out}\left[\left(\sum_{i=0}^{M_L} w_{i,L} h_i^{(j)}\right) + b_k\right] \qquad (3)$$

where $k = [1,2,\ldots,N_2]$ and $f_{act}^{out}$ is a linear transfer function or also known as *purelin*, its mathematical expression is given in Equation (4)

$$purelin(x) = x \qquad (4)$$

To train the FFNNP, several optimisation algorithms can be used to minimise the performance function (also known as loss/cost function) [32,38–40]. These algorithms use either the Jacobian of the network errors or the gradient of the network performance. Both Jacobian and gradient are computed via the back-propagation algorithm which is an efficient computational trick for calculating derivatives inside the deep feed-forward NNs. In this work, we made a comparison study among the four major used algorithms including the Levenberg–Marquardt (LM), Bayesian regularization (BR), BFGS quasi-Newton and Scaled conjugate gradient (SCG). For each training algorithm, the following basic system training parameters are used: maximum number of epochs = 5000, learning rate = 0.01, performance goal = 0, time of training = Infinity. All these parameters were checked for different number of neurons and hidden layers as presented in Table 1.

Table 1. Mean squared error of different FFNNP structures/algorithms.

| Training Algorithms | Hidden Layers | MSE/Time(s) | Number of Neurons for Each Hidden Layer | | | | | | | |
|---|---|---|---|---|---|---|---|---|---|---|
| | | | 1 | 5 | 10 | 15 | 20 | 25 | 30 | 35 |
| LM | 1 | MSE | 0.0241 | 0.0052 | 0.0025 | 0.0016 | 0.0017 | 0.0015 | 0.0015 | 0.0014 |
| | | Time | 1.9520 | 9.3870 | 17.1920 | 6.9760 | 6.6720 | 24.9600 | 21.1340 | 16.2700 |
| | 2 | MSE | 0.0248 | 0.0036 | 0.0017 | 0.0014 | 0.0014 | 0.0012 | 0.0013 | 0.0012 |
| | | Time | 8.1680 | 8.1740 | 4.4090 | 31.4030 | 29.4810 | 90.1620 | 54.3320 | 235.6880 |
| | 3 | MSE | 0.0244 | 0.0017 | 0.0015 | 0.0012 | 0.0012 | 0.0011 | 0.0011 | 0.0012 |
| | | Time | 6.9090 | 10.7930 | 6.9720 | 38.5620 | 43.3770 | 304.0570 | 193.9730 | 346.6590 |
| BR | 1 | MSE | 0.0242 | 0.0106 | 0.0022 | 0.0015 | 0.0015 | 0.0015 | 0.0014 | 0.0014 |
| | | Time | 4.3540 | 6.5850 | 33.0220 | 20.9350 | 40.8840 | 69.8870 | 225.6650 | 268.2380 |
| | 2 | MSE | 0.0243 | 0.0022 | 0.0014 | 0.0011 | 0.0010 | 0.0009 | 0.0008 | 0.0008 |
| | | Time | 41.7 | 6.5 | 132.6 | 260.6 | 657.3 | 1583.5 | 2954.5 | 5438.6 |
| | 3 | MSE | 0.0243 | 0.0015 | 0.0012 | 0.0009 | 0.0008 | 0.0007 | 0.0006 | 0.0006 |
| | | Time | 41.4 | 67.6 | 126 | 518.2 | 1064.3 | 4741.2 | 6936.3 | 13217.4 |
| BFG | 1 | MSE | 0.0245 | 0.0082 | 0.0065 | 0.0036 | 0.0030 | 0.0029 | 0.0022 | 0.0023 |
| | | Time | 2.2560 | 2.4880 | 1.8110 | 4.7750 | 5.3230 | 7.8430 | 13.4960 | 7.9190 |
| | 2 | MSE | 0.0245 | 0.0088 | 0.0024 | 0.0017 | 0.0016 | 0.0020 | 0.0019 | 0.0017 |
| | | Time | 1.6280 | 2.2800 | 8.2930 | 26.7120 | 25.1810 | 20.8690 | 66.0140 | 223.0960 |
| | 3 | MSE | 0.0251 | 0.0048 | 0.0053 | 0.0019 | 0.0017 | 0.0016 | 0.0018 | 0.0018 |
| | | Time | 1.6 | 8.1 | 15.6 | 27.8 | 85.0 | 278.1 | 601.0 | 1125.1 |
| SCG | 1 | MSE | 0.0258 | 0.0145 | 0.0100 | 0.0090 | 0.0070 | 0.0052 | 0.0096 | 0.0070 |
| | | Time | 1.2110 | 1.1140 | 1.5290 | 2.4660 | 2.8820 | 5.4720 | 2.0470 | 4.3250 |
| | 2 | MSE | 0.0261 | 0.0204 | 0.0317 | 0.0040 | 0.0041 | 0.0023 | 0.0051 | 0.0027 |
| | | Time | 1.0180 | 1.0500 | 0.7660 | 5.9290 | 6.2720 | 11.5320 | 5.1960 | 27.5590 |
| | 3 | MSE | 0.0261 | 0.0082 | 0.0061 | 0.0055 | 0.0028 | 0.0025 | 0.0024 | 0.0027 |
| | | Time | 1.5150 | 4.1530 | 4.7230 | 6.3070 | 12.4290 | 21.7610 | 24.7390 | 42.3530 |

The performance of each training algorithm was measured via the mean squared error ($mse$) which is given in Equation (5), where $y_i^*$ is the desired output (target), $y_i$ is the actual (predicted) output, and $N$ is the number of dataset.

$$F = mse = \frac{1}{N}\sum_{i=0}^{N}(e_i)^2 = \frac{1}{N}\sum_{i=0}^{N}(y_i^* - y_i)^2 \qquad (5)$$

The best performance, in terms of training time and mean squared error $mse$, of each algorithm is tinted with green colour (Table 1). The predicted output results as well as the error that corresponds to the best performance (green cells) for each training algorithm are respectively shown in Figures 7 and 8.

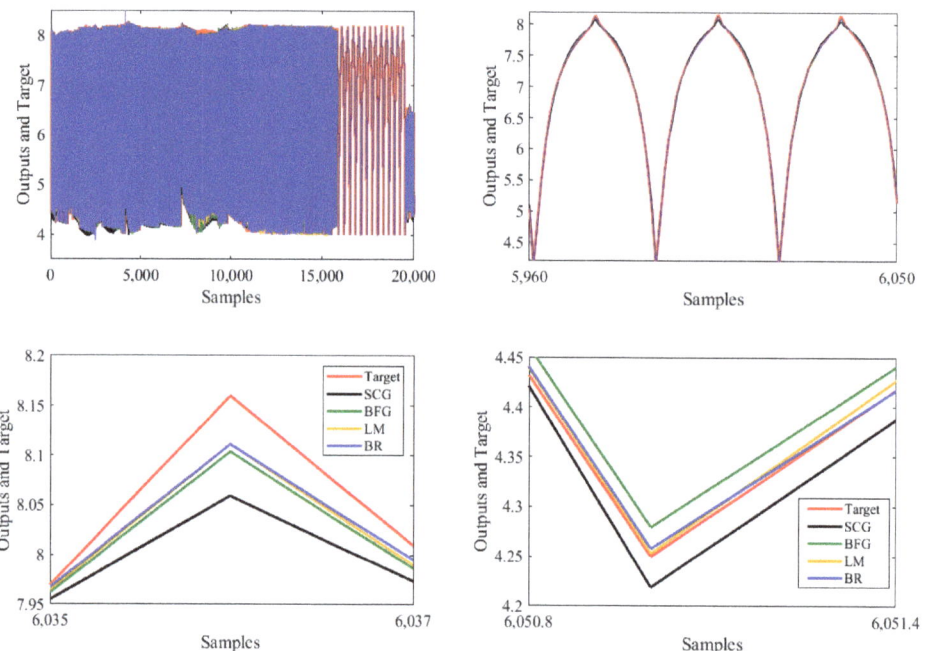

**Figure 7.** Predicted output results when using SCG, BFG, LM and BR.

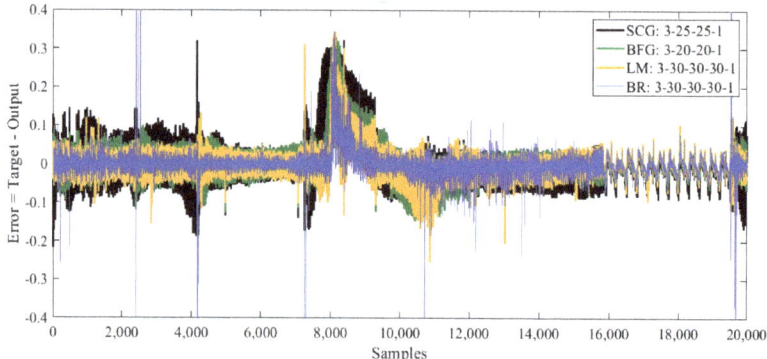

**Figure 8.** Obtained training errors when using SCG, BFG, LM and BR.

According to these figures, it is clear that the BR training algorithm with the structure of 3 hidden layers and 30 neurons for each predicts the best output results in terms of accuracy, where the SCG shows the worst predicted results in comparison with the rest of the algorithms. In terms of time, the SCG shows the fastest training since it takes only around 11 s to predict the output while the BR needs around 6930 s. However, although the BR takes around 2 h for the training, it finally provides a highly accurate model which is one of the main goals of this study.

Figure 9 shows the regression plots which were used to validate the performance of the obtained trained model. According to this figure, it is clear that the predicted model is characterised by high accuracy since most of the data points fall along a 45 degree line, where the output is equal to the target. The goodness of the model also can be analysed via the R values which ranged between 0 (lowest accuracy) and 1 (ideal model). In our case, the accuracy of the obtained model is proven by the following R values: training, R = 0.99974, test, R = 0.99735 and all, R = 0.99938.

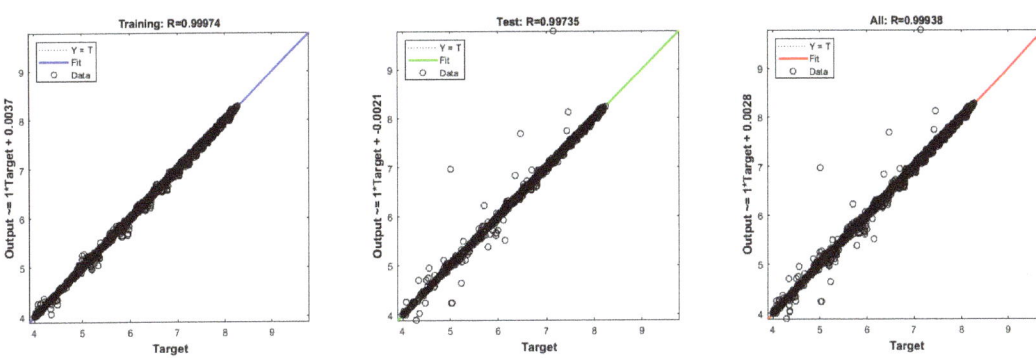

**Figure 9.** Performance analysis of the predicted model.

### 2.2. PEMFC Control with ANN-PID

#### 2.2.1. Control Design

Although PID control is one of the most used controllers in industries, it still suffers from systems sufficient nonlinearity which make its constant parameters not optimal in each operating moment. This is due to the difficulties of determining the parameters which are usually tuned via the conventional trial and error method. As a solution, we have designed a self-adaptive PID based feed-forward artificial neural network (ANN-PID) aiming to avoid parameters manual tuning. The input of the ANN-PID controller is the error $e(k)$ which is achieved from the difference between the desired and actual PEMFC stack currents, and the output is the duty cycle signal $u(k)$. The error is decomposed into three variables $x_i$ ($i$ = [1,2,3]) similarly to the conventional PID, but they will be respectively associated with three weights $w_i$ which are self-tuned via the Hebb supervised learning rule method. The implementation of the of ANN-PID in the hardware system is explained in Figure 10.

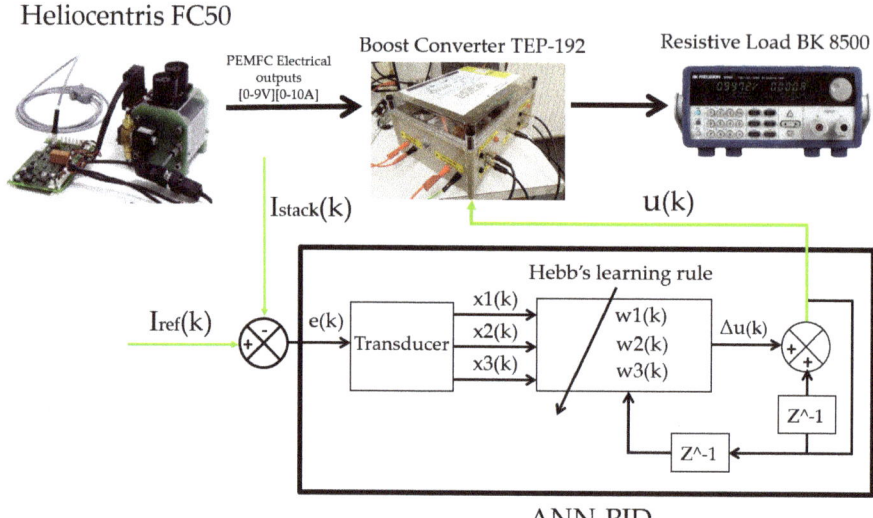

**Figure 10.** Implementation of ANN-PID on Heliocentris FC50 hardware system.

The output of the ANN-PID controller $\Delta u(k)$ is given in Equation (6), where $k$ is a positive parameter determined by the user, and the values of the three inputs $x_i$ are given in Equation (7).

$$\Delta u(k) = k \sum_{i=1}^{3} x_i(k) w_i(k) = k(x_1(k)w_1(k) + x_2(k)w_2(k) + x_3(k)w_3(k)) \quad (6)$$

$$x_i(k) = \begin{cases} x_1(k) = \Delta e(k) \\ x_2(k) = e(k) \\ x_3(k) = \Delta e(k) - \Delta e(k-1) \end{cases} \quad (7)$$

The biological origin of Hebb's supervised learning was established from a neuroscience perspective: when two neurons are activated simultaneously, the link intensity (also called plasticity) is proportional to the multiplication of their stimulation [41,42]. Therefore, this concept can be translated mathematically for the adjustment of the PID parameters ($k_p$, $k_i$ and $k_d$) which can be obtained through a neural settlement of Equation (8) as Equation (8) shows, where $\eta_i$ are learning rates that correspond to $w_i(k)$ [43].

$$w_i(k) = w_i(k-1) + \eta_i x_i(k) u(k-1) e(k) \quad (8)$$

Recently, it has been found that the weight values used for PID online regulation are mainly related to $e(k)$ and $\Delta e(k)$ [44]. Hence, the inputs $x_i(k)$ of Equation (8) can be replaced by $e(k) + \Delta e(k)$. Finally, the running algorithm of the control law can be expressed as Equation (9).

$$\begin{cases} w_i(k) = w_i(k-1) + \eta_i [e(k) + \Delta e(k)] u(k-1) e(k) \\ w'_i(k) = \dfrac{w_i(k)}{\sum_{i=1}^{3} |w_i(k)|} \\ u(k) = u(k-1) + K \sum_{i=1}^{3} w'_i(k) x_i(k) \end{cases} \quad (9)$$

2.2.2. Metrics Used for Control Performance Improvement

To achieve high tracking performance, the minimisation of the integral of the absolute error (IAE), the root mean squared error (RMSE) and the relative root mean squared error (RRMSE), which are described by Equation (10), have been used to adjust and tune the gains of the controller, whereas the values can be determined by taking into account the error reduction in real time.

$$\begin{cases} IAE = \sum_{i=1}^{N} |e_i| \Delta t \\ RMSE = \sqrt{\dfrac{1}{N} \sum_{i=1}^{N} (e_i)^2} \\ RRMSE = \sqrt{\sum_{i=1}^{N} (e_i)^2 / \sum_{i=1}^{N} (r_i)} \times 100\% \end{cases} \quad (10)$$

where $N$, $e_i$ and $r_i$ are, respectively, an observation data length time for the calculation, the tracking error and the reference along the $i$-th *sample*.

## 3. Results and Discussion

### 3.1. Comparison between the Experiment and Simulation Results

The $I_{stack}$-$V_{stack}$ and $I_{stack}$-$P_{stack}$ polarisation curves of the simulated and real model are presented in Figure 11. According to this figure, it is clear that the predicted model succeeded in providing the same results obtained by the real fuel cell system. It should be noted that the temperature in the experiment curves has an error around $\pm 0.5\,°C$ since it is difficult to make experiments at constant temperatures. One other variable factor that also should be taken into account is the input Hydrogen pressure which is controlled by the manufacture. However, although these two variable factors can differ the predicted results from the real ones, only slight deviations occurred.

### 3.2. Effect of Temperature and Humidity on the PEM Fuel Cell Stack Performance

The effects of the operation temperature on the polarisation curves for a low, medium and high humidification (fans power are set at 10%, 50% and 100%) are presented in Figure 11. At low humidification (Figure 11a,b), by varying the temperature from 25 °C to 43 °C the stack performance improves from $T = 25\,°C$ until $T = 31\,°C$ and then deteriorates for temperatures up to 31 °C. The improvement of the performance from $T = 25\,°C$ until $T = 31\,°C$ can be explained by the enhancement of the conductivity of the membrane which leads to reducing the activation loss. However, for temperatures above 31 °C the membrane starts to dry due to the lack of water content which leads as a consequence to decrease the performance of the stack. At medium humidification (Figure 11c,d), the stack performance improves with increasing temperature. However, for higher temperatures only slight improvements occurred since the membrane requires an additional amount of water content. Regarding the last case at which the membrane is 100% humidified (Figure 11e,f), the stack performance is largely improved with increasing the temperature from $T = 25\,°C$ until $T = 39\,°C$. It is noticed that even for higher temperatures the stack performance is still improving and this is due to the well humidification provided by the fans. It should be noted that although the high humidification has a positive effect on the stack performance for higher temperatures, it also has a negative effect for lower temperatures. Hence, according to Figure 11b,f and for a low temperature equal to 25 °C, it is clear that the stack performance for low humidification ($P_{max} = 33\,W$) is better than the one obtained by high humidification ($P_{max} = 28\,W$).

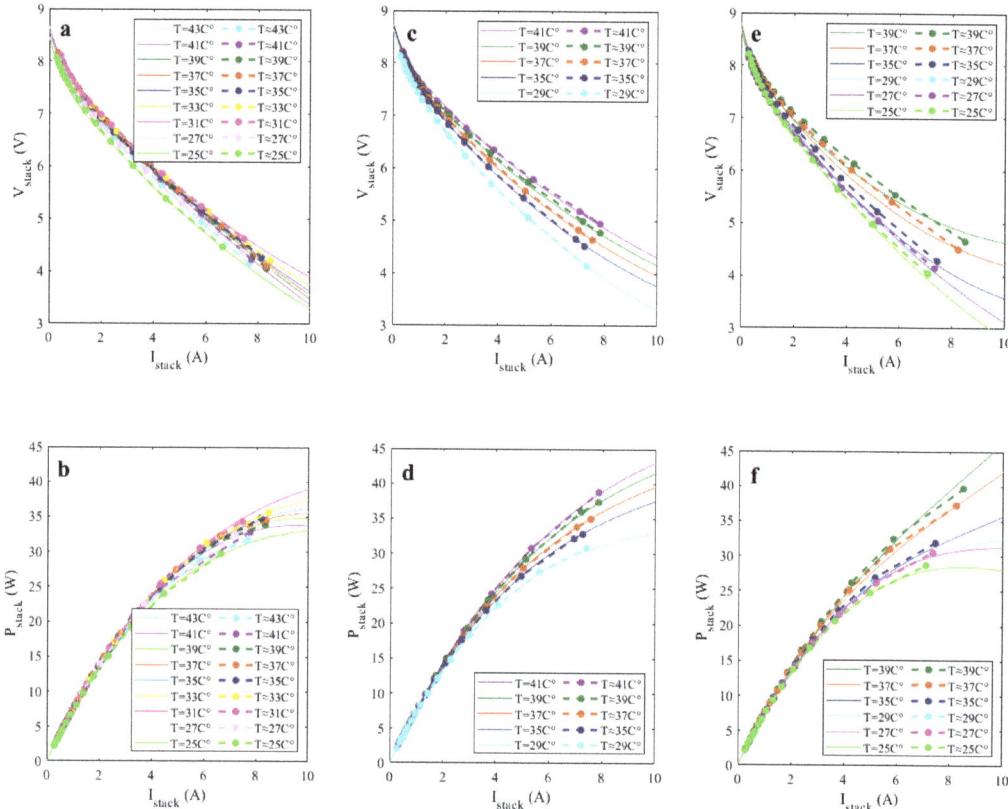

**Figure 11.** Simulation and experiment results (simulation: continuous line; experiment: dashed line); (**a**): $I_{stack}$-$V_{stack}$ polarisation curves when Fans Power = 10%; (**b**): $I_{stack}$-$P_{stack}$ polarisation curves when Fans Power = 10%; (**c**): $I_{stack}$-$V_{stack}$ polarisation curves when Fans Power = 50%; (**d**): $I_{stack}$-$P_{stack}$ polarisation curves when Fans Power = 50%; (**e**): $I_{stack}$-$V_{stack}$ polarisation curves when Fans Power = 100%; (**f**): $I_{stack}$-$P_{stack}$ polarisation curves when Fans Power = 100%.

*3.3. Control Results*

To keep the fuel cell operating at an adequate power point, PID and NN-PID are used. First, the controllers were designed and tested via the the predicted model so as to determine their coefficients. Then, they were implemented on the PEMFC hardware system using the Matlab/Simulink™ graphical interface. To test the performance of the PID and the NN-PID, two load variations respectively from 20 Ω to 50 Ω and from 50 Ω to 20 Ω are applied during the experiments. The obtained results are clearly presented in Figures 12 and 13.

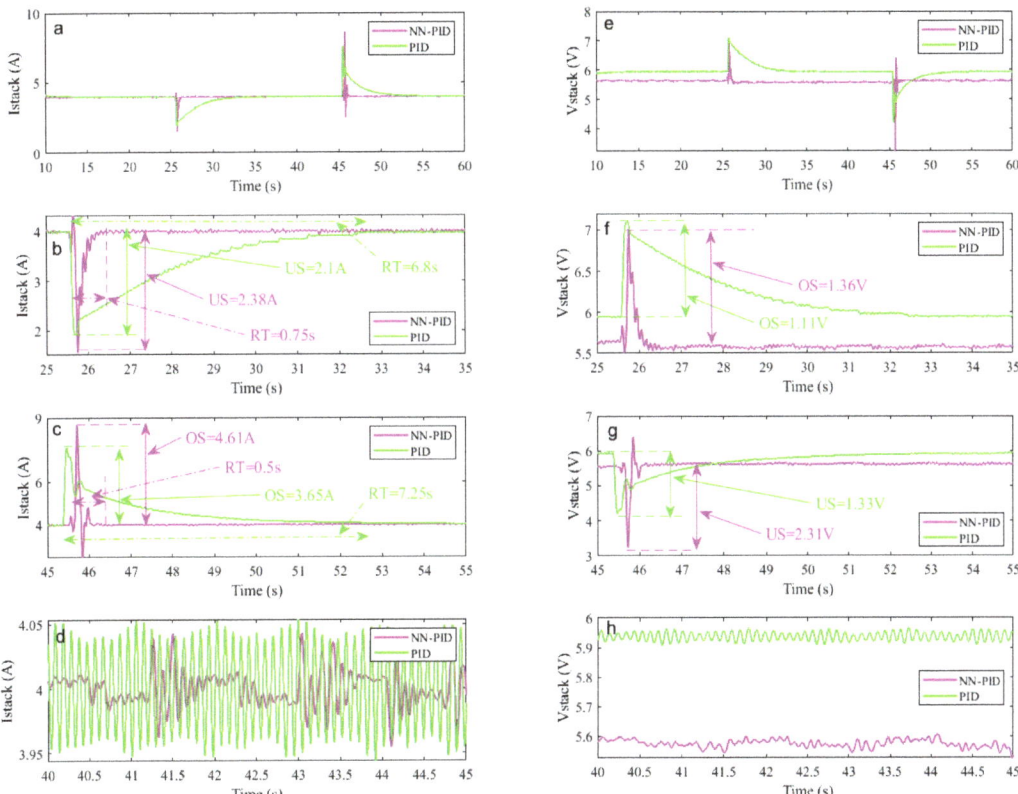

**Figure 12.** PID and NN-PID experimental results; (**a**–**d**): PEMFC stack current signal; (**e**–**h**): PEMFC stack voltage signal.

The waveforms of the stack current $I_{stack}$ and voltage $V_{stack}$ are presented in Figure 12a–d and Figure 12e–h, where (b–d) and (f–h) are respectively the zoom in of a and e. The stack power $P_{stack}$ is shown in Figure 13a–d (b–d are the zoom in of a); whereas the duty cycle and the boost converter output signals (current, voltage and power) are exhibited in Figure 13e–h. According to these graphs, it is clear that both PID and NN-PID succeeded in driving the PEMFC to operate at an adequate power point even when experiencing large load variation. However, although the PID track the reference, slow motion at each load variation occurred. It takes around 6.8 s and 7.25 s to converge to the desired value (response time) respectively for the first and second load variation; whereas the NN-PID requires only 0.75 s and 0.5 s for the same load variations. Regarding the overshoots and undershoots, the PID shows an undershoot current of 2.1 A, an overshoot voltage of 1.11 V and an undershoot power of 10.21 W for the first load variation and an overshoot current of 3.65 A, an undershoot voltage of 1.33 V and an overshoot power of 8.58 W for the second load variation is displayed. On the other hand, the application of the NN-PID performs an undershoot current of 2.38 A, an overshoot voltage of 1.36 A and an undershoot power of 11.18 W for the first load variation and an overshoot current of 4.61 A, an undershoot voltage of 2.31 V and an overshoot power of 7.2 W for the second load variation. It should be noted that both experiments are made at different temperature since it is difficult to keep the fuel cell operating at a constant temperature. Since the stack current is forced via the controllers to follow the reference, the temperature effect of each experiment on the stack performance appears in the stack voltage as presented in Figure 12e–h. The steady state error of current, voltage and power for both PID and NN-PID are respectively shown in (d) and (h) of Figure 12 and (d) of Figure 13. According to these results, it is clear that the NN-PID provides better results in terms of accuracy since it reduces the amplitude of ripples from 0.1 A to less than 0.01 A. Therefore, although the PID shows slightly lower overshoots in comparison with the NN-PID, this latter provides significantly higher performance in terms of response time and steady state error.

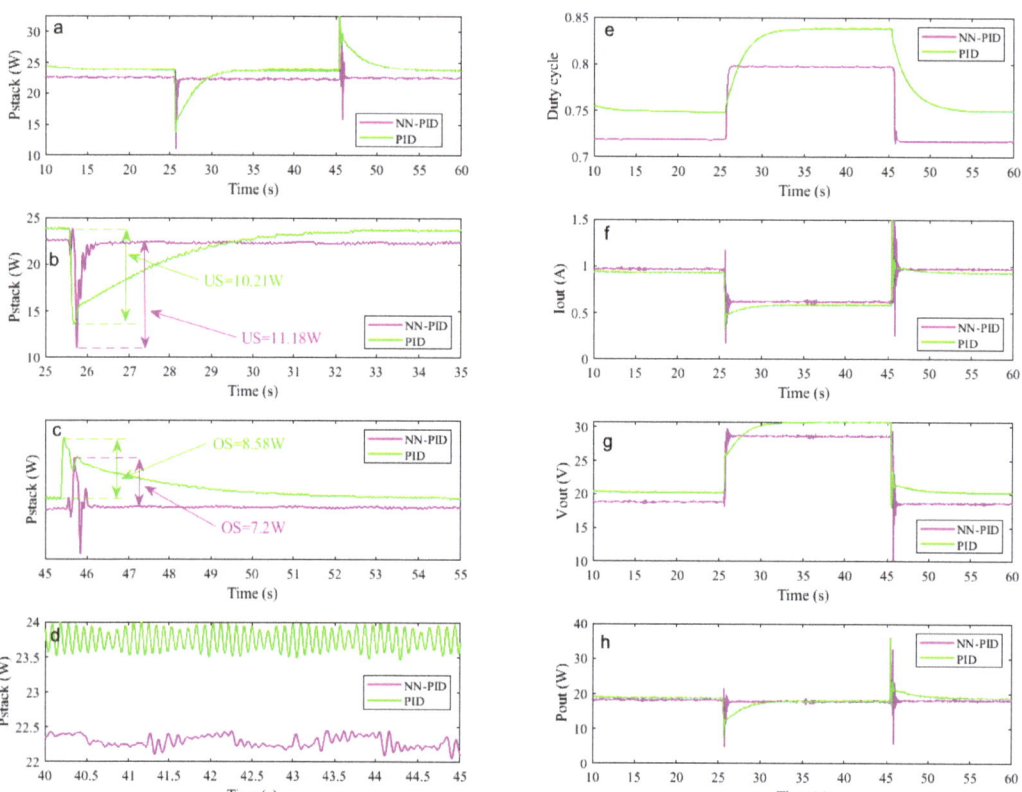

**Figure 13.** PID and NN-PID experimental results; (**a–d**): PEMFC stack power signal; (**e**): Duty cycle signal (**f**): Boost converter output current; (**g**): Boost converter output voltage; (**h**): Boost converter output power.

Finally, Table 2 summarises the results of the metrics used to measure the control demeanour. It can be seen that the NN-PID achieved a better outcome in terms of the IAE since it gathered a lower value in comparison to the PID, which represents 62.8% of performance increment. In regards to the accuracy, the trend is still favourable for the NN-PID which is in contrast to the PID of 93.6%. The same situation is seen in the RRMSE since the NN-PID provided a higher improvement as 0.344% was obtained where, in comparison, the PID achieved 5.38%.

**Table 2.** Comparison of the different metrics.

| IAE | | RMSE | | RRMSE (%) | |
|---|---|---|---|---|---|
| NN-PID | PID | NN-PID | PID | NN-PID | PID |
| 0.0049 | 0.0132 | 0.0138 | 0.2154 | 0.3440 | 5.3857 |

## 4. Conclusions

This paper presented an analysis of a commercial Heliocentris FC50 PEM fuel cell system; the objective was to model and control the device via the application of a deep machine learning based artificial neural network. Due to its several input variations, such as stack temperature, humidity and oxygen, which results in nonlinearities and high model complexity, extensive tests with various ANN parameters were required to predict an efficient model.

Since the ANN model requires a large dataset, an efficient automatic method was designed to simplify and facilitate the data collection. This was obtained by generating a triangular signal which varies the duty cycle of the power converter that was inserted between the stack and the load. An experimental dataset composed of 20,512 samples over a wide operating range (different operating current, temperature and fan power) of a commercial stack was recorded and saved for the training process.

Different structures of feed-forward neural network perceptron with backpropagation learning rule were tested to predict the performance of the Heliocentries FC50 fuel cell system. A comparison study including various ANN parameters such as the training algorithm, the number of hidden layers and the number of neurons at each layer was made to obtain the highest accurate model. Finally, an accurate model composed of 3 hidden layers and 90 neurons trained by BR algorithm was used for a comparison study with the real results. On the other hand, the predicted model also was adopted for determining the parameters of the NN-PID control method.

The effect of temperature on the PEM fuel cell stack performance was studied for low, medium and high humidification. At low humidification, it was obtained that the performance of the stack improves for low temperatures (from $T = 25\,°C$ until $T = 31\,°C$) and deteriorates for temperatures up to 31 °C. At medium and high humidifications, it was obtained that the stack performance improves with increasing temperature. However, the effect of temperature is clearly pronounced at higher humidification since the increase of temperature results in a large increase in the stack performance.

At last, two controllers were designed and performed to keep the fuel cell stack operating point at an adequate power stage. Results have demonstrated that both PID and NN-PID have succeeded in driving the stack operation to the desired power point even when experiencing large load variation. However, comparison results have shown high-performance tracking in terms of response time, and steady state error was obtained via the application of the proposed NN-PID control method.

Through this research, future trends for modelling and control of PEM fuel cell systems were analysed and will be the goal of the forthcoming studies. Other types of ANN such as recurrent neural network (RNN) can be an option to improve the performance of the model. Regarding the control method, robust and adaptive controls tuned via neural approach can be also an efficient trend to improve the tracking performance.

**Author Contributions:** Conceptualisation, M.D., C.N. and O.B.; methodology, M.D. and C.N.; software, M.D. and C.N.; validation, M.D.; formal analysis, M.D.; investigation, M.D., C.N. and O.B.; resources, O.B.; writing—original draft preparation, M.D. and C.N.; writing—review and editing, M.D., C.N. and O.B.; supervision, O.B.; project administration, O.B. All authors have read and agreed to the published version of the manuscript.

**Funding:** This research was funded by the Basque Government, Diputación Foral de Álava and UPV/EHU, respectively, through the projects EKOHEGAZ (ELKARTEK KK-2021/00092), CONAVANTER and GIU20/063.

**Institutional Review Board Statement:** Not applicable.

**Informed Consent Statement:** Not applicable.

**Data Availability Statement:** Not applicable.

**Acknowledgments:** The authors would like to express their gratitude to the UPV/EHU, the Basque Government and the Diputación Foral de Álava.

**Conflicts of Interest:** The authors declare no conflict of interest.

## Abbreviations

The following abbreviations are used in this manuscript:

| | |
|---|---|
| FC | fuel cell |
| PEM | proton exchange membrane |
| ML | machine learning |
| ANN | artificial neural network |
| PEMFC | polymer electrolyte membrane fuel cell |
| SOFCs | solid oxide fuel cells |
| PAFCs | phosphoric acid fuel cells |
| MCFCs | molten carbonate fuel cells |
| MEA | membrane electrode assembly |
| SVM | support vector machine |
| MLP | multilayer perceptron |
| ANN-PID | artificial neural network proportional integral-derivative |
| PID | proportional integral derivative |
| FFNNP | feed-forward neural network perceptron |
| LM | levenberg-marquardt |
| BR | bayesian regularization |
| SCG | scaled conjugate gradient |
| MSE | mean squared error |
| BFGS | broyden fletcher goldfarb shanno |
| IAE | integral of the absolute error |
| RMSE | root mean squared error |
| RRMSE | relative root mean squared error |
| RNN | recurrent neural network |

## References

1. Li, D.; Li, S.; Ma, Z.; Xu, B.; Lu, Z.; Li, Y.; Zheng, M. Ecological Performance Optimization of a High Temperature Proton Exchange Membrane Fuel Cell. *Mathematics* **2021**, *9*, 1332. [CrossRef]
2. Mahapatra, M.K.; Singh, P. Chapter 24-Fuel Cells: Energy Conversion Technology. In *Future Energy*, 2nd ed.; Letcher, T.M., Ed.; Elsevier: Boston, MA, USA, 2014; pp. 511–547. [CrossRef]
3. Kadyk, T.; Winnefeld, C.; Hanke-Rauschenbach, R.; Krewer, U. Analysis and Design of Fuel Cell Systems for Aviation. *Energies* **2018**, *11*, 375. [CrossRef]
4. Oldenbroek, V.; Smink, G.; Salet, T.; van Wijk, A.J. Fuel Cell Electric Vehicle as a Power Plant: Techno-Economic Scenario Analysis of a Renewable Integrated Transportation and Energy System for Smart Cities in Two Climates. *Appl. Sci.* **2020**, *10*, 143. [CrossRef]
5. Xing, H.; Stuart, C.; Spence, S.; Chen, H. Fuel Cell Power Systems for Maritime Applications: Progress and Perspectives. *Sustainability* **2021**, *13*, 1213. [CrossRef]
6. Wang, Y.; Chen, K.S.; Mishler, J.; Cho, S.C.; Adroher, X.C. A review of polymer electrolyte membrane fuel cells: Technology, applications, and needs on fundamental research. *Appl. Energy* **2011**, *88*, 981–1007. [CrossRef]
7. Weber, A.Z.; Balasubramanian, S.; Das, P.K. Chapter 2-Proton Exchange Membrane Fuel Cells. In *Fuel Cell Engineering*; Sundmacher, K., Ed.; Academic Press: Cambridge, MA, USA, 2012; Volume 41, pp. 65–144. [CrossRef]
8. Ji, M.; Wei, Z. A Review of Water Management in Polymer Electrolyte Membrane Fuel Cells. *Energies* **2009**, *2*, 1057–1106. [CrossRef]
9. Han, J.; Charpentier, J.F.; Tang, T. An Energy Management System of a Fuel Cell/Battery Hybrid Boat. *Energies* **2014**, *7*, 2799–2820. [CrossRef]
10. Fang, L.; Di, L.; Ru, Y. A Dynamic Model of PEM Fuel Cell Stack System for Real Time Simulation. In Proceedings of the 2009 Asia-Pacific Power and Energy Engineering Conference, Wuhan, China, 27–31 March 2009; pp. 1–5. [CrossRef]
11. Saadi, A.; Becherif, M.; Aboubou, A.; Ayad, M. Comparison of proton exchange membrane fuel cell static models. *Renew. Energy* **2013**. [CrossRef]
12. Amphlett, J.C.; Baumert, R.M.; Mann, R.F.; Peppley, B.A.; Roberge, P.R.; Harris, T.J. Performance Modeling of the Ballard Mark IV Solid Polymer Electrolyte Fuel Cell: I . Mechanistic Model Development. *J. Electrochem. Soc.* **1995**, *142*, 1–8. [CrossRef]
13. Kandidayeni, M.; Macias, A.; Boulon, L.; Trovão, J.P.F. Online Modeling of a Fuel Cell System for an Energy Management Strategy Design. *Energies* **2020**, *13*, 3713. [CrossRef]
14. Je, L.; Dicks, A. Proton exchange membrane fuel cells. *Fuel Cell Syst. Explain.* **2013**, *18*, 67–119. [CrossRef]

15. Kim, J.; Lee, S.; Srinivasan, S.; Chamberlin, C. Modeling of Proton Exchange Membrane Fuel Cell Performance with an Empirical Equation. *J. Electrochem. Soc.* **1995**, *142*, 2670–2674. [CrossRef]
16. Pathapati, P.; Xue, X.; Tang, J. A new dynamic model for predicting transient phenomena in a PEM fuel cell system. *Renew. Energy* **2005**, *30*, 1–22. [CrossRef]
17. Ansari, S.; Khalid, M.; Kamal, K.; Abdul Hussain Ratlamwala, T.; Hussain, G.; Alkahtani, M. Modeling and Simulation of a Proton Exchange Membrane Fuel Cell Alongside a Waste Heat Recovery System Based on the Organic Rankine Cycle in MATLAB/SIMULINK Environment. *Sustainability* **2021**, *13*, 1218. [CrossRef]
18. Rubio, G.A.; Agila, W.E. A Fuzzy Model to Manage Water in Polymer Electrolyte Membrane Fuel Cells. *Processes* **2021**, *9*, 904. [CrossRef]
19. Moaveni, B.; Rashidi Fathabadi, F.; Molavi, A. Fuzzy control system design for wheel slip prevention and tracking of desired speed profile in electric trains. *Asian J. Control* **2020**, 1–13. [CrossRef]
20. Napole, C.; Barambones, O.; Calvo, I.; Derbeli, M.; Silaa, M.; Velasco, J. Advances in Tracking Control for Piezoelectric Actuators Using Fuzzy Logic and Hammerstein-Wiener Compensation. *Mathematics* **2020**, *8*, 2071. [CrossRef]
21. Tian, Y.; Zou, Q.; Han, J. Data-Driven Fault Diagnosis for Automotive PEMFC Systems Based on the Steady-State Identification. *Energies* **2021**, *14*, 1918. [CrossRef]
22. Shao, J.; Liu, X.; He, W. Kernel Based Data-Adaptive Support Vector Machines for Multi-Class Classification. *Mathematics* **2021**, *9*, 936. [CrossRef]
23. Napole, C.; Barambones, O.; Derbeli, M.; Calvo, I.; Silaa, M.; Velasco, J. High-Performance Tracking for Piezoelectric Actuators Using Super-Twisting Algorithm Based on Artificial Neural Networks. *Mathematics* **2021**, *9*, 244. [CrossRef]
24. Nanadegani, F.S.; Lay, E.N.; Iranzo, A.; Salva, J.A.; Sunden, B. On neural network modeling to maximize the power output of PEMFCs. *Electrochim. Acta* **2020**, *348*, 136345. [CrossRef]
25. Qin, Y.; Duan, H. Single-Neuron Adaptive Hysteresis Compensation of Piezoelectric Actuator Based on Hebb Learning Rules. *Micromachines* **2020**, *11*, 84. [CrossRef] [PubMed]
26. Vt, S.E.; Shin, Y.C. Radial basis function neural network for approximation and estimation of nonlinear stochastic dynamic systems. *IEEE Trans. Neural Netw.* **1994**, *5*, 594–603.
27. Li, Y.; Qiang, S.; Zhuang, X.; Kaynak, O. Robust and adaptive backstepping control for nonlinear systems using RBF neural networks. *IEEE Trans. Neural Netw.* **2004**, *15*, 693–701. [CrossRef] [PubMed]
28. Priddy, K.L.; Keller, P.E. *Artificial Neural Networks: An Introduction*; SPIE Press: Bellingham, WA, USA, 2005; Volume 68.
29. Derbeli, M.; Charaabi, A.; Barambones, O.; Napole, C. High-Performance Tracking for Proton Exchange Membrane Fuel Cell System PEMFC Using Model Predictive Control. *Mathematics* **2021**, *9*, 1158. [CrossRef]
30. Derbeli, M.; Barambones, O.; Farhat, M.; Ramos-Hernanz, J.A.; Sbita, L. Robust high order sliding mode control for performance improvement of PEM fuel cell power systems. *Int. J. Hydrogen Energy* **2020**, *45*, 29222–29234. [CrossRef]
31. Derbeli, M.; Barambones, O.; Ramos-Hernanz, J.A.; Sbita, L. Real-time implementation of a super twisting algorithm for PEM fuel cell power system. *Energies* **2019**, *12*, 1594. [CrossRef]
32. Karim, H.; Niakan, S.R.; Safdari, R. Comparison of neural network training algorithms for classification of heart diseases. *IAES Int. J. Artif. Intell.* **2018**, *7*, 185. [CrossRef]
33. Falcão, D.; Pires, J.C.M.; Pinho, C.; Pinto, A.; Martins, F.G. Artificial neural network model applied to a PEM fuel cell. In Proceedings of the IJCCI 2009: Proceedings of the International Joint Conference on Computational Intelligence, Funchal, Portugal, 5–7 October 2009.
34. Dao, D.V.; Adeli, H.; Ly, H.B.; Le, L.M.; Le, V.M.; Le, T.T.; Pham, B.T. A sensitivity and robustness analysis of GPR and ANN for high-performance concrete compressive strength prediction using a Monte Carlo simulation. *Sustainability* **2020**, *12*, 830. [CrossRef]
35. Nhu, V.H.; Hoang, N.D.; Duong, V.B.; Vu, H.D.; Bui, D.T. A hybrid computational intelligence approach for predicting soil shear strength for urban housing construction: A case study at Vinhomes Imperia project, Hai Phong city (Vietnam). *Eng. Comput.* **2020**, *36*, 603–616. [CrossRef]
36. Arabi, M.; dehshiri, A.; Shokrgozar, M. Modeling transportation supply and demand forecasting using artificial intelligence parameters (Bayesian model). *Istraz. I Proj. Za Privredu* **2018**, *16*, 43–49. [CrossRef]
37. Mudunuru, V. Comparison of activation functions in multilayer neural networks for stage classification in breast cancer. *Neural Parallel Sci. Comput. Arch.* **2016**, *24*, 83–96.
38. Kumar, D.A.; Murugan, S. Performance Analysis of MLPFF Neural Network Back Propagation Training Algorithms for Time Series Data. In Proceedings of the 2014 World Congress on Computing and Communication Technologies, Trichirappalli, India, 27 February–1 March 2014; pp. 114–119. [CrossRef]
39. Sharma, B.; Venugopalan, K. Comparison of Neural Network Training Functions for Hematoma Classification in Brain CT Images. *IOSR J. Comput. Eng.* **2014**, *16*, 31–35. [CrossRef]
40. Shende, K.V.; Kumar, M.R.; Kale, K. Comparison of Neural Network Training Functions for Prediction of Outgoing Longwave Radiation over the Bay of Bengal. In *Computing in Engineering and Technology*; Springer: Berlin/Heidelberg, Germany, 2020; pp. 411–419.
41. Meng, F.; Hu, Y.; Ma, P.; Zhang, X.; Li, Z. Practical Control of a Cold Milling Machine using an Adaptive PID Controller. *Appl. Sci.* **2020**, *10*, 2516. [CrossRef]

42. Magotra, A.; Kim, J. Improvement of Heterogeneous Transfer Learning Efficiency by Using Hebbian Learning Principle. *Appl. Sci.* **2020**, *10*, 5631. [CrossRef]
43. Napole, C.; Barambones, O.; Calvo, I.; Velasco, J. Feedforward Compensation Analysis of Piezoelectric Actuators Using Artificial Neural Networks with Conventional PID Controller and Single-Neuron PID Based on Hebb Learning Rules. *Energies* **2020**, *13*, 3929. [CrossRef]
44. Liang, Y.; Xu, S.; Hong, K.; Wang, G.; Zeng, T. Neural network modeling and single-neuron proportional–integral–derivative control for hysteresis in piezoelectric actuators. *Meas. Control* **2019**, *52*, 1362–1370. [CrossRef]

Article

# "Holographic Implementations" in the Complex Fluid Dynamics through a Fractal Paradigm

Alexandra Saviuc [1,*], Manuela Gîrțu [2], Liliana Topliceanu [3], Tudor-Cristian Petrescu [4] and Maricel Agop [5,6]

[1] Faculty of Physics, Alexandru Ioan Cuza University of Iași, 700506 Iasi, Romania
[2] Department of Mathematics and Informatics, Vasile Alecsandri University of Bacau, 600114 Bacau, Romania; girtum@yahoo.com
[3] Faculty of Engineering, Vasile Alecsandri University of Bacau, 600115 Bacau, Romania; lili@ub.ro
[4] Department of Structural Mechanics, Gheorghe Asachi Technical University of Iasi, 700050 Iasi, Romania; tudor.petrescu@tuiasi.ro
[5] Department of Physics, Gheorghe Asachi Technical University of Iași, 700050 Iasi, Romania; m.agop@yahoo.com
[6] Romanian Scientists Academy, 050094 Bucharest, Romania
* Correspondence: iuliana.saviuc@gmail.com

Citation: Saviuc, A.; Gîrțu, M.; Topliceanu, L.; Petrescu, T.-C.; Agop, M. "Holographic Implementations" in the Complex Fluid Dynamics through a Fractal Paradigm. *Mathematics* 2021, 9, 2273. https://doi.org/10.3390/math9182273

Academic Editors: Catalin I. Pruncu and Efstratios Tzirtzilakis

Received: 8 July 2021
Accepted: 14 September 2021
Published: 16 September 2021

**Publisher's Note:** MDPI stays neutral with regard to jurisdictional claims in published maps and institutional affiliations.

**Copyright:** © 2021 by the authors. Licensee MDPI, Basel, Switzerland. This article is an open access article distributed under the terms and conditions of the Creative Commons Attribution (CC BY) license (https://creativecommons.org/licenses/by/4.0/).

**Abstract:** Assimilating a complex fluid with a fractal object, non-differentiable behaviors in its dynamics are analyzed. Complex fluid dynamics in the form of hydrodynamic-type fractal regimes imply "holographic implementations" through velocity fields at non-differentiable scale resolution, via fractal solitons, fractal solitons–fractal kinks, and fractal minimal vortices. Complex fluid dynamics in the form of Schrödinger type fractal regimes imply "holographic implementations", through the formalism of Airy functions of fractal type. Then, the in-phase coherence of the dynamics of the complex fluid structural units induces various operational procedures in the description of such dynamics: special cubics with SL(2R)-type group invariance, special differential geometry of Riemann type associated to such cubics, special apolar transport of cubics, special harmonic mapping principle, etc. In such a manner, a possible scenario toward chaos (a period-doubling scenario), without concluding in chaos (nonmanifest chaos), can be mimed.

**Keywords:** differentiability; fractal hydrodynamic regimes; fractal Schrödinger regimes; fractal soliton; fractal kink; "holographic implementations"; cubics; apolar transport; harmonic mapping principle; period doubling scenario

## 1. Introduction

Common models used to describe the dynamics in complex fluids, are founded on a mix of basic theories, derived primarily from physics and computer simulations [1–3]. In such a context, their description implies both computational simulations based on specific algorithms [2], as well as developments on fundamental theories. With respect to models developed on fundamental theories, the following classes can be distinguished:

(i) A class of models developed on spaces with integer dimension—i.e., differentiable models (for example, Navier–Stokes systems, etc.) [1–3];

(ii) Another class of models developed on spaces with non-integer dimensions, which is clearly defined by means of fractional derivatives [4,5]—i.e., non-differentiable models, with examples including the fractal models [6];

(iii) Expanding the previous class of models, new developments have been made based on Scale Relativity Theory. In such a context, the dynamics of any complex fluid can be developed on monofractal manifolds (theory of Nottale, in the fractal dimension $D_f$ = 2) [7], or on the multifractal manifolds (as in the case of the Fractal Theory of Motion) [8,9].

Both in the context of Scale Relativity Theory in the sense of Nottale [7], as well as in the one of Fractal Theory of Motion [8,9], the fundamental hypothesis is the following: assuming that any type of complex fluid is assimilated to a fractal object, said dynamics can be analyzed using motions of the structural units of any complex fluid, on fractal curves.

Such a hypothesis may be illustrated by considering the following scenario: between two successive interactions of the structural units belonging to any complex system, the trajectory of the complex fluid's structural unit is a straight line. This straight line becomes non-differentiable in the impact point. From such a perspective, taking into account that all interaction points construct an uncountable set of points, it can be stated that the trajectories of the complex fluid's structural units become fractal curves. Given the diversity of the structural units which compose any complex fluid and the diversity of interactions taking place between them, extrapolating the preceding argument for any type of complex fluid, it results that it can be assimilated to a fractal in the general sense of Mandelbrot [6].

All these considerations imply that, in the description of complex fluid dynamics, instead of "working" with a single variable (regardless of its nature, i.e., velocity, density, etc.) governed through a non-differentiable function, it is necessary to "work" just with approximations of this function (i.e., mathematical function was given by averaging them on various scale resolutions). From such a perspective, it results that any mathematical variable purposed to characterize the complex fluid dynamics will act as the limit of a class of functions. Thus, said variable will be non-differentiable for null scale resolutions and differentiable otherwise [7–9]. To put it differently, from a mathematical point of view, these variables can be explained through fractal functions, i.e., functions dependent not only on spatial and temporal coordinates, but also on the scale resolution.

Because for a large temporal scale resolution when referring to the inverse of the highest Lyapunov exponent [10,11], the deterministic trajectories of any structural unit belonging to a complex fluid can be substituted by a "class" of virtual trajectories, such that the notion of a definite trajectory can be supplanted by the one of probability density. Considering all of the above, the fractality expressed by means of stochasticity, in the depiction of the dynamics of complex fluid, becomes operational in the fractal paradigm through the Fractal Theory of Motion [8,9].

In this context, the present study was directed to the modeling of the behavior of complex fluid dynamics. A mathematical model was created considering the complex fluid as a fractal object, and its dynamics were analyzed in the framework of Scale Relativity Theory [7–9].

## 2. Mathematical Model

The complex fluid is a collection of entities (or structured units) that, by means of their interactions, relationships, or dependencies construct a unified total. In what follows, the complex fluid will be assimilated with a fractal. Then, Scale Relativity Theory in the form of Fractal Theory of Motion becomes operational through the scale covariant derivative [8,9]:

$$\frac{d\hat{F}}{dt} = \left[\partial_t + \hat{V}^l \partial_l + \frac{1}{4}(dt)^{(\frac{2}{D_f})-1} D^{lp} \partial_l \partial_p\right] F, \qquad (1)$$

where

$$\begin{aligned}
\hat{V}^l &= V_D^l - V_F^l \\
D^{lp} &= d^{lp} - i\hat{d}^{lp} \\
d^{lp} &= \lambda_+^l \lambda_+^p - \lambda_-^l \lambda_-^p \\
\hat{d}^{lp} &= \lambda_+^l \lambda_+^p + \lambda_-^l \lambda_-^p
\end{aligned} \qquad (2)$$

$\partial_t = \frac{\partial}{\partial t}$, $\partial_l = \frac{\partial}{\partial x^l}$, $\partial_l \partial_p = \frac{\partial}{\partial x^l}\frac{\partial}{\partial x^p}$, $i = \sqrt{-1}$, $l, p = 1, 2, 3$

In relations (2), the meaning of the variables and parameters are as follows:

- $x^l$ is the fractal spatial coordinate;
- $t$ is the non-fractal time having the role of an affine parameter of the motion curves;
- $\hat{V}^l$ is the complex velocity;
- $V_D^l$ is the differential velocity independent on the scale resolution;
- $V_F^l$ is the non-differentiable velocity dependent on the scale resolution;
- $dt$ is the scale resolution;
- $D_f$ is the fractal dimension of the movement curve;
- $D^{lp}$ is the constant tensor associated with the differentiable–non-differentiable transition;
- $\lambda_+^l \left( \lambda_+^p \right)$ is the constant vector associated with the backward differentiable–non-differentiable dynamic processes;
- $\lambda_-^l \left( \lambda_-^p \right)$ is the constant vector associated with the forward differentiable–non-differentiable dynamic processes;
- $F$ is a fractal function.

Many modes, and as such, an equally varied choice of definitions of fractal dimensions exist. More precisely, the fractal dimension of the Kolmogorov type and the fractal dimension of Hausdorff–Besikovitch type are the most frequently used [6,10,11]. Choosing one of the above fractal dimensions in the description of any complex fluid dynamics, the value of the fractal dimension must be constant and arbitrary in any dynamical analysis. For instance: $D_f < 2$ for correlative processes in complex fluid dynamics, $D_f > 2$ for non-correlative processes in said dynamics, etc. [10,11].

Accepting the functionality of the scale covariance principle, which refers to applying the operator (1) to the complex velocity field (2), for the case of free motions, the geodesics equation on fractal space takes the following form [8,9]:

$$\frac{d\hat{V}^i}{dt} = \partial_t \hat{V}^i + \hat{V}^l \partial_l \hat{V}^i + \frac{1}{4}(dt)^{(\frac{2}{D_f})-1} D^{lk} \partial_l \partial_k \hat{V}^i = 0, \qquad (3)$$

This means that the fractal acceleration, $\partial_t \hat{V}^i$, the fractal convection, $\hat{V}^l \partial_l \hat{V}^i$ and the fractal dissipation, $D^{lk} \partial_l \partial_k \hat{V}^i$, achieve their equilibrium at any point of the fractal curve.

If the fractalization is achieved by Markov-type stochastic processes (see Introduction and [6–9]), then:

$$\lambda_+^i \lambda_+^l = \lambda_-^i \lambda_-^l = 2\lambda \delta^{il}, \qquad (4)$$

In (4), $\lambda$ is a coefficient linked to the differentiable-non-differentiable transition and $\delta^{il}$ is Kronecker's pseudo-tensor. In these conditions, the geodesics Equation (3) becomes:

$$\frac{d\hat{V}^i}{dt} = \partial_t \hat{V}^i + \hat{V}^l \partial_l \hat{V}^i - i\lambda(dt)^{(\frac{2}{D_f})-1} \partial^l \partial_l \hat{V}^i = 0 \qquad (5)$$

## 3. Dynamics of Complex Fluids in the Form of Hydrodynamic—Type Fractal "Regimes"

The division of the complex fluid's dynamics on scale resolutions implies, through (5), both the conservation law of the specific momentum at differentiable scale resolution:

$$\frac{\partial V_D^i}{dt} = \partial_t V_D^i + V_D^l \partial_l V_D^i - \left[ V_F^l + \lambda(dt)^{(\frac{2}{D_f})-1} \partial_l \right] \partial^l V_F^i = 0, \qquad (6)$$

and also the conservation laws of the specific momentum at non-differentiable scale resolutions:

$$\frac{\partial V_F^i}{dt} = \partial_t V_F^i + V_D^l \partial_l V_F^i + \left[ V_F^l + \lambda(dt)^{(\frac{2}{D_f})-1} \partial_l \right] \partial^l V_D^i = 0, \qquad (7)$$

From (6), it results that the specific force:

$$f_F^i = \left[V_F^l + \lambda(dt)^{(\frac{2}{D_f})-1}\partial_l\right]\partial^l V_F^i, \tag{8}$$

induced by the velocity fields $V_F^i$. This becomes a "measure" of non-differentiability of motion curves of complex fluid entities.

In the case of stationary complex fluid dynamics $(\partial_t V_D^i = 0, \partial_t V_F^i = 0)$, the conservation laws (6), (7) become:

$$V_D^l \partial_l V_D^i - \left[V_F^l + \lambda(dt)^{(\frac{2}{D_f})-1}\partial_l\right]\partial^l V_F^i = 0, \tag{9}$$

$$V_D^l \partial_l V_F^i + \left[V_F^l + \lambda(dt)^{(\frac{2}{D_f})-1}\partial_l\right]\partial^l V_D^i = 0, \tag{10}$$

while, in the static case $(\partial_t V_D^i = 0, V_D^i = 0, \partial_t V_F^i = 0)$ these take the form:

$$\left[V_F^l + \lambda(dt)^{(\frac{2}{D_f})-1}\partial_l\right]\partial^l V_F^i = 0, \tag{11}$$

The result (11) specifies that, although at differentiable scale resolution, the complex fluid dynamics are absent while, at the non-differentiable scale resolution, the complex fluid dynamics can be "dictated" by the hydrodynamic fractal- type equations:

$$V_F^l \partial_l V_F^i + \lambda(dt)^{(\frac{2}{D_f})-1}\partial_l \partial^l V_F^i = 0 \tag{12}$$

$$\partial_l V_F^l = 0 \tag{13}$$

Equation (13) corresponds to the complex fluid incompressibility at the non-differentiable scale resolution (i.e., the states' density $\rho$ at the non-differentiable scale resolution is constant).

Generally, it is difficult to obtain an analytical solution for the previous equation system, taking into account its non-linear nature. However, it is still possible to obtain an analytic solution in the case of plane symmetry (for example, in $(x, y)$ coordinates) of the complex fluid dynamics. In order to obtain such a solution, in what follows, the method described in [12] will be used. Let it be considered the equations system (12) and (13) in the form:

$$U_0 \partial_x U_0 + V_0 \partial_y U_0 = \sigma_0 \partial_{yy}^2 U_0, \tag{14}$$

$$\partial_x U_0 + \partial_y V_0 = 0, \tag{15}$$

where:

$$V_{Fx} = U_0(x,y), \quad V_{Fy} = V_0(x,y), \quad \sigma_0 = \lambda(dt)^{(\frac{2}{D_f})-1} \tag{16}$$

Imposing now the following conditions:

$$\lim_{y \to 0} V_0(x,y) = 0, \quad \lim_{y \to 0} \frac{\partial U_0}{\partial y} = 0, \quad \lim_{y \to \infty} U_0(x,y) = 0, \tag{17}$$

and considering constant flux moment per unit of depth:

$$Q = \rho \int_{-\infty}^{+\infty} U_0^2 dy = const., \tag{18}$$

the velocity fields as the solution of the equations system (14) and (15), take the form:

$$U_0 = \frac{1.5\left(\frac{Q}{6\rho}\right)^{\frac{2}{3}}}{(\sigma_0 x)^{\frac{1}{3}}} \operatorname{sech}^2\left[\frac{0.5y\left(\frac{Q}{6\rho}\right)^{\frac{1}{3}}}{(\sigma_0 x)^{\frac{2}{3}}}\right], \qquad (19)$$

$$V_0 = \frac{1.9\left(\frac{Q}{6\rho}\right)^{\frac{2}{3}}}{(\sigma_0 x)^{\frac{1}{3}}} \left\{ \frac{y\left(\frac{Q}{6\rho}\right)^{\frac{1}{3}}}{(\sigma_0 x)^{\frac{2}{3}}} \operatorname{sech}^2\left[\frac{0.5y\left(\frac{Q}{6\rho}\right)^{\frac{1}{3}}}{(\sigma_0 x)^{\frac{2}{3}}}\right] - \tanh\left[\frac{0.5y\left(\frac{Q}{6\rho}\right)^{\frac{1}{3}}}{(\sigma_0 x)^{\frac{2}{3}}}\right] \right\}, \qquad (20)$$

The previous can be simplified greatly through the use of non-dimensional variables:

$$X = \frac{x}{x_0},\ Y = \frac{y}{y_0},\ U = \frac{U_0}{w_0},\ V = \frac{V_0}{w_0}, \qquad (21)$$

and non-dimensional parameters:

$$\mu = \frac{\sigma_0}{v_0},\ v_0 = \frac{y_0^{\frac{3}{2}}}{x_0}\left(\frac{Q}{6\rho}\right)^{\frac{1}{2}},\ w_0 = \frac{1}{(y_0)^{\frac{1}{2}}}\left(\frac{Q}{6\rho}\right)^{\frac{1}{2}}, \qquad (22)$$

where $x_0$, $y_0$, $w_0$, and $v_0$ represent specific lengths, specific velocity, and "fractal degree" of the complex fluid dynamics. In these conditions, the normalized velocity fields become:

$$U = \frac{1.5}{(\mu X)^{\frac{1}{3}}} \operatorname{sech}^2\left[\frac{0.5Y}{(\mu X)^{\frac{2}{3}}}\right], \qquad (23)$$

$$V = \frac{1.9}{(\mu X)^{\frac{1}{3}}} \left\{ \frac{Y}{(\mu X)^{\frac{2}{3}}} \operatorname{sech}^2\left[\frac{0.5Y}{(\mu X)^{\frac{2}{3}}}\right] - \tanh\left[\frac{0.5Y}{(\mu X)^{\frac{2}{3}}}\right] \right\}, \qquad (24)$$

Any of the above relations describe the non-linear character of the velocity fields. This character can be explained through the fractal soliton (i.e., soliton depending on scale resolution) for the velocity field across the Ox axis, respectively "mixtures" of fractal soliton-fractal kink (i.e., kink dependent on scale resolution), for the velocity fields across the Oy axis. The specificities in the complex fluid dynamics are "explained" in Figures 1a–d and 2a–d. Details on the soliton, kink, and other classical non-linear solutions are given in [10,11].

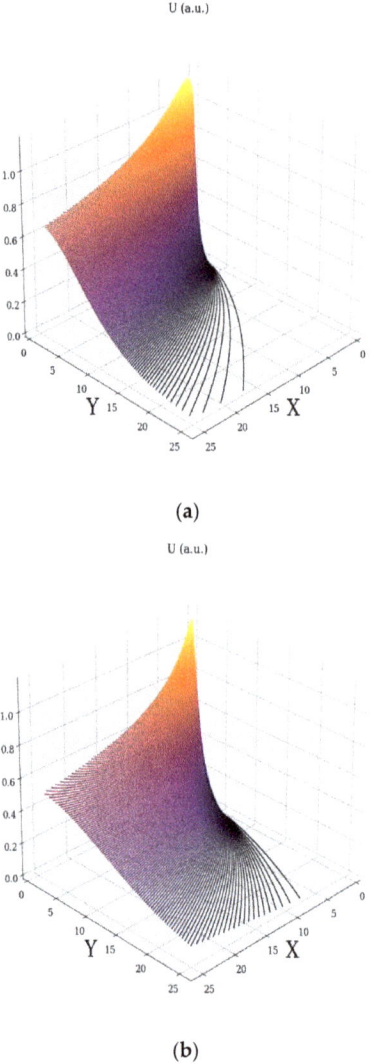

(a)

(b)

**Figure 1.** *Cont.*

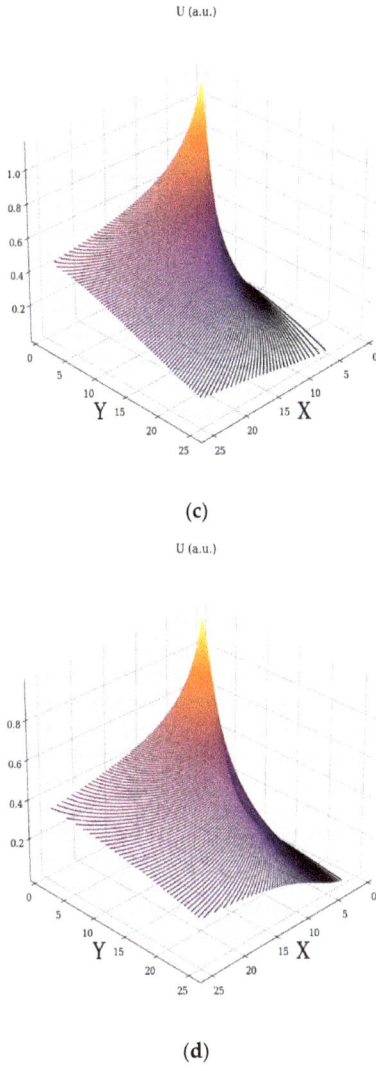

(c)

(d)

**Figure 1.** Non-dimensional velocity field $U$ for different fractal degree: (**a**) $\mu = 0.5$; (**b**) $\mu = 1$; (**c**) $\mu = 1.5$; (**d**) $\mu = 3$.

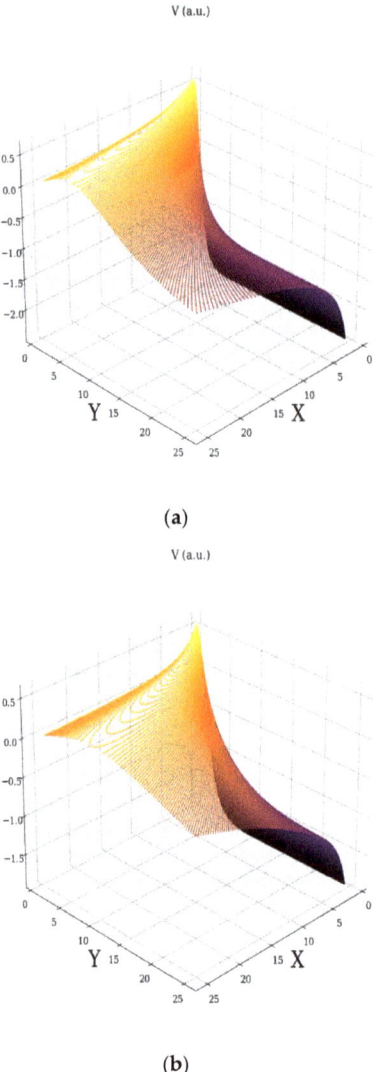

(a)

(b)

**Figure 2.** *Cont.*

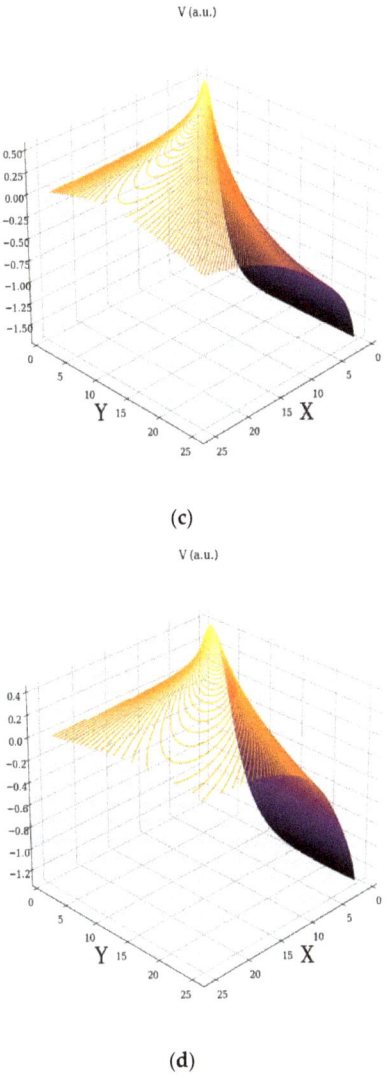

**Figure 2.** Non-dimensional velocity field $V$ for different fractal degree: (**a**) $\mu = 0.5$; (**b**) $\mu = 1$; (**c**) $\mu = 1.5$; (**d**) $\mu = 3$.

The velocity fields (23) and (24) induce the fractal minimal vortex (Figure 3a–d).

$$\Omega = \left(\frac{\partial U}{\partial Y} - \frac{\partial V}{\partial Y}\right) = \frac{0.57Y}{(\mu X)^2} + \frac{0.63\mu}{(\mu X)^{\frac{4}{3}}}\tanh\left[\frac{0.5Y}{(\mu X)^{\frac{2}{3}}}\right] + \frac{1.9Y}{(\mu X)^2}\operatorname{sech}^2\left[\frac{0.5Y}{(\mu X)^{\frac{2}{3}}}\right] - \frac{0.57Y}{(\mu X)^2}\tanh^2\left[\frac{0.5Y}{(\mu X)^{\frac{2}{3}}}\right] - \left[\frac{1.5}{\mu X} + \frac{1.4Y^2}{X(\mu X)^{\frac{5}{3}}}\right]\operatorname{sech}^2\left[\frac{0.5Y}{(\mu X)^{\frac{2}{3}}}\right]\tanh\left[\frac{0.5Y}{(\mu X)^{\frac{2}{3}}}\right], \quad (25)$$

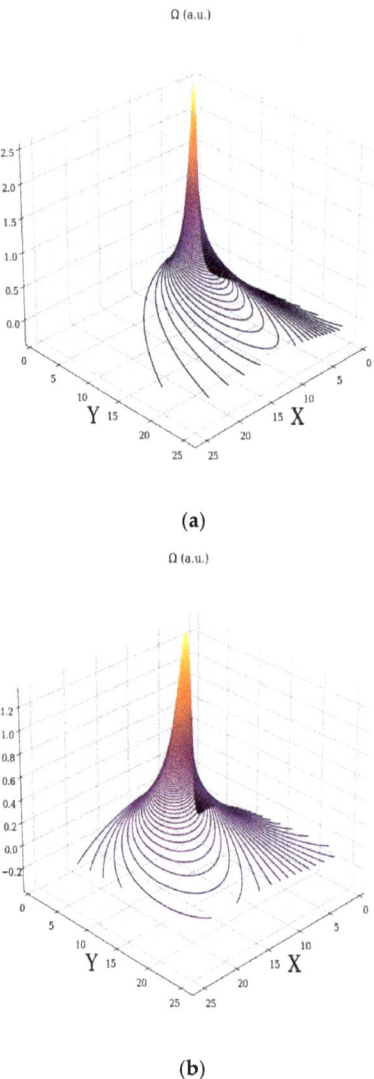

(a)

(b)

**Figure 3.** Cont.

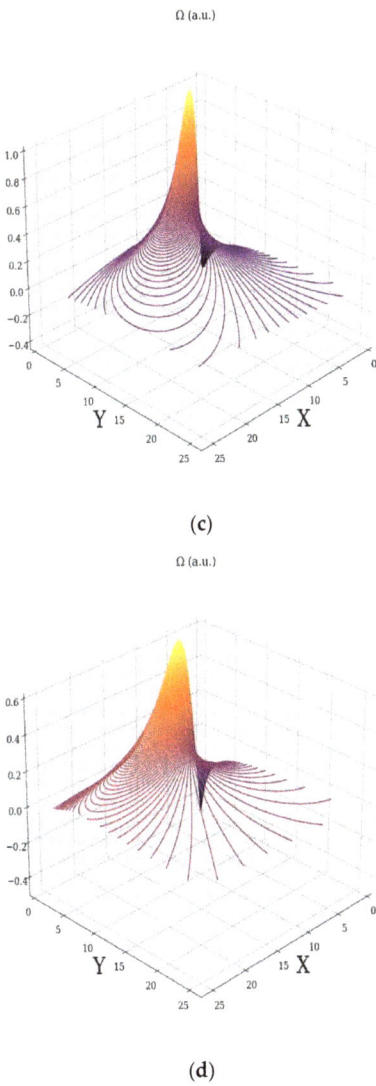

(c)

(d)

**Figure 3.** Fractal minimal vortex field $\Omega$ for different fractal degree: (**a**) $\mu = 0.5$ (**b**) $\mu = 1$; (**c**) $\mu = 1.5$; (**d**) $\mu = 3$.

This previous result was used to specify the fact that the turbulence sources may be induced by fractal vortices. As long as the complex fluid is not constrained externally, fractal vortices do not manifest themselves. Phrasing it differently, they are "virtual" fractal vortices and manifest as "virtual" turbulence sources. In the presence of an external constraint, they become "real" and the turbulence mechanism is triggered. Essentially, the discussion revolves around "holographic implementation" of turbulences in the complex fluid dynamics. It is reminded that, since the dynamics of complex fluid entities are described by continuous but non-differentiable curves, curves which exhibit the property of self-similarity in every one of its points, these can be viewed as a holographic mechanism (every part reflects the whole) of dynamics description. It is noted that the previous choice of the fractality degree (i.e., the scale resolution, type of motion curve through its fractal dimension) can generally cover various types of dynamics found in complex fluids. Moreover, it is noted that the previous Figures were obtained in a Python programming environment.

## 4. Dynamics of Complex Fluids in the Form of Schrödinger-Type "Regimes"

In the case of irrotational motions of the complex fluid structural units, the complex velocity field $\hat{V}^i$ from (2) becomes:

$$\hat{V}^i = -2i\lambda(dt)^{(\frac{2}{D_f})-1}\partial^i \ln \Psi \tag{26}$$

where $\ln \Psi$ is the fractal scalar potential of the velocity fields and $\Psi$ is a fractal state function.

Then, substituting (26) in (5), the geodesics Equation (5) becomes (for details on the method, see [7–9]):

$$\lambda^2(dt)^{(\frac{4}{D_f})-2}\partial^l\partial_l\Psi + i\lambda(dt)^{(\frac{2}{D_f})-1}\partial_t\Psi = 0 \tag{27}$$

Relation (27) is a Schrödinger equation of fractal type. As a consequence, different dynamics of any complex fluids can be explained as Schrödinger-type fractal "regimes". In the particular case of the dynamics of structural units belonging to the complex fluid, on Peano-type curves ($D_f \to 2$) at Compton scale ($\lambda = h/4\pi m_0$, where $h$ is Planck's constant and $m_0$ is the rest mass of the structural unit belonging to the complex fluid), (27) becomes the standard Schrödinger equation from quantum mechanics.

The solution of the one-dimensional Schrödinger equation of fractal type can be written in the form (for details see [13,14]):

$$\Psi(x,t) = \frac{1}{\sqrt{t}}exp\left(i\frac{x^2}{4\mu t}\right), \\ \mu = \lambda(dt)^{(\frac{2}{D_f})-1} \tag{28}$$

and is defined, of course, up to an arbitrary multiplicative constant.

As such, the general solution of Equation (27) can be written as a linear superposition of the form:

$$\Psi(x,t) = \frac{1}{\sqrt{t}}\int_{-\infty}^{+\infty} u(y)exp\left[i\frac{(x-y)^2}{4\mu t}\right]dy \tag{29}$$

Now, if $u(y)$ is an Airy function of fractal type, then $\Psi(x,t)$ retains this property, in the sense that its amplitude is an Airy function of fractal type. Indeed, in this case, there will be:

$$u(y) \equiv Ai(y) = \frac{1}{2\pi}\int_{-\infty}^{+\infty} exp\left[i\left(\frac{\omega^3}{3}+\omega y\right)\right]d\omega \tag{30}$$

in such a way as the state function (29) will be written in the form:

$$\Psi(x,t) = \frac{1}{2\pi\sqrt{t}}\int_{-\infty}^{+\infty} exp\left\{i\left[\frac{\omega^3}{3}+\omega y+\frac{(x-y)^2}{4ut}\right]\right\}dyd\omega \tag{31}$$

If, at first, the integration will be carried out after $y$, up to a multiplicative constant, the results is:

$$\Psi(x,t) = \frac{1}{2\pi}\int_{-\infty}^{+\infty} exp\left[i\left(\frac{\omega^3}{3}+\omega x-\mu t\omega^2\right)\right]d\omega \tag{32}$$

The final result is obtained based on a special relationship developed in [13,14] and it is:

$$\Psi(x,t) = \left[Ai\left(kx - v^2 t^2\right)\right] \exp\left[ivt\left(kx - \frac{2}{3}v^2 t^2\right)\right] \quad (33)$$

with

$$v = k^2 \mu \quad (34)$$

In these conditions, if $\Psi$ is chosen in the form:

$$\Psi(x,t) = A(x,t)\exp[i\phi(x,t)] \quad (35)$$

where $A(x,t)$ is an amplitude and $\Phi(x,t)$ is a phase, by identifying in (33) the amplitude and the phase, there will be:

$$\begin{aligned} A(x,t) &= Ai(kx - v^2 t^2), \\ \phi(x,t) &= vt\left(kx - \tfrac{2}{3}v^2 t^2\right) \end{aligned} \quad (36)$$

By substituting (35) in (27), by means of direct calculation, the following relation is checked:

$$i\partial_t \Psi + \mu \partial_l \partial^l \Psi = -\left[\partial_t \phi + \mu(\partial_l \phi)^2 - \mu\frac{\partial_l \partial^l A}{A}\right] + \frac{i}{2A^2}\left[\partial_t A^2 + 2\mu \partial_l\left(A^2 \partial^l \phi\right)\right] \quad (37)$$

Now, the "specific constraints" necessary for $\Psi$ to be a solution of the non-stationary differential Equation (37) will be reducible to the differential equations:

$$\begin{aligned} \partial_t \phi + \mu\left(\partial_l \phi \partial^l \phi\right) &= \mu\frac{\partial_l \partial^l A}{A} \\ \partial_t A^2 + 2\mu\left(A^2 \partial_l \phi\right) &= 0 \end{aligned} \quad (38)$$

The first of these equations is the Hamilton–Jacobi equation of fractal type, while the second equation is the continuity equation of fractal type. From here, the correspondence with the hydrodynamic model of fractal type, pertaining to scale relativity [7–9], becomes evident based on the substitutions:

$$\begin{aligned} V_D^i &= \mu \partial^i \phi, \\ \rho &= A^2 \end{aligned} \quad (39)$$

where $V_D^i$ is the differential component of the velocity field and $\rho$ is the density of states. In this condition, the conservation law of fractal type of the specific momentum:

$$\partial_t V_D^i + V_D^l \partial_l V_D^i = -\partial^i Q \quad (40)$$

and respectively, the conservation law of the density of states of fractal type:

$$\partial_t \rho + \partial^l\left(\rho V_D^l\right) = 0 \quad (41)$$

can be found.

The specific potential of fractal type:

$$Q = -\mu^2 \frac{\partial_l \partial^l \sqrt{\rho}}{\sqrt{\rho}} \quad (42)$$

through the induced specific force of fractal type:

$$f^i = -\partial^i Q = -\mu^2 \partial^i \left( \frac{\partial_l \partial^l \sqrt{\rho}}{\sqrt{\rho}} \right) \tag{43}$$

becomes a measure of the fractal degree pertaining to the motion curves.

Now, through (36), the in-phase coherence of the structural unit dynamics for any complex fluid implies the condition

$$\Phi(x, t) = vt \left( kx - \frac{2}{3} v^2 t^2 \right) = const \tag{44}$$

or, moreover, in the notations:

$$-\frac{2}{3} v^3 t^3 = a_0 X^3, \ a_1 = 0, \ vkxt = 3a_2, \ a_3 = const.$$

the cubic equation:
$$a_0 X^3 + 3a_1 X^2 + 3a_2 X + a_3 = 0 \tag{45}$$

If (45) has real roots [14,15]:
$$\begin{aligned} X_1 &= \frac{h + \bar{h}k}{1+k}, \\ X_2 &= \frac{h + \varepsilon \bar{h}k}{1+\varepsilon k}, \\ X_3 &= \frac{h + \varepsilon^2 \bar{h}k}{1+\varepsilon^2 k} \end{aligned} \tag{46}$$

with $h, \bar{h}$ the roots of Hessian, and $\varepsilon \equiv \left( -1 + i\sqrt{3} \right)/2$ the cubic root of unity $(i = \sqrt{-1})$, the values of variables $h, \bar{h}$, and $k$ can be "scanned" by a simple transitive group with real parameters. This group can be revealed through Riemann-type spaces associated with the previous cubic. The basis of this approach is the fact that the simply transitive group with real parameters [14,15]:

$$X_l \leftrightarrow \frac{aX_l + b}{cX_l + d}, \ l = 1, 2, 3 \ a, b, c, d \in R \tag{47}$$

where $X_l$ are the roots of the cubic (45), induces the simply transitive group in the quantities $h, \bar{h}$, and $k$, whose actions are:

$$\begin{aligned} h &\leftrightarrow \frac{ah + b}{ch + d}, \\ \bar{h} &\leftrightarrow \frac{a\bar{h} + b}{c\bar{h} + d}, \\ k &\leftrightarrow \frac{c\bar{h} + d}{ch + d} k \end{aligned} \tag{48}$$

The structure of this group is typical of SL(2R), i.e.,

$$\begin{aligned}{} [B^1, B^2] &= B^1, \\ [B^2, B^3] &= B^3, \\ [B^3, B^1] &= -2B^2 \end{aligned} \tag{49}$$

where $B^l$ are the infinitezimal generators of the group:

$$\begin{aligned} B^1 &= \frac{\partial}{\partial h} + \frac{\partial}{\partial \bar{h}} \\ B^2 &= h \frac{\partial}{\partial h} + \bar{h} \frac{\partial}{\partial \bar{h}} \\ B^3 &= h^2 \frac{\partial}{\partial h} + \bar{h}^2 \frac{\partial}{\partial \bar{h}} + \left( h - \bar{h} \right) k \frac{\partial}{\partial k} \end{aligned} \tag{50}$$

and admit the absolute invariant differentials

$$\begin{aligned} \omega^1 &= \frac{dh}{(h-\bar{h})k} \\ \omega^2 &= -i\left(\frac{dk}{k} - \frac{dh+d\bar{h}}{h-\bar{h}}\right) \\ \omega^3 &= -\frac{kd\bar{h}}{h-\bar{h}} \end{aligned} \tag{51}$$

and the 2-form (the metric):

$$ds^2 = \left(\frac{dk}{k} - \frac{dh+d\bar{h}}{h-\bar{h}}\right)^2 - 4\frac{dhd\bar{h}}{\left(h-\bar{h}\right)^2} \tag{52}$$

In real terms

$$h = u+iv, \; \bar{h} = u+iv, \; k = e^{i\theta} \tag{53}$$

and for

$$\begin{aligned} \Omega^1 &= \omega^2 = d\theta + \frac{du}{v} \\ \Omega^2 &= \cos\theta \frac{du}{v} + \sin\theta \frac{dv}{v} \\ \Omega^3 &= -\sin\theta \frac{du}{v} + \cos\theta \frac{dv}{v}, \end{aligned} \tag{54}$$

the connection with Poincaré representation of the Lobachevsky plane can be obtained. Indeed, the metric is a three-dimensional Lorentz structure:

$$ds^2 = -\left(\Omega^1\right)^2 + \left(\Omega^2\right)^2 + \left(\Omega^3\right)^2 = -\left(d\theta + \frac{du}{v}\right)^2 + \frac{du^2+dv^2}{v^2} \tag{55}$$

This metric reduces to that of Poincaré, in cases where $\Omega^1 \equiv 0$, which defines the variable $\theta$ as the "angle of parallelism" of the hyperbolic planes (the connection). In fact, recalling that

$$\frac{dk}{k} - \frac{dh+d\bar{h}}{h-\bar{h}} = 0 \leftrightarrow d\theta = -\frac{du}{v} \tag{56}$$

represents the connection form of the hyperbolic plane, the relationship (54) then represents general Bäcklund transformations in that plane. In such a conjecture, it is noted that, if the temporal cubic is assumed to have distinct roots, the condition (56) is satisfied, if, and only if, the differential forms $\Omega^1$ is null.

Therefore, for the metric (55) with restriction (56), the relation becomes:

$$ds^2 = \frac{dhd\bar{h}}{\left(h-\bar{h}\right)^2} = \frac{du^2+dv^2}{v^2} \tag{57}$$

The parallel transport of the hyperbolic plane actually represents the apolar transport of the cubics (45).

Such a metric approach allows harmonic mappings from the usual space to the hyperbolic one (space associated to the dynamics of the complex fluid), through the functional (for details see [14–16]):

$$J = \frac{1}{2}\iiint d^3X \left[\frac{\partial_l h \partial^l \bar{h}}{\left(h-\bar{h}\right)^2}\right] \tag{58}$$

where the usual notation $\partial_l$ denotes the gradient and $d^3X$ is the elementary volume.

In the case of the synchronization of dynamics of any complex fluid structural units, i.e., in-phase coherence through the condition (44), the Euler equations corresponding to the functional (58) is:

$$\left(h - \bar{h}\right)\nabla(\nabla h) = 2(\nabla h)^2 \tag{59}$$

which admits

$$h = i\frac{\cosh\chi - \sinh\chi e^{-i\bar{\Omega}}}{\cosh\chi + \sinh\chi e^{-i\bar{\Omega}}}, \quad \chi = \frac{\psi}{2} \tag{60}$$

as a solution, as long as $\chi$ (and thus $\psi$) are solutions of a Laplace-type equation for the free space.

Therefore, space-time "synchronization modes" in phase and amplitude of the complex fluid structural units imply group invariances of a $SL(2R)$ type. Then, period doubling emerges as a natural behavior in the complex fluid dynamics (see Figure 4a–c where $r = \tanh\chi$, $|h| \equiv Amplitude$ and $\bar{\Omega} = \Omega t$ at various scale resolutions, given by means of the maximum value of $\Omega$, i.e., $\Omega_{max}$).

As it can be observed in Figure 4a–c, the natural transition of a complex fluid is to evolve from a normal period doubling state towards damped oscillating and strong modulated dynamics. The complex fluid never reaches a chaotic state, but it permanently evolves towards that state. There is a periodicity to the whole series of transitions, the system evolves through period doubling, damped oscillations even reaching in some cases an intermittent state (the damped oscillations, intermittent states, etc. will be analyzed by us in a future paper), but it never reaches a pure chaotic state. The evolution of the systems sees a "jump" into a period doubling oscillation state and the transition resumes towards a quasi-chaotic state.

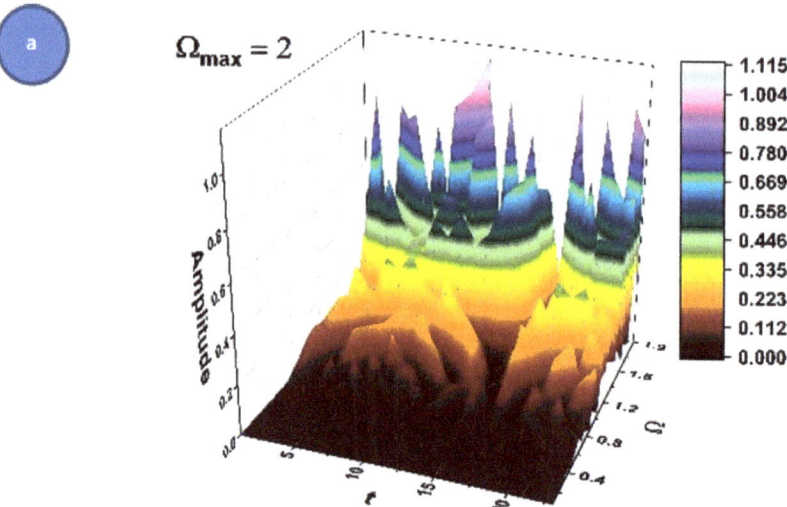

**Figure 4.** Cont.

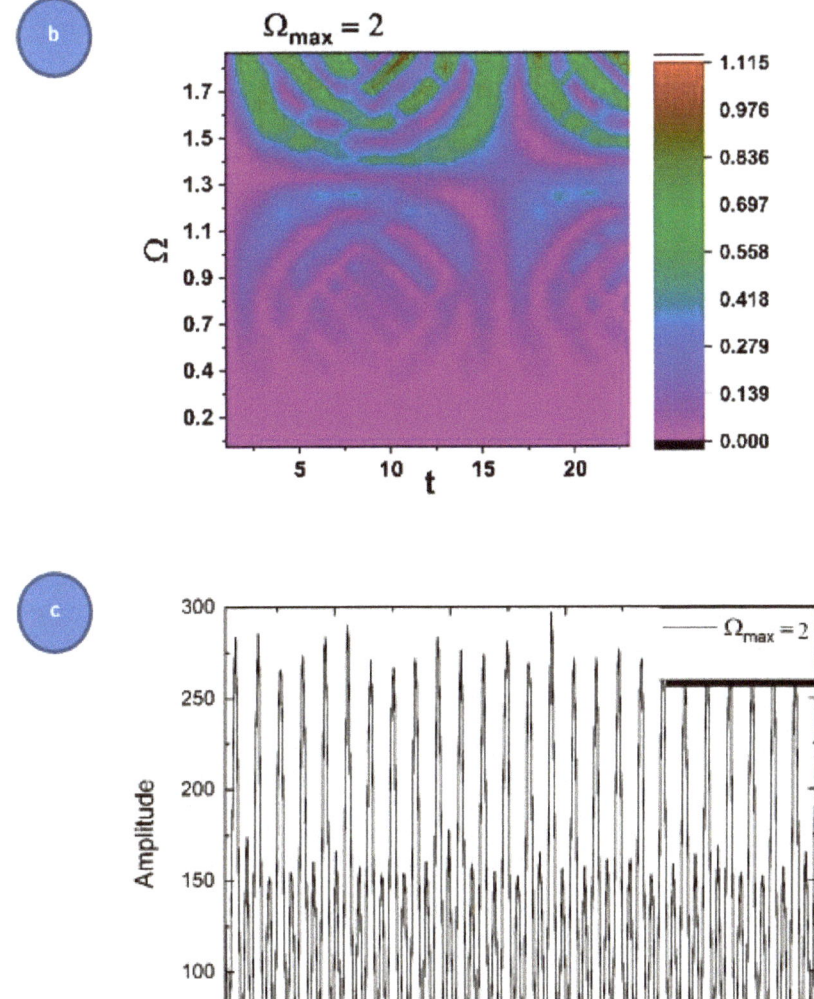

**Figure 4.** (a–c) A period doubling (a–c) "synchronization mode" of complex fluid structural units (3D, contour plot, and time-series) for the scale resolution given by $\Omega_{max} = 2$.

The Bifurcation Map is presented (Figure 5) where again it is observed that the complex fluid starts from a steady state (double period state) and evolves towards a chaotic one ($\Omega_{max} = 2$) but it never reaches that state. For each periodic transition scenario, it is possible to observe the system swiping through all the previously mentioned dynamic states. Therefore, there is an overall periodicity with a continuous increase in oscillation amplitude.

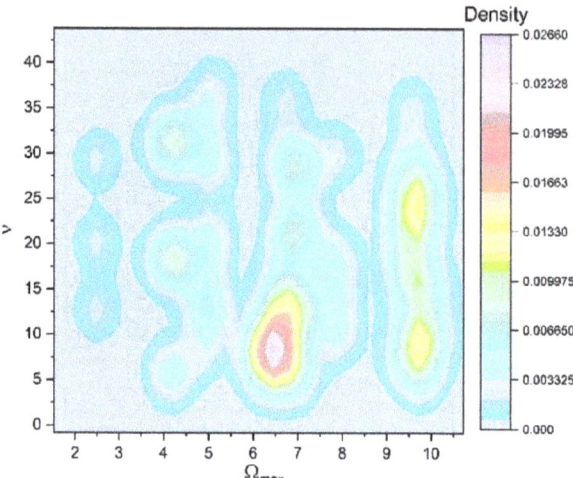

**Figure 5.** The oscillation frequency of the complex system as a function of a scale resolution chosen by $\Omega_{max}$ bifurcation map.

Let it be noted that the mathematical formalism of the Fractal Theory of Motion implies various operational procedures (invariance groups, harmonic mappings, groups isomorphism, embedding manifolds, etc.) with quite a number of applications in complex fluid dynamics [17,18].

## 5. Conclusions

Assimilating the complex fluid with a fractal, different dynamics at various scale resolutions are analyzed. Therefore, the following conclusions may be stated:

(i) The complex fluid dynamics, in the form of hydrodynamic-type fractal regimes, specify velocity fields at non-differentiable scale resolution, in the form of fractal solitons, fractal solitons–fractal kinks, and fractal minimal vortices;

(ii) The fractal vortices can be linked to turbulence sources in complex fluid dynamics at non-differentiable scale resolutions. So long as they are not acted upon by any external constraint, fractal vortices are virtual and non-manifest. However, the presence of external constraints radically changes the complex fluid dynamics, in the sense that the vortices will manifest as a real turbulences. Since the dynamics of complex fluid entities are described by continuous but non-differentiable curves (which exhibit the property of self-similarity in every one of their points), these can be viewed as a holographic mechanism (every part reflects the whole) in the description of complex fluid dynamics;

(iii) The description of the complex fluid dynamics in the form of Schrödinger type-fractal regimes imply "holographic implementations", through the formalism of Airy functions of fractal type. From such a perspective, the in-phase coherence of the dynamics of the complex fluid structural units induces various operational procedures in the description of such dynamics: special cubics with SL(2R)-type group invariance, special differential geometry of Riemann type associated to such cubics, special apolar transport of cubics, special harmonic mapping principle, etc. Referring to the special harmonic principle, in-phase coherence allows harmonic mappings from the usual space to the hyperbolic one, situation in which the period doubling "synchronization mode" among the structural units of a complex fluid becomes functional. In such a manner, a possible scenario toward chaos (period doubling scenario), without concluding in chaos (nonmanifest chaos), can be mimed.

**Author Contributions:** Conceptualization, A.S. and M.A.; methodology, M.G. and L.T.; investigation, A.S., T.-C.P., and M.A.; resources, T.-C.P. and M.G.; writing—original draft preparation, A.S., T.-C.P., and M.A.; writing—review and editing, M.A. and T.-C.P. All authors have read and agreed to the published version of the manuscript.

**Funding:** This research received no external funding.

**Institutional Review Board Statement:** Not applicable.

**Informed Consent Statement:** Not applicable.

**Data Availability Statement:** Data will be available on request.

**Acknowledgments:** The Authors would like to thank both the editor and the reviewers for their valuable feedback, regarding the quality of this paper.

**Conflicts of Interest:** The authors declare no conflict of interest.

## References

1. Saramito, P. *Complex Fluids—Modeling and Alghorithms*; Springer: Berlin/Heidelberg, Germany, 2016.
2. Hou, T.Y.; Liu, C.; Liu, J.G. *Multi-Scale Phenomena in Complex Fluids: Modeling, Analysis and Numerical Simulations*; World Scientific Publishing Company: Singapore, 2009.
3. Deville, M.T.; Gatski, B. *Mathematical Modeling for Complex Fluids and Flows*; Springer: Berlin/Heidelberg, Germany, 2012.
4. Băceanu Diethelm, K.; Scalas, E.; Trujillo, H. *Fractional Calculus, Models and Numerical Methods*; World Scientific: Singapore, 2016.
5. Ortigueria, M.D. *Fractional Calculus for Scientists and Engineers*; Springer: Berlin/Heidelberg, Germany, 2011.
6. Mandelbrot, B.B. *Fractal and Chaos*; Springer: Berlin/Heidelberg, Germany, 2004.
7. Nottale, L. *Scale Relativity and Fractal Space-Time: A New Approach to Unifying Relativity and Quantum Mechanics*; Imperial College Press: London, UK, 2011.
8. Merches, I.; Agop, M. *Differentiability and Fractality in Dynamics of Physical Systems*; World Scientific: Hackensack, NJ, USA, 2016.
9. Agop, M.; Paun, V.P. On the new perspectives of fractal theory. In *Fundaments and Applications*; Romanian Academy Publishing House: Bucharest, Romania, 2017.
10. Strogatz, S.H. *Nonlinear Dynamics and Chaos*, 2nd ed.; CRC Press: Boca Raton, FL, USA, 2015.
11. Cristescu, C.P. Nonlinear dynamics and chaos. In *Theoretical Fundaments and Applications*; Romanian Academy Publishing House: Bucharest, Romania, 2008.
12. Schlichting, H. *Boundary-Layer Theory*; Springer: Berlin, Germany, 2018.
13. Vallée, O.; Soares, M. *Airy Functions and Applications to Physics*; London Imperial College Press: London, UK, 2010.
14. Mazilu, N.; Agop, M.; Merches, I. *Scale Transitions as Foundations of Physics*; World Scientific: Singapore, 2021.
15. Mazilu, N.; Agop, M. Skyrmions. In *A Great Finishing Touch to Classical Newtonian Philosophy*; Ny Nova Science Publishers C: New York, NY, USA, 2012.
16. Xi, Y. *Geometry of Harmonics Maps*; Springer: New York, NY, USA, 2018.
17. Irimiciuc, S.A.; Chertopalov, S.; Lancok, J.; Craciun, V. Langmuir Probe Technique for Plasma Characterization during Pulsed Laser Deposition Process. *Coatings* **2021**, *11*, 762. [CrossRef]
18. Schrittwieser, R.W.; Ionita, C.; Teodorescu-Soare, C.T.; Vasilovici, O.; Gurlui, S.; Irimiciuc, S.A.; Dimitriu, D.G. Spectral and electrical diagnosis of complex space-charge structures excited by a spherical grid cathode with orifice. *Phys. Scr.* **2017**, *92*, 044001. [CrossRef]

*Article*

# Reliability Simulation of Two Component Warm-Standby System with Repair, Switching, and Back-Switching Failures under Three Aging Assumptions

Kiril Tenekedjiev [1,2,*], Simon Cooley [1], Boyan Mednikarov [2], Guixin Fan [1] and Natalia Nikolova [1,2]

[1] Australian Maritime College, University of Tasmania, 1 Maritime Way, Launceston, TAS 7250, Australia; smcooley@utas.edu.au (S.C.); gfan@utas.edu.au (G.F.); Natalia.Nikolova@utas.edu.au or natalianik@gmail.com (N.N.)
[2] Nikola Vaptsarov Naval Academy—Varna, 73 V. Drumev Str., 9002 Varna, Bulgaria; bobmednikarov@abv.bg
* Correspondence: Kiril.Tenekedjiev@utas.edu.au or Kiril.Tenekedjiev@fulbrightmail.org; Tel.: +61-3-6324-9724

**Citation:** Tenekedjiev, K.; Cooley, S.; Mednikarov, B.; Fan, G.; Nikolova, N. Reliability Simulation of Two Component Warm-Standby System with Repair, Switching, and Back-Switching Failures under Three Aging Assumptions. *Mathematics* **2021**, *9*, 2547. https://doi.org/10.3390/math9202547

Academic Editors: Maria Luminița Scutaru and Catalin I. Pruncu

Received: 23 August 2021
Accepted: 6 October 2021
Published: 11 October 2021

**Publisher's Note:** MDPI stays neutral with regard to jurisdictional claims in published maps and institutional affiliations.

**Copyright:** © 2021 by the authors. Licensee MDPI, Basel, Switzerland. This article is an open access article distributed under the terms and conditions of the Creative Commons Attribution (CC BY) license (https://creativecommons.org/licenses/by/4.0/).

**Abstract:** We analyze the influence of repair on a two-component warm-standby system with switching and back-switching failures. The repair of the primary component follows a minimal process, i.e., it experiences full aging during the repair. The backup component operates only while the primary component is being repaired, but it can also fail in standby, in which case there will be no repair for the backup component (as there is no indication of the failure). Four types of system failures are investigated: both components fail to operate in a different order or one of two types of switching failures occur. The reliability behavior of the system is investigated under three different aging assumptions for the backup component during warm-standby: full aging, no aging, and partial aging. Four failure and repair distributions determine the reliability behavior of the system. We analyzed two cases—in the First Case, we utilized constant failure rate distributions. In the Second Case, we applied the more realistic time-dependent failure rates. We used three methods to identify the reliability characteristics of the system: analytical, numerical, and simulational. The analytical approach is limited and only viable for constant failure rate distributions i.e., the First Case. The numerical method integrates simultaneous Algebraic Differential Equations. It produces a solution in the First Case under any type of aging, and in the Second Case but only under the assumption of full aging in warm-standby. On the other hand, the developed simulation algorithms produce solutions for any set of distributions (i.e., the First Case and the Second Case) under any of the three aging assumptions for the backup component in standby. The simulation solution is quantitively verified by comparison with the other two methods, and qualitatively verified by comparing the solutions under the three aging assumptions. It is numerically proven that the full aging and no aging solutions could serve as bounds of the partial aging case even when the precise mechanism of partial aging is unknown.

**Keywords:** state probability functions; partial aging in standby; Monte Carlo simulation; qualitative and quantitative verification of simulation model

## 1. Introduction

Assessing the reliability of a system is a key engineering task that has economic and safety implications. Having a better understanding of failure/repair rates of system components is a tool to design highly reliable systems and conduct repair operations at adequate cost levels while complying with adequate and reasonable maintenance schedules. A common approach for improving the reliability is to provide redundancy for excessively failing components. The redundant components may operate simultaneously in a sense that the system will never fail if at least one of the parallel components operates [1]. Another possibility is to design a "*k* out of *n*" configuration where all *n* components are in operation and the system is not failing if at least *k* of them operate properly [2]. However, the standby arrangement is the simplest, cheapest and the most utilized one; the system operates with some of its components (called primary) whereas the redundant components (called backup components) are in standby, but when a primary component fails, they take its place [3]. In this paper, we will focus on a two-component system with a standby arrangement where the backup components may fail either while in standby, or during operation after some imperfect switching mechanism has put those online. The switching mechanism can be continuous type when it actively monitors the primary component and makes its own decisions, but it can malfunction at any time [4]. However, in this paper, we will treat exclusively the widespread mechanisms that can fail only on demand when the switching is needed [5]. According to the failure intensities of the backup component, such systems are classified as cold-standby, warm-standby, and hot-standby [6]. In the hot-standby system, the intensity of failures of the backup component is the same during standby and operation, whereas in the cold-standby, there is no failure in standby. We will concentrate on the two-component warm-standby systems where there are failures of the backup component in standby, but with smaller intensity compared to its operational mode.

Additionally, the reliability of a system with backup components depends on the way of aging of the backup components while in standby. Previous works have identified three types of aging of the backup component during warm-standby: full aging, no aging and partial aging [7]. The full aging assumption means that the component changes its failure/repair rate during standby as if it is operational. Under the no aging assumption, the component does not change its failure/repair rate during standby. The partial aging assumption models the intermediate situation where the backup component experiences some wear during standby, but at a slower rate than if operational.

If some components of the system are deemed repairable, the system can be brought to its full operational capacity by replacing parts or by making adjustments [8]. In most works, the focus is on single-component repairable systems under various repair activities. A detailed discussion on how such tasks can be approached with modern statistical tools is offered in [9]. There are different types of repairs that can be adopted depending on objectives. The first possibility is the so-called perfect repair (a.k.a. as-good-as-new (AGAN) repair), where the primary component fails and it is replaced or restored to its original or good-as-new condition [10]. Minimal repair restores the device to the condition it was in immediately before the failure [11] (pp. 226–227). There may also be intermediate types of repairs (e.g., the partial perfect repair procedures mentioned in [8]). In the current work, the focus is on the case of minimal repair of the special type worse-than-old (WTO) [12]. The assumption is that during this repair, the non-repaired elements of the primary unit age as if the latter was operational.

In this paper, we will investigate the effects that adding repair and back-switching failures to a two-component warm-standby system with switching mechanism has on the reliability of the system. Our goal is to analyze how this affects the system reliability under different aging assumptions in standby. In such a system, the primary component begins operation, and when it fails, the system will try to activate the backup component, but a switching failure is possible. When the backup component is operational, the primary component undergoes minimal repair. If the latter finishes before the backup component fails, the system will try to activate back the primary component, but again a back-switching failure is possible. However, it is possible that the backup component will fail in standby. In that case, there will be no repair for the backup component since there will be no indication of the failure. The system is considered to have failed when either both components fail to operate at any given time, or when, after primary component failure, the switching mechanism fails on demand to switch the system operation to the secondary component (switching failure), or when, after a successful repair, the switching mechanism fails on demand to switch the system operation back to the primary component (back-switching failure). The primary component undergoes minimal repair, i.e., the primary component experiences full aging during the repair. The reliability behavior of the system will be investigated under three different aging assumptions for the backup component during standby: full aging, no aging and partial aging. Only mechanical aging of the components will be considered, which excludes any influence by some software aging (for discussion of the latter topic see [13,14]).

The focus of our investigation would be a two-component warm-standby system with repair, switching, and back-switching failures, denoted as 2SBRSBF. Our study concentrates on the characteristics of the uptime of the 2SBRSBF. We only consider repairs of the failed primary component of a working system. The repair of the failed system, which relates to downtime, is an important component of system availability, but it is outside of the scope of our paper (for elaborate case study of data center availability using Markovian modeling, see [14]). In [15], the causes of system and component failures were classified as technological failures, natural disaster failures, and man-made disasters (e.g., terrorism). In our study, we will consider only the technological failure of 2SBRSBF since the other two types tend to cause dependent component failures, which is outside of the scope of our study. Here, the standby mode of the 2SBRSBF is defined as a situation, where the primary component is working properly, and the backup component is fully operational, but its failure will not affect the normal operation of the system at this moment (in [16] such component configuration is classified as "active/cold-standby").

The reliability behavior of the 2SBRSBF depends on four distributions: the failure and the repair distributions of the primary component, the failure distributions of the backup component in operation and in standby. We will analyze two cases for those distributions. In the First Case, all distributions will be with constant failure/repair rates. In the Second Case, the more realistic time-dependent failure/repair rates will be applied.

We will use three methods to identify the reliability characteristics of the 2SBRSBF: analytical, numerical, and simulational. The analytical approach is applicable for the First Case distributions. We will develop novel analytical solutions for the state probability functions in the case of exponential distributions. The numerical method creates and integrates simultaneous Ordinary Differential Equations (ODEs) for 2SBRSBF. This method is applicable for any set of First Case distributions and for Second Case distributions under the assumption of full aging in standby. However, there are no simultaneous ODEs that describe the behavior of 2SBRSBF with time-dependent distributions (i.e., Second Case) under no aging or partial aging assumptions in standby. To facilitate the simulational solution, we will introduce a novel method to generate failure times of the backup component in standby under the assumptions of full aging, no aging, or partial aging. Using this method, we will modify and generalize the algorithm from [17] to simulate the behavior of 2SBRSBF and to calculate its most important reliability characteristics. That algorithm will produce a novel simulation solution for any set of distributions (i.e., the First Case and the Second Case) under any of the three aging assumptions for the backup component in standby. The proposed algorithm will be validated quantitively by comparing with the analytical and with the numerical solutions (if those exist) as well as quantitatively by comparing with the full aging results.

In what follows, Section 2 summarizes the state-of-the-art in the field and outlines the contributions of our paper. Section 3 will setup the problem for reliability characteristics assessment of a 2SBRSBF function. In Section 4, we present a novel analytical solution of the formulated problem in the case of distributions with constant failure/repair rate. A numerical solution will be identified in Section 5 where a system of four simultaneous deferential algebraic equations will model the 2SBRSBF in the case of full aging of the backup component during standby. In Section 6, the same problem will be solved with simulation which can be used with any distributions under three different assumptions about the aging mechanism of the backup component. Section 7 contains the results of three numerical examples, where we validate the proposed simulational algorithm quantitatively (by comparing with the analytical and the numerical solutions when those exist) and qualitatively (by checking whether the effects of no aging and partial aging correspond to the logically expected ones). Section 8 concludes the paper.

## 2. Related Works and Contributions of the Paper

Although the publications about warm-backup system reliability are growing recently, they are rare in comparison with reliability studies of cold-backup and hot-backup system, since the realistic models of the former tend to be more elaborate [18]. In [19] (pp. 113–115), the analytical solution for two-component warm-standby system with switching failure (2SBSF) was developed. The switching mechanism fails on demand. The failure distributions were considered exponential. Hence, no aging effect was taken into account. Explicit formulae were derived for the reliability of the system and for all state probability functions. In [20] (pp. 167–170), a model of a two-component warm-standby system (2SB) with arbitrary failure distributions was proposed. Although no particular simulation algorithm was developed, general advice was given on how to acquire the reliability function and the state probability functions using Monte Carlo simulation and how to deal with different aging assumptions. In [21], a model of a 2SB system was proposed under general standby, which generalizes the three special cases of warm-, cold- and hot-standby. The failure distributions can be arbitrary. The aging effects are accounted for using a pre-specified virtual aging function. An integral equation, connecting the failure rates and the virtual aging function with the reliability of the system was proposed. In [6], these results from [21] were generalized to solve the problem of allocation of redundancy that includes two independent and one generalized standby component. The reliability and the state probability functions of a generic two-component standby system under full aging, no aging, and partial aging were identified with a simulation algorithm in [22] using arbitrary failure distributions. That solution is verified with analytical and numerical special cases. The results from this work were expanded in [17] to model the 2SBSF, but some numerical problems connected with random variate generation and arbitrary failure rate calculations were resolved.

The majority of the above models consider aging effects, but none of them has repairs.

A 14-states model of two dissimilar warm-standby subsystems in series with repair were discussed in [23]. The failure distributions are exponential, and the system is with constant repair rates. The type of repair is AGAN. The problem of aging is not considered. Some analytical steady-state characteristics of the system are provided using Laplace transforms. Those characteristics for two-component warm-standby system with repair (2SBR) can be obtained as special cases from the results in the paper. The work [4] performs reliability analysis for a two-component warm-standby system with repair and switching failures (2SBRSF). The failure and repair rates are constant. The switching mechanism is of continuous type and has its own failure distribution. This leads to a possibility of repairing the failed backup unit while the primary component operates. All failure and repair distributions are exponential. Any failure of the switch leads to system failure. The repairs are AGAN and no aging is considered. The system has 10 states. The reliability and the state probability functions of the system were identified with a numerical algorithm as a solution of an ODE system. Another interesting two-identical-component standby system is given in [24]. The type of standby is difficult to determine since the failure in standby mode is deterministic and happens after surpassing a pre-specified time. The failure distribution of the operating unit is exponential, but the repair rate is arbitrary. There is no switching failure, but the switching mechanism inspects the failed standby unit and decides whether to replace it or to repair it. No aging is considered in this model. Some steady-state measures of reliability are obtained using semi-Markov models. In [25], the authors propose a system with $m$ identical components working in parallel with $s$ components in warm-standby. The system includes a service station that can also fail and be repaired. There are no switching failures, and all failure and repair distributions of the components are exponential. The failure and repair distributions of the service station are also exponential. The repairs are AGAN, and no aging is considered. The reliability and the state probability functions of the system were approximated using symbolic computer software. The work [18] presents a system of $n$ components in series with one component in warm-standby. There are neither switching failures nor aging considerations. The failure distributions are exponential, but the repair distributions are arbitrary. The system is also subjected to non-repairable failures. Some reliability and availability steady-state characteristics of the system are derived using Laplace transforms. In [26], the authors discuss a three identical component warm-standby system. Initially, the primary component is working, and the other two are in standby. The failure of the operating unit and the repairs are with random distributions, however in standby there is constant failure rate. The repairs are AGAN, there are no switching failures, and no aging is considered. An integral equation, connecting the failure rates with the reliability of the system is proposed.

The models with repair discussed above do not consider any aging effects.

In Table 1, we summarized seven characteristics for each of the above-discussed 12 papers plus the current work. The information in Table 1 highlights the novelty of our work against the discussed state-of-the-art studies in the literature. The contributions of our study can be outlined as follows:

1. We shall formulate a novel model of 2SBRSBF containing three operational states and four system failure substates. The switching mechanism will fail on demand and the repair of the primary unit will be WTO. This warm-standby system will utilize arbitrary failure and repair distributions and will have three types of aging modes of the backup component in warm-standby—no aging, full aging, and partial aging.
2. We shall create a novel six-attribute procedure, which gives numerically stable estimates of the equivalent age of the backup unit under any of the three aging assumptions.
3. We shall formulate 11 properties of the event chain (EC) describing the 2SBRSBF that can happen during the normal exploitation of the system.
4. We shall develop a novel algorithm to generate a random EC for the 2SBRSBF, which satisfies the EC properties in step 3 above.
5. We shall propose a simulation algorithm to calculate the state probability functions and the rest of the reliability characteristics of a 2SBRSBF in their dynamics.
6. We shall develop a novel analytical solution of the 2SBRSBF when the failure and the repair rates are constant. We will prove that the solution is real for any constant failure/repair rates and switching mechanism failure probabilities.
7. We shall develop a numerical solution of the 2SBRSBF under the assumption of full aging of the backup component in warm-standby. The procedure will use a semi-explicit system of four simultaneous differential algebraic equations (DAEs) with differential index 1, singular constant mass matrix, and Jacobian matrix depending only on the time. The main novelty is the calculation of stable approximations of the failure/repair rates at any moment of time.
8. We shall verify quantitatively the results from the simulation procedure using analytical and numerical solutions in special cases of the 2SBRSBF. The solutions in the three aging modes will serve as qualitative validation of the simulation solution.

Table 1. Overview of the state-of-the-art publications in the warm-standby area.

| Reference | Arbitrary Failure Distribution | Arbitrary Repair Distribution | Switching Failure | Aging | Repair | Repair Type | Dynamic Solution |
|---|---|---|---|---|---|---|---|
| [19] | no | N/A | yes | no | no | N/A | yes |
| [20] (pp. 167–170) | yes | N/A | no | yes | no | N/A | yes |
| [21] | yes | N/A | no | yes | no | N/A | yes |
| [6] | yes | N/A | no | yes | no | N/A | yes |
| [22] | yes | N/A | no | yes | no | N/A | yes |
| [17] | yes | N/A | yes | yes | no | N/A | yes |
| [23] | no | no | no | no | yes | AGAN | no |
| [4] | no | no | yes | no | yes | AGAN | yes |
| [24] | no | yes | no | no | yes | AGAN | no |
| [25] | no | no | no | no | yes | AGAN | yes |
| [18] | no | yes | no | no | yes | AGAN | no |
| [26] | yes | yes | no | no | yes | AGAN | yes |
| Current study | yes | yes | yes | yes | yes | WTO | yes |

## 3. States, Transition Rates, and Distributions

The dynamics of a 2SBRSBF system can be determined by its transition between several possible states [27]. The 2SBRSBF has four major states, but State 4 (where the 2SBRSBF system is not operational) is subdivided into 4 substates, called types.

In State 1, the primary component operates, the backup component is fully operational but is in standby. Sooner or later, one of the two components will fail:

(A) If the primary component fails, the system will attempt a transit to State 2, where the backup component operates and the primary component is under repair. However, if the switching device fails to operate properly, we observe the so-called switching failure on demand resulting in transition to State 4, where the 2SBRSBF system is not operational (type $a$ system failure).

(B) If the backup component fails in standby, the system will transit to State 3 where the primary component operates but the backup component is not operational. There will be no indication whether the system is in State 1 or in State 3, so no maintenance decision will be made in those two states.

In State 2 sooner or later either the primary component will be repaired, or the backup component will fail. Then one of the following two events will occur:

(A) If the primary component is repaired, the system will try a transit to State 1. However, if the switching device fails to operate properly, we observe the so-called back-switching failure resulting in transition to State 4, where the 2SBRSBF system is not operational (type $b$ system failure).

(B) If the backup component fails in operation, the system will transit to State 4 where the 2SBRSBF system is not operational (type $c$ system failure).

In State 3, sooner or later, the primary component will fail and there will be no operational backup component to take over. The system will transit to State 4 where the 2SBRSBF system is not operational (type $d$ system failure).

The State 4, where the 2SBRSBF system is not operational, is irreversible in our model regardless of the type of the system failure.

The described system is partially observable since we will not know whether the system is in State 1 or in State 3, but State 4 and State 2 are observable. At the same time, 2SBRSBF is controllable by three trivial event-driven decisions: (a) when the primary component fails, attempt to move to State 2, by switching to the backup unit; (b) when the backup unit is in operation, start repairing the primary component; (c) when the primary component is repaired, attempt to move to State 1, by back-switching to the primary unit.

The state function $P_g(t)$ (for $g$ = 1,2,3,4) measures the probability of the 2SBRSBF to be in State $g$ at time $t$ (for $t \geq 0$). Since the system will be in one state and in one state only at any non-negative time moment $t$, then:

$$P_1(t) + P_2(t) + P_3(t) + P_4(t) = 1, \text{ for } t \in [0, \infty) \tag{1}$$

The 2SBRSBF system starts in fully operational mode so initially it will be in State 1:

$$P_1(0) = 1 \text{ and } P_2(0) = P_3(0) = P_4(0) = 0 \tag{2}$$

If the four state functions are identified, then the 2SBRSBF system is quantitatively described and we can calculate all its reliability characteristics. The reliability of the system is the sum of the first three state probabilities (i.e., the probability not to be in State 4):

$$R_{sys}(t) = P_1(t) + P_2(t) + P_3(t) = 1 - P_4(t), \text{ for } t \in [0, \infty) \tag{3}$$

The mean time to failure (MTTF) of the 2SBRSBF system can be calculated as:

$$MTTF_{sys} = \int_0^\infty R_{sys}(t)dt \tag{4}$$

The time for which the reliability of the system will be $\alpha$ is known as $\alpha$-design life ($t_{des,\alpha}$). It can be identified as the unique solution of Equation (5) in the domain $t_{des,\alpha} \in (0, \infty)$:

$$R_{sys}(t_{des,\alpha}) = \alpha, \text{ for } \alpha \in (0,1) \tag{5}$$

The median ($Median_{sys}$), the B1 life ($B1\_life$), the B10-life ($B10\_life$), and the interquartile range ($IQR_{sys}$) of the 2SBRSBF system reliability can be easily estimated using Equation (5) respectively as $t_{des,0.5}, t_{des,0.99}, t_{des,0.9}$, and $t_{des,0.25} - t_{des,0.75}$ [20] (pp. 87–88).

To identify the four required state functions of the 2SBRSBF system, we need to know:

- The probability, $p_f$, for switching failure on demand.
- The probability, $p_r$, for back-switching failure on demand.
- The probability density function (PDF), $f_1(t)$, of the failure distribution for the primary component in operation.
- The PDF, $f_2(t)$, of the failure distribution for the standby component in operation.
- The PDF, $f_3(t)$, of the failure distribution for the standby component in standby.
- The PDF, $f_4(t)$, of the repair distribution for the primary component.

Each of the four PDFs, $f_k(t)$, (for $k$ = 1, 2, 3, 4) can be transformed into four alternative forms: a cumulative distribution function (CDF), $F_k(t)$, a failure/repair rate, $\lambda_k(t)$ (as shown in [28]), a complementary CDF, or $R_k(t)$, and an inverse CDF, i.e., $F_k^{-1}(p)$. The five forms $f_k(t), F_k(t), \lambda_k(t), R_k(t)$, and $F_k^{-1}(p)$ contain the same information and are equivalent. In the ideal world the domain of the first four functions and the range of the last one will be $t \in [0, \infty)$ where $t$ can be interpreted as time. However, this is not always the case—those failure and repair distributions are based on information about the behavior of the components. The first step is to summarize the available information in several nodes of the CDF. If the reliability information is in the form of fully observed or multiply sensor data, then we can produce an empirical distribution, using either the Kaplan-Meier product limit estimator method [29] (see the function *ecdf.m* in [30], which embodies the method) or the invertible ECDF estimator with maximum count of nodes [31], or any other modern method. If the information is in the form of expert knowledge, then we can extract subjective quantiles using the triple bisection method [32] as described in [33]. The second step is to fit a parametric distribution of some type to the nodes of the CDF identified in the first step. The work [20] (p. 399) gives several reasons to use parametric distributions rather than empirical ones, with the most important one being that empirical distributions can only be trusted at the beginning of the failure/repair process. Regardless of the method utilized to identify the parameters in the second step (least square, maximum likelihood estimation, Bayesian estimation, etc.), it is quite possible that some of the derived parametric distributions would have substantial support for negative values of the argument $t$. For purely pragmatic reasons, we assume that for each $k$, we are given only procedures to calculate $f_k(t), F_k(t)$, and $F_k^{-1}(p)$. Such numerical procedures exist in almost any software package. For example, the Statistics and Machine Learning Toolbox in MATLAB contains the *pdf.m*, *cdf.m*, and *icdf.m* which calculate the PDF, the CDF, and the inverse CDF values for any distribution object created by the *makedist.m* [30]. The latter can choose a wide variety of parametrical 1D distributions with arbitrary specified parameters. Unluckily, some of those parametrical distributions are defined over the whole real axis (e.g., the normal distribution, or the extreme value distribution). Traditionally, no numerical procedures are given for estimating the values of $\lambda_k(t)$ and $R_k(t)$, which have to be approximated using $f_k(t), F_k(t), F_k^{-1}(p)$. In this paper, any of the procedures $f_k(t), F_k(t)$,

$R_k(t)$, $\lambda_k(t)$, $F_k^{-1}(p)$ will be called the $k$th original distribution since the five of them describe in alternative form the uncertainty of a real continuous variable:

$$\begin{cases} \text{(a) } f_k(t), \text{ for } k = 1,2,3,4 \text{ with Domain } t \in (-\infty, +\infty) \\ \text{(b) } F_k(t) = \int_{-\infty}^{t} f_k(t)dt, \text{ for } k = 1,2,3,4 \text{ with Domain } t \in (-\infty, +\infty) \\ \text{(c) } \lambda_k(t) = f_k(t)/[1 - F_k(t)], \text{ for } k = 1,2,3,4 \text{ with Domain } t \in (-\infty, +\infty) \\ \text{(d) } R_k(t) = 1 - F_k(t), \text{ for } k = 1,2,3,4 \text{ with Domain } t \in (-\infty, +\infty) \\ \text{(e) } F_k^{-1}(p), \text{ for } k = 1,2,3,4 \text{ with Domain } p \in [0,1] \end{cases} \quad (6)$$

Here, $R_k(t)$ from Equation (6) a is aka original reliability/repair function when the real argument $t$ is non-negative and can be interpreted as time. In our problem, the argument $t$ would be most often the time (or other suitable non-negative variable, e.g., mileage), so we will use the original distribution in Equation (6) a–e to approximate their truncated versions which take the form of conditional distributions provided that the failure/repair has not happened till time 0:

$$\begin{cases} \text{(a)} f_{k,trun}(t) = f_k(t|0) = f_k(t)/R_k(0), \text{for} k = 1,2,3,4 \text{with Domain} t \in [0, +\infty) \\ \text{(b)} F_{k,trun}(t) = F_k(t|0) = 1 - R_k(t)/R_k(0), \text{for} k = 1,2,3,4 \text{with Domain} t \in [0, +\infty) \\ \text{(c)} \lambda_{k,trun}(t) = \lambda_k(t|0) = f_{k,trun}(t)/[1 - F_{k,trun}(t)], \text{for} k = 1,2,3,4 \text{with Domain} t \in [0, +\infty) \\ \text{(d)} R_{k,trun}(t) = R_k(t|0) = 1 - F_{k,trun}(t), \text{for} k = 1,2,3,4 \text{with Domain} t \in [0, +\infty) \\ \text{(e)} F_{k,trun}^{-1}(p) = F_k^{-1}(p|0), \text{for} k = 1,2,3,4 \text{with Domain} p \in [0,1] \end{cases} \quad (7)$$

In this paper, any of the functions $f_{k,trun}(t)$, $F_{k,trun}(t)$, $R_{k,trun}(t)$, $\lambda_{k,trun}(t)$, $F_{k,trun}^{-1}(p)$ will be called the $k$th truncated distribution, since the five of them describe in alternative forms the uncertainty of a real non-negative continuous variable which can be interpreted as time. The $R_{k,trun}(t)$ from Equation (7) d is aka truncated reliability/repair function. Let us concentrate on the 2SBRSBF system at time $t$:

- The rate for transitioning between State 1 and State 2 will depend on $P_1(t)$, on $p_f$, and on the conditional failure distribution $f_1(\tau | t)$ (failure density of the primary component in operation, given that it has not failed till time $t$). The reason is that any possible previous repairs of the primary component were from minimal type which equates to the full aging assumption for the primary component during repair and any failure will behave like a first failure at time $t$.
- The rate for transitioning between State 1 and State 4 (type $a$ system failure) will depend on $P_1(t)$ and on the conditional failure distribution $f_1(\tau | t)$ since the same arguments made for the State 1–State 2 transition apply.
- The rate for transitioning between State 3 and State 4 (type $d$ system failure) will depend on $P_3(t)$, on $p_f$, and on the conditional failure distribution $f_1(\tau | t)$ since the same arguments made in the State 1–State 2 transition apply.
- The rate for transitioning between State 2 and State 1 will depend on $P_2(t)$, on $p_r$, and on the conditional repair distribution $f_4(\tau | t)$ (repair density of the primary component, given that the repair starts at time $t$). The reason is that any possible previous repairs of the primary component were from minimal type, which equates to the full aging assumption for the primary component during operation and any repair will look like a first repair at time $t$.
- The rate for transitioning between State 2 and State 4 (type $b$ system failure) will depend on $P_2(t)$, on $p_r$, and on the conditional repair distribution $f_4(\tau | t)$ since the same arguments made in the State 2–State 1 transition apply.
- The rate for transitioning between State 1 and State 3 will depend on $P_1(t)$ and on the conditional failure distribution $f_3(\tau | t)$ (failure density of the backup component in standby, given that it has not failed till time $t$). The backup component is never repaired until there is a system failure, which suggests that the failure rate in standby should depend only on the time the system operates but not on the backup component history of utilization (alternating between operational and standby modes).

- The rate for transitioning between State 2 and State 4 (type $c$ system failure) will depend on $P_2(t)$ and on the conditional failure distribution $f_2(\tau \mid t_{age})$ (failure density of the backup component in operation, given that it has not failed till time $t_{age}$). Here $t_{age}$ is the equivalent aging of the backup component in operation. It depends on the type of aging and possibly on the backup component history of utilization (alternating between operational and standby modes).

The four state functions of 2SBRSBF system can be identified using computer simulation in the above setup for any set of distributions and aging assumptions during standby. However, for verification purposes, two alternative solution methods can be developed for some special cases of the 2SBRSBF system. This approach was successfully applied in [34] for verification of a novel simulation-based optimization algorithm used in redundancy allocati on problems using Markovian models as special cases.

If we have a set of First Case distributions, then all state transitions will depend on the absolute densities, rather than from conditional ones. The reason is that the exponential distributions have no memory, and hence any aging assumptions are irrelevant. Then the probabilities for transitioning between the states depend only on the current state of the system, but not on the history describing how the system turns out to be in the current state. This means that the 2SBRSBF system with First Case distributions degenerates to a Markov model [27] (more precisely to a partially observable Markov decision process [35]). Such Markov model can be conveniently visualized with the Rate Diagram (RD) [20] (pp. 155–170) shown in Figure 1a. Using that RD, we will derive an analytical solution for the four state probability functions of the 2SBRSBF system with First Case distributions.

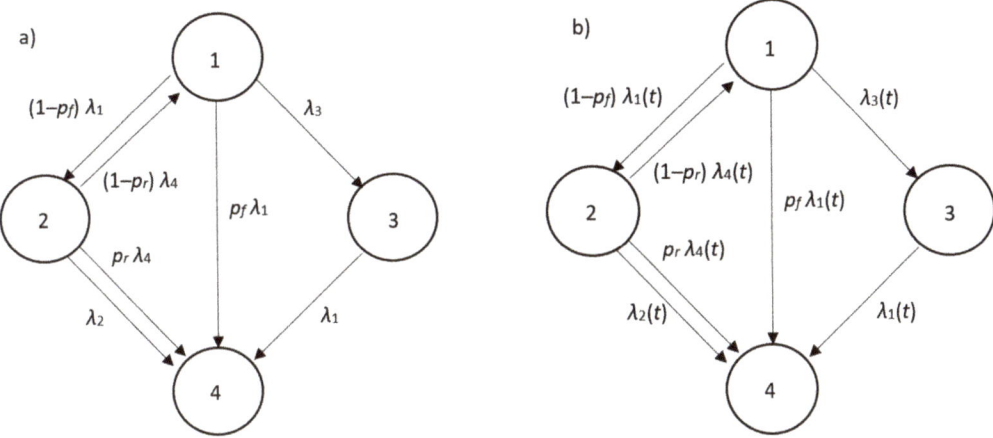

**Figure 1.** Rate diagram for 2SBRSBF with: (**a**) First Case distributions; (**b**) Second Case distributions with full aging in standby.

If we have a system with full aging assumption during standby, then the equivalent aging of the backup component in operation rate $t_{age}$, described above in the transitioning between State 2 and State 4 (type $c$ system failure) will be simply the current time $t$. The reason is that the backup component is assumed to age during standby in the same fashion as in operation, which shows that any failure of the backup component during operation will behave like a first failure at time $t$. This means that the 2SBRSBF system with full aging assumption degenerates to a semi-Markov model where the transition probabilities depend not only on the current state but also on the current time [36]. The semi-Markov model can be conveniently visualized with the Generalized Rate Diagram (GRD) shown in Figure 1b [37] (pp. 521–526). Using the GRD, we can describe the 2SBRSBF system with simultaneous ODEs. This is possible because the failure/repair rate of any distribution,

$F(t)$, at time $t^*$ coincides with the failure/repair rate of the conditional distribution $F(\tau | T)$ at the same time $t^*$, if $T \leq t^*$. This trivial fact is proven in Appendix A. The derived Cauchi problem can be solved numerically. Obviously, such solution exists also for the First Case distribution, which will allow the comparison of the analytical and the numerical solutions.

Neither the analytical, nor the numerical solutions can be derived for the cases of the Second Case distribution under the assumptions of no aging and partial aging since general aging effects cannot be described by any Markovian or semi-Markovian model and there is no system of ODE which fully quantifies the reliability behavior of 2SBRSBF unless when the primary component is subjected to full aging in standby (see [14,36]).

## 4. Analytical Solution

This solution is applicable only for First Case distributions, where the failure/repair rates are constant. The rate diagram in Figure 1a can be represented as a system of three ODEs from Equation (8) about the first three state probability functions [38]:

$$\begin{cases} \frac{dP_1}{dt}(t) = -(\lambda_1 + \lambda_3)P_1(t) + (1 - p_r)\lambda_4 P_2(t) \\ \frac{dP_2}{dt}(t) = \left(1 - p_f\right)\lambda_1 P_1(t) - (\lambda_4 + \lambda_2)P_2(t) \\ \frac{dP_3}{dt}(t) = \lambda_3 P_1(t) - \lambda_1 P_3(t) \end{cases} \quad (8)$$

The initial conditions are given in Equation (2). After solving Equation (8), the last probability function, $P_4(t)$, can be estimated from Equation (3) as the complement to 1 of the other state probability functions. The analytical solution of 2SBRSBF with First Case distributions can be described as: "set the constants from Equation (9) and form the state probability functions from Equation (10)" (see Appendix B for the proof).

$$\begin{aligned} &K = (\lambda_1 + \lambda_2 + \lambda_3 + \lambda_4)/2; C = (\lambda_1 + \lambda_3)(\lambda_2 + \lambda_4) - \left(1 - p_f\right)(1 - p_r)\lambda_1\lambda_4 \\ &s_1 = -K + \sqrt{K^2 - C}; s_2 = -K - \sqrt{K^2 - C} \\ &A_1 = \frac{s_1 + \lambda_2 + \lambda_4}{s_1 - s_2}; B_1 = \frac{s_2 + \lambda_2 + \lambda_4}{s_2 - s_1}; A_2 = \frac{(1 - p_f)\lambda_1}{s_1 - s_2}; B_2 = \frac{(1 - p_f)\lambda_1}{s_2 - s_1} \\ &A_3 = \frac{\lambda_3(s_1 + \lambda_2 + \lambda_4)}{(s_1 - s_2)(\lambda_1 + s_1)}; B_3 = \frac{\lambda_3(s_2 + \lambda_2 + \lambda_4)}{(s_2 - s_1)(\lambda_1 + s_2)}; C_3 = \frac{\lambda_3(\lambda_1 + \lambda_2 + \lambda_4)}{(\lambda_1 - s_1)(\lambda_1 - s_2)} \end{aligned} \quad (9)$$

$$\text{Domain}: t \in [0, \infty)$$
$$\begin{cases} P_1(t) = A_1 e^{s_1 t} - B_1 e^{s_2 t} \\ P_2(t) = A_2 e^{s_1 t} - B_2 e^{s_2 t} \\ P_3(t) = A_3 e^{s_1 t} - B_3 e^{s_2 t} + C_3 e^{-\lambda_1 t} \\ P_4(t) = 1 - (A_1 + A_2 + A_3)e^{s_1 t} + (B_1 + B_2 + B_3)e^{s_2 t} - C_3 e^{-\lambda_1 t} \end{cases} \quad (10)$$

The reliability of the system from Equation (11) and its MTTF from Equation (12) are derived as special cases of Equations (3) and (4):

$$R_{sys}(t) = (A_1 + A_2 + A_3)e^{s_1 t} - (B_1 + B_2 + B_3)e^{s_2 t} + C_3 e^{-\lambda_1 t}, \text{ for } t \in [0, \infty) \quad (11)$$

$$MTTF_{sys} = -(A_1 + A_2 + A_3)/s_1 + (B_1 + B_2 + B_3)/s_2 + C_3/\lambda_1 \quad (12)$$

## 5. Numerical Solution

This solution is applicable for any Second Case distribution with full aging of the backup component in standby and for any First Case distribution. The GRD in Figure 1b can be represented as a system of four simultaneous DAEs from Equation (16) about the four state probability functions, $P_g(t)$ ($g$ = 1,2,3,4). The system of DAEs will be numerically integrated from 0 to $t_{end}$, where the latter will be selected sufficiently large, so $R_{sys}(t_{end}) \approx 0$ (< 0.01). The main numerical difficulty in solving Equation (16) is to advise a procedure for stable approximation of the failure/repair rates, $\lambda_k(t)$ ($k$ = 1,2,3,4), at any $t \in [0, t_{end}]$. That problem is far from trivial since sometimes $F_k(t)$ is so close to 1, that the denominator of Equation (7) turns into 0. For each of the four distributions, using the original inverse CDF function, we can calculate the time $t_{\lambda,k}$, where the denominator of Equation (7) equals to 100 times the machine epsilon ($\epsilon$):

$$t_{\lambda,k} = F_k^{-1}(1 - 100\varepsilon), \text{ for } k = 1,2,3,4 \tag{13}$$

The approximated failure/repair rate, $\lambda_{k,a}(t)$ ($k$ = 1,2,3,4) equals to Equation (7) if its denominator is greater than $100\epsilon$ or equals the failure/repair rate at $t_{\lambda,k}$ otherwise:

$$\lambda_{k,a} = \begin{cases} f_k(t)/[1 - F_k(t)] & , t \in [0, t_{\lambda,k}] \\ f_k(t_{\lambda,k})/[1 - F_k(t_{\lambda,k})] & , t \in (t_{\lambda,k}, \infty) \end{cases}, \text{ where } k = 1,2,3,4 \tag{14}$$

Equation (14) produces numerically stable approximations of the failure/repair rates at any non-negative time not greater than $t_{end}$. This is true even when a distribution is truncated which means that $F_k(0) > 0$ and only its part in the non-negative domain has to be used. Then, according to Appendix A, the value of the failure/repair rate for any non-negative time will be the same as that of the non-truncated distribution since the truncated distribution can be represented as a conditional nontruncated one:

$$F_{k,trun}(t) = F_k(t|T_0 = 0) = 1 - [1 - F_k(t)]/[1 - F_k(0)], \ t \geq 0 \tag{15}$$

Now, we can write the DAE system corresponding to Figure 1b:

$$\begin{cases} \frac{dP_1}{dt}(t) = -[\lambda_{1,a}(t) + \lambda_{3,a}(t)]P_1(t) + (1 - p_r)\lambda_{4,a}(t)P_2(t) \\ \frac{dP_2}{dt}(t) = \left(1 - p_f\right)\lambda_{1,a}(t)P_1(t) - [\lambda_{4,a}(t) + \lambda_{2,a}(t)]P_2(t) \\ \frac{dP_3}{dt}(t) = \lambda_{3,a}(t)P_1(t) - \lambda_{1,a}(t)P_3(t) \\ 0 = P_1(t) + P_2(t) + P_2(t) + P_2(t) - 1 \end{cases} \tag{16}$$

The dependent variables can be organized in a 4D vector: $\vec{y}(t) = [P_1(t), P_2(t), P_3(t), P_4(t)]^T$. The DAE from Equation (16) is semi-explicit with differential index 1. It has a singular constant mass matrix:

$$M(t, \vec{y}) = \begin{pmatrix} 1 & 0 & 0 & 0 \\ 0 & 1 & 0 & 0 \\ 0 & 0 & 1 & 0 \\ 0 & 0 & 0 & 0 \end{pmatrix} \tag{17}$$

The Jacobian matrix of the RHS of Equation (16) depends only on the time $t$:

$$J(t, \vec{y}) = \begin{bmatrix} -\lambda_{1,a}(t) - \lambda_{3,a}(t) & (1 - p_r)\lambda_{4,a}(t) & 0 & 0 \\ (1 - p_r)\lambda_{1,a}(t) & -\lambda_{4,a}(t) - \lambda_{2,a}(t) & 0 & 0 \\ \lambda_{3,a}(t) & 0 & -\lambda_{1,a}(t) & 0 \\ 1 & 1 & 1 & 1 \end{bmatrix} \tag{18}$$

The initial conditions given in Equation (2) together with Equations (16) and (17) form a Cauchy problem:

$$M(t, \vec{y})\vec{y}'(t) = \vec{f}(t, \vec{y}(t)) \text{ with } \vec{y}_{ini} = \vec{y}(0) = [P_1(0), P_2(0), P_3(0), P_4(0)]^T = [1,0,0,0]^T \tag{19}$$

Here, $\vec{y}'(t) = [dP_1(t)/dt, dP_2(t)/dt, dP_3(t)/dt, dP_4(t)/dt]^T$, $M(t, \vec{y})$ is the mass matrix (17), and the 4D $\vec{f}(t, \vec{y}(t))$ is the RHS of Equation (16). The problem from Equation (19) can be numerically integrated (e.g., using ode15s.m from MATLAB [39]) at 2000 evenly distributed time points from 0 to $t_{end}$:

$$t_i = (i - 1)t_{end}/1999, \text{ for } i = 1,2,\ldots,2000 \tag{20}$$

The reliability function and the $MTTF_{sys}$ can be calculated approximating Equations (3) and (4) as:

$$R_{sys}(t_i) = 1 - P_4(t_i), \text{ for } i = 1, 2, \ldots, 2000 \qquad (21)$$

$$MTTF_{sys} = \left[ R_{sys}(t_1) + R_{sys}(t_{2000}) + 2 \sum_{i=2}^{1999} R_{sys}(t_i) \right] t_{end} / 1999 \qquad (22)$$

## 6. Simulation Solution

This solution is applicable for any set of distribution (First Case or Second Case) and for any type of aging of the backup component in standby (full aging, no aging, or partial aging). Any simulation uses multiple pseudo-realities to study the system in question. The information from each generated pseudo-reality will be kept in an EC, whose definition and properties will be discussed in Section 6.1. In Section 6.2 we will concentrate on the development of specific functions generating random time intervals for the 2SBRSBF system. Those functions will be used in Section 6.3 where an algorithm will be developed to generate a random EC describing the 2SBRSBF system. In Section 6.4 we will extract the information in the generated ECs to calculate the state probability functions and the rest of the reliability characteristics of a 2SBRSBF system.

### 6.1. Definition and Properties of the Event Chains for 2SBRSBF

In the simulational solution, we generate a large count $N$ of pseudo-realities in which we observe the behavior of the 2SBRSBF system from time 0 to system failure or to time $t_{end}$ whichever comes first. As in the numerical solution (described in Section 5) the constant $t_{end}$ is selected sufficiently large, so $R_{sys}(t_{end}) < 0.01$. The pseudo-realities are described with the ECs introduced in [22] where the EC of the $j$th pseudo-reality is defined as the set:

$$EC_j = \{ [timepsr_j(k), statepsr_j(k)] | k = 1, 2, \cdots, q_j \} \qquad (23)$$

The notation in Equation (23) shows that the $j$th pseudo-reality contains $q_j$ state transitions (called events) where the $k$th consecutive event which happened at time $timepsr_j(k)$ is a transition to state/substate $timepsr_j(k)$. The latter is coded either with 1, 2, and 3 respectively for State 1, State 2, and State 3, or with 40, 41, 42, and 43 respectively for system failure type $b$, type $a$, type $c$, and type $d$ (all of them denoting State 4). Any EC for a 2SBRSBF system should have the following properties:

**p1)** It contains at least one event: $q_j \geq 1$.
**p2)** The events happen at strictly increasing times: $timepsr_j(k) < timepsr_j(k + 1)$ for $k = 1, 2, \ldots, (q_j - 1)$.
**p3)** The initial event is at time zero: $timepsr_j(1) = 0$.
**p4)** The final event happens before $t_{end}$: $timepsr_j(q_j) < t_{end}$.
**p5)** The simulation starts with fully operational system: $statepsr_j(1) = 1$.
**p6)** Whenever a system failure is observed the simulation ends: if $statepsr_j(b) > 3$, then $q_j = b$.
**p7)** Whenever the State 3 is observed either it is the last event, or the next event is the system failure type $d$: if $statepsr_j(b) = 3$, then either $q_j = b$, or $q_j = (b + 1)$ and $statepsr_j(q_j) = 43$.
**p8)** The State 3 and the State 4 (in all its substates) can happen only once: #[$statepsr_j(k) = 3$] $\leq 1$, #[$statepsr_j(k) = 40$] $\leq 1$, #[$statepsr_j(k) = 41$] $\leq 1$, #[$statepsr_j(k) = 42$] $\leq 1$, #[$statepsr_j(k) = 43$] $\leq 1$.
**p9)** The State 1 and State 2 alternate in the beginning of the EC including to the $h$th event and neither one happens later: $statepsr_j(k) = 1$ if and only if $k$ is odd and $k \leq h$, whereas $statepsr_j(k) = 2$ if and only if $k$ is even and $k \leq h$.
**p10)** There could be maximum two events after $h$: $h \leq q_j \leq (h + 2)$.
**p11)** If there are events after the $h$th one, they are either a transition to State 3 or a transition to State 4 (in all its substates): $statepsr_j(k) \geq 3$ for all $k > h$ and $k \leq q_j$.
**p12)** The State 2 can be observed only on an even position and the previous event is always a transition to State 1: if $statepsr_j(b) = 2$, then $b$ is even and $statepsr_j(b - 1) = 1$.

**p13)** The State 3 can be observed only on an even position and the previous event is always a transition to State 1: if $statepsr_j(b) = 3$, then $b$ is even and $statepsr_j(b-1) = 1$.

The formulated EC properties will facilitate the generation of time-period variates presented in Section 6.2. The algorithm described in Section 6.3 will generate ECs with the formulated EC properties. The latter will be used in Section 6.4 to prove the methods for extracting reliability information from the generated set of ECs for 2SBRSBF system.

*6.2. Generating Times Periods Using Conditional Distributions from 2SBRSBF*

As discussed in Section 3, to simulate an EC of a 2SBRSBF system we need to generate random time-periods complying with the conditional failure distributions $f_1(\tau \mid t), f_3(\tau \mid t)$, and $f_2(\tau \mid t_{age})$ and with the conditional repair distribution $f_4(\tau \mid t)$, where $t$ and $t_{age}$ are non-negative values.

We do not know which of the four original distributions, $f_k(t)$ ($k = 1,2,3,4$), are defined only in the non-negative domain and which are defined in the entire real axes so we need to substitute them with their truncated distributions, $f_{k,trunc}(t) = f_k(t \mid 0)$ for $k = 1,2,3,4$. Noting that if the first condition is met, then $f_{k,trunc}(t) = f_k(t \mid 0) = f_k(t)$ ($k = 1,2,3,4$), and we can safely work only with truncated distributions. So, strictly speaking, we need to generate time-period variates from the conditional truncated distributions $f_{1,trun}(\tau \mid t), f_{3,trun}(\tau \mid t)$, $f_{2,trun}(\tau \mid t_{age})$, and $f_{4,trun}(\tau \mid t)$. However, for any $k$ it is true that:

$$f_{k,trun}(\tau|t) = \frac{f_{k,trun}(\tau+t)}{R_{k,trun}(t)} = \frac{f_k(\tau+t|0)}{R_k(t|0)} = \frac{f_k(\tau+t+0)/R_k(0)}{R_k(t+0)/R_k(0)} = \frac{f_k(\tau+t)}{R_k(t)} = f_k(\tau|t) \quad (24)$$

According to Equation (24) the conditional truncated distributions coincide with the conditional original distributions. In case $t$ and $t_{age}$ are known entities we can generate random time-period variates as special cases of the Practical Indirect Sampling Method from Conditional CDF (PISMCF) [17] where the algorithm is motivated, formalized, illustrated, and proven. On its basis we can define a three-attribute procedure, PISMCF(.), which generates numerically stable random time interval variate, $\Delta\tau$, from a given conditional CDF, $F(t \mid T_{surv})$, where $T_{surv}$ is a non-negative real number representing the time of survival:

$$\Delta\tau = \text{PISMCF}\left(F(.), F^{-1}(.), T_{surv}\right) \quad (25)$$

In Equation (25), $F(.)$ is the unconditional CDF which can express $F(t \mid T_{surv})$ using Equation (26):

$$1 - F(t|T_{surv}) = \frac{1 - F(t + T_{surv})}{1 - F(T_{surv})} \quad (26)$$

The second argument, $F^{-1}(.)$, of the PISMCF procedure from Equation (25) being the inverse CDF, can be used to estimate the time $t_\lambda$ where the denominator of Equation (26) is 100 machine epsilons ($\epsilon$):

$$t_\lambda = F^{-1}(1 - 100\epsilon) \quad (27)$$

In short, the algorithm for estimating Equation (25) is: (a) Calculate $t_\lambda$ using Equation (27); (b) if $T_{surv} < t_\lambda$, then set $T_{cut} = T_{surv}$, else set $T_{cut} = t_\lambda$; (c) Generate $RD$ as a uniformly distributed variate in the unit interval (0,1); (d) estimate $p_{RD} = 1 - RD\,[1 - F(T_{cut})]$; (e) Set $\Delta\tau = F^{-1}(p_{RD})$.

Let us assume that while performing the simulation of the $j$th pseudo-reality for the 2SBRSBF system we have observed only the first $k_{cur}$ state events. The simulation probably will continue and therefore, the EC is yet incomplete:

$$EC_j^{inc} = \{[timepsr_j(k), statepsr_j(k)] \mid k = 1, 2, \cdots, k_{cur}\} \quad (28)$$

Then, the current state of the system is $s_{cur} = statespr_j(k_{cur})$ and the simulational time is $T_{cur} = timespr_j(k_{cur}) < t_{end}$ (see EC property p3). The incomplete EC in Equation (28) is never empty since $k_{cur} \geq 1$ (see EC properties p1, p3, and p5).

If $s_{cur} > 3$, we do not need to generate any time-period variates since it shows a system failure, i.e., end of the simulation in the $j$th pseudo-reality (see EC properties p8, p11, and p6).

If $s_{cur}$ is 1, we need to generate two possible time-period variates: the time to failure of the primary unit, $\Delta\tau_{1,fp}$, and the time to failure in standby of the backup unit, $\Delta\tau_{1,fb}$. Using Equation (25):

$$\Delta\tau_{1,fp} = \text{PISMCF}\left(F_1(.), F_1^{-1}(.), T_{cur}\right) \tag{29}$$

$$\Delta\tau_{1,fb} = \text{PISMCF}\left(F_3(.), F_3^{-1}(.), T_{cur}\right) \tag{30}$$

If $s_{cur}$ is 3, we do not need to generate any time-period variate since the possible time to failure of the primary unit is known to be $\left(\Delta\tau_{1,fp} - \Delta\tau_{1,fb}\right)$, where $\Delta\tau_{1,fp}$ and $\Delta\tau_{1,fb}$ are generated in the previous State 1 (see EC properties p9 and p13).

If $s_{cur} > 3$, we do not need to generate any time-period variate since we have observed a system failure of some type which means that the simulation in the $j$th pseudo-reality should stop and therefore $q_j = k_{cur}$ (see EC properties p6, p8, and p11).

If $s_{cur}$ is 2, we need to generate two possible time-period variates: the time to repair of the primary unit, $\Delta\tau_{2,rp}$, and the time to failure in operation of the backup unit, $\Delta\tau_{2,fb}$. Using Equation (25):

$$\Delta\tau_{2,rp} = \text{PISMCF}\left(F_4(.), F_4^{-1}(.), T_{cur}\right) \tag{31}$$

$$\Delta\tau_{2,fb} = \text{PISMCF}\left(F_2(.), F_2^{-1}(.), t_{age}\right) \tag{32}$$

If the 2SBRSBF operates with First Case distributions, the equivalent age, $t_{age}$, of the backup unit when it starts operation at time $T_{cur}$ is rather irrelevant since $F_2(.)$ is the CDF of an exponential distribution. Then we can compare the state probability functions derived by the simulational solution with the same acquired, on one hand, from numerical solution with the DAE system from Equation (16) according to the RD in Figure 1b and on the other hand with the analytical solution from Equations (9)–(12) according to the RD in Figure 1a.

If, however, the 2SBRSBF operates with Second Case distributions, then in order to use Equation (32), we have to determine the equivalent age, $t_{age}$, at time $T_{cur}$. Since we need $\Delta\tau_{2,fb}$ only when the system is State 2, it follows that $k_{cur}$ is even (see EC property p9). From the beginning of the $j$th pseudo-reality up to time $T_{cur}$, the backup component has been in standby ($k_{cur}/2$) times when the primary component was operating till its failure (see EC property p9). Up to $T_{cur}$, the backup component has never failed in standby when the primary component was in operation, i.e., during the compound time interval with overall positive length $T_{sb}$ (see EC properties p6, p8, and p9). The latter time length can be defined using Equation (28) as:

$$T_{sb} = \sum_{i=1}^{k_{cur}/2}\left[timepsr_j(2i) - timepsr_j(2i-1)\right] \tag{33}$$

On the other hand, the backup component has been in operation ($k_{cur}/2 - 1$) times, when the primary component was in successful repair (see EC property p9). Up to $T_{cur}$, the backup component has never failed during operation when the primary component was successfully repaired, i.e., during the compound time interval with overall non-negative length $T_{oper}$ time (see EC properties p7, p8, and p9). The latter time length can be estimated by noting that up to $T_{cur}$ the 2SBRSBF system is either in State 1 or in State 2:

$$T_{oper} = T_{cur} - T_{sb} \tag{34}$$

The non-negative value of $t_{age}$ will be the sum of the backup component operation time, $T_{oper}$, with the equivalent operating time $T_{equ}$ with which the backup component would age during the standby time $T_{sb}$:

$$t_{age} = T_{oper} + T_{equ} \qquad (35)$$

The equivalent operating time $T_{equ}$ depends on aging in standby mechanism under which the 2SBRSBF system functions. There are three alternative assumptions for the nature of this aging in standby mechanism: full aging, no aging, or partial aging.

The full aging assumption accepts that the aging of the backup component during standby is the same as during operation (see Figure 2a):

$$T_{equ} = T_{sb} \Rightarrow t_{age} = T_{oper} + T_{sb} = T_{cur} - T_{sb} + T_{sb} = T_{cur} \qquad (36)$$

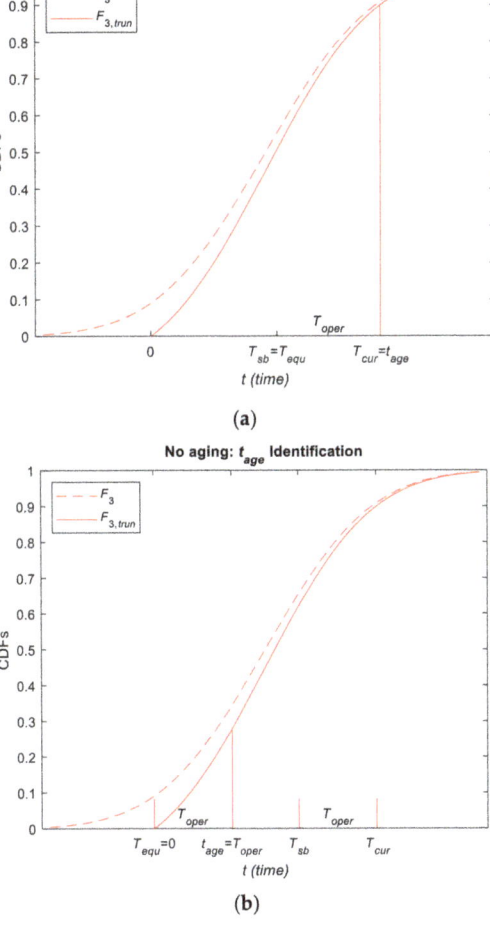

(a)

(b)

**Figure 2.** Cont.

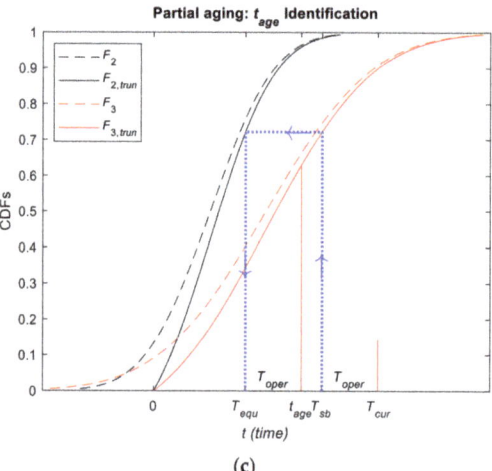

(c)

**Figure 2.** Identification of the equivalent aging time for different aging assumptions: (**a**) under full aging; (**b**) under no aging; (**c**) under partial aging.

Under the full aging assumption, a 2SBRSBF can be described with the DAE system from Equation (16) according to the RD in Figure 1b, which allows us to acquire numerical solution for Second Case distributions. The numerical solution can be compared with simulational state probability functions.

The no aging assumption accepts that the backup component during standby never ages (see Figure 2b):

$$T_{equ} = 0 \Rightarrow t_{age} = T_{oper} + 0 = T_{cur} - T_{sb} \tag{37}$$

Under the no aging assumption, a 2SBRSBF cannot be described with a DAE system since no RD adequately reflects the reliability behavior of the 2SBRSBF. For Second Case distributions, the only possible solution is the simulational one.

The partial aging assumption accepts that the backup component in standby ages to the same reliability as the backup component in operation during the equivalent operating time $T_{sb}^{equ}$ (see Figure 2c):

$$F_{2,trun}\left(T_{sb}^{equ}\right) = F_{3,trun}(T_{sb}) \tag{38}$$

Equation (38) in simplified form was firstly proposed in [22], where it was successfully tested for Two-Component Standby System with Failures in Standby. In a real 2SBRSBF system, the failures of the backup component will be more frequent during operation than during standby which means that $F_{2,trun}(t) \geq F_{3,trun}(t)$ for any non-negative time $t$. This inequality, together with Equation (38), assures that practically always $T_{equ} \in [0, T_{sb}]$. Applying Equation (15) twice to Equation (38) we get:

$$T_{equ} = \begin{cases} F_2^{-1}(p^{equ}) & \text{if } p^{equ} < 1 - 100\varepsilon \\ t_{\lambda,2} & \text{if } p^{equ} \geq 1 - 100\varepsilon \end{cases}, \text{ where } p^{equ} = 1 - \frac{1 - F_2(0)}{1 - F_3(0)}[1 - F_3(T_{sb})] \tag{39}$$

In Equation (39), $t_{\lambda,2}$ is calculated with Equation (13) for $k = 2$, therefore it uses the ideas in Equation (14) for stable approximation of the equivalent operating time, $T_{equ}$, at any $T_{cur} \in [0, t_{end}]$ for arbitrary incomplete EC from Equation (23) describing the behavior of a 2SBRSBF.

Under the partial aging assumption, the 2SBRSBF cannot be described with a DAE system since no RD adequately reflects the reliability behavior of the 2SBRSBF similarly to the no aging assumption. Again, for Second Case distributions, the only possible solution is the simulational one.

We combined Equations (33)–(37) into a six-attribute procedure, TAGEASS(.), which gives numerically stable estimates for the equivalent age, $t_{age}$, of the backup unit under any of the three aging assumptions:

$$t_{age} = \text{TAGEASS}\left(F_2(.), F_2^{-1}(.), F_3(.), EC_j^{inc}, T_{cur}, Flag_A\right) \tag{40}$$

In Equation (40), the variable $Flag_A$ is 1, 2 or 3, respectively when the 2SBRSBF operates under the full aging, no aging, or partial aging assumptions. Then Equation (40) can be estimated using Algorithm 1.

---

**Algorithm 1** Equivalent Age Estimation in the $j$th Pseudo-Reality for 2SBRSBF

1) Calculate the total standby time of the backup component, $T_{sb}$, using (33).
2) Calculate the total operational time of the backup component, $T_{oper}$, using (34).
3) If $Flag_A$ = 1 (full aging assumption), then calculate the equivalent operating time, $T_{equ}$, using (36).
4) If $Flag_A$ = 2 (no aging assumption), then calculate the equivalent operating time, $T_{equ}$, using (37).
5) If $Flag_A$ = 3 (partial aging assumption), then:
    5.1) Calculate the positive constant, $t_{\lambda,2}$, using (13) with $k$ = 2.
    5.2) Calculate the probability, $p^{equ}$, using the second part of (39).
    5.3) Calculate the equivalent operating time, $T^{equ}$, the first part of (39)
6) Calculate the equivalent age of the backup component, $t_{age}$, using (35).

---

### 6.3. Event Chain Generation for 2SBRSBF

After developing the procedures for random time-period generation in Section 6.2, we may simulate an EC for the $j$th pseudo-reality of a 2SBRSBF system which satisfies all EC properties defined in Section 6.1.

The following is given:

(1) For each $k$ = 1, 2, 3, 4, the original CDFs, $F_k(t)$, defined for any real argument.
(2) For each $k$ = 1, 2, 3, 4 the original inverse CDFs $F_k^{-1}(p)$, defined for any $p$ belonging to the unit interval.
(3) The probability, $p_f$, for switching failure on demand.
(4) The probability, $p_r$, for back-switching failure on demand.
(5) The value of the $Flag_A$, which determines under which aging assumption the 2SBRSBF operates.
(6) The positive final simulation time, $t_{end}$, such that $R_{sys}(t_{end}) \approx 0 \ (< 0.01)$.
(7) The consecutive number, $j$, of the pseudo-reality.

It is easy to demonstrate that any EC generated by Algorithm 2 satisfies all EC properties formulated in Section 6.1.

The event chain for the $j$th pseudo-reality, $EC_j$ can be calculated using Algorithm 2.

**Algorithm 2** Generation of the Event Chain for the *j*th Pseudo-Reality of 2SBRSBF

1) Initiate the incomplete event chain, $EC_j^{inc}$:
   1.1) Set, $T_{cur}$ =0 (the current system time is zero)
   1.2) Set, $k_{cur}$ = 1 (the current count of events is one)
   1.3) Set, $timepsr_j(k_{cur}) = T_{cur}$ (the time of the first event is zero)
   1.4) Set, $statepsr_j(k_{cur})$ = 1 (the system starts from State 1)
2) If $statepsr_j(k_{cur})$ = 1 (the system is currently in State 1), then:
   2.1) Estimate, $\Delta\tau_{1,fp}$ = PISMCF$(F_1(.), F_1^{-1}(.), T_{cur})$ (the possible time to failure of the primary unit).
   2.2) Estimate, $\Delta\tau_{1,fb}$ = PISMCF$(F_3(.), F_3^{-1}(.), T_{cur})$ (the possible time to standby failure of the backup unit).
   2.3) If $t_{end} \leq T_{cur} + \Delta\tau_{1,fp}$ and $t_{end} \leq T_{cur} + \Delta\tau_{1,fb}$ (the end of simulation comes first), then:
      2.3.1) Set $q_j = k_{cur}$ (the last event count in $EC_j$)
      2.3.2) Set $EC_j = EC_j^{inc}$ (the final $EC_j$)
      2.3.3) Stop the Algorithm
   2.4) If $\Delta\tau_{1,fp} \leq \Delta\tau_{1,fb}$ (the primary unit is failing first), then:
      2.4.1) Set $k_{cur} = k_{cur}$ +1 (new event)
      2.4.2) Set $T_{cur} = T_{cur} + \Delta\tau_{2,fb}$ (new current system time)
      2.4.3) Set $timepsr_j(k_{cur}) = T_{cur}$ (the time of the new event)
      2.4.4) Generate *RN* as an evenly distributed number in the unit interval (check which is the new state)
         2.4.4.1) If $RN > p_f$, then $statepsr_j(k_{cur})$ = 2 (i.e., no switching failure, move to State 2)
         A. If $RN \leq p_f$, then $statepsr_j(k_{cur})$ = 41 (i.e., switching failure, move to State 4, type *a*)
   2.5) If $\Delta\tau_{1,fp} > \Delta\tau_{1,fb}$ (the backup unit is failing first), then:
      2.5.1) Set $k_{cur} = k_{cur}$ + 1 (new event)
      2.5.2) Set $T_{cur} = T_{cur} + \Delta\tau_{1,fb}$ (new current system time)
      2.5.3) Set $timepsr_j(k_{cur}) = T_{cur}$ (the time of the new event)
      2.5.4) Set $statepsr_j(k_{cur})$ = 3 (move to State 2)
3) If $statepsr_j(k_{cur})$ = 2 (the system is currently in State 2), then:
   3.1) Estimate, $\Delta\tau_{2,rp}$ = PISMCF$(F_4(.), F_4^{-1}(.), T_{cur})$ (the possible time to repair of the primary unit).
   3.2) Estimate $t_{age}$ = TAGEASS$\left(F_2(.), F_2^{-1}(.), F_3(.), EC_j^{inc}, T_{cur}, Flag_A\right)$ (the equivalent age of the backup unit)
   3.3) Estimate, $\Delta\tau_{2,fb}$ = PISMCF$(F_2(.), F_2^{-1}(.), t_{age})$ (the possible time to operational failure of the backup unit).
   3.4) If $t_{end} \leq T_{cur} + \Delta\tau_{2,rp}$ and $t_{end} \leq T_{cur} + \Delta\tau_{2,fb}$ (the end of simulation comes first), then:
      3.4.1) Set $q_j = k_{cur}$ (the last event count in $EC_j$)
      3.4.2) Set $EC_j = EC_j^{inc}$ (the final $EC_j$)
      3.4.3.) Stop the Algorithm
   3.5) If $\Delta\tau_{2,rp} \leq \Delta\tau_{2,fb}$ (the primary unit is repaired first), then:
      3.5.1) Set $k_{cur} = k_{cur}$ + 1 (new event)
      3.5.2) Set $T_{cur} = T_{cur} + \Delta\tau_{2,rp}$ (new current system time)
      3.5.3) Set, $timepsr_j(k_{cur}) = T_{cur}$ (the time of the new event)
      3.5.4) Generate *RN* as an evenly distributed number in the unit interval (check which is the new state)
         3.5.4.1) If $RN > p_r$, then $statepsr_j(k_{cur})$ =1 (no back-switching failure, move to State 1)
         3.5.4.2) If $RN \leq p_r$, then $statepsr_j(k_{cur})$ = 40 (back-switching failure, move to State 4, type *b*)
   3.6) If $\Delta\tau_{2,rp} > \Delta\tau_{2,fb}$ (the backup unit is failing first), then:
      3.6.1) Set, $k_{cur} = k_{cur}$ + 1 (new event)
      3.6.2) Set, $T_{cur} = T_{cur} + \Delta\tau_{2,fb}$ (new current system time)
      3.6.3) Set, $timepsr_j(k_{cur}) = T_{cur}$ (the time of the new event)
      3.6.4) Set, $statepsr_j(k_{cur})$ =42 (move to State 4, type *c*)
4) If $statepsr_j(k_{cur})$ = 3 (the system is currently in State 3), then:
   4.1) If $t_{end} \leq T_{cur} + \Delta\tau_{1,fp} - \Delta\tau_{1,fb}$ (the end of simulation comes first), then:
      4.1.1) Set, $q_j = k_{cur}$ (the last event count in $EC_j$)
      4.1.2) Set, $EC_j = EC_j^{inc}$ (the final $EC_j$)
      4.1.3) Stop the Algorithm
   4.2) If $t_{end} > T_{cur} + \Delta\tau_{1,fp} - \Delta\tau_{1,fb}$ (the primary unit is failing first), then:
      4.2.1) Set, $k_{cur} = k_{cur}$+1 (new event)
      4.2.2) Set, $T_{cur} = T_{cur} + \Delta\tau_{1,fp} - \Delta\tau_{1,fb}$ (new current system time)
      4.2.3) Set, $timepsr_j(k_{cur}) = T_{cur}$ (the time of the new event)
      4.2.4) Set, $statepsr_j(k_{cur})$ = 43 (switching failure, move to State 4, type *d*)
5) If $statepsr_j(k_{cur}) > 3$ (the system is currently in State 4), then:
   5.1) Set, $q_j = k_{cur}$ (the last event count in $EC_j$)
   5.2) Set, $EC_j = EC_j^{inc}$ (the final $EC_j$)
   5.3) Stop the Algorithm
6) Go to Step 2 (try a next transition)

### 6.4. Extracting Reliability Information from the Simulated ECs

Let $N$ be a large positive integer representing the count of the randomly simulated pseudo-realities. Using Algorithm 2, we can simulate $EC_j$, for $j = 1,2,\ldots,N$. In this section, we will extract the reliability information from the simulated ECs, approach which is the essence of any Monte Carlo simulation [37] (pp. 290–294).

Let us calculate the state probability functions at the 2000 evenly distributed time points from 0 to $t_{end}$ given in Equation (20). For a given $EC_j$ we can estimate the state, $St_{i,j}$, at each of the time points $t_i$:

$$St_{i,j} = \begin{cases} statepsr_j(k) \text{ if } timepsr_j(k) \leq t_i < timepsr_j(k+1), \text{ for } k < q_j \\ statepsr_j(q_j) \text{ if } timepsr_j(q_j) \leq t_i \leq t_{end} \end{cases}, \text{ where } \begin{matrix} i = 1,2,\ldots,2000 \\ j = 1,2,\ldots,N \end{matrix} \quad (41)$$

From Equation (41) it is easy to estimate the values of the first three state probability functions at the time point, $t_i$:

$$P_g(t_i) = \frac{1}{N} \#(S_{i,j} = g | j = 1,2,\ldots,N), \text{ where } g = 1,2,3 \text{ and } i = 1,2,\ldots,2000 \quad (42)$$

In Equation (42) the $\#(S_{i,j} = g | j = 1,2,\ldots,N)$ is the count of all states at the time point $t_i$ which are equal to $g$.

The fourth state probability function can be estimated using Equation (1) as:

$$P_4(t_i) = 1 - P_3(t_i) - P_2(t_i) - P_3(t_i), \text{ for } i = 1,2,\ldots,2000 \quad (43)$$

The reliability function and the $MTTF_{sys}$ can be approximated with Equations (21) and (22). According to the ES property p1, the reliability in Equation (22) has decreasing nodes:

$$R_{sys}(t_i) \geq R_{sys}(t_{i+1}), \text{ for } i = 1,2,\ldots,1999 \quad (44)$$

One way to identify the $\alpha$-design life, $t_{des,\alpha}$ for given $\alpha$ is to transform the nodes, $\{[t_i, R_{sys}(t_i)] | i = 1,2,\ldots,2000\}$, of the system reliability from Equation (22) into strictly decreasing purged nodes $\left\{\left[t_i^{pu}, R_{sys}^{pu}\left(t_i^{pu}\right)\right] \Big| i = 1,2,\ldots,n^{pu}\right\}$ where:

$$R_{sys}^{pu}\left(t_i^{pu}\right) > R_{sys}^{pu}\left(t_{i+1}^{pu}\right), \text{ for } i = 1,2,\ldots,n^{pu} \quad (45)$$

Such a purging procedure is proposed in [17], where the algorithm is motivated, formalized, illustrated, and proven. In short, it runs in the steps summarized in Algorithm 3.

---
**Algorithm 3** Purging Algorithm

---
(a) Identify the time of the first purged node $\left[t_1^{pu}, R_{sys}^{pu}(t_1^{pu}) = 1\right]$ as the greatest $t_i$ for which $R_{sys}(t_i) = 1$;
(b) Substitute all internal nodes with equal reliability with one purged node in the center of the horizontal platform;
(c) Identify the time of the last purged node $\left[t_{n^{pu}}^{pu}, R_{sys}^{pu}(t_{n^{pu}}^{pu})\right]$ as the smallest $t_i$ for which $R_{sys}(t_i) = R_{sys}(t_{2000})$.

---

Having the strictly decreasing purged system reliability function, we can identify the $\alpha$-design life, $t_{des,\alpha}$ for any $\alpha \in \left[R_{sys}^{pu}\left(t_{n^{pu}}^{pu}\right), 1\right]$:

$$t_{des,\alpha} = t_i^{pu} + \left[R_{sys}^{pu}\left(t_i^{pu}\right) - \alpha\right] \frac{t_{i+1}^{pu} - t_i^{pu}}{R_{sys}^{pu}\left(t_i^{pu}\right) - R_{sys}^{pu}\left(t_{i+1}^{pu}\right)}, \text{ for } R_{sys}^{pu}\left(t_i^{pu}\right) \geq \alpha > R_{sys}^{pu}\left(t_{i+1}^{pu}\right) \quad (46)$$

As discussed in Section 3, the reliability numerical characteristics $Median_{sys}$, $B1\_life$, $B10\_life$, and $IQR_{sys}$ can be estimated as $t_{des,0.5}$, $t_{des,0.99}$, $t_{des,0.9}$, and $t_{des,0.25} - t_{des,0.75}$ respectively by applying Equation (46) five times.

The simulational solution is universal and exists even when the numerical and analytical solutions are impossible. Even when the numerical and the analytical solutions exist, the simulational solution can provide richer reliability information.

For example, it is obvious that the 2SBRSBF system will have 100% chance to ever be in the State 1. It is also clear that if tend is correctly selected, then the 2SBRSBF system will have more than 99% chance to ever be in the State 4. However, it is interesting to know the chance, $P_{2,ever}$, for the 2SBRSBF system to ever be in the State 2, since that probability will help us plan the resources needed for the repair of the primary unit. Similarly, the chance, $P_{3,ever}$, for the 2SBRSBF system to ever be in the State 3 is important, since that will show us the prevalence of the failure in standby of the backup unit. So, for a given 2SBRSBF system, we can estimate the chances, $P_{g,ever}$, for $g = 1,2,3$:

$$P_{g,ever} = \frac{100}{N} \#(\exists i, \text{ that } S_{i,j} = g | j = 1, 2, \ldots, N), \text{ where } g = 1, 2, 3 \qquad (47)$$

In Equation (47), $\#(\exists i, \text{ that } S_{i,j} = g | j = 1, 2, \ldots, N)$ is the count of pseudo-realities in which State $g$ can be found at least once. Similarly, for a given 2SBRSBF system we can estimate the chance, $P_{4,ever}$ as:

$$P_{4,ever} = \frac{100}{N} \#(\exists i, \text{ that } S_{i,j} > 3 | j = 1, 2, \ldots, N) \qquad (48)$$

In Equation (48), $\#(\exists i, \text{ that } S_{i,j} > 3 | j = 1, 2, \ldots, N)$ is the count of pseudo-realities in which State 4 (system failure) can be found at least once.

As another example for reliability information, which can be acquired neither with the numerical, nor with the analytical solution, can be found in the four conditional chances, $P_{g,ever}^{cond}$ (for $g = 40, 41, 42, 43$), of the 2SBRSBF system to have respectively type $b$, type $a$, type $c$, or type $d$ system failure, provided that system has failed:

$$P_{g,ever}^{cond} = 100 \frac{\#\left(S_{i,q_j} = g | j = 1, 2, \ldots, N\right)}{NP_{4,ever}/100}, \text{ where } g = 40, 41, 42, 43 \qquad (49)$$

The information in Equation (49) allows to identify the types of system failures which dominate the 2SBRSBF system. That knowledge will increase the efficiency of the reliability improvement measures. Equations (42), (47)–(49) use the frequentist interpretation of probability [40] (pp. 42–43).

Knowing how to simulate an EC for the $j$th pseudo-reality of a 2SBRSBF system, allows us to develop the simulational solution of a given 2SBRSBF system. We have the following given:

(1) For each $k = 1, 2, 3, 4$, the original CDFs, $F_k(t)$, defined for any real argument.
(2) For each $k = 1, 2, 3, 4$, the original inverse CDFs $F_k^{-1}(p)$, defined for any $p$ belonging to the unit interval.
(3) The probability, $p_f$, for switching failure on demand.
(4) The probability, $p_r$, for back-switching failure on demand.
(5) The value of the $Flag_A$, which determines under which aging assumption the 2SBRSBF operates.

The proposed algorithm in [17] uses simulation to find the reliability characteristics of a two-component standby systems with switching failures and aging in standby. The simulational solution for 2SBRSBF system can be obtained through a generalization of that algorithm, which is formalized as Algorithm 4 below.

With the formulation of Algorithm 4 the universal simulational solution for a 2SBRSBF system is complete.

**Algorithm 4** Simulational Solution of a 2SBRSBF System

1) Select the count $N$ of pseudo-realities to be simulated as a large integer.
2) Select the final simulation time, $t_{end}$, as a positive real number.
3) Set, $j = 1$ (initiate the consecutive number of the simulated pseudo-reality)
4) Generate the $EC_j$, using Algorithm 2.
5) Set, $j = j + 1$ (move to next pseudo-reality).
6) If $j \leq N$, then go to Step 4 (repeat the EC generation $N$ times).
7) Estimate 2000 equally spaced times, $t_i$, in the closed interval $[0, t_{end}]$ using Equation (20).
8) Estimate the states, $St_{i,j}$, using Equation (41).
9) Estimate the first three state probability functions, $P_g(t_i)$ (for $g = 1, 2, 3$) at the time points $t_i$ using Equation (42).
10) Estimate the fourth state probability function, $P_4(t_i)$, at the time point $t_i$ using Equation (43).
11) Estimate the system reliability function, $R_{sys}(t_i)$ at the time point $t_i$ using Equation (21).
12) Estimate the system mean time to failure, $MTTF_{sys}$ using Equation (22).
13) Estimate the nodes, $\{[t_i^{pu}, R_{sys}^{pu}(t_i^{pu})] | i = 1, 2, \ldots, n^{pu}\}$, of the invertible reliability function using Algorithm 3.
14) Estimate the design lives, $t_{des,0.5}$, $t_{des,0.99}$, $t_{des,0.9}$, $t_{des,0.25}$ and $t_{des,0.75}$ using Equation (46) five times.
15) Set the median time, $Median_{sys} = t_{des,0.5}$.
16) Set the B1 life, $t_{des,0.99}$.
17) Set the B10 life, $t_{des,0.9}$.
18) Set the interquartile range, $IQR = t_{des,0.25} - t_{des,0.75}$.
19) Estimate the first three unconditional chances, $P_{g,ever}$ (for $g = 1, 2, 3$) using Equation (47).
20) Estimate the fourth unconditional chance, $P_{4,ever}$ using Equation (48).
21) Estimate the conditional chances, $P_{g,ever}^{cond}$ (for $g$ = 40, 41, 42, 43) using Equation (49).

## 7. Illustrative Examples

### 7.1. Examples Setup

We shall analyze three Illustrative Examples. In all of them, the probability for switching failure is $p_f = 0.12$, whereas the probability for back-switching failure is $p_r = 0.03$. The ratio between those values is plausible for the following reasons. If the switching is successful, it means that the switching device operated properly. Then a back-switching failure is less probable since it will be demanded shortly afterwards (the repair time of the primary component is much smaller than its failure time).

In Example 1, any of the four original distributions has a constant failure/repair rate $\lambda_k$ shown in Table 2 (for $k$ = 1,2,3,4). The PDFs of the original exponential distributions are:

$$f_k(t) = \lambda_k e^{-\lambda_k t}, \text{ for } t \geq 0 \text{ where } k = 1, 2, 3, 4 \tag{50}$$

**Table 2.** Description of the original distributions in Example 1.

| Component—Event—Mode | Distribution | Parameters |
|---|---|---|
| Primary Component Failure | Exponential | $\lambda 1$ = 0.0005 failures/h |
| Backup Component Failure in operation | Exponential | $\lambda 2$ = 0.0008 failures/h |
| Backup Component Failure in standby | Exponential | $\lambda 3$ = 0.00025 failures/h |
| Primary Component Repair | Exponential | $\lambda 4$ = 0.008 failures/h |

The PDFs, the reliability/repair functions, and the failure/repair rates of the truncated distributions from Equation (50) are plotted in Figure 3. Example 1 will illustrate the behavior of the 2SBRSBF system with First Case distributions. Here, the original and the truncated distributions coincide.

In Example 2, the original distributions are as follows:

(1) a Rayleigh distribution with shape parameter $b_1$ [41] for the failures of the primary component:

$$f_1(t) = (t/b_1)e^{-0.5(t/b_1)^2}, \text{ for } t \geq 0 \tag{51}$$

(2) a normal distribution with mean value $\mu_2$ h and standard deviation $\sigma_2$ h [42] for the failures of the backup component in operation:

$$f_2(t) = \frac{1}{\sqrt{2\pi}\sigma_2} e^{-0.5(t-\mu_2)^2/(\sigma_2)^2}, \text{ for } t \in (-\infty, +\infty) \tag{52}$$

(3) a Weibull distribution with a scale parameter $\theta_3$ h and a shape parameter $\beta_3$ [43] for the failures of the backup component in standby:

$$f_3(t) = \frac{\beta_3}{\theta_3}\left(\frac{t}{\theta_3}\right)^{\beta_3-1} e^{-(t/\theta_3)^{\beta_3}}, \text{ for } t \geq 0 \quad (53)$$

(4) a lognormal distribution with median time $t_{med,4}$ h and shape parameter $s_4$ [44] for the repairs of the primary component:

$$f_4(t) = \frac{1}{\sqrt{2\pi}s_4 t} e^{-0.5\ln^2(t/t_{med,4})/(s_4)^2}, \text{ for } t \geq 0 \quad (54)$$

The original distribution Example 2 are described in Table 3. The PDFs, the reliability/repair functions, and the failure/repair rates of the truncated distributions from Equations (51)–(54) are plotted in Figure 4. Example 2 will illustrate the behavior of the 2SBRSBF system with Second Case distributions where the failures of the backup component in operation have an Increasing Failure Rate (IFR). Such a typical situation can occur when the operational failure is caused mainly by high wearing in the backup component [11] (pp. 73–75). Here, the original and the truncated distributions coincide except for the $f_2(t)$ and $f_{2,trunc}(t)$.

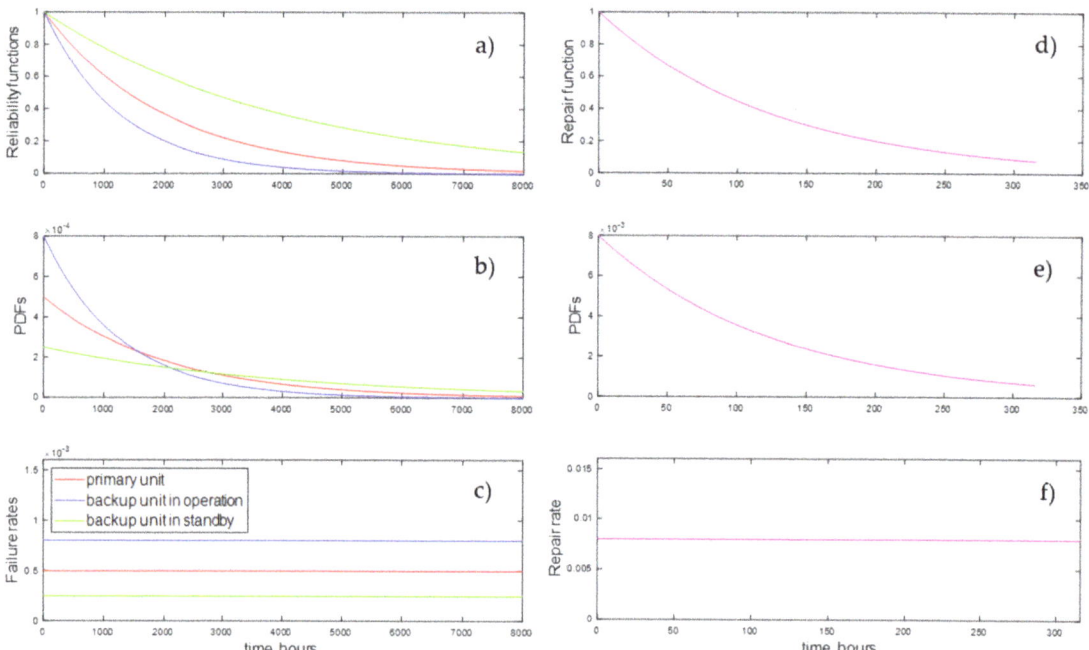

**Figure 3.** The truncated distributions in Example 1. The three failure distributions are shown in section (**a**–**c**), whereas the repair distribution is shown in section (**d**–**f**). The reliability/repair functions, the PDFs and the failure/repair rates are given respectively in the first (sections (**a**,**d**)), the second (section (**b**,**e**)), and the third row (sections (**c**,**f**)).

**Table 3.** Description of the original distributions in Example 2.

| Component—Event—Mode | Distribution | Parameters |
|---|---|---|
| Primary Component Failure | Rayleigh | $b_1 = 1600$ h |
| Backup Component Failure in operation | normal | $\mu_2 = 1000$ h and $\sigma_2 = 900$ h |
| Backup Component Failure in standby | Weibull | $\theta_3 = 4500$ h and $\beta_3 = 2.2$ |
| Primary Component Repair | lognormal | $t_{med,4} = 90$ h and $s_4 = 0.8$ |

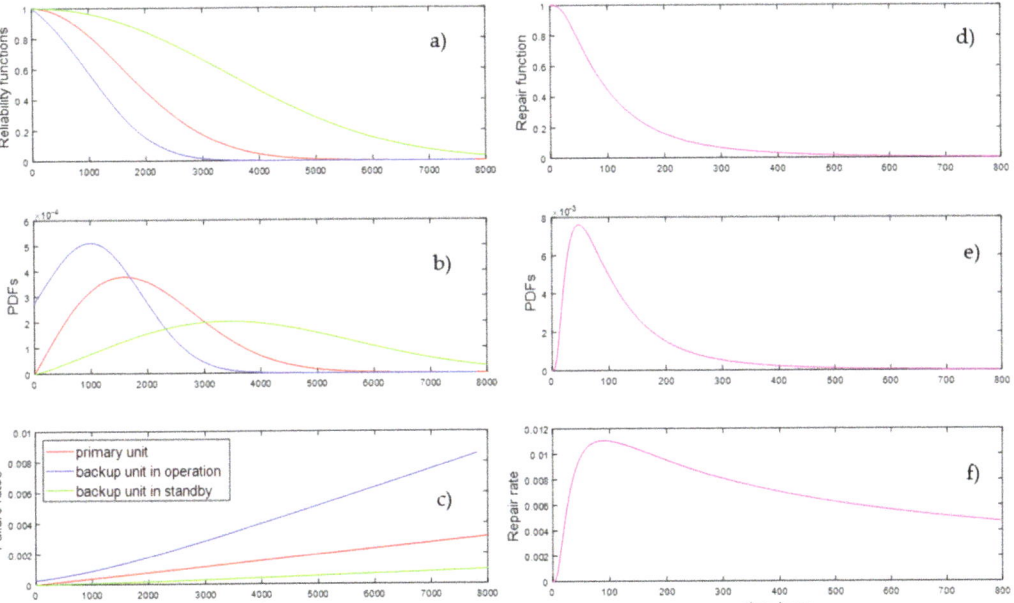

**Figure 4.** The truncated distributions in Example 2. The three failure distributions are shown in section (**a**–**c**), whereas the repair distribution is shown in section (**d**–**f**). The reliability/repair functions, the PDFs and the failure/repair rates are given respectively in the first (sections (**a**,**d**)), the second (section (**b**,**e**)), and the third row (sections (**c**,**f**)).

In Example 3 the distributions are the same as in Example 2, except for the second type, which changes to:

2) a lognormal distribution with median time $t_{med,2}$ h and shape parameter $s_2$ for the failures of the backup component in operation:

$$f_2(t) = \frac{1}{\sqrt{2\pi}s_2 t} e^{-0.5\ln^2(t/t_{med,2})/(s_2)^2}, \text{ for } t \geq 0 \quad (55)$$

The original distribution Example 3 are described in Table 4. The PDFs, the reliability/repair functions, and the failure/repair rates of the truncated distributions from Equations (51), (53)–(55) are plotted in Figure 5. Example 3 will illustrate the behavior of the 2SBRSBF system with Second Case distributions where the failures of the backup component in operation have a Decreasing Failure Rate (DFR). Such an atypical situation can occur when the operational failure is caused mainly by high child mortality in the backup component [11] (pp. 73–75). Here, the original and the truncated distributions coincide.

**Table 4.** Description of the original distributions in Example 3.

| Component—Event—Mode | Distribution | Parameters |
|---|---|---|
| Primary Component Failure | Rayleigh | $b_1 = 1600$ h |
| Backup Component Failure in operation | lognormal | $t_{med,2} = 537$ h and $s_2 = 1.3$ |
| Backup Component Failure in standby | Weibull | $\theta_3 = 4500$ h and $\beta_3 = 2.2$ |
| Primary Component Repair | lognormal | $t_{med,4} = 90$ h and $s_4 = 0.8$ |

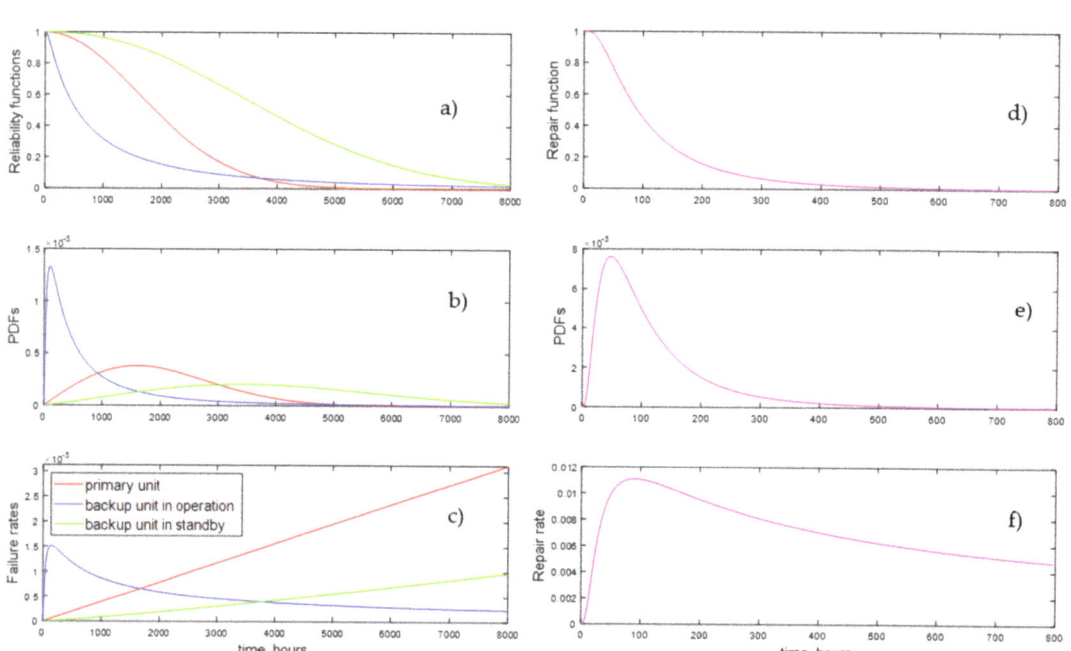

**Figure 5.** The truncated distributions in Example 3. The three failure distributions are shown in section (**a**–**c**), whereas the repair distribution is shown in section (**d**–**f**). The reliability/repair functions, the PDFs and the failure/repair rates are given respectively in the first (sections (**a**,**d**)), the second (section (**b**,**e**)), and the third row (sections (**c**,**f**)).

### 7.2. Example 1 Solution

Since in Example 1, we are dealing with First Case distributions, the type of aging has no effect on the reliability performance of the 2SBRSBF system. The simulation solution was obtained by Algorithm 4 with $N = 10{,}000$ pseudo-realities for time from 0 to $t_{end} = 20{,}000$ h. Four typical pseudo-realities are shown in Table 5 where the different types of system failures are demonstrated. The four state probability functions are shown in Figure 6a–d, respectively. The system reliability function is depicted in Figure 7. The simulation reliability at $t_{end}$ was negligible (as required $R_{sys}(20{,}000) = 0.0024 < 0.01$) which justifies the selection of $t_{end}$. Important simulation numerical characteristics of the 2SBRSBF reliability can be found in Table 6. The chances of some events of interest (described in Section 6.4) can be found in Table 7. It is revealing so see that the backup component has approximately 69% chance to endure failure in standby (State 3). Another useful fact is that the switching failures (Type *a*) are more frequent than the backup component failures in operation (Type *c*) (17% vs. 11% conditional chance). That fact suggests that it is easier to improve the reliability by upgrading the switching mechanism than by upgrading the backup unit.

**Table 5.** Four typical pseudo-realities from Example 1.

| 1 | Time 0.0 h: Start of the simulation. The primary component operates, the backup component is ready. | Time 0.0 h: Start of the simulation. The primary component operates, the backup component is ready. | Time 0.0 h: Start of the simulation. The primary component operates, the backup component is ready. | Time 0.0 h: Start of the simulation. The primary component operates, the backup component is ready. |
|---|---|---|---|---|
| 2 | Time 1378.3 h: The primary component fails in operation. The primary component under repair, the backup component operates. | Time 1753.2 h: The primary component fails in operation. The primary component under repair, the backup component operates. | Time 2016.9 h: The primary component fails in operation. The primary component under repair, the backup component operates. | Time 2348.6 h: The primary component fails in operation. The primary component under repair, the backup component operates. |
| 3 | Time 1467.6 h: The primary component successfully repaired. The primary component operates, the backup component is ready. | Time 1821.4 h: The primary component successfully repaired. The primary component operates, the backup component is ready. | Time 2042.9 h: The primary component successfully repaired. The primary component operates, the backup component is ready. | Time 2406.5 h: The primary component successfully repaired. The primary component operates, the backup component is ready. |
| 4 | Time 2099.6 h: The primary component fails in operation. Switching failure. Type $a$ system failure (switching failure). | Time 4321.5 h: The primary component fails in operation. The primary component under repair, the backup component operates. | Time 8168.7 h: The primary component fails in operation. The primary component under repair, the backup component operates. | Time 3057.8 h: The backup component fails in standby. The primary component operates, the backup component failed in standby. |
| 5 |  | Time 4460.6 h: The primary component successfully repaired. Back-Switching failure. Type $b$ system failure (back-switching failure). | Time 8288.8 h: The backup component fails in operation. Type $c$ system failure (backup component failure during primary repair). | Time 3712.4 h: The primary component fails in operation. Type $d$ system failure (standby failure+ primary failure). |

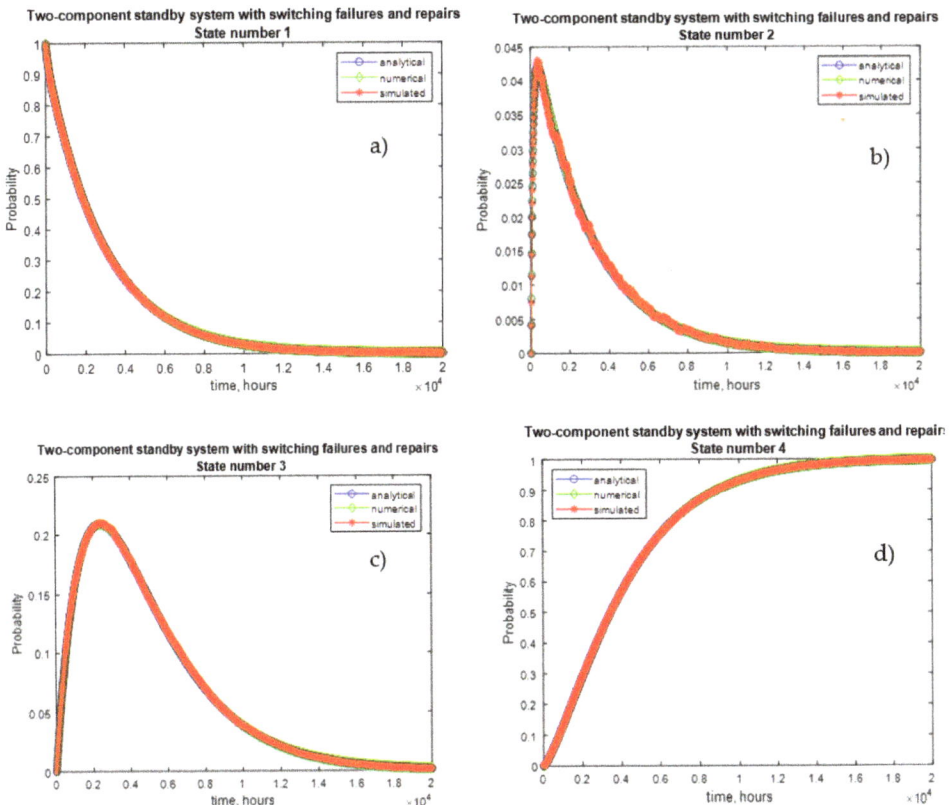

**Figure 6.** State probability functions for Example 1 (with states 1 through 4 given in sections (**a–d**) respectively) from the analytical, numerical and simulation solution.

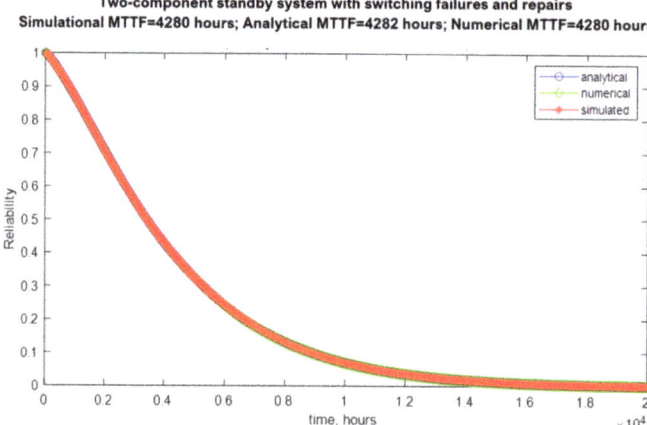

**Figure 7.** Reliability functions for Example 1 from the analytical, numerical and simulation solution.

**Table 6.** Reliability characteristics of the 2SBRSBF from Example 1.

| Count of pseudo-realities | 100,000 |
|---|---|
| Simulation time | $2.000 \times 10^{+4}$ h |
| Mean value (Simulation) | $4.294 \times 10^{+3}$ h |
| Median | $3.418 \times 10^{+3}$ h |
| Interquartile range | $4.187 \times 10^{+3}$ h |
| B10 life | $7.922 \times 10^{+2}$ h |
| B1 life | $1.174 \times 10^{+2}$ h |
| Mean value (Analytical) | $4.282 \times 10^{+3}$ h |
| Mean value (Numerical) | $4.280 \times 10^{+3}$ h |

**Table 7.** Chances for events of interest in % for Example 1.

| | | |
|---|---|---|
| Unconditional Chance for State 1 to happen | 100.00% | The primary component operates, the backup component is ready |
| Unconditional Chance for State 2 to happen | 58.66% | The primary component under repair, the backup component operates |
| Unconditional Chance for State 3 to happen | 68.98% | The primary component operates, the backup component failed in standby |
| Unconditional Chance for State 4 to happen | 99.76% | System failure |
| Conditional chance for type $a$ failure to happen | 16.68% | Switching failure |
| Conditional chance for type $b$ failure to happen | 3.25% | Back-switching failure |
| Conditional chance for type $c$ failure to happen | 11.07% | Backup component failure during primary repair |
| Conditional chance for type $d$ failure to happen | 69.00% | Standby failure + primary failure |

The simulation results were verified by comparison with the precise analytical solution (see Section 4). According to Table 6, the precise analytical MTTF is 4282 h, whereas the simulational MTTF is estimated as 4294 h, which contains less than 0.3% error.

Also, the simulational results were verified by comparison with the numerical solution (see Section 5), which, as seen from Figures 6 and 7, produced undistinguishable curves from the simulational state probabilities and the simulational reliability function. According to Table 6, the numerical MTTF is 4280 h, whereas the simulational MTTF is estimated as 4294 h. The numerical solution for Example 1 (as well as in Examples 2 and 3) was derived by solving the index-1 DAE system described in Section 5 with the MATLAB multistep procedure *ode15s.m*. The software successfully integrated the DAE system from 0 to $t_{end}$ = 20,000 h using variable-step method of variable order from 1 to 5 [45].

As seen from Figures 6 and 7, the analytical and the numerical solutions produce undistinguishable curves from the simulational state probabilities and the simulational reliability function. The observed overlap is an essential part of the verification of the presented simulation algorithm: in the case of exponential distribution, the model is Markovian, where the analytical, the numerical, and the simulational solutions should practically coincide.

### 7.3. Example 2 Solution

Since Example 2 deals with Second Case distributions, the type of aging has an effect on the reliability performance of the 2SBRSBF system. Three simulational solutions were obtained by repeatedly using Algorithm 4 with $N$ = 10,000 pseudo-realities for the three aging assumptions: full aging, no aging and patrial aging of the backup component in standby. Each of those solutions was estimated for time from 0 to $t_{end}$ = 8000 h. The three sets of four state probability functions are shown in Figure 8a–d, respectively. The three system reliability functions are depicted in Figure 9. The simulational reliabilities at $t_{end}$ were negligible and much lower than 0.01 (for full aging-$R_{sys}$(8000) = 0; for no aging-$R_{sys}$(8000) = 0.0025; for partial aging-$R_{sys}$(8000) = 0.0003) which justifies the selection of $t_{end}$.

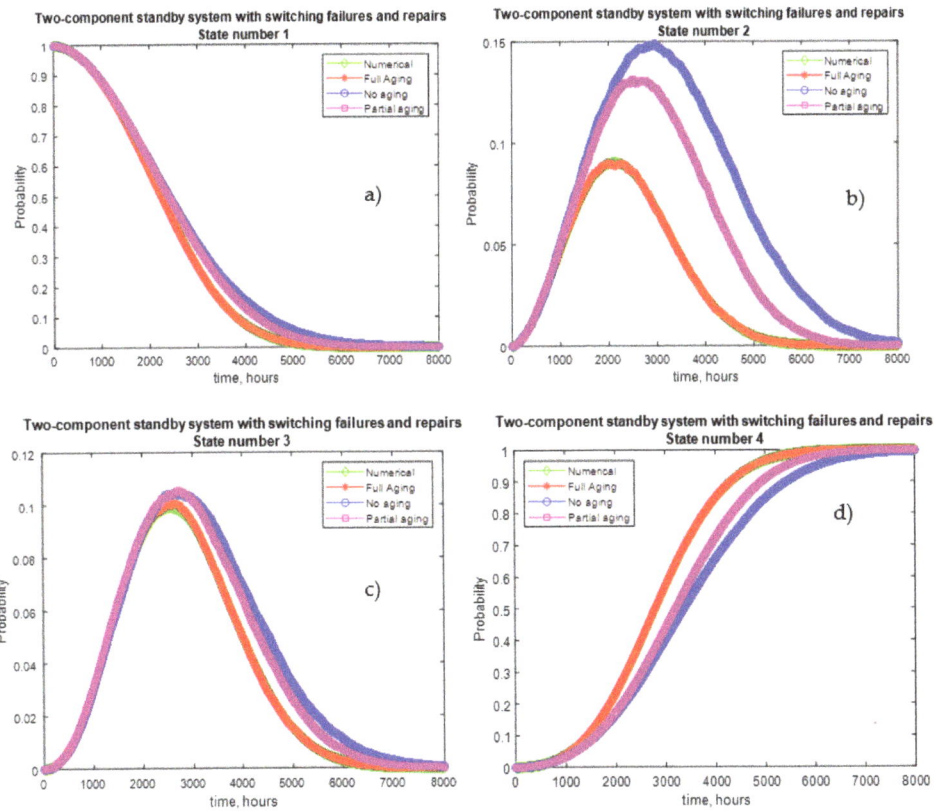

**Figure 8.** State probability functions for Example 2 (with states 1 through 4 given in sections (**a**–**d**) respectively) under the three aging assumptions.

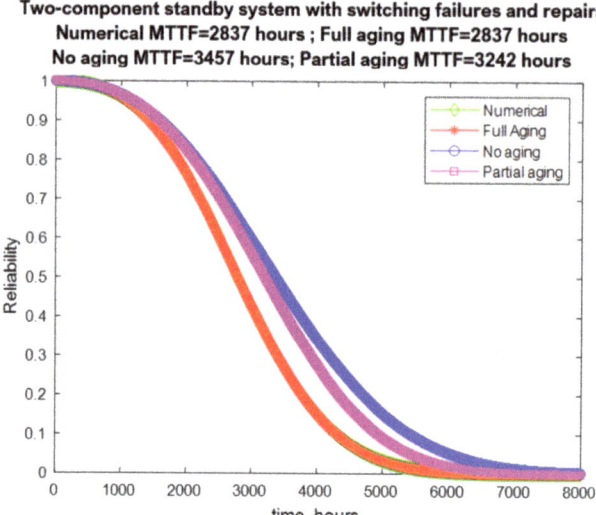

**Figure 9.** Reliability functions for Example 2 under the three aging assumptions.

Important simulational numerical characteristics of the 2SBRSBF reliabilities can be found in Table 8 for the three types of aging. The chances of some events of interest (described in Section 6.4) can be found in Table 9 for each of the three aging assumptions. It is revealing to see that the backup component has between 31% and 41% chance to endure failure in standby (State 3) depending on the aging model. An interesting dynamic is observed in the conditional chances of observing the different types of failure. At full aging, the backup component failures during primary repair (Type $c$) have more than 50% chance, whereas the primary component failures after failure in standby (Type $d$) constitute only around 30% of the system failures. At no aging, the backup failures in operation are less likely and, therefore, the primary component failures after backup failure in standby (Type $d$) are more frequent than the backup component failures during primary repair (Type $c$) (41% vs. 36% conditional chance). At the same time, Type $c$ and Type $d$ system failures are marginally the same at partial aging of the backup component in standby (37% vs. 42% conditional chance). Those facts suggest that to increase the reliability of the 2SBRSBF it is of paramount importance correctly to identify the aging mechanism of the backup unit during standby.

**Table 8.** Reliability characteristics of the 2SBRSBF from Example 2 under the three aging assumptions.

|  | Full Aging | No Aging | Partial Aging |
|---|---|---|---|
| Count of pseudo-realities | 100,000 | 100,000 | 100,000 |
| Simulation time | $8.000 \times 10^{+3}$ h | $8.000 \times 10^{+3}$ h | $8.000 \times 10^{+3}$ h |
| Mean value (Simulation) | $2.837 \times 10^{+3}$ h | $3.457 \times 10^{+3}$ h | $3.242 \times 10^{+3}$ h |
| Median | $2.783 \times 10^{+3}$ h | $3.364 \times 10^{+3}$ h | $3.199 \times 10^{+3}$ h |
| Interquartile range | $1.531 \times 10^{+3}$ h | $2.051 \times 10^{+3}$ h | $1.785 \times 10^{+3}$ h |
| B10 life | $1.419 \times 10^{+3}$ h | $1.596 \times 10^{+3}$ h | $1.581 \times 10^{+3}$ h |
| B1 life | $5.536 \times 10^{+2}$ h | $5.735 \times 10^{+2}$ h | $5.818 \times 10^{+2}$ h |
| Mean value (Analytical) | NA | NA | NA |
| Mean value (Numerical) | $2.837 \times 10^{+3}$ h | NA | NA |

Table 9. Chances in % for events of interest for Example 2 under the three aging assumptions.

| | Full Aging | No Aging | Partial Aging | |
|---|---|---|---|---|
| Unconditional Chance for State 1 to happen | 100.00% | 100.00% | 100.00% | The primary component operates, the backup component is ready |
| Unconditional Chance for State 2 to happen | 71.76% | 71.78% | 72.00% | The primary component under repair, the backup component operates |
| Unconditional Chance for State 3 to happen | 30.43% | 40.78% | 37.13% | The primary component operates, the backup component failed in standby |
| Unconditional Chance for State 4 to happen | 100.00% | 99.75% | 99.97% | System failure |
| Conditional chance for type $a$ failure to happen | 15.56% | 20.11% | 18.19% | Switching failure |
| Conditional chance for type $b$ failure to happen | 1.85% | 3.30% | 2.70% | Back-switching failure |
| Conditional chance for type $c$ failure to happen | 52.16% | 35.75% | 41.99% | Backup component failure during primary repair |
| Conditional chance for type $d$ failure to happen | 30.43% | 40.85% | 37.12% | Standby failure + primary failure |

In Example 2, the distribution of the backup component failures in operation has an IFR (see the blue line in Figure 5c), indicating that the wear out is the most likely reason for those failures. This is by far the most widespread case in the engineering practice where the backup component operates at the rear end of the bathtub curve [46]. Then, the severity of the aging should increase the failure incidence of the backup component in operation and subsequently should decrease the reliability. As expected, the system reliability function is the best at no-aging and worst at full aging (see Figure 9 for 1500–5500 h). The MTTF increases from 2837 h at full aging, through 3242 h at partial aging, to 3457 h at no aging, which corresponds to substantial 21% improvement. Similar behavior can be observed in the median, B10 life, and at the B1 life (see Table 8). Another expected result is that the state probability functions for partial aging are between the state probability functions for no aging and full aging (see Figure 8). The real distinction between the three curves can be seen in State 2 probability function (Figure 8b) which is very sensitive to the aging mode. The observed forms of the State 2 probability functions are justifiable since the severity of aging increases the incidence of failure of the operational backup unit, which moves the system to State 4 and decreases the probability of the 2SBRSBF to be in State 2. All the above can serve as a qualitative validation of Algorithm 4 for simulating the reliability behavior of the 2SBRSBF system.

Also, the simulational results were quantitatively verified by comparison with the numerical solution (as described in Section 7.2), which, as seen from Figures 8 and 9, produced undistinguishable curves from the simulational state probabilities and the simulational reliability function in the case of full aging of the backup component during standby. This overlap is an important result: under the full-aging assumption the model is semi-Markovian where the numerical, and the simulational solutions should practically coincide. According to Table 8, the numerical MTTF and the simulational MTTF at full aging are estimated to be equal (2837 h). Note that the analytical solution is impossible to be derived in Example 2 since the failure/repair rates are not constant.

*7.4. Example 3 Solution*

Since Example 3 deals with Second Case distributions, similarly to Example 2, the type of aging has effect on the reliability performance of the 2SBRSBF system. Three simulational solutions were obtained by repeatedly using Algorithm 4 with $N$ = 10,000 pseudo-realities for the three aging assumptions: full aging, no aging and patrial aging of the backup component in standby. Each of those solutions was estimated for time from 0 to $t_{end}$ = 12,000 h. The three sets of four state probability functions are shown in Figure 10a–d, respectively. The three system reliability functions are depicted in Figure 11. The simulational reliabilities at $t_{end}$ were negligible and lower than 0.01 (for full aging-$R_{sys}$(12000) = 0.0062; for no aging-$R_{sys}(12,000) = 0.0011$; for partial aging-$R_{sys}(12,000) = 0.002$) which justifies the selection of $t_{end}$.

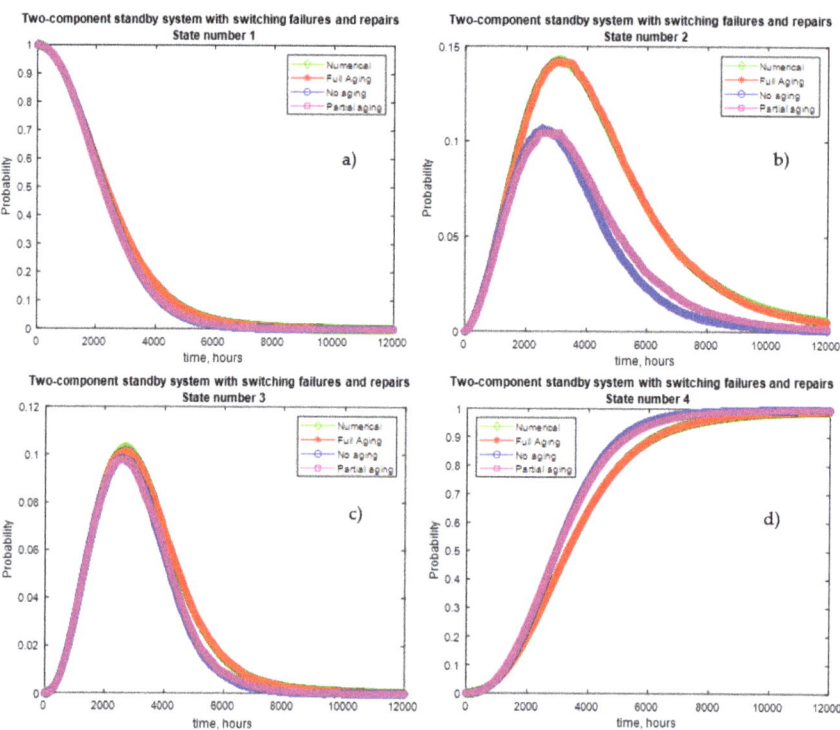

**Figure 10.** State probability functions for Example 3 (with states 1 through 4 given in sections (**a**–**d**) respectively) under the three aging assumptions.

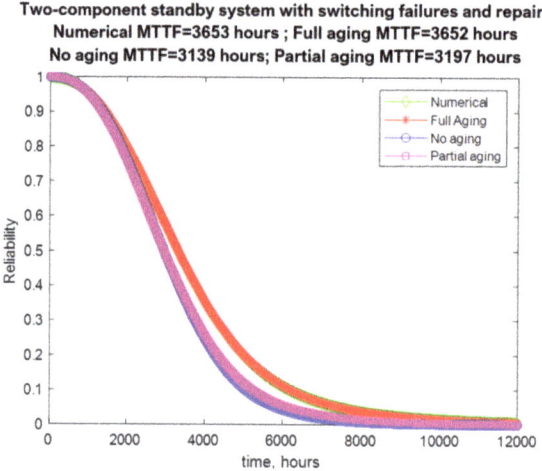

**Figure 11.** Reliability functions for Example 3 under the three aging assumptions.

Important simulational numerical characteristics of the 2SBRSBF reliabilities can be found in Table 10 for the three types of aging. The chances of some events of interest (described in Section 6.4) can be found in Table 11 for each of the three aging assumptions.

**Table 10.** Reliability characteristics of the 2SBRSBF from Example 3 under the three aging assumptions.

|  | Full Aging | No Aging | Partial Aging |
|---|---|---|---|
| Count of pseudo-realities | 100,000 | 100,000 | 100,000 |
| Simulation time | $1.200 \times 10^{+4}$ h | $1.200 \times 10^{+4}$ h | $1.200 \times 10^{+4}$ h |
| Mean value (Simulation) | $3.652 \times 10^{+3}$ h | $3.139 \times 10^{+3}$ h | $3.197 \times 10^{+3}$ h |
| Median | $3.317 \times 10^{+3}$ h | $2.957 \times 10^{+3}$ h | $2.957 \times 10^{+3}$ h |
| Interquartile range | $2.400 \times 10^{+3}$ h | $1.924 \times 10^{+3}$ h | $2.015 \times 10^{+3}$ h |
| B10 life | $1.432 \times 10^{+3}$ h | $1.364 \times 10^{+3}$ h | $1.336 \times 10^{+3}$ h |
| B1 life | $4.870 \times 10^{+2}$ h | $5.031 \times 10^{+2}$ h | $4.999 \times 10^{+2}$ h |
| Mean value (Analytical) | NA | NA | NA |
| Mean value (Numerical) | $3.653 \times 10^{+3}$ h | NA | NA |

**Table 11.** Chances in % for events of interest for Example 3 under the three aging assumptions.

|  | Full Aging | No Aging | Partial Aging |  |
|---|---|---|---|---|
| Unconditional Chance for State 1 to happen | 100.00% | 100.00% | 100.00% | The primary component operates, the backup component is ready |
| Unconditional Chance for State 2 to happen | 71.75% | 71.83% | 71.80% | The primary component under repair, the backup component operates |
| Unconditional Chance for State 3 to happen | 44.31% | 35.71% | 36.55% | The primary component operates, the backup component failed in standby |
| Unconditional Chance for State 4 to happen | 99.38% | 99.89% | 99.80% | System failure |
| Conditional chance for type $a$ failure to happen | 21.53% | 17.73% | 18.31% | Switching failure |
| Conditional chance for type $b$ failure to happen | 3.88% | 2.48% | 2.68% | Back-switching failure |
| Conditional chance for type $c$ failure to happen | 30.04% | 44.06% | 42.39% | Backup component failure during primary repair |
| Conditional chance for type $d$ failure to happen | 44.56% | 35.74% | 36.62% | Standby failure + primary failure |

In Example 3, the distribution of the backup component failures in operation has an DFR (see the blue line in Figure 5c), indicating that the child mortality is the most likely reason for those failures. This is a very rare case in the engineering practice where the backup component operates at the front end of the bathtub curve. Then, the severity of the aging should decrease the failure incidence of the backup component in operation and subsequently should increase the reliability. As expected, the system reliability function is the worst at no-aging and best at full aging (see Figure 11 for 2000–8000 h). The MTTF increases from 3139 h at no aging, through 3187 h at partial aging, to 3625 h at full aging, which corresponds to noticeable 16% improvement. Similar behavior can be observed in the median, B10 life, and at the B1 life (see Table 10). Another expected result is that the state probability functions for partial aging are between the state probability functions for no aging and full aging (see Figure 10). The real distinction between the three curves can be seen in State 2 probability function (Figure 10b) which is very sensitive to the aging mode. The observed forms of the State 2 probability functions are justifiable since the severity of aging decreases the incidence of failure of the operational backup unit, which moves the system to State 4 and increases the probability of the 2SBRSBF to be in State 2. All the above can serve as a qualitative validation of Algorithm 4 for simulating the reliability behavior of the 2SBRSBF system.

A partial overlap between the no aging simulation solution and the partial aging simulation solution can be spotted in Figures 10 and 11. The same can also be observed in Figures 8 and 9 to a lesser extent. Those partial overlaps reflect the fact that for almost all realistic distribution sets, under the applied method, the solution of partial aging is much closer to the solution with no aging assumption than to the solution with full aging assumption.

Again, the simulational results were quantitatively verified by comparison with the numerical solution (as described in Section 7.2) which as seen from Figures 10 and 11 produced undistinguishable curves from the simulational state probabilities and the simulational reliability function in the case of full aging of the backup component during standby (for comment on the observed overlap see Section 7.3). According to Table 10, the numerical MTTF and the simulational MTTF at full aging are estimated to be virtually equal (3653 h vs. 3652 h, respectively). Note that the analytical solution is impossible to be derived in Example 2 since the failure/repair rates are not constant.

## 8. Conclusions

In this paper, we investigated the reliability effect of introducing a primary component minimal repair in a two-component standby system with switching failures and aging in warm-standby. A novel analytical solution was derived for distributions with constant failure/repair rates. Under a full aging assumption of the backup component during standby, an index-1 DAE system of four simultaneous equations with constant mass singular matrix was proposed and solved to numerically approximate the state probability functions and system reliability. A universal simulational algorithm was designed to solve the 2SBFSR system under three types of aging. That algorithm generates pseudo-realities with ECs, which satisfy the newly formulated EC properties for the 2SBFSR system. Novel function to assess the equivalent age of the backup component under arbitrary aging mechanisms was proposed and utilized during the EC generation. The system has a stable operation with any type of distribution. There is a significant practical benefit in the ability of the user to write their own distribution functions, which reflect several modes of failure during operation, several modes of failure during warm-standby, and several modes of repair.

Three numerical examples were elaborated to validate quantitatively and qualitatively the simulational solution. To model the 2SBRSBF system with partial aging in standby, we assumed that that the backup component in standby ages to the same reliability as the backup component in operation. That is a logical and plausible hypothesis that allows to produce a tractable aging model whose results can be treated as best estimate. Even if the real aging mechanism is different the numerical examples show that the partial aging results always will be bounded by the full aging and the no-aging results. That fact allows the designers and the maintenance staff to correctly assess the effect of alternative measures aiming at improving the system reliability even if the precise aging in standby mechanism is known.

Although our model may look too specific and simplified, it is easily scalable. The demonstrated methodology can easily be applied to multiple-component warm-standby system with random configuration. We have not given such an example for purely volume constraints in this work. Any different aging assumptions can be incorporated by modifying Algorithm 1 (hence the function TAGEASS). All aspects and elements of such a multi-component warm-standby system can be found in 2SBRSBF. In such a way, our model is suitable for applications in industrial systems, manufacturing, design of ship electrical and propulsion systems, power plants, etc.

As a direction for future studies, we may study the ways to adapt our procedures to the case of perfect repair [10] and intermediate repair [8], as this work only analyzed the case of minimal repair.

**Author Contributions:** Conceptualization, K.T. and N.N.; methodology, K.T., N.N. and S.C.; software, K.T. and S.C.; validation, S.C. and G.F.; investigation, K.T., B.M. and G.F.; data curation, B.M. and S.C.; writing—original draft, K.T., S.C. and B.M.; writing—review and editing, N.N., G.F. and B.M.;

visualization, K.T., N.N. and S.C. All authors have read and agreed to the published version of the manuscript.

**Funding:** This research received no external funding.

**Institutional Review Board Statement:** Not applicable.

**Informed Consent Statement:** Not applicable.

**Data Availability Statement:** Data sharing not applicable. No new data were created or analyzed in this study. Data sharing is not applicable to this article.

**Conflicts of Interest:** The authors declare no conflict of interest.

**Appendix A**

**Given:** Let the random variable $T$ be the time to failure (or repair) of a component. Also, let the random event $A(T_0)$ be that the component is operational at, or has not been repaired up to, time $T_0$. In fact, $T$ is the deterministic time $T_0$ plus the random time period till the next event (failure or repair). This definition of $T$ is true only for Appendix A. Then:

- The unconditional Cumulative Distribution Function (CDF) of $T$ is $F(t)$, for $t \in [0, \infty)$.
- The unconditional Probability Density Function (PDF) of $T$ is $f(t)$, for $t \in [0, \infty)$.
- The unconditional reliability (repair) function of $T$ is $R(t) = 1 - F(t)$, for $t \in [0, \infty)$.
- The unconditional failure (repair) rate of $T$ is $\lambda(t) = \frac{f(t)}{R(t)}$, for $t \in [0, \infty)$.
- The conditional CDF of $T$ if $A(T_0)$, is $F_{cond}(\tau|T_0)$, for $\tau = (t - T_0) \in [0, \infty)$.
- The conditional PDF of $T$ if $A(T_0)$, is $f_{cond}(\tau|T_0) = \frac{dF_{cond}(\tau|T_0)}{d\tau}$, for $\tau = (t - T_0) \in [0, \infty)$.
- The conditional reliability (repair) function of $T$ if $A(T_0)$, is $R_{cond}(\tau|T_0) = 1 - F_{cond}(\tau|T_0)$, for $\tau = (t - T_0) \in [0, \infty)$.
- The conditional failure (repair) rate of $T$ if $A(T_0)$, is $\lambda_{cond}(\tau|T_0) = \frac{f_{cond}(\tau|T_0)}{R_{cond}(\tau|T_0)}$, for $\tau = (t - T_0) \in [0, \infty)$.

**Prove:** The unconditional and the conditional failure (repair) rate are equal for any $t* \geq T_0$, i.e., $\lambda(t*) = \lambda_{cond}(t* - T_0|T_0)$, for $t* \in [T_0, \infty)$.

**Proof.** The unconditional $R(t)$ and $f(t)$ are given in Figure A1a,c. The relationship between these functions is:

$$f(t) = \frac{dF(t)}{dt} = \frac{d[1 - R(t)]}{dt} = -\frac{dR(t)}{dt} \text{ for } t \in [0, \infty) \quad (A1)$$

Similarly, the conditional $R_{cond}(\tau|T_0)$ and $f_{cond}(\tau|T_0)$ are given in Figure A1b,d. The relationship between these functions is:

$$f_{cond}(\tau|T_0) = \frac{dF_{cond}(\tau|T_0)}{d\tau} = \frac{d[1 - R_{cond}(\tau|T_0)]}{d\tau} = -\frac{dR_{cond}(\tau|T_0)}{d\tau} \text{ for } \tau = (t - T_0) \in [0, \infty) \quad (A2)$$

According to [11] (p. 72), the value of the conditional $R_{cond}(\tau|T_0)$ can be expressed as the ratio of two unconditional values of $R(t)$:

$$R_{cond}(\tau|T_0) = \frac{R(\tau + T_0)}{R(T_0)} \text{ for } \tau = (t - T_0) \in [0, \infty) \quad (A3)$$

The interdependency between Figure A1a,b illustrates Equation (A3). The constant $R(T_0)$ is the height of the red vertical line in Figure A1a.

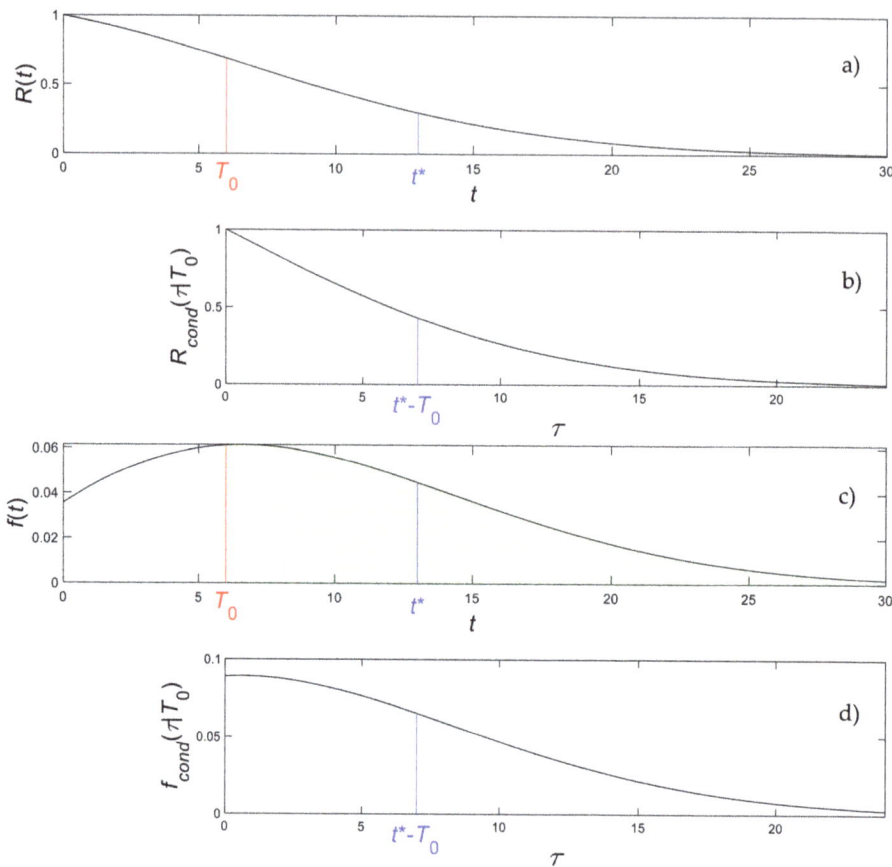

**Figure A1.** A generic distribution described by: (**a**) unconditional reliability; (**b**) conditional reliability; (**c**) unconditional density; (**d**) conditional density.

Let us take the first derivative about $\tau$ from Equation (A3) and multiply both sides by negative 1. Then,

$$-\frac{dR_{cond}(\tau|T_0)}{d\tau} = -\frac{d}{d\tau}\frac{R(\tau+T_0)}{R(T_0)} \quad \text{for } \tau = (t-T_0) \in [0,\infty) \quad \text{(A4)}$$

Let us simplify the RHS of Equation (A4) using Equation (A1) and utilizing that $\tau = (t-T_0)$:

$$\begin{aligned}
-\frac{d}{d\tau}\frac{R(\tau+T_0)}{R(T_0)} &= -\frac{1}{R(T_0)}\frac{dR(\tau+T_0)}{d\tau} = -\frac{1}{R(T_0)}\frac{dR(\tau+T_0)}{d(\tau+T_0)}\frac{d(\tau+T_0)}{d\tau} \\
&= -\frac{1}{R(T_0)}\frac{dR(t)}{dt}\frac{dt}{d\tau} = \frac{1}{R(T_0)}f(t)\frac{d(\tau+T_0)}{d\tau} \\
&= \frac{f(\tau+T_0)}{R(T_0)}(1) = \frac{f(\tau+T_0)}{R(T_0)}
\end{aligned} \quad \text{(A5)}$$

According to Equation (A2), the LHS of Equation (A4) is $f_{cond}(\tau|T_0)$. Then, from Equations (A4) and (A5) it follows that:

$$f_{cond}(\tau|T_0) = \frac{f(\tau+T_0)}{R(T_0)} \quad \text{for } \tau = (t-T_0) \in [0,\infty) \quad \text{(A6)}$$

The interdependency between Figure A1c,d illustrates Equation (A6). The constant $R(T_0)$ is the area of the green patch in Figure A1c, since from Equation (A1) it follows that
$$R(T_0) = \int_{T_0}^{\infty} f(t)dt.$$
The conditional failure (repair) rate of $T$ if $A(T_0)$ can be transformed using Equations (A3) and (A6):

$$\lambda_{cond}(\tau|T_0) = \frac{f_{cond}(\tau|T_0)}{R_{cond}(\tau|T_0)} = \frac{f(\tau+T_0)}{R(T_0)} \div \frac{R(\tau+T_0)}{R(T_0)} = \frac{f(\tau+T_0)}{R(\tau+T_0)} \text{ for } \tau = (t - T_0) \in [0, \infty) \quad (A7)$$

Let's select a time point $t* \geq T_0$. The unconditional failure (repair) rate of $T$ at time $t*$ is:

$$\lambda(t*) = \frac{f(t*)}{R(t*)} \quad (A8)$$

The nominator and the denominator in Equation (A8) are respectively the heights of the blue lines in Figure A1a,c. The conditional time $\tau$ is simply the time $t$ delayed with $T_0$ (i.e., $t = \tau + T_0$).

From here, the relative time moment $\tau*$ which coincides with time $t*$ is:

$$\tau* = t* - T_0 \quad (A9)$$

Equation (A9) is illustrated by the transition from Figure A1a to Figure A1b, and in the transition from Figure A1c to Figure A1d.

The value of $\lambda_{cond}(\tau|T_0)$ at relative time point $\tau*$ can be easily calculated from Equation (A7) utilizing Equations (A8) and (A9):

$$\begin{aligned} \lambda_{cond}(t* - T_0|T_0) &= \frac{f_{cond}(t*-T|T_0)}{R_{cond}(t*-T|T_0)} = \lambda_{cond}(\tau*|T_0) \\ &= \frac{f(\tau*+T_0)}{R(\tau*+T_0)} = \frac{f(t*)}{R(t*)} = \lambda(t*) \end{aligned}, \text{ for } t* \in [T_0, \infty) \quad (A10)$$

□

**Appendix B**

Given: Let $\lambda_1$, $\lambda_2$, $\lambda_3$, and $\lambda_4$ be real positive constants, whereas $p_r$ and $p_f$ are real positive constants less than 1. The real functions $P_1(t)$, $P_2(t)$, and $P_3(t)$ are defined in the Domain $t \in [0, \infty)$ and satisfy the system from Equation (A11) of three simultaneous ordinary differential equations. The initial conditions of the functions are given in Equation (A12).

$$\begin{cases} \frac{dP_1}{dt}(t) = -(\lambda_1 + \lambda_3)P_1(t) + (1-p_r)\lambda_4 P_2(t) \\ \frac{dP_2}{dt}(t) = (1-p_f)\lambda_1 P_1(t) - (\lambda_4 + \lambda_2)P_2(t) \\ \frac{dP_3}{dt}(t) = \lambda_3 P_1(t) - \lambda_1 P_3(t) \end{cases} \quad (A11)$$

$$P_1(0) = 1, P_2(0) = 0, P_3(0) = 0 \quad (A12)$$

Find:
(a) The solution of the initial-value problem for $P_1(t)$, $P_2(t)$, and $P_3(t)$ in the Domain $t \in [0, \infty)$.
(b) The functions $P_4(t) = 1 - P_1(t) - P_2(t) - P_3(t)$ and $R_{sys}(t) = 1 - P_4(t)$ in the Domain $t \in [0, \infty)$.
(c) The quantity $MTTF_{sys} = \int_0^{\infty} R_{sys}(t)dt$.

Solution:
(a) Taking Laplace transformation [47] (pp. 331–335) of the three equations in Equation (A11) yields a system of three algebraic equations about the Laplace transforms $Y_1(s)$, $Y_2(s)$,

and $Y_3(s)$ of the functions $P_1(t)$, $P_2(t)$, and $P_3(t)$, where $s$ is a complex number known as frequency:

$$\begin{cases} sY_1(s) - P_1(0) = -(\lambda_1 + \lambda_3)Y_1(s) + (1 - p_r)\lambda_4 Y_2(s) \\ sY_2(s) - P_2(0) = (1 - p_r)\lambda_1 Y_1(s) - (\lambda_4 + \lambda_2)Y_2(s) \\ sY_3(s) - P_3(0) = \lambda_3 Y_1(s) - \lambda_1 Y_3(s) \end{cases} \quad (A13)$$

Substituting Equation (A12) into Equation (A13) and simplifying gives:

$$\begin{cases} (s + \lambda_1 + \lambda_3)Y_1(s) - (1 - p_r)\lambda_4 Y_2(s) = 1 \\ -\left(1 - p_f\right)\lambda_1 Y_1(s) + (s + \lambda_2 + \lambda_4)Y_2(s) = 0 \\ -\lambda_3 Y_1(s) + (\lambda_1 + s)Y_3(s) = 0 \end{cases} \quad (A14)$$

The first two equations in Equation (A14) can be solved for $Y_1(s)$, $Y_2(s)$ using the Cramer's rule [48]:

$$Y_1(s) = \frac{s + \lambda_2 + \lambda_4}{(s + \lambda_1 + \lambda_3)(s + \lambda_2 + \lambda_4) - (1 - p_r)\left(1 - p_f\right)\lambda_1 \lambda_4} \quad (A15)$$

$$Y_2(s) = \frac{\left(1 - p_f\right)\lambda_1}{(s + \lambda_1 + \lambda_3)(s + \lambda_2 + \lambda_4) - (1 - p_r)\left(1 - p_f\right)\lambda_1 \lambda_4} \quad (A16)$$

The denominator in both Equations (A15) and (A16) is a quadratic polynomial with real coefficients 1, $K$, and $C$:

$$(s + \lambda_1 + \lambda_3)(s + \lambda_2 + \lambda_4) - (1 - p_r)\left(1 - p_f\right)\lambda_1 \lambda_4 = s^2 + 2Ks + C \quad (A17)$$

where the real constants $K$ and $C$ are:

$$\begin{aligned} K &= (\lambda_1 + \lambda_2 + \lambda_3 + \lambda_4)/2 \\ C &= (\lambda_1 + \lambda_3)(\lambda_2 + \lambda_4) - \left(1 - p_f\right)(1 - p_r)\lambda_1 \lambda_4 \end{aligned} \quad (A18)$$

We will prove that the discriminant, $\Delta$, of the quadratic polynomial Equation (A17) is always positive:

$$\begin{aligned} \Delta &= (2K)^2 - 4(1)C = [2(\lambda_1 + \lambda_2 + \lambda_3 + \lambda_4)/2]^2 - 4\left[(\lambda_1 + \lambda_3)(\lambda_2 + \lambda_4) - \left(1 - p_f\right)(1 - p_r)\lambda_1 \lambda_4\right] \\ &= (\lambda_1 + \lambda_2 + \lambda_3 + \lambda_4)^2 - 4(\lambda_1 + \lambda_3)(\lambda_2 + \lambda_4) + 4\left(1 - p_f\right)(1 - p_r)\lambda_1 \lambda_4 \\ &\quad (\lambda_1 + \lambda_2 + \lambda_3 + \lambda_4)^2 - 4(\lambda_1 + \lambda_3)(\lambda_2 + \lambda_4) + 4(1 - 1)(1 - p_r)\lambda_1 \lambda_4 \\ &= (\lambda_1 + \lambda_2 + \lambda_3 + \lambda_4)^2 - 4(\lambda_1 + \lambda_3)(\lambda_2 + \lambda_4) = [(\lambda_1 + \lambda_3) + (\lambda_2 + \lambda_4)]^2 - 4(\lambda_1 + \lambda_3)(\lambda_2 + \lambda_4) \\ &= (\lambda_1 + \lambda_3)^2 + (\lambda_2 + \lambda_4)^2 + 2(\lambda_1 + \lambda_3)(\lambda_2 + \lambda_4) - 4(\lambda_1 + \lambda_3)(\lambda_2 + \lambda_4) \\ &= (\lambda_1 + \lambda_3)^2 + (\lambda_2 + \lambda_4)^2 - 2(\lambda_1 + \lambda_3)(\lambda_2 + \lambda_4) = [(\lambda_1 + \lambda_3) - (\lambda_2 + \lambda_4)]^2 \geq 0 \\ &\Rightarrow \Delta > 0 \end{aligned} \quad (A19)$$

In Equation (A19) we used that $4\left(1 - p_f\right)(1 - p_r)\lambda_1 \lambda_4 > 0$ since $\left(1 - p_f\right) > 0$, $(1 - p_r) > 0$, $\lambda_1 > 0$, and $\lambda_4 > 0$. From Equation (A19) it follows that the roots $s_1$ the $s_2$ of the quadratic polynomial Equation (A19) are always real and different:

$$s_{1,2} = \left(-2K \pm \sqrt{\Delta}\right)/2 = \left(-2K \pm \sqrt{4K^2 - 4C}\right)/2 = -K \pm \sqrt{K^2 - C} \quad (A20)$$

In Equation (A20) we assume that $s_1 > s_2$ (i.e., $s_1 = -K + \sqrt{K^2 - C}$ and $s_2 = -K - \sqrt{K^2 - C}$). It can easily be seen that the constants $s_1$ the $s_2$ are always negative. Using the quadratic factorization formula together with Equation (A17) the denominator in both Equations (A15) and (A16) can be factored to:

$$s^2 + 2Ks + C = 1(s - s_1)(s - s_2) = (s - s_1)(s - s_2) \tag{A21}$$

From Equations (A15)–(A17), and (A21), $Y_1(s)$, $Y_2(s)$ can be simplified to:

$$Y_1(s) = \frac{s + \lambda_2 + \lambda_4}{(s - s_1)(s - s_2)} \tag{A22}$$

$$Y_2(s) = \frac{(1 - p_f)\lambda_1}{(s - s_1)(s - s_2)} \tag{A23}$$

Substituting Equation (A22) in the third equation of Equation (A14) we can find $Y_3(s)$:

$$Y_3(s) = \frac{\lambda_3 Y_1(s)}{(\lambda_1 + s)} = \frac{\lambda_3 (s + \lambda_2 + \lambda_4)}{(s - s_1)(s - s_2)(\lambda_1 + s)} \tag{A24}$$

The identified $Y_1(s)$, $Y_2(s)$, and $Y_3(s)$ are rational fractions according to Equations (A22)–(A24). To facilitate the inverse Laplace transform, those rational fractions can be subjected to a partial fraction decomposition [49] (pp. 533–540):

$$Y_1(s) = \frac{s + \lambda_2 + \lambda_4}{(s - s_1)(s - s_2)} = \frac{A_1}{(s - s_1)} + \frac{B_1}{(s - s_2)} \tag{A25}$$

The constants $A_1$ and $B_1$ in Equation (A25) are:

$$A_1 = \frac{s_1 + \lambda_2 + \lambda_4}{s_1 - s_2} \text{ and } B_1 = \frac{s_2 + \lambda_2 + \lambda_4}{s_2 - s_1} \tag{A26}$$

$$Y_2(s) = \frac{(1 - p_f)\lambda_1}{(s - s_1)(s - s_2)} = \frac{A_2}{(s - s_1)} + \frac{B_2}{(s - s_2)} \tag{A27}$$

The constants $A_2$ and $B_2$ in Equation (A27) are:

$$A_2 = \frac{(1 - p_f)\lambda_1}{s_1 - s_2} \text{ and } B_2 = \frac{(1 - p_f)\lambda_1}{s_2 - s_1} \tag{A28}$$

$$Y_3(s) = \frac{\lambda_3(s + \lambda_2 + \lambda_4)}{(s - s_1)(s - s_2)(\lambda_1 + s)} = \frac{A_3}{(s - s_1)} + \frac{B_3}{(s - s_2)} + \frac{C_3}{(\lambda_1 + s)} \tag{A29}$$

The constants $A_3$, $B_3$, and $C_3$ in Equation (A29) are:

$$A_3 = \frac{\lambda_3(s_1 + \lambda_2 + \lambda_4)}{(s_1 - s_2)(\lambda_1 + s_1)}, B_3 = \frac{\lambda_3(s_2 + \lambda_2 + \lambda_4)}{(s_2 - s_1)(\lambda_1 + s_2)}, \text{ and } C_3 = \frac{\lambda_3(\lambda_1 + \lambda_2 + \lambda_4)}{(\lambda_1 - s_1)(\lambda_1 - s_2)} \tag{A30}$$

Now, we can apply the inverse Laplace transform over Equations (A25), (A27), and (A29) and find the solutions $P_1(t)$, $P_2(t)$, and $P_3(t)$ of the stated initial-value problem:

$$\begin{cases} \text{Domain}: t \in [0, \infty) \\ P_1(t) = A_1 e^{s_1 t} - B_1 e^{s_2 t} \\ P_2(t) = A_2 e^{s_1 t} - B_2 e^{s_2 t} \\ P_3(t) = A_3 e^{s_1 t} - B_3 e^{s_2 t} + C_3 e^{-\lambda_1 t} \end{cases} \tag{A31}$$

(b) Using the Equation (A31), the required functions can be simplified to:

$$P_4(t) \begin{aligned} &\text{Domain}: t \in [0, \infty) \\ &= 1 - P_1(t) - P_2(t) - P_3(t) \\ &= 1 - \left(A_1 e^{s_1 t} - B_1 e^{s_2 t}\right) - \left(A_2 e^{s_1 t} - B_2 e^{s_2 t}\right) - \left(A_3 e^{s_1 t} - B_3 e^{s_2 t} + C_3 e^{-\lambda_1 t}\right) \\ &= 1 - A_1 e^{s_1 t} + B_1 e^{s_2 t} - A_2 e^{s_1 t} + B_2 e^{s_2 t} - A_3 e^{s_1 t} + B_3 e^{s_2 t} - C_3 e^{-\lambda_1 t} \\ &= 1 - (A_1 + A_2 + A_3) e^{s_1 t} + (B_1 + B_2 + B_3) e^{s_2 t} - C_3 e^{-\lambda_1 t} \end{aligned}$$ (A32)

$$R_{sys}(t) \begin{aligned} &\text{Domain}: t \in [0, \infty) \\ &= 1 - P_4(t) \\ &= 1 - \left[1 - \left(A_1 e^{s_1 t} - B_1 e^{s_2 t}\right) - \left(A_2 e^{s_1 t} - B_2 e^{s_2 t}\right) - \left(A_3 e^{s_1 t} - B_3 e^{s_2 t} + C_3 e^{-\lambda_1 t}\right)\right] \\ &= 1 - 1 + A_1 e^{s_1 t} - B_1 e^{s_2 t} + A_2 e^{s_1 t} - B_2 e^{s_2 t} + A_3 e^{s_1 t} - B_3 e^{s_2 t} + C_3 e^{-\lambda_1 t} \\ &= (A_1 + A_2 + A_3) e^{s_1 t} - (B_1 + B_2 + B_3) e^{s_2 t} + C_3 e^{-\lambda_1 t} \end{aligned}$$ (A33)

(c) The required improper integral for $MTTF_{sys}$ when the integrand is given by Equation (A33) can be calculated using the following formula:

$$\int_0^\infty e^{-at} dt = \frac{1}{a} \text{ where } a > 0$$ (A34)

Then,

$$\begin{aligned} MTTF_{sys} &= \int_0^\infty R_{sys}(t) dt = \int_0^\infty (A_1 + A_2 + A_3) e^{s_1 t} - (B_1 + B_2 + B_3) e^{s_2 t} + C_3 e^{-\lambda_1 t} dt \\ &= (A_1 + A_2 + A_3) \int_0^\infty e^{s_1 t} dt - (B_1 + B_2 + B_3) \int_0^\infty e^{s_2 t} dt + C_3 \int_0^\infty e^{-\lambda_1 t} dt \\ &= -(A_1 + A_2 + A_3)/s_1 + (B_1 + B_2 + B_3)/s_2 + C_3/\lambda_1 \end{aligned}$$ (A35)

In the derivation shown in Equation (A35) we applied Equation (A34) three times since $s_1 < 0$, $s_2 < 0$, and $(-\lambda_1) < 0$.

## References

1. Hausken, K. Strategic defense and attack for series and parallel reliability systems. *Eur. J. Oper. Res.* **2008**, *186*, 856–881. [CrossRef]
2. Aghaei, M.; Hamadani, A.Z.; Ardakan, M.A. Redundancy allocation problem for k-out-of-n systems with a choice of redundancy strategies. *J. Ind. Eng. Int.* **2017**, *13*, 81–92. [CrossRef]
3. Amari, S.V.; Dill, G. A new method for reliability analysis of standby systems. In Proceedings of the 2009 Annual Reliability and Maintainability Symposium, Fort Worth, TX, USA, 26–29 January 2009; pp. 417–422. [CrossRef]
4. Yuan, L.; Meng, X.-Y. Reliability analysis of a warm standby repairable system with priority in use. *Appl. Math. Model.* **2011**, *35*, 4295–4303. [CrossRef]
5. Ardakan, M.A.; Rezvan, M.T. Multi-objective optimization of reliability–redundancy allocation problem with cold-standby strategy using NSGA-II. *Reliab. Eng. Syst. Saf.* **2018**, *172*, 225–238. [CrossRef]
6. Li, X.; Zhang, Z.; Wu, Y. Some new results involving general standby systems. *Appl. Stoch. Model. Bus. Ind.* **2009**, *25*, 632–642. [CrossRef]
7. Kwiatuszewska-Sarnecka, B. Reliability improvement of large multi-state series-parallel systems. *Int. J. Autom. Comput.* **2006**, *3*, 157–164. [CrossRef]
8. Yang, Q.; Zhang, N.; Hong, Y. Reliability Analysis of Repairable Systems with Dependent Component Failures Under Partially Perfect Repair. *IEEE Trans. Reliab.* **2013**, *62*, 490–498. [CrossRef]
9. Lindqvist, B.H. On the Statistical Modeling and Analysis of Repairable Systems. *Stat. Sci.* **2006**, *21*, 532–551. [CrossRef]
10. Zhang, Y.L. A geometric-process repair-model with good-as-new preventive repair. *IEEE Trans. Reliab.* **2002**, *51*, 223–228. [CrossRef]
11. Modarres, M.; Kaminskiy, M.; Krivtsov, V. *Reliability Engineering and Risk Analysis: A Practical Guide*, 2nd ed.; CRC Press: Boca Raton, FL, USA, 2010; pp. 72–227. ISBN 9780849392474.
12. Badía, F.; Berrade, M.; Cha, J.H.; Lee, H. Optimal replacement policy under a general failure and repair model: Minimal versus worse than old repair. *Reliab. Eng. Syst. Saf.* **2018**, *180*, 362–372. [CrossRef]
13. Bao, Y.; Sun, X.; Trivedi, K. A Workload-Based Analysis of Software Aging, and Rejuvenation. *IEEE Trans. Reliab.* **2005**, *54*, 541–548. [CrossRef]
14. Nguyen, T.A.; Kim, D.S.; Park, J.S. A Comprehensive Availability Modeling and Analysis of a Virtualized Servers System Using Stochastic Reward Nets. *Sci. World J.* **2014**, *2014*, 1–18. [CrossRef]

15. Nguyen, T.A.; Kim, D.S.; Park, J.S. Availability modeling and analysis of a data center for disaster tolerance. *Futur. Gener. Comput. Syst.* **2016**, *56*, 27–50. [CrossRef]
16. Loveland, S.; Dow, E.M.; Lefevre, F.; Beyer, D.; Chan, P.F. Leveraging virtualization to optimize high-availability system configurations. *IBM Syst. J.* **2008**, *47*, 591–604. [CrossRef]
17. Tenekedjiev, K.; Nikolova, N.; Fan, G.; Symes, M.; Nguyen, O. Simulation algorithms to assess the impact of aging on the reliability of standby systems with switching failures. In *Advances in Intelligent Systems Research and Innovation*; Book Series: Studies in Systems, Decision and Control; Sgurev, V.V., Jotsov, J., Kacprzyk, J., Eds.; Springer Nature: New York, NY, USA, 2021; Chapter 21; pp. 463–496. [CrossRef]
18. Wells, C.E. Reliability analysis of a single warm-standby system subject to repairable and nonrepairable failures. *Eur. J. Oper. Res.* **2014**, *235*, 180–186. [CrossRef]
19. Ebeling, C.E. *An Introduction to Reliability and Maintainability Engineering*, 2nd ed.; Waveland Press Inc.: Long Grove, IL, USA, 2010; pp. 113–115. ISBN 1-57766-625-9.
20. Ebeling, C.E. *An Introduction to Reliability and Maintainability Engineering*, 3rd ed.; Waveland Press Inc.: Long Grove, IL, USA, 2019; pp. 87–399. ISBN 978-1478637349.
21. Cha, J.H.; Mi, J.; Yun, W.Y. Modelling a general standby system and evaluation of its performance. *Appl. Stoch. Model. Bus. Ind.* **2007**, *24*, 159–169. [CrossRef]
22. Nikolova, N.; Fan, G.; Symes, M.; Tenekedjiev, K. Simulating State-Dependent Systems with Partial Aging in Standby. In Proceedings of the IEEE 10th International Conference on Intelligent Systems (IS'2020), Varna, Bulgaria, 28–30 August 2020; pp. 51–60.
23. Yang, L. Reliability Model for Warm Standby System under Consideration of Replace Time. *Int. J. Hybrid. Inf. Technol.* **2016**, *9*, 135–146. [CrossRef]
24. Bhardwaj, R.K.; Kaur, K.; Malik, S.C. Reliability indices of a redundant system with standby failure and arbitrary distribution for repair and replacement times. *Int. J. Syst. Assur. Eng. Manag.* **2017**, *8*, 423–431. [CrossRef]
25. Wang, K.-H.; Ke, J.-B.; Lee, W.-C. Reliability and sensitivity analysis of a repairable system with warm standbys and R unreliable service stations. *Int. J. Adv. Manuf. Technol.* **2007**, *31*, 1223–1232. [CrossRef]
26. Srinivasan, S.K.; Subramanian, R. Reliability analysis of a three unit warm standby redundant system with repair. *Ann. Oper. Res.* **2006**, *143*, 227–235. [CrossRef]
27. Maciel, P.R.M.; Dantas, J.R.; Júnior, R.d.S.M. Markov chains and stochastic Petri nets for availability and reliability modeling. In *Reliability Engineering: Methods and Applications*; Ram, M., Ed.; CRC Press: Boca Raton, FL, USA, 2020; pp. 127–151.
28. Zhao, X.; Nakagawa, T. An Overview on Failure Rates in Maintenance Policies. In *Reliability Engineering*; CRC Press: Boca Raton, CL, USA, 2019; pp. 166–196.
29. Kishore, J.; Goel, M.; Khanna, P. Understanding survival analysis: Kaplan-Meier estimate. *Int. J. Ayurveda Res.* **2010**, *1*, 274–278. [CrossRef]
30. MATLAB. *MATLAB R2019a and Statistics and Machine Learning Toolbox 11.5*.; The MathWorks Inc.: Natick, MA, USA, 2019.
31. Nikolova, N.; Toneva, D.; Tsonev, Y.; Burgess, B.; Tenekedjiev, K. Novel Methods to Construct Empirical CDF for Continuous Random Variables using Censor Data. In Proceedings of the IEEE 10th International Conference on Intelligent Systems (IS'2020), Varna, Bulgaria, 28–30 August 2020; pp. 61–68.
32. Fuzzy Rationality in Quantitative Decision Analysis. *J. Adv. Comput. Intell. Intell. Inform.* **2005**, *9*, 65–69. [CrossRef]
33. Nikolova, N.D.; Dimitrakicv, D.; Tenekedjiev, K.I. Fuzzy rationality in the elicitation of subjective quantiles. In Proceedings of the Second International IEEE Conference on Intelligent Systems IS'2004, Varna, Bulgaria, 22–24 June 2004; Volume 3, pp. 32–34.
34. Attar, A.; Raissi, S.; Khalili-Damghani, K. A simulation-based optimization approach for free distributed repairable multi-state availability-redundancy allocation problems. *Reliab. Eng. Syst. Saf.* **2017**, *157*, 177–191. [CrossRef]
35. Williams, J.D.; Young, S. Partially observable Markov decision processes for spoken dialog systems. *Comput. Speech Lang.* **2007**, *21*, 393–422. [CrossRef]
36. Distefano, S.; Trivedi, K.S. Non-Markovian State-Space Models in Dependability Evaluation. *Qual. Reliab. Eng. Int.* **2013**, *29*, 225–239. [CrossRef]
37. Birolini, A. *Reliability Engineering: Theory and Practice*, 8th ed.; Springer: New York, NY, USA, 2017; pp. 290–526.
38. Darling, R.; Norris, J. Differential equation approximations for Markov chains. *Probab. Surv.* **2008**, *5*, 37–79. [CrossRef]
39. MATLAB. *MATLAB R2019a*; The MathWorks: Natick, MA, USA, 2019.
40. Nikolaidis, E.; Mourelatos, Z.P.; Pandey, V. *Design Decisions under Uncertainty with Limited Information*; CRC Press: Boca Raton, CA, USA, 2011; pp. 42–43. ISBN 9781138115095.
41. Merovci, F.; Elbatal, I. Weibull Rayleigh distribution: Theory and applications. *Appl. Math. Inf. Sci.* **2015**, *9*, 1–11. [CrossRef]
42. Horrace, W.C. Moments of the truncated normal distribution. *J. Prod. Anal.* **2015**, *43*, 133–138. [CrossRef]
43. Kızılersü, A.; Kreer, M.; Thomas, A.W. The Weibull distribution. *Significance* **2018**, *15*, 10–11. [CrossRef]
44. Mouri, H. Log-normal distribution from a process that is not multiplicative but is additive. *Phys. Rev. E* **2013**, *88*, 042124. [CrossRef]
45. Shampine, L.F.; Reichelt, M.W.; Kierzenka, J.A. Solving Index-1 DAEs in MATLAB and Simulink. *SIAM Rev.* **1999**, *41*, 538–552. [CrossRef]
46. Jiang, R. A new bathtub curve model with a finite support. *Reliab. Eng. Syst. Saf.* **2013**, *119*, 44–51. [CrossRef]

47. James, G.; Dyke, P. *Advanced Modern Engineering Mathematics*, 5th ed.; Pearson Education: Hoboken, NJ, USA, 2018; pp. 331–335. ISBN 9781292174341.
48. Habgood, K.; Arel, I. A condensation-based application of Cramer's rule for solving large-scale linear systems. *J. Discret. Algorithms* **2012**, *10*, 98–109. [CrossRef]
49. Stewart, J. *Calculus, Metric Version*, 8th ed.; Cengage Learning: Boston, MA, USA, 2015; pp. 533–540. ISBN 9781473742437.

Article

# Lie-Group Modeling and Numerical Simulation of a Helicopter

Alessandro Tarsi [1] and Simone Fiori [2,*]

[1] School of Automation Engineering, Alma Mater Studiorum—Università di Bologna, Viale del Risorgimento 2, I-40136 Bologna, Italy; S1082358@studenti.univpm.it
[2] Department of Information Engineering, Marches Polytechnic University, Brecce Bianche Rd., I-60131 Ancona, Italy
* Correspondence: s.fiori@staff.univpm.it

**Abstract:** Helicopters are extraordinarily complex mechanisms. Such complexity makes it difficult to model, simulate and pilot a helicopter. The present paper proposes a mathematical model of a fantail helicopter type based on Lie-group theory. The present paper first recalls the Lagrange–d'Alembert–Pontryagin principle to describe the dynamics of a multi-part object, and subsequently applies such principle to describe the motion of a helicopter in space. A good part of the paper is devoted to the numerical simulation of the motion of a helicopter, which was obtained through a dedicated numerical method. Numerical simulation was based on a series of values for the many parameters involved in the mathematical model carefully inferred from the available technical literature.

**Keywords:** Lagrange–d'Alembert principle; non-conservative dynamical system; Euler–Poincaré equation; helicopter model; Lie group

**Citation:** Tarsi, A.; Fiori, S. Lie-Group Modeling and Numerical Simulation of a Helicopter. *Mathematics* **2021**, *9*, 2682. https://doi.org/10.3390/math9212682

Academic Editors: Maria Luminița Scutaru and Catalin I. Pruncu

Received: 17 September 2021
Accepted: 13 October 2021
Published: 22 October 2021

**Publisher's Note:** MDPI stays neutral with regard to jurisdictional claims in published maps and institutional affiliations.

**Copyright:** © 2021 by the authors. Licensee MDPI, Basel, Switzerland. This article is an open access article distributed under the terms and conditions of the Creative Commons Attribution (CC BY) license (https://creativecommons.org/licenses/by/4.0/).

## 1. Introduction

Conventional helicopters are built with two propellers that can be arranged as two coplanar rotors both providing upward thrust, but spinning in opposite directions in order to balance the torques exerted upon the body of the helicopter, or as one main rotor providing thrust and a smaller side rotor oriented laterally and counteracting the torque produced by the main rotor, as shown in the Figure 1. Helicopters with no tail rotors ('notar') use a jet of compressed air to compensate for the unwanted yawing of the fuselage.

**Figure 1.** Eurocopter EC 135, with a fantail assembly tail rotor (reproduced from https://en.wikipedia.org/wiki/Tail_rotor accessed on 21 April 2021).

Controls on a helicopter are numerous. Considering a rigid rotor system, the attitude and the position of a helicopter are mainly controlled through two systems, called the *collective control* system and *cyclic control* system. The power exerted by the rotors is usually constant, in fact, the blades are designed to operate at a specific rotational speed. However, it is possible to slightly vary the engine power using the *throttle control*, whereas the direction the aircraft nose points, the yaw angle, could be changed using the *pedals control*. A summary of helicopter controls is given in the following.

**Collective control**: The *collective control* is used to increase or decrease the total thrust generated by the rotors. This technique is adopted in the main rotor and in the tail rotor. To grow (to reduce) the thrust it is necessary to increase (to decrease) the angle of attack $\alpha_c$ of all blades. This angle is in each instant the same for all the blades. An example of the usage of the *collective control* is illustrated in Figure 2.

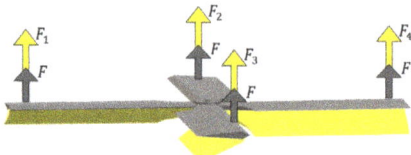

**Figure 2.** *Collective control* changes the angle of attack of all blades at the same time. The main rotor is in the grey position, horizontal to the ground, if not actuated. A maneuver of the collective control brings the blades to rotate independently to the yellow configuration. The force generated by each propeller is represented by $F$ in the standard configuration and by $F_1 = F_2 = F_3 = F_4$ in the collective controlled case.

**Cyclic control**: The *cyclic control* is distinctive of the main rotor. To tilt the body of a helicopter forward and backwards (pitch) or sideways (roll), a pilot must alter the angle of attack of the main rotor blades cyclically during rotation, as illustrated in Figure 3. In particular, controlling the angle of attack of the blades in such a way that the forward half of the rotor disk exerts more (less) thrust than the backward half makes the helicopter pitch upward (downward). Generally, to vary the attitude of a helicopter it is necessary to modify the angle of the thrust exerted by the main rotor, which is generated by the rotation of the blades, hence it is necessary to create different amounts of thrust at different points in the cycle. Where a greater (smaller) amount of thrust is necessary the blade increases (decrease) its angle. Two angles, namely $\alpha_p$ and $\alpha_r$, are used to indicate the direction of the thrust vector generated by the main rotor.

**Pedals control**: Because of momentum conservation, the rotation of the main rotor causes a rotation of the body of the helicopter in the opposite direction: as the engine turns the main rotor system in a counterclockwise direction, the helicopter fuselage turns clockwise. The amount of torque is directly related to the amount of engine power being used to turn the main rotor system. The unwanted yawing of the fuselage may be balanced by controlling the thrust of the tail rotor, as illustrated in Figure 4. The anti-torque pedals change the tail rotor collective angle of attack $\alpha_c^T$. The yaw angle variation depends upon variations of the tail rotor thrust or variations on the main rotor thrust. The *pedals control* is used for heading changes while hovering, but also to maintain the actual helicopter nose direction.

**Actuators**: The mentioned pilot control systems are actuated through a series of devices that are briefly described in the following:

- The *cyclic control* and the *collective control* of the main rotor work through a complex mechanical system called 'swash-plate', whose functioning is illustrated, e.g., at page 272 of the manual [1]. The swash-plate is composed of two parts, one that is tight with the rotor mast and one that can rotate with the main rotor. Each blade is strictly connected with the swash-plate revolving part using a rod. This causes a variation of the angle of attack of the blade when the swash-plate changes position. The swash-plate manages the cyclic and collective angles and sets up constraints in their ranges. The *collective control* causes a movement upward or downward of the swash-plate on the rotor mast, therefore all the blades increase or decrease their angle of attack

simultaneously. The *cyclic control* changes the attitude of the swash-plate. This causes a changing of the angle of attack that is different in every part of the rotation cycle.
- The tail rotor actuator is called a "pitch change spider" and, similarly to the swash-plate, it is used to vary all the angles of attack of the blades simultaneously. A figure at page 272 of the manual [1] illustrates the functioning of the pitch change spider. Helicopters, usually, possess a stabilizer that reduces the noise of the wind, providing an easier use of the yaw pedals. The pitch change spider also sets up the constraints for the range of variation of its angle of attack $\alpha_c^T$.

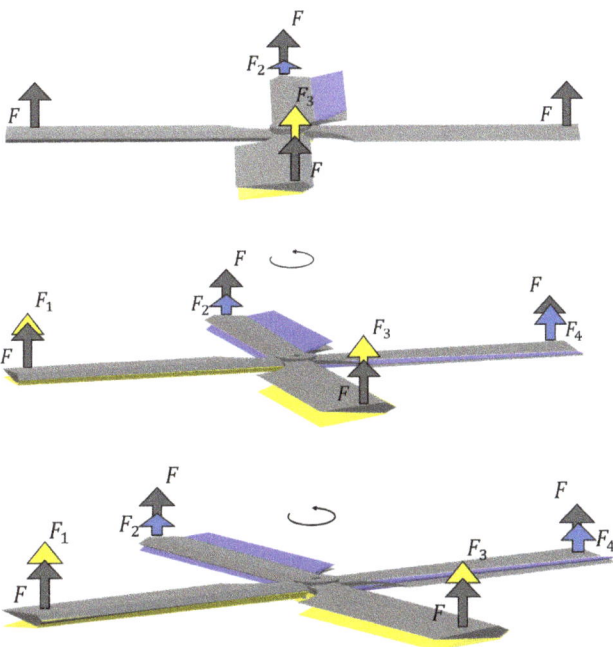

**Figure 3.** Series of frames representing the rotation of the main rotor actuated by the *Cyclic control*. The artwork shows the forces exerted by each blade during a rotation. The grey arrow denotes the force F produced when the blade is horizontal, whereas the yellow (blue) arrows denote the forces produced when the angle taken by the blade is such that the thrust is stronger (weaker) than F.

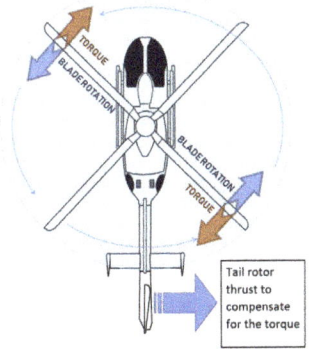

**Figure 4.** Anti-torque effect of the tail rotor.

***Throttle***: The throttle controls the power of the engine which is connected to both rotors by the transmission. The throttle setting must maintain enough engine power to keep the rotor speed within the limits where the rotor produces enough lift for flight. The throttle changes the blades' angular velocity in a range of few values percentage. Helicopters possess only a gear to drive both the main rotor and tail rotor, hence increasing the speed of the main rotor causes an increase in the tail rotor speed. More throttle means more speed and hence a larger value of thrust. The angular velocity of the rotors is usually reported in percentage for a more intuitive perception, the value of 100% is the typical one under standard conditions.

A helicopter is an extraordinarily complicated machine, whose functioning is based on a number of mechanical devices whose actions interact intricately to one another. Such complex design make its modeling and control by a pilot a fascinating challenge. The main challenge encountered during the present research work was to design a mathematical model that, on one hand, is able to capture the essential features of a helicopter, hence being sufficiently accurate to predict its behavior and, on the other hand, to be simple enough for the result to be mathematically tractable.

In the present paper, we derive, through the Euler–Poincaré formalism, the mathematical model of a simplified helicopter. The model concerns a helicopter with a principal rotor and a tail rotor. More accurate (and mathematically complicated) aircraft models are available in the specialized literature [2–4]. The structure of the present paper may be outlined as follows:

- Section 2 presents a summary of definitions and properties regarding Lie groups, such as the tools used in this research to formalize the mathematical model of a helicopter, i.e., tangent bundle, Lie algebra and exponential map. Moreover, this section introduces a system of differential equations that are used to describe the motion of a helicopter.
- Section 3 introduces the structure of the helicopter, a reference system and the structure of forces used to complete the mathematical model, as the thrust of the rotors. In addition, this section outlines a derivation of the equations of motion starting from a Lagrangian function.
- Section 4 presents a numerical scheme to simulate on a computing platform the system of equations determined in Section 3 using a forward Euler (*fEul*) method tailored to the Lie group of rotations.
- Section 5 introduces a helicopter type and shows the values of the parameters required to perform simulation analysis. These values are presented in tables and figures and have been gathered (and calculated) from data-sheets.
- Section 6 illustrates eight simulation results. Each simulation is particularly focused on a specific response, i.e., pitch response and roll response.
- Section 7 concludes this paper with a recapitulation of the obtained results and an overview of the key points of the performed analysis.

We would like to mention that the scientific literature about system modeling (including mechanical system modeling) is rich in inventions. A few alternative techniques to the more traditional equation-based modeling and control are bond graph modeling utilized, e.g., in [5] to design a Kalman filter observer for an industrial back-support exoskeleton; closed loop identification and frequency domain analysis, utilized in [6] to determine a dynamic model of a quadrotor prototype; deep neural networks, used in [7] to predict the remaining useful life (RUL) of aircraft gas turbine engines. The present authors are not familiar enough with the mentioned techniques to judge their advantages or disadvantages in relation to the proposed modeling method, which arises as a more elaborated version of the familiar Euler–Lagrange formalism (except for the neural-network modeling approach that provides an approximated, data-derived model in contrast to exact models).

## 2. The Lagrange–d'Alembert–Pontryagin Principle and the Forced Euler–Poincaré Equation

In this paper, we consider non-conservative non-linear dynamical systems whose state space $\mathbb{G}$ possesses the mathematical structure of a *Lie group*.

### 2.1. Definition and Properties

Let us recapitulate the following definitions and properties [8,9] (see also [10,11] for a non-strictly mathematical viewpoint):

**Matrix Lie group**: A smooth matrix manifold $\mathbb{G}$ that is also an algebraic group is termed a matrix Lie group. A matrix group is a matrix set endowed with an associative binary operation, termed *group multiplication* which, for any two elements $g, h \in \mathbb{G}$, is denoted by $gh$, endowed with the property of closure, an *identity element* with respect to the multiplication, denoted by $e$, such that $eg = ge = g$ for any $g \in \mathbb{G}$, and an *inversion* operation, denoted by $g^{-1}$, with respect to multiplication, such that $g^{-1}g = gg^{-1} = e$ for any $g \in \mathbb{G}$. A *left translation* $L : \mathbb{G} \times \mathbb{G} \to \mathbb{G}$ is defined as $L_g(h) := g^{-1}h$. An instance of matrix Lie group is $SO(3) := \{R \in \mathbb{R}^{3 \times 3} \mid R^\top R = RR^\top = I_3, \det(R) = +1\}$, where the symbol $^\top$ denotes matrix transposition and the quantity $I_3$ represents a $3 \times 3$ identity matrix.

**Tangent bundle and its metrization**: Given a point $g \in \mathbb{G}$, the tangent space to $\mathbb{G}$ at $g$ will be denoted as $T_g\mathbb{G}$. The tangent bundle associated with a manifold-group $\mathbb{G}$ is denoted by $T\mathbb{G}$ and plays the role of phase-space for a dynamical system whose state-space is $\mathbb{G}$. The inner product of two tangent vectors $\xi, \eta \in T_g\mathbb{G}$ is denoted by $\langle \xi, \eta \rangle_g$. A smooth function $F : \mathbb{G} \to \mathbb{G}$ induces a linear map $dF_g : T_g\mathbb{G} \to T_{F(g)}\mathbb{G}$ termed *pushforward map*. For a matrix Lie group, the pushforward map $d(L_g)_h : T_h\mathbb{G} \to T_{L_g(h)}\mathbb{G}$ associated with a left translation is $d(L_g)_h(\eta) := g^{-1}\eta$, with $\eta \in T_h\mathbb{G}$.

**Lie algebra**: The tangent space $\mathfrak{g} := T_e\mathbb{G}$ to a Lie group at the identity is termed *Lie algebra*. The Lie algebra is endowed with *Lie brackets*, denoted as $[\cdot, \cdot] : \mathfrak{g} \times \mathfrak{g} \to \mathfrak{g}$, and an *adjoint endomorphism* $\mathrm{ad}_\xi \eta := [\xi, \eta]$. The Lie algebra associated with the group $SO(3)$ is $\mathfrak{so}(3) := \{\xi \in \mathbb{R}^{3\times 3} \mid \xi + \xi^\top = 0\}$. On a matrix Lie algebra, the Lie brackets coincide with matrix commutator, namely $[\xi, \eta] := \xi\eta - \eta\xi$. The matrix commutator in $\mathfrak{so}(3)$ is an anti-symmetric bilinear form, namely $[\xi, \eta] + [\eta, \xi] = 0$. A pushforward map $d(L_g)_g : T_g\mathbb{G} \to \mathfrak{g}$ is denoted as $dL_g$ for brevity. Given a smooth function $\ell : \mathfrak{g} \to \mathbb{R}$, for a matrix Lie group one may define the *fiber derivative* of $\ell$, $\frac{\partial \ell}{\partial \xi} \in \mathfrak{g}$, at $\xi \in \mathfrak{g}$ as the unique algebra element such that $\left\langle \frac{\partial \ell}{\partial \xi}, \eta \right\rangle_e = \mathrm{tr}((\mathbf{J}_\xi \ell)^\top \eta)$ for any $\eta \in \mathfrak{g}$, where $\mathbf{J}_\xi \ell$ denotes the Jacobian matrix of the function $\ell$ with respect to the matrix $\xi$. (Notice that $\mathbf{J}_\xi \ell$ is a formal Jacobian, namely a matrix of partial derivatives with respect to each entry of the matrix $\xi$ without any regard of the internal structure of the matrix $\xi$ itself.)

**Exponential map**: Given a point $g \in \mathbb{G}$ and a tangent vector $v \in T_g\mathbb{G}$, the *exponential* maps $g$ to a point $\exp_g(v)$, namely, it flows the point $g$ along a geodesic line departing from $g$ with initial direction $v$. On a matrix Lie group endowed with the Euclidean metric, it holds that $\exp_g(v) = g\mathrm{Exp}(g^{-1}v)$, where 'Exp' denotes a matrix exponential.

### 2.2. The Euler–Poincaré Equations

The Lagrange–d'Alembert–Pontryagin (LDAP) principle is one of the fundamental concepts in mathematical physics to describe the time-evolution of the state of a physical system and to handle non-conservative external forces. The state-variables of the system are subjected to holonomic constraints, which are embodied in the structure of the state Lie group $\mathbb{G}$. These external forces often arise as control actions designed with the aim of driving the physical system into a predefined state [12]. Let $\Lambda : T\mathbb{G} \to \mathbb{R}$ denote a Lagrangian function and $F : T\mathbb{G} \to T\mathbb{G}$ a generalized force field. (A generalized force field is generally taken as a smooth map from $T\mathbb{G}$ to its dual $T^\star\mathbb{G}$ or, for left-invariant force fields, from an algebra $\mathfrak{g}$ to its dual $\mathfrak{g}^\star$. We adopt a non-standard definition because it eases the notation and is more easily translated into implementation). The LDAP principle affirms that a dynamical system follows a trajectory $g : [a, b] \to \mathbb{G}$ such that:

$$\delta \int_a^b \Lambda(g(t), \dot{g}(t)) \, dt + \int_a^b \langle F(g(t), \dot{g}(t)), \delta g(t) \rangle_{g(t)} \, dt = 0, \tag{1}$$

The leftmost integral is called *action* and the symbol $\delta$ denotes variation, namely the change of the action value from a trajectory $g$ to a trajectory that is infinitely close to $g$, whose point-by-point change is denoted as $\delta g$. The variation *vanishes at endpoints* and is elsewhere *arbitrary*. In the above expression, an over-dot (as in $\dot{g}$) denotes derivation with respect to the parameter $t$. The vanishing of the first term alone is called principle of stationary action. The rightmost integral represents the total work achieved by the force field $F$ due to the variation.

A variational formulation is based on a smooth family of curves $g : U \subset \mathbb{R}^2 \to \mathbb{G}$, where each element is denoted as $g(t, \varepsilon)$. The index $\varepsilon$ selects a curve in the family, and the index $t$ individuates a point over this curve. All the curves in the family depart from the same initial point and arrive at the same endpoint, namely, $g(a, \varepsilon)$ and $g(b, \varepsilon)$ are constant with respect to $\varepsilon$. The variations in (1) are defined as

$$\delta \int_a^b \Lambda(g, \dot{g}) dt := \int_a^b \frac{\partial}{\partial \varepsilon} \Lambda(g(t, \varepsilon), \dot{g}(t, \varepsilon)) \, dt \bigg|_{\varepsilon=0}, \quad \delta g(t) := \frac{\partial g(t, \varepsilon)}{\partial \varepsilon} \bigg|_{\varepsilon=0}. \tag{2}$$

The following result, enunciated directly for matrix Lie groups, is of prime importance, as it relates a variation of velocity to velocity of variation.

**Lemma 1** ([13]). *Given a smooth function $g : U \subset \mathbb{R}^2 \to \mathbb{G}$ on a matrix Lie group, define:*

$$\xi(t, \varepsilon) := g^{-1}(t, \varepsilon) \frac{\partial g(t, \varepsilon)}{\partial t}, \quad \eta(t, \varepsilon) := g^{-1}(t, \varepsilon) \frac{\partial g(t, \varepsilon)}{\partial \varepsilon}. \tag{3}$$

*A variation of a trajectory induces a variation of its velocity field given by*

$$\frac{\partial \xi}{\partial \varepsilon} = \dot{\eta} + \mathrm{ad}_\xi \eta. \tag{4}$$

Assuming that the Lagrangian as well as the generalized force field $F$ are left invariant, we may write $\Lambda(g, \dot{g}) = \ell(g^{-1}\dot{g})$ and $g^{-1}F(g, \dot{g}) = f(g^{-1}\dot{g})$, where $\ell : \mathfrak{g} \to \mathbb{R}$ and $f : \mathfrak{g} \to \mathfrak{g}$ denote a *reduced Lagrangian* and a *reduced force field*, respectively. In addition, if the inner product is left-invariant, it holds that

$$\langle F(g, \dot{g}), \delta g \rangle_g = \langle f(g^{-1}\dot{g}), g^{-1}\delta g \rangle_e. \tag{5}$$

Therefore, the LDAP principle (1) reduces to

$$\delta \int_a^b \ell(g^{-1}\dot{g}) \, dt + \int_a^b \langle f(g^{-1}\dot{g}), g^{-1}\delta g \rangle_e \, dt = 0, \tag{6}$$

where it is legitimate to replace $g^{-1}\dot{g}$ with $\xi$ and $g^{-1}\delta g$ with $\eta$ and then set $\varepsilon$ to 0.

By means of the Lemma 1, the variational formulation of the reduced LDAP principle may be recast in a differential form.

**Theorem 1** ([13]). *Let $\xi := g^{-1}\dot{g}$ and $\eta := g^{-1}\delta g$. The solution of the integral Lagrange–d'Alembert equation (6) under perturbations of the form $\frac{\partial \xi}{\partial \varepsilon} = \dot{\eta} + \mathrm{ad}_\xi \eta$, which vanishes at endpoints, satisfies the Euler–Poincaré equation*

$$\frac{d}{dt} \frac{\partial \ell}{\partial \xi} = \mathrm{ad}_\xi^\star \left( \frac{\partial \ell}{\partial \xi} \right) + f, \tag{7}$$

where ad$^\star$ denotes the adjoint (The adjoint $\omega^\star$ of an operator $\omega : \mathfrak{g} \to \mathfrak{g}$ with respect to an inner product $\langle \cdot, \cdot \rangle$ satisfies $\langle \omega(\xi), \eta \rangle = \langle \xi, \omega^\star(\eta) \rangle$.) of the operator ad with respect to the inner product of $\mathfrak{g}$.

The complete system of differential equations then read

$$\begin{cases} \dot{g} = g\xi, \\ \frac{d}{dt}\frac{\partial \ell}{\partial \xi} = \mathrm{ad}^\star_\xi\left(\frac{\partial \ell}{\partial \xi}\right) + f. \end{cases} \quad (8)$$

The above equations may be used to describe the rotational component of motion of a flying object such as a helicopter or a drone. The forcing term takes into account several *external* driving phenomena, such as:

**Energy dissipation**: Energy dissipation is due, e.g., to friction with air particles. For instance, a linear dissipation term represents aerodynamic drag.

**Control actions**: Other than dissipation (which is often neglected in simplistic models), the forcing term depends on the problem under investigation. It might serve to incorporate into the equations control terms aimed, for instance, at stabilizing the motion or to drive a dynamical system [14].

### 2.3. Particular Case: Euclidean Space

In order to clarify the physical meaning of the Euler–Poincaré equations, let us recall the classical version of these equations for the space $\mathbb{R}^n$, which is also instrumental in describing the translational component of motion of a flying device. The principle (1) on $\mathbb{R}^n$, endowed with the Euclidean inner product, reads:

$$\delta \int_a^b \Lambda(p(t), \dot{p}(t)) \, dt + \int_a^b f(p(t), \dot{p}(t))^\top \delta p(t) \, dt = 0, \quad (9)$$

where $\Lambda : \mathbb{R}^n \times \mathbb{R}^n \to \mathbb{R}$ denotes a Lagrangian function, $p = p(t)$ a trajectory in $\mathbb{R}^n$ and $f : \mathbb{R}^n \times \mathbb{R}^n \to \mathbb{R}^n$ a non-conservative force field. Upon computing the variation, we obtain

$$\int_a^b \left( \left(\frac{\partial \Lambda}{\partial p}\right)^\top \delta p + \left(\frac{\partial \Lambda}{\partial \dot{p}}\right)^\top \delta \dot{p} + f^\top \delta p \right) dt = 0. \quad (10)$$

Integrating by parts the second term and recalling that the variations vanish at the endpoints, we obtain

$$\int_a^b \left( \frac{\partial \Lambda}{\partial p} - \frac{d}{dt}\frac{\partial \Lambda}{\partial \dot{p}} + f \right)^\top \delta p \, dt = 0. \quad (11)$$

Since the variation $\delta p$ is arbitrary, the dynamics of the variable $p$ is governed by the Euler–Lagrange equation

$$\frac{d}{dt}\frac{\partial \Lambda}{\partial \dot{p}} = \frac{\partial \Lambda}{\partial p} + f \quad (12)$$

where the quantity $q := \frac{\partial \Lambda}{\partial \dot{p}}$ is usually termed linear momentum.

## 3. Mathematical Model of a Helicopter

This section introduces a helicopter model based on the Lie group $\mathbb{G} := SO(3)$ of the 3-dimensional rotations $R$.

Since, in the state space $\mathbb{G} := SO(3)$, it holds that $(dL_R)^{-1}(\dot{\xi}) = R\xi$ and $\mathrm{ad}^\star_\xi \eta = -\mathrm{ad}_\xi \eta$ [13], the Euler–Poincaré equations read

$$\begin{cases} \dot{R} = R\xi, \\ \frac{d}{dt}\frac{\partial \ell}{\partial \xi} = -\mathrm{ad}_\xi\left(\frac{\partial \ell}{\partial \xi}\right) + \tau, \end{cases} \quad (13)$$

where $\tau$ denotes the resultant of all external *mechanical torques*. In this context, the state variable $R \in SO(3)$ denotes the *attitude* of a rigid body (i.e., its orientation with respect to a earth-fixed reference frame) and the state-variable $\xi \in \mathfrak{so}(3)$ denotes its *instantaneous angular velocity*. Moreover, the quantity $\mu := \frac{\partial \ell}{\partial \xi}$ represents an angular momentum and the second Euler–Poincaré equation reads $\dot{\mu} = [\mu, \xi] + \tau$, which is a generalization of the well-known angular momentum theorem, where the term $[\mu, \xi]$ represents the inertial torque due to the internal mass unbalance of a body.

It is convenient to define the operator $[\![ \cdot ]\!] : \mathbb{R}^3 \to \mathfrak{g}$ as:

$$x := \begin{bmatrix} x_1 \\ x_2 \\ x_3 \end{bmatrix} \mapsto [\![x]\!] := \begin{bmatrix} 0 & -x_3 & x_2 \\ x_3 & 0 & -x_1 \\ -x_2 & x_1 & 0 \end{bmatrix}. \tag{14}$$

Since any skew-symmetric matrix in $\mathfrak{so}(3)$ may be written as in (14), it is convenient to define a basis of $\mathfrak{so}(3) = \text{span}(\xi_x, \xi_y, \xi_z)$ as follows:

$$\xi_x := \begin{bmatrix} 0 & 0 & 0 \\ 0 & 0 & -1 \\ 0 & 1 & 0 \end{bmatrix}, \xi_y := \begin{bmatrix} 0 & 0 & 1 \\ 0 & 0 & 0 \\ -1 & 0 & 0 \end{bmatrix}, \xi_z := \begin{bmatrix} 0 & -1 & 0 \\ 1 & 0 & 0 \\ 0 & 0 & 0 \end{bmatrix}. \tag{15}$$

In order to shorten some relations, it is also convenient to introduce the *matrix anti-commutator* $\{A, B\} := AB + BA$. Moreover, some relations take advantage of the skew-symmetric projection $\{\!\{ \cdot \}\!\} : \mathbb{R}^{3 \times 3} \to \mathfrak{so}(3)$, defined as $\{\!\{A\}\!\} := \frac{1}{2}(A - A^\top)$. It also pays to define the 'diag' operator as $\text{diag}(a, b, c) := \begin{bmatrix} a & 0 & 0 \\ 0 & b & 0 \\ 0 & 0 & c \end{bmatrix}$.

In the present setting, we equip the algebra $\mathfrak{so}(3)$ with the canonical metric $\langle \xi, \eta \rangle_e := \text{tr}(\xi^\top \eta)$. With this choice, the fiber derivative of a scalar function $\ell : \mathfrak{so}(3) \to \mathbb{R}$ takes a special form.

**Lemma 2** ([15]). *The fiber derivative of a scalar function $\ell : \mathfrak{so}(3) \to \mathbb{R}$ under the canonical metric takes the form*

$$\frac{\partial \ell}{\partial \xi} = \frac{1}{2}(\mathbf{J}_\xi \ell - \mathbf{J}_\xi^\top \ell) \in \mathfrak{so}(3). \tag{16}$$

It is immediate to verify that the fiber derivative corresponds to the orthogonal projection of the Jacobian into the algebra $\mathfrak{g}$, namely $\frac{\partial \ell}{\partial \xi} = \{\!\{\mathbf{J}_\xi \ell\}\!\}$. Moreover, it is convenient to recall a property of the matrix 'trace' operator, namely the cyclic permutation property $\text{tr}(ABC) = \text{tr}(BCA) = \text{tr}(CAB)$ for any square conformable matrices $A, B, C$.

Modeling a complex object to obtain the differential equations that describe its rotational and translational dynamics consists essentially in:

- Defining a Lagrangian function $\ell$ on the basis of the kinetic and potential energy of its components, which accounts for the geometrical and mechanical features of each component;
- Computing the total mechanical torque $\tau$ exerted by the moving parts on the body of the complex object.

These descriptors, for a helicopter, are evaluated in the next subsections.

### 3.1. Model of a Helicopter with a Single Principal Rotor and a Tail Rotor

In order to formalize the behavior of a helicopter into a mathematical model, let us fix an inertial (earth) reference frame $\mathcal{F}_E$. Further, it is necessary to establish a body-fixed reference frame $\mathcal{F}_B$, as shown in Figure 5: the origin of the reference frame $\mathcal{F}_B$ is located at the center of gravity of the helicopter and the three axes coincide with its principal inertia axes. The thrust $\varphi_m$ exerted by the principal rotor appears at the tip of the helicopter's body, which is located along the $z$-axis at a distance $D_m$ from the center of gravity, whereas the thrust $\varphi_t$ exerted by the tail rotor appears at the tail of the helicopter's body, which is located along the $-x$ axis at a distance $D_t$ from the center of gravity.

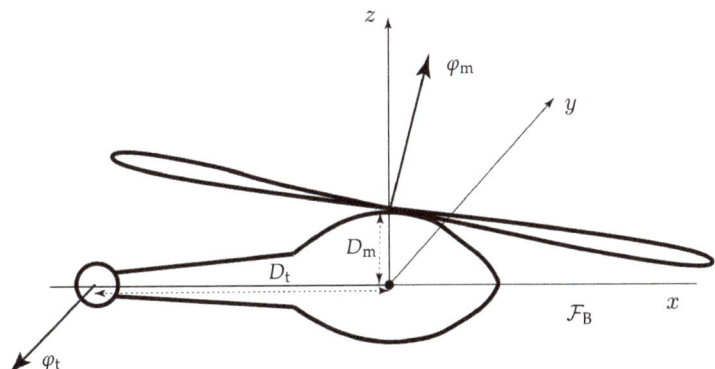

**Figure 5.** Schematic of a helicopter with a principal rotor and a tail rotor (adapted from [12]). (The principal rotor to center of mass distance $D_m$ and the tail rotor to center of mass distance $D_t$ are expressed in meters (m).)

Furthermore, the term $\frac{1}{2}u_m$ represents the intensity of the thrust exerted by the main rotor, while $\frac{1}{2}u_t$ denotes the thrust exerted by the tail rotor, both expressed in Newtons (N). Considering the total thrust $\varphi := \varphi_t + \varphi_m$ as a vector, a *collective control* management of the main rotor results in a change of the thrust intensity exerted, namely a change in $u_m$, whereas a *cyclic control* management changes the direction of the lift exerted, therefore the pitch angle $\alpha_p$ (in radians (rad)) and the sideways roll angle $\alpha_r$ (in radians). The expressions of the thrusts (from [12]) and of their moment arms in the helicopter's body-fixed frame $\mathcal{F}_B$ are given by

$$\varphi_m := \tfrac{1}{2}u_m \begin{bmatrix} \sin\alpha_p \cos\alpha_r \\ -\sin\alpha_r \\ \cos\alpha_p \cos\alpha_r \end{bmatrix}, \; b_m := \begin{bmatrix} 0 \\ 0 \\ D_m \end{bmatrix}, \qquad (17)$$

$$\varphi_t := \tfrac{1}{2}u_t \begin{bmatrix} 0 \\ -1 \\ 0 \end{bmatrix}, \; b_t := \begin{bmatrix} -D_t \\ 0 \\ 0 \end{bmatrix}. \qquad (18)$$

The vector $\frac{2\varphi_m}{u_m}$ may be regarded as the unit normal to the *rotor disk* [2]. A further forcing term to account for the resistance of the air during forward vertical motion is described in Section 3.4. Concerning the thrust generated by the principal rotor, we may notice what follows:

- Whenever $\alpha_r = \alpha_p = 0$, the thrust takes the expression $\tfrac{1}{2}u_m \begin{bmatrix} 0 \\ 0 \\ 1 \end{bmatrix}$, namely, only the z-component is non-null and the thrust is vertical;

- Whenever $\alpha_p = 0$ and $\alpha_r \neq 0$, the thrust takes the expression $\tfrac{1}{2}u_m \begin{bmatrix} 0 \\ -\sin\alpha_r \\ \cos\alpha_r \end{bmatrix}$, namely, the x-component is null and the thrust belongs to the y–z plane, as shown in Figure 6, hence it may only produce a rotation along the x-axis, which corresponds to pure rolling. (*Remark:* The right-hand law defines the positive angle variation.)

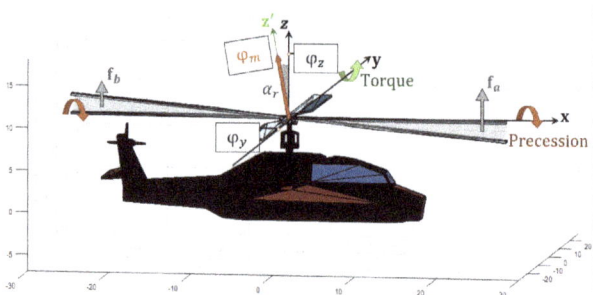

**Figure 6.** Illustration of a positive variation of the angle of attack along the $x$-axis, where $\varphi_y = -\frac{1}{2}u_m \sin \alpha_r$ and $\varphi_z = \frac{1}{2}u_m \cos \alpha_r$ are the projections of the thrust vector $\varphi_m$ along the $y$- and $z$-axis, respectively. The maneuver to control rolling assigns to the blades an angle such that a greater amount of force is produced in the positive $x$-axis, namely $f_a$, with respect to the force in the negative $x$-axis, namely $f_b$. Those two vector forces produce a torque and a precession rotation due to the gyroscopic effect.

- whenever $\alpha_r = 0$ and $\alpha_p \neq 0$, the thrust takes the expression $\frac{1}{2}u_m \begin{bmatrix} \sin \alpha_p \\ 0 \\ \cos \alpha_p \end{bmatrix}$, namely,

the $y$-component is null and the thrust belongs to the $x$–$z$ plane, as shown in Figure 7, hence it may only produce a rotation along the $y$-axis, which corresponds to pure pitching.

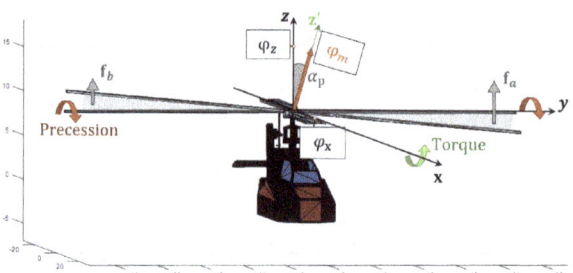

**Figure 7.** Illustration of a positive variation of the angle of attack along the $y$-axis, where $\varphi_x = \frac{1}{2}u_m \sin \alpha_p$ and $\varphi_z = \frac{1}{2}u_m \cos \alpha_p$ are the projections of the main rotor thrust $\varphi_m$ to the $x$- and $z$-axis, respectively. The maneuver to control the pitching assigns to the blades an angle such that a larger amount of force is produced in the positive $y$-axis, namely $f_a$, with respect to the force along the negative $y$-axis, namely $f_b$. Those two vector forces produce a mechanical torque and a precession of the fuselage.

Notice that the inclination of the blades influences the thrust and the torque acting on the fuselage, but does not influence directly the roll and the pitch attitude of the helicopter. Further, notice that the thrust $\frac{1}{2}u_m$ does not distribute equally across the three directions of space and, in particular, that a change in the angles of attack of the blades weakens the vertical component of the thrust: when a helicopter tilts, it tends to fall, unless the thrust is compensated by the pilot. It is also worth noticing that the total thrust $\varphi$ acting on the fuselage has a $y$ component that depends on the tail rotor thrust. This component causes the translation of the helicopter in the direction of $\varphi_t$: this is called drift effect (or translation tendency). The mechanical torque exerted by the two rotors on the helicopter's fuselage, expressed in N·m, is termed *active torque* and is given by

$$\tau_{\mathcal{A}} := \tfrac{1}{2}\left(\varphi_m b_m^\top - b_m \varphi_m^\top + \varphi_t b_t^\top - b_t \varphi_t^\top\right)$$
$$= \frac{1}{2}\begin{bmatrix} 0 & -D_t u_t & D_m u_m \sin\alpha_p \cos\alpha_r \\ D_t u_t & 0 & -D_m u_m \sin\alpha_r \\ -D_m u_m \sin\alpha_p \cos\alpha_r & D_m u_m \sin\alpha_r & 0 \end{bmatrix}. \quad (19)$$

The mechanical torque due to the drag of the principal rotor, namely the resultant of the torque that tends to make the helicopter spin as a counter-reaction to the spinning of the rotor, expressed in N·m, may be quantify by

$$\tau_{\mathcal{D}} := -\tfrac{1}{2}\gamma u_m \tilde{\zeta}_z, \quad (20)$$

where $\gamma > 0$ is termed *air drag* coefficient (whose measurement unit is meters) and represents the efficacy with which the air surrounding the helicopter pushes the rotor as a reaction of its spinning. According to the canonical basis (15), the total mechanical torque $\tau := \tau_{\mathcal{A}} + \tau_{\mathcal{D}}$ may be decomposed as $\tau = \tau_x \tilde{\zeta}_x + \tau_y \tilde{\zeta}_y + \tau_z \tilde{\zeta}_z$, with

$$\begin{cases} \tau_x = \tfrac{1}{2} D_m u_m \sin\alpha_r, \\ \tau_y = \tfrac{1}{2} D_m u_m \sin\alpha_p \cos\alpha_r, \\ \tau_z = \tfrac{1}{2} D_t u_t - \tfrac{1}{2}\gamma u_m. \end{cases} \quad (21)$$

The component $\tau_x$ is responsible for the rolling of the helicopter (plane y–z), the component $\tau_y$ is responsible for the pitching of the helicopter (plane x–z). The component $\tau_z$ is responsible for the control of the yawing of the helicopter (plane x–y): to prevent the spinning of the aircraft, it is necessary to control the thrust $u_t$ of the tail rotor in such a way that $D_t u_t - \gamma u_m \approx 0$. During hovering, the vertical component of the total thrust needs to balance the weight force of the helicopter. A further torque component is introduced in Section 3.3 to account for friction-type resistance during fast yawing. According to the specialized literature (see, e.g., [16]), the maximum value of the thrust $u_m$ of the main rotor (in Newtons) may be computed by the expression

$$u_m := \tfrac{1}{2} C_u \rho A (l_\mathcal{R} \Omega_m)^2, \quad (22)$$

where $C_u$ is a (dimensionless) thrust coefficient that represents the efficiency of the rotor, $\rho$ represents the density of the air at a given temperature and altitude in kg·m$^{-3}$, $A$ denotes the area of the rotor disk, in m$^2$, which contributes to generating the thrust, $l_\mathcal{R}$ represents the radius of the rotor disk (namely, the length of each blade) in meters and $\Omega_m$ denotes the angular velocity of the rotor in rad·s$^{-1}$. In fact, the product $l_\mathcal{R}\Omega_m$ denotes the tip velocity of a blade. Such thrust may be expressed compactly as a quadratic function of the rotor speed as $u_m = \beta_u \Omega_m^2$. Further, the mechanical power (in Watts) that the engine transfers to the rotor is given by

$$w := \tfrac{1}{2} C_w \rho A (l_\mathcal{R} \Omega_m)^2 \Omega_m, \quad (23)$$

where $C_w$ denotes a (dimensionless) power coefficient. Such power may be expressed as a cubic function of the rotor speed, namely $w = \beta_w \Omega_m^3$. The main rotor disk area $A$ changes its value thanks to *collective control* and consequently to $\alpha_c$. In fact, such value is related to the portion of each blade that pushes the helicopter, for instance, upward. In order to describe correctly the area of the disk that contributes to generating thrust, it is assumed that $A = \pi l_\mathcal{R}^2 \sin\alpha_c$; therefore, if the blades are considered with no thickness, no built-in twists and to be perfectly horizontal, namely in the earth inertial reference's x–y plane, then when $\alpha_c = 0$ the helicopter has no thrust. Instead, when all blades take an angle of attack $\alpha_c > 0$ the thrust is no longer null and the turning of the blades produces a vertical thrust that tends to counteract the helicopter's weight force. The Equation (22) becomes:

$$u_m := \tfrac{1}{2} C_u \rho \pi l_\mathcal{R}^4 \Omega_m^2 \sin\alpha_c, \quad (24)$$

with $\alpha_c \in [\alpha_{c,\min}, \alpha_{c,\max}]$. The minimum and the maximum value of the thrust depend on the range of the angle of attack of the principal rotors blades, whereas the range of the angle of attack is related to the shape and the built-in twist of the blades, besides the swash-plate rods mobility. The power coefficient $C_w$ is related to the thrust coefficient $C_u$ by the relationship

$$C_w = \frac{C_u^{3/2}}{\sqrt{2}}. \tag{25}$$

The mechanical power $w$ absorbed by the helicopter's engine at the reference speed of 100% is usually provided by data-sheets. Considering $w$ as known, it is possible to calculate the power and the thrust coefficients, that otherwise would have to be measured through experiments. The value of the first coefficient, following the Equation (23), is

$$C_w = \frac{2w}{\rho A (l_\mathcal{R} \Omega_m)^2 \Omega_m} \tag{26}$$

Consequently it is possible to find the value of $C_u$ using the Equation (25). The expression (24) holds for the main rotor, while a similar expression may describe the thrust exerted by the tail rotor. The equation below is based on tail rotor characteristics:

$$u_t := \tfrac{1}{2} C_u^T \rho \pi l_\mathcal{T}^4 \Omega_t^2 \sin \alpha_c^T, \tag{27}$$

where $C_u^T$ denotes the thrust coefficient of the tail rotor and $l_\mathcal{T}$ denotes the length of the tail rotor's blades. The drag coefficient is generally unknown, but it is possible to estimate its value by assuming that the helicopter hovering and that the mechanical torque of the tail rotor balances the undesired drag torque, which would tend to make the helicopter yaw. Indeed, in hovering condition, with the tail rotor's blades collective angle at a value set to a half of its interval range, namely $\alpha_{c,\mathrm{mid}}^T := \frac{\alpha_{c,\min}^T + \alpha_{c,\max}^T}{2}$ (see Table 1), and at 100% of the tail rotor speed, the helicopter should have no yawing. The drag coefficient could hence be determined by imposing the condition $e_z^\top \tau = 0$, where $e_z := [0\ 0\ 1]^\top$, which leads to the expression

$$\gamma = D_t \frac{C_u l_\mathcal{R}^4 \Omega_m^2 \sin \alpha_c}{C_u^T l_\mathcal{T}^4 \Omega_t^2 \sin \alpha_{c,\mathrm{mid}}^T}. \tag{28}$$

The numerical values of these (as well as others) parameters will be specified in Section 5.

**Table 1.** Tail rotor collective angle range, tail and main rotors weight, speed ([1], pages 303, 254 and 157), tail rotor speed ([17], page 3) and cycling angle range ([18], page 11).

|  | Weight | Speed 100% | Collective Angle | Cyclic Angle | |
|---|---|---|---|---|---|
|  | [kg] | [RPM] | min ÷ max [deg] | Longitudinal [deg] | Lateral [deg] |
| Main rotor | 277.2 [1] | 395 | 11 ÷ 31 [2] | −21.8 ÷ 21.8 | −15 ÷ 15 |
| Tail rotor | 8.2 | 3584 | −16.8 ÷ 34.2 | − | − |

[1] The main rotor weight is the result of the addition of various components that compose the entire main rotor. These values have been taken from [19] (page 3) which is the technical data-sheet of the helicopter AS350B3 also known as H125, that is the lower level helicopter by the same manufacturer. The values taken have not been modified because the model is supposed to be similar. The final weight is calculated by the sum of: anti-vibration device (28.4 kg), main rotor mast (55.7 kg), rotor hub (57.5 kg) and four blades (4 × 33.9 kg = 135.6 kg). [2] As stated in [20] (page 57) the value of the collective angle could vary in the range −5 ÷ 15 degrees and the negative angle could be necessary to achieve zero lift if blades have a built-in axial twist. From Reference [1] (page 200), we know that the EC135 P2+ helicopter has a positive twist of 16 degrees in the region where the pitch control cuff joins the airfoil section. This provides the airfoil section with a corresponding preset pitch angle. Using the Equation (24), the collective angles range becomes 11 ÷ 31 degrees, the minimum angle and the maximum angle to modify the intensity of the thrust generated, respectively.

## 3.2. Lagrangian Function Associated to the Helicopter Model

To complete the present description of a helicopter motion dynamics, it is necessary to write explicitly the Lagrangian function of a helicopter, which coincides with its kinetic energy minus its potential energy, both expressed in the inertial reference frame $\mathcal{F}_E$.

**Kinetic energy of the fuselage**: The position of the center of gravity of the helicopter in the inertial reference frame $\mathcal{F}_E$ at time $t$ is denoted as $p(t)$. The position of each infinitesimal volume of the body (fuselage) in the body-fixed frame $\mathcal{F}_B$ is denoted by $s$. Since the helicopter's fuselage is rigid, the position of each volume element, at time $t$, is $p(t) + R(t)s$, where $R(t) \in SO(3)$ denotes a rotation matrix that takes the body-fixed frame $\mathcal{F}_B$ to coincide with $\mathcal{F}_E$. The kinetic energy of the helicopter's body $\mathcal{B}$ with respect to the inertial reference frame $\mathcal{F}_E$ may be written as

$$\ell_\mathcal{B} := \frac{1}{2}\int_\mathcal{B} \left\|\frac{d(p+Rs)}{dt}\right\|^2 dm = \frac{1}{2}\int_\mathcal{B} \|\dot{p} + \dot{R}s\|^2 dm, \tag{29}$$

where $dm$ denotes the mass content of each infinitesimal volume. Recalling that $\dot{R} = R\hat{\zeta}$, with $\hat{\zeta} \in \mathfrak{so}(3)$, we get:

$$\begin{aligned}\ell_\mathcal{B} &= \frac{1}{2}\int_\mathcal{B} \operatorname{tr}((\dot{p}+\dot{R}s)(\dot{p}+\dot{R}s)^\top) dm = \frac{1}{2}\int_\mathcal{B} \operatorname{tr}(\dot{p}\dot{p}^\top + R\hat{\zeta}ss^\top\hat{\zeta}^\top R^\top + 2\dot{p}s^\top\dot{R}) dm \\ &= \tfrac{1}{2}M_\mathcal{B}\|\dot{p}\|^2 + \tfrac{1}{2}\operatorname{tr}(R\hat{\zeta}\hat{J}_\mathcal{B}\hat{\zeta}^\top R^\top) + M_\mathcal{B}\operatorname{tr}(\dot{p}c_\mathcal{B}^\top \dot{R}),\end{aligned} \tag{30}$$

where the cancellation is due to the cyclic permutation property of the trace operator and to the defining property of rotations ($R^\top R = I_3$). The constant quantities that appear in the expression (30) are defined as follows

$$M_\mathcal{B} := \int_\mathcal{B} dm > 0,\ c_\mathcal{B} := \frac{1}{M_\mathcal{B}}\int_\mathcal{B} s\, dm \in \mathbb{R}^3,\ \hat{J}_\mathcal{B} := \int_\mathcal{B} ss^\top dm \in \mathbb{R}^{3\times 3}. \tag{31}$$

The quantity $M_\mathcal{B}$ denotes the total mass of the helicopter's fuselage. The matrix $\hat{J}_\mathcal{B}$ denotes a *non-standard inertia tensor* [21]. The *standard inertia tensor* of the helicopter's body is defined as

$$J_\mathcal{B} := \int_\mathcal{B} [\![s]\!][\![s]\!]^\top dm. \tag{32}$$

(Refer to (14) for this notation.) These inertia tensors are related by the following result:

**Lemma 3** ([21]). *The non-standard moment of inertia $\hat{J}$ of a body is related to its standard moment of inertia $J$ by the relationship $\hat{J} = \tfrac{1}{2}\operatorname{tr}(J)I_3 - J$.*

The standard and non-standard moment of inertia constitute two different ways of quantifying the inertia of a rigid body and differ only by their trace. Their difference is particularly evident in bodies with symmetries, as the ones treated within the present exposition.

Assuming that the shape of the fuselage may be assimilated to an ellipsoid, its standard inertial tensor takes the form:

$$J_\mathcal{B} = \begin{bmatrix} \frac{M_\mathcal{B}(b^2+c^2)}{5} & 0 & 0 \\ 0 & \frac{M_\mathcal{B}(a^2+c^2)}{5} & 0 \\ 0 & 0 & \frac{M_\mathcal{B}(a^2+b^2)}{5} \end{bmatrix}, \tag{33}$$

where $a, b, c$ denote the semi-axes lengths ($a$ refers to the $x$-axis, $b$ refers to the $y$-axis and $c$ refers to the $z$-axis). The non-standard inertial tensor of the fuselage reads

$$\hat{J}_B = \begin{bmatrix} \frac{M_B a^2}{5} & 0 & 0 \\ 0 & \frac{M_B b^2}{5} & 0 \\ 0 & 0 & \frac{M_B c^2}{5} \end{bmatrix}. \tag{34}$$

Since the origin of the reference frame $\mathcal{F}_B$ coincides with the center of gravity of the aircraft, not of the fuselage alone, in general it holds that the center of mass of the fuselage $c_B \neq 0$, therefore

$$\ell_B = \tfrac{1}{2} M_B \|\dot{p}\|^2 + \tfrac{1}{2} \mathrm{tr}(\xi \hat{J}_B \xi^\top) + M_B \mathrm{tr}(\dot{p} c_B^\top \xi^\top R^\top). \tag{35}$$

**Kinetic energy of the principal rotor**: The position of the center of gravity of the principal rotor with respect to the reference frame $\mathcal{F}_B$ is individuated by the vector $b_m$ defined in (17). A reference frame $\mathcal{F}_R$ whose $z$-axis coincides with the $z$-axis of the reference frame $\mathcal{F}_B$ is associated with the rotor. Hence the position of each volume element in the principal rotor $\mathcal{R}$ at time $t$ in the inertial reference frame $\mathcal{F}_E$ takes the expression $p(t) + R(t)(b_m + R_m(t)s)$, where $R_m \in \mathrm{SO}(3)$ denotes the instantaneous orientation matrix of the principal rotor (a rotation that aligns the rotor-fixed reference frame $\mathcal{F}_R$ to the body-fixed reference frame $\mathcal{F}_B$) and $s$ denotes the position of a point of the rotor in a rotor-fixed reference frame. The matrix $R_m$ represents a rotation about the $z$-axis of the reference frame $\mathcal{F}_R$, hence it takes the form $\begin{bmatrix} \cos\theta_m & -\sin\theta_m & 0 \\ \sin\theta_m & \cos\theta_m & 0 \\ 0 & 0 & 1 \end{bmatrix}$, therefore $\dot{R}_m = \zeta_m R_m$, where $\zeta_m = \Omega_m \zeta_z$ and $\theta_m$ indicates the rotation angle of the main rotor. The time-derivative of the position of each volume element is

$$\tfrac{d}{dt}[p + R(b_m + R_m s)] = \dot{p} + \dot{R}(b_m + R_m s) + R\dot{R}_m s = \dot{p} + R\xi b_m + R(\xi + \zeta_m)R_m s. \tag{36}$$

The angular velocity matrix $\zeta_m \in \mathfrak{so}(3)$ of the principal rotor is controlled by the pilot and is hence a known quantity (although, as already underlined, most helicopters are designed to keep a fixed rotor speed). The kinetic energy per mass element $dm$ of the principal rotor $\mathcal{R}$ may be written as

$$\tfrac{1}{2} \mathrm{tr}([\dot{p} + \dot{R} b_m + R(\xi + \zeta_m) R_m s][\dot{p} + \dot{R} b_m + R(\xi + \zeta_m) R_m s]^\top) = \\ \tfrac{1}{2} \|\dot{p}\|^2 + \tfrac{1}{2} \mathrm{tr}(R\xi b_m b_m^\top \xi^\top R^\top) + \tfrac{1}{2} \mathrm{tr}(R(\xi + \zeta_m) R_m s s^\top R_m^\top (\xi + \zeta_m)^\top R^\top) + \\ \mathrm{tr}(\dot{p} b_m^\top \xi^\top R^\top) + \mathrm{tr}(\dot{p} s^\top R_m^\top (\xi + \zeta_m)^\top R^\top) + \mathrm{tr}(R\xi b_m s^\top R_m^\top (\xi + \zeta_m)^\top R^\top). \tag{37}$$

The kinetic energy of the principal rotor $\mathcal{R}$ in the earth frame $\mathcal{F}_E$ may thus be written as

$$\ell_\mathcal{R} = \tfrac{1}{2} M_\mathcal{R} \|\dot{p}\|^2 + \tfrac{1}{2} M_\mathcal{R} \mathrm{tr}(\xi b_m b_m^\top \xi^\top) + \tfrac{1}{2} \mathrm{tr}((\xi + \zeta_m) R_m \hat{J}_\mathcal{R} R_m^\top (\xi + \zeta_m)^\top) + \\ M_\mathcal{R} \mathrm{tr}(\dot{p} b_m^\top \xi^\top R^\top) + M_\mathcal{R} \mathrm{tr}(\dot{p} c_\mathcal{R}^\top R_m^\top (\xi + \zeta_m)^\top R^\top) + M_\mathcal{R} \mathrm{tr}(\xi b_m c_\mathcal{R}^\top R_m^\top (\xi + \zeta_m)^\top), \tag{38}$$

where

$$M_\mathcal{R} := \int_\mathcal{R} dm > 0, \ \hat{J}_\mathcal{R} := \int_\mathcal{R} s s^\top dm \in \mathbb{R}^{3\times 3} \text{ and } c_\mathcal{R} := \frac{1}{M_\mathcal{R}} \int_\mathcal{R} s\, dm \in \mathbb{R}^3. \tag{39}$$

In order to simplify the expression (38), we may assume that the principal rotor is perfectly symmetric about its center of mass, which implies that $c_\mathcal{R} = 0$. Moreover, we may assume that the principal rotor may be schematized as two rods of mass $\tfrac{1}{2} M_\mathcal{R}$ each and length $2\,l_\mathcal{R}$, one along the $x$ axis and one along the $y$-axis, spinning around the $z$-axis, therefore:

$$J_\mathcal{R} = \begin{bmatrix} j_\mathcal{R} & 0 & 0 \\ 0 & j_\mathcal{R} & 0 \\ 0 & 0 & 2j_\mathcal{R} \end{bmatrix} \text{ that is } \hat{J}_\mathcal{R} = j_\mathcal{R}\text{diag}(1,1,0), \tag{40}$$

by Lemma 3, with $j_\mathcal{R} := \frac{1}{12}\frac{M_\mathcal{R}}{2}(2l_\mathcal{R})^2 = \frac{1}{6}M_\mathcal{R} l_\mathcal{R}^2$. (Refer to the beginning of the present section for the notation used.) A consequence is that the expression $R_\text{m}\hat{J}_\mathcal{R} R_\text{m}^\top$ simplifies to $\hat{J}_\mathcal{R}$; therefore, the kinetic energy of the principal rotor is given by

$$\ell_\mathcal{R} = \tfrac{1}{2}M_\mathcal{R}\|\dot{p}\|^2 + \tfrac{1}{2}M_\mathcal{R}\text{tr}(\xi b_\text{m} b_\text{m}^\top \xi^\top) + \tfrac{1}{2}\text{tr}((\xi+\xi_\text{m})\hat{J}_\mathcal{R}(\xi+\xi_\text{m})^\top) + M_\mathcal{R}\text{tr}(\dot{p}b_\text{m}^\top \xi^\top R^\top). \tag{41}$$

Rearranging these terms shows that the kinetic energy of the principal rotor may be written equivalently as the quadratic form

$$\ell_\mathcal{R} = \tfrac{1}{2}M_\mathcal{R}\|\dot{p} + R\xi b_\text{m}\|^2 + \tfrac{1}{2}\text{tr}((\xi+\xi_\text{m})\hat{J}_\mathcal{R}(\xi+\xi_\text{m})^\top), \tag{42}$$

where the first term represents the translational kinetic energy of the center of mass of the principal rotor in the reference system $\mathcal{F}_\text{E}$, whereas the second term represents the rotational kinetic energy of the principal rotor in the reference system $\mathcal{F}_\text{E}$.

**Kinetic energy of the tail rotor**: The position of the tail rotor with respect to the reference frame $\mathcal{F}_\text{B}$ is individuated by the vector $b_\text{t}$ defined in (18), hence the position of each point in the tail rotor $\mathcal{T}$ at time $t$ is given by $p(t) + R(t)(b_\text{t} + R_\text{t}(t)s)$, where $R_\text{t} \in SO(3)$ denotes the instantaneous orientation matrix of the tail rotor with respect to a body-fixed reference frame $\mathcal{F}_\text{B}$ and $s$ denotes the position of a point of the tail rotor in a rotor-fixed reference frame. In this case, it holds that

$$\tfrac{d}{dt}[p + R(b_\text{t} + R_\text{t}s)] = \dot{p} + \dot{R}(b_\text{t} + R_\text{t}s) + R\dot{R}_\text{t}s = \dot{p} + R\xi b_\text{t} + R(\xi+\xi_\text{t})R_\text{t}s, \tag{43}$$

where $\dot{R}_\text{t} = \xi_\text{t} R_\text{t}$. The angular velocity matrix $\xi_\text{t} \in \mathfrak{so}(3)$ of the principal rotor is controlled by the pilot and is hence to be held as a known quantity. Since the instantaneous axis of rotation of the tail rotor is fixed and coincides to the $-y$ axis, the angular matrix $\xi_\text{t}$ takes the explicit expression

$$\xi_\text{t} := -\Omega_\text{t}\xi_y = \begin{bmatrix} 0 & 0 & -\Omega_\text{t} \\ 0 & 0 & 0 \\ \Omega_\text{t} & 0 & 0 \end{bmatrix}, \tag{44}$$

where $\Omega_\text{t}$ denotes the instantaneous rotation speed of the tail rotor.

The kinetic energy of the tail rotor $\mathcal{T}$ in the earth frame $\mathcal{F}_\text{E}$ has an expression which is derived in a similar manner to (38) and may be written as

$$\ell_\mathcal{T} = \tfrac{1}{2}M_\mathcal{T}\|\dot{p}\|^2 + \tfrac{1}{2}M_\mathcal{T}\text{tr}(\xi b_\text{t} b_\text{t}^\top \xi^\top) + \tfrac{1}{2}\text{tr}((\xi+\xi_\text{t})R_\text{t}\hat{J}_\mathcal{T} R_\text{t}^\top (\xi+\xi_\text{t})^\top) + \\ M_\mathcal{T}\text{tr}(\dot{p}b_\text{t}^\top \xi^\top R^\top) + M_\mathcal{T}\text{tr}(\dot{p}c_\mathcal{T}^\top R_\text{t}^\top (\xi+\xi_\text{t})R^\top) + M_\mathcal{T}\text{tr}(\xi b_\text{t} c_\mathcal{T}^\top R_\text{t}^\top (\xi+\xi_\text{t})^\top), \tag{45}$$

where

$$M_\mathcal{T} := \int_\mathcal{T} dm > 0,\ \hat{J}_\mathcal{T} := \int_\mathcal{T} ss^\top dm \in \mathbb{R}^{3\times 3} \text{ and } c_\mathcal{T} := \frac{1}{M_\mathcal{T}}\int_\mathcal{T} s\,dm \in \mathbb{R}^3. \tag{46}$$

In order to simplify the expression (45), we may assume that the tail rotor is perfectly symmetric about its own center of mass $c_\mathcal{T}$, which implies that $c_\mathcal{T} = 0$. Moreover, we assume that the tail rotor may be schematized as a full disk of mass $M_\mathcal{T}$ and radius $l_\mathcal{T}$, laying over the $x$–$z$ plane, spinning around the $y$-axis, namely that

$$J_\mathcal{T} = \begin{bmatrix} j_\mathcal{T} & 0 & 0 \\ 0 & 2j_\mathcal{T} & 0 \\ 0 & 0 & j_\mathcal{T} \end{bmatrix}, \text{ that is, } \hat{J}_\mathcal{T} = j_\mathcal{T}\text{diag}(1,0,1), \tag{47}$$

by Lemma 3, with $j_\mathcal{T} := \frac{1}{4} M_\mathcal{T} l_\mathcal{T}^2$. Since

$$R_t = \begin{bmatrix} \cos\theta_t & 0 & -\sin\theta_t \\ 0 & 1 & 0 \\ \sin\theta_t & 0 & \cos\theta_t \end{bmatrix}, \quad (48)$$

direct calculations show that $R_t \hat{J}_\mathcal{T} R_t^\top = \hat{J}_\mathcal{T}$; therefore, the kinetic energy of the tail rotor is given by

$$\ell_\mathcal{T} = \tfrac{1}{2} M_\mathcal{T} \|\dot{p}\|^2 + \tfrac{1}{2} M_\mathcal{T} \mathrm{tr}(\xi b_t b_t^\top \xi^\top) + \tfrac{1}{2} \mathrm{tr}((\xi + \xi_t) \hat{J}_\mathcal{T} (\xi + \xi_t)^\top) + M_\mathcal{T} \mathrm{tr}(\dot{p} b_t^\top \xi^\top R^\top). \quad (49)$$

Rearranging terms shows that the kinetic energy of the tail rotor may be written equivalently as

$$\ell_\mathcal{T} = \tfrac{1}{2} M_\mathcal{T} \|\dot{p} + R\xi b_t\|^2 + \tfrac{1}{2} \mathrm{tr}((\xi + \xi_t) \hat{J}_\mathcal{T} (\xi + \xi_t)^\top), \quad (50)$$

where the first term represents the translational kinetic energy of the center of mass of the tail rotor and the second term represents the rotational kinetic energy of the tail rotor, both expressed in the reference frame $\mathcal{F}_E$.

**Potential energy associated with a helicopter model**: The potential energy associated with the helicopter is $(M_\mathcal{B} + M_\mathcal{R} + M_\mathcal{T}) \bar{g} e_z^\top p$, where the scalar $\bar{g}$ denotes gravitational acceleration.

**Lagrangian function associated with a helicopter model**: The Lagrangian function associated with a helicopter model is hence obtained by gathering the kinetic energies (35), (41), (49) and the potential energy and defining the total Lagrangian as

$$\begin{aligned}
\ell_\mathcal{H} &:= \ell_\mathcal{B} + \ell_\mathcal{R} + \ell_\mathcal{T} - (M_\mathcal{B} + M_\mathcal{R} + M_\mathcal{T}) \bar{g} e_z^\top p \\
&= \tfrac{1}{2} M_\mathcal{B} \|\dot{p}\|^2 + \tfrac{1}{2} \mathrm{tr}(\xi \hat{J}_\mathcal{B} \xi^\top) + M_\mathcal{B} \mathrm{tr}(\dot{p} c_\mathcal{B}^\top \xi^\top R^\top) + \\
&\quad \tfrac{1}{2} M_\mathcal{R} \|\dot{p}\|^2 + \tfrac{1}{2} M_\mathcal{R} \mathrm{tr}(\xi b_m b_m^\top \xi^\top) + \tfrac{1}{2} \mathrm{tr}((\xi + \xi_m) \hat{J}_\mathcal{R} (\xi + \xi_m)^\top) + M_\mathcal{R} \mathrm{tr}(\dot{p} b_m^\top \xi^\top R^\top) + \\
&\quad \tfrac{1}{2} M_\mathcal{T} \|\dot{p}\|^2 + \tfrac{1}{2} M_\mathcal{T} \mathrm{tr}(\xi b_t b_t^\top \xi^\top) + \tfrac{1}{2} \mathrm{tr}((\xi + \xi_t) \hat{J}_\mathcal{T} (\xi + \xi_t)^\top) + M_\mathcal{T} \mathrm{tr}(\dot{p} b_t^\top \xi^\top R^\top) - \\
&\quad (M_\mathcal{B} + M_\mathcal{R} + M_\mathcal{T}) \bar{g} e_z^\top p.
\end{aligned}$$

The expression of the Lagrangian $\ell_\mathcal{H}$ contains several similar terms and may be rewritten compactly as

$$\begin{aligned}
\ell_\mathcal{H} &= \tfrac{1}{2} M_\mathcal{H} \|\dot{p}\|^2 + \tfrac{1}{2} \mathrm{tr}(\xi \hat{J}_\mathcal{H} \xi^\top) + M_\mathcal{H} \mathrm{tr}(\dot{p} c_\mathcal{H}^\top \xi^\top R^\top) + \\
&\quad \tfrac{1}{2} \mathrm{tr}((\xi + \xi_m) \hat{J}_\mathcal{R} (\xi + \xi_m)^\top) + \tfrac{1}{2} \mathrm{tr}((\xi + \xi_t) \hat{J}_\mathcal{T} (\xi + \xi_t)^\top) - M_\mathcal{H} \bar{g} e_z^\top p,
\end{aligned} \quad (51)$$

where the following placeholders have been made use of

$$M_\mathcal{H} := M_\mathcal{B} + M_\mathcal{R} + M_\mathcal{T}, \quad \hat{J}_\mathcal{H} := \hat{J}_\mathcal{B} + M_\mathcal{R} b_m b_m^\top + M_\mathcal{T} b_t b_t^\top, \quad c_\mathcal{H} := \tfrac{1}{M_\mathcal{H}} (M_\mathcal{B} c_\mathcal{B} + M_\mathcal{R} b_m + M_\mathcal{T} b_t). \quad (52)$$

Since the origin of the body-fixed reference frame was taken at the center of gravity of the helicopter, it holds that $c_\mathcal{H} = 0$, therefore the helicopter's Lagrangian takes the final expression

$$\ell_\mathcal{H}(\dot{p}, \xi, p) = \tfrac{1}{2} M_\mathcal{H} \|\dot{p}\|^2 - \tfrac{1}{2} \mathrm{tr}(\hat{J}_\mathcal{H} \xi^2) - \tfrac{1}{2} \mathrm{tr}(\hat{J}_\mathcal{R} (\xi + \xi_m)^2) - \tfrac{1}{2} \mathrm{tr}(\hat{J}_\mathcal{T} (\xi + \xi_t)^2) - M_\mathcal{H} \bar{g} e_z^\top p, \quad (53)$$

where we have used the Lie-algebra property that $\xi^\top = -\xi$ and the cyclic permutation property of the trace operator. The Lagrangian (53) is a function of the variables $\dot{p}$, $\xi$ and $p$.

### 3.3. Rotational Component of Motion

The rotational component of motion, which governs the evolution of the Lie-algebra variable $\xi$, is described by the Euler–Poincaré equations (13) applied to the Lagrangian function (53) and to the rotors-generated mechanical torque (19).

As a first step in the determination of a Lie-group differential description of the rotational component of motion, it is necessary to compute the fiber derivative of the Lagrangian $\ell_\mathcal{H}$. The Jacobian of the Lagrangian at a point $\xi$ may be computed easily by the property:

$$\ell_\mathcal{H}(\xi + \Delta\xi) - \ell_\mathcal{H}(\xi) = \mathrm{tr}(\Delta\xi^\top \mathbf{J}_\xi \ell_\mathcal{H}) + \text{higher-order terms in } \Delta\xi, \tag{54}$$

where $\Delta\xi$ denotes an arbitrary perturbation. It is essential to recall that, while evaluating the Jacobian, the matrix $\xi$ is to be considered as unconstrained (namely, not an element of $\mathfrak{g}$). Straightforward calculations yield

$$\mathbf{J}_\xi \ell_\mathcal{H} = -\frac{1}{2}\left(\{\xi, \hat{J}_\mathcal{H}\}^\top + \{\xi + \xi_m, \hat{J}_\mathcal{R}\}^\top + \{\xi + \xi_t, \hat{J}_\mathcal{T}\}^\top\right). \tag{55}$$

Plugging the above expression into the relation (16) and recalling that inertia tensors are symmetric matrices, one gets the angular momentum

$$\frac{\partial \ell_\mathcal{H}}{\partial \xi} = \{\{\mathbf{J}_\xi \ell_\mathcal{H}\}\} = \frac{1}{2}(\{\xi, \hat{J}_\mathcal{H}\} + \{\xi + \xi_m, \hat{J}_\mathcal{R}\} + \{\xi + \xi_t, \hat{J}_\mathcal{T}\}). \tag{56}$$

It pays to recall that the anti-commutator is a bilinear form, hence, upon defining

$$\hat{J}_\mathcal{H}^\star := \hat{J}_\mathcal{H} + \hat{J}_\mathcal{R} + \hat{J}_\mathcal{T}, \tag{57}$$

the angular momentum (56) may be simplified to

$$\mu := \frac{\partial \ell_\mathcal{H}}{\partial \xi} = \frac{1}{2}(\{\xi, \hat{J}_\mathcal{H}^\star\} + \{\xi_m, \hat{J}_\mathcal{R}\} + \{\xi_t, \hat{J}_\mathcal{T}\}). \tag{58}$$

The angular momentum $\mu$ represents the 'quantity of rotational motion' of the helicopter as it is proportional to the inertia and to the rotational speed of its components. The time-derivative of the angular momentum may be rewritten as

$$\dot{\mu} = \frac{d}{dt}\frac{\partial \ell_\mathcal{H}}{\partial \xi} = \frac{1}{2}(\{\dot{\xi}, \hat{J}_\mathcal{H}^\star\} + \{\dot{\xi}_m, \hat{J}_\mathcal{R}\} + \{\dot{\xi}_t, \hat{J}_\mathcal{T}\}), \tag{59}$$

and direct calculations lead to

$$-\mathrm{ad}_\xi\left(\frac{\partial \ell_\mathcal{H}}{\partial \xi}\right) = \left[\frac{\partial \ell_\mathcal{H}}{\partial \xi}, \xi\right] = \frac{1}{2}[\hat{J}_\mathcal{H}^\star, \xi^2] + \frac{1}{2}[\{\xi_m, \hat{J}_\mathcal{R}\} + \{\xi_t, \hat{J}_\mathcal{T}\}, \xi]. \tag{60}$$

The term $\dot{\mu}$ represents the rate of change of the angular momentum that is to be equated to the total torque acting on the helicopter.

To take into account energy dissipation due to friction between the helicopter and the air molecules during rotation of the helicopter along the vertical direction, which tends to brake the motion of the helicopter, the equation governing the rotational motion may be completed by introducing a non-conservative force proportional to the helicopter rotation speed along the z-axis. The resulting Euler–Poincaré equation for the helicopter model reads

$$\{\dot{\xi}, \hat{J}_\mathcal{H}^\star\} = [\hat{J}_\mathcal{H}^\star, \xi^2] + [\{\xi_m, \hat{J}_\mathcal{R}\} + \{\xi_t, \hat{J}_\mathcal{T}\}, \xi] - \{\dot{\xi}_m, \hat{J}_\mathcal{R}\} - \{\dot{\xi}_t, \hat{J}_\mathcal{T}\} + 2\tau - \beta_r \langle \xi, \xi_z \rangle \xi_z, \tag{61}$$

where $\beta_r \geq 0$ is a coefficient that quantifies the braking action of the air around the helicopter during fast yawing.

### 3.4. Translational Component of Motion

The translational component of motion obeys the Euler–Lagrange equation (12) written in the inertial (earth) reference frame $\mathcal{F}_E$. In this case, the non-conservative force field is given by the total thrust $\varphi_m + \varphi_t$ rotated of a quantity $R$ to express it in the earth frame $\mathcal{F}_E$, therefore, the Euler–Lagrange equation reads:

$$\frac{d}{dt}\frac{\partial \ell_\mathcal{H}}{\partial \dot{p}} = \frac{\partial \ell_\mathcal{H}}{\partial p} + R(\varphi_m + \varphi_t). \tag{62}$$

Notice that

$$\frac{d}{dt}\frac{\partial \ell_\mathcal{H}}{\partial \dot{p}} = M_\mathcal{H}\ddot{p}, \quad \frac{\partial \ell_\mathcal{H}}{\partial p} = -M_\mathcal{H}\bar{g}e_z. \tag{63}$$

To take into account energy dissipation due to friction between the helicopter and the air molecules, that tends to brake the motion of the helicopter, the equation governing the translational motion may be completed by introducing a non-conservative force proportional to the helicopter speed. Ultimately, the equation that describes the translational motion of a helicopter may be written as follows:

$$M_\mathcal{H}\ddot{p} = R(\varphi_m + \varphi_t) - M_\mathcal{H}\bar{g}e_z - B\dot{p}, \tag{64}$$

where $B := \mathrm{diag}(\beta_h, 0, \beta_v)$. The non-negative coefficients $\beta_h$ and $\beta_v$ quantify the braking action on the helicopter, which is more pronounced along the vertical direction than horizontally, due to the helicopter's shape. Focusing on the Equation (64), it is clear that when the helicopter fuselage is horizontal, namely $R = I_3$, the tail rotor influences the horizontal component of the second derivative of the position $p$. The tail rotor term when the helicopter is tilted ($R \neq I_3$) causes an additional difficulty in controlling the position of the helicopter.

*3.5. Explicit State-Space Form of the Equations of Motion*

In order to write the equations of motion in an explicit form, we start off with a few important simplifications.

- The terms related to the principal rotors may be rewritten explicitly as follows. The term $\{\Omega_m \zeta_z, \hat{J}_\mathcal{R}\} = j_\mathcal{R}\Omega_m\{\zeta_z, \mathrm{diag}(1,1,0)\} = 2j_\mathcal{R}\Omega_m\zeta_z$. Likewise, the term $\{\dot{\Omega}_m\zeta_z, \hat{J}_\mathcal{R}\} = 2j_\mathcal{R}\dot{\Omega}_m\zeta_z$.
- The terms related to the tail rotors may be rewritten explicitly by noticing that the term $\{-\Omega_t\zeta_y, \hat{J}_\mathcal{T}\} = -j_\mathcal{T}\Omega_t\{\zeta_y, \mathrm{diag}(1,0,1)\} = -2j_\mathcal{T}\Omega_t\zeta_y$. Likewise, the term $\{-\dot{\Omega}_t\zeta_y, \hat{J}_\mathcal{T}\} = -2j_\mathcal{T}\dot{\Omega}_t\zeta_y$.
- The constant $\hat{J}_\mathcal{H}^\star = \hat{J}_B + M_\mathcal{R}b_m b_m^\top + M_\mathcal{T}b_t b_t^\top + \hat{J}_\mathcal{R} + \hat{J}_\mathcal{T}$. Notice that $b_m b_m^\top = D_m^2 \mathrm{diag}(0,0,1)$ and $b_t b_t^\top = D_t^2 \mathrm{diag}(1,0,0)$. In addition, recall that the reference frame $\mathcal{F}_B$ has been chosen with the orthogonal axes coincident with the principal axes of inertia of the fuselage itself, hence the tensor $\hat{J}_B$ is diagonal. As a consequence, the total helicopter's non-standard inertia tensor is diagonal, namely $\hat{J}_\mathcal{H}^\star = \mathrm{diag}(j_x, j_y, j_z)$.
- As a last observation, the quantity $\{\dot{\xi}, \hat{J}_\mathcal{H}^\star\}$ may be written equivalently as $S\dot{\xi}S$, where $S := \mathrm{diag}(s_x, s_y, s_z)$, with

$$s_x := \sqrt{\frac{(j_x + j_y)(j_x + j_z)}{j_y + j_z}}, \quad s_y := \sqrt{\frac{(j_y + j_x)(j_y + j_z)}{j_x + j_z}}, \quad s_z := \sqrt{\frac{(j_z + j_x)(j_z + j_y)}{j_x + j_y}}. \tag{65}$$

The equations of motion of the helicopter model taken into consideration in the present paper may be written explicitly as

$$\begin{cases} \dot{R} = R\xi, \\ \dot{\xi} = S^{-1}([\hat{J}_{\mathcal{H}}^\star, \xi^2] + 2[j_\mathcal{R}\Omega_m\xi_z - j_\mathcal{T}\Omega_t\xi_y, \xi] - 2j_\mathcal{R}\Omega_m\xi_z + 2j_\mathcal{T}\Omega_t\xi_y + 2\tau - \beta_r\langle\xi,\xi_z\rangle\xi_z)S^{-1}, \\ \tau := \frac{1}{2}D_m u_m \sin\alpha_r \xi_x + \frac{1}{2}D_m u_m \sin\alpha_p \cos\alpha_r \xi_y + \frac{1}{2}(D_t u_t - \gamma u_m)\xi_z, \\ \ddot{p} = \frac{1}{M_\mathcal{H}}R\varphi - \bar{g}e_z - \frac{1}{M_\mathcal{H}}B\dot{p}, \\ \varphi := \begin{bmatrix} \frac{1}{2}u_m \sin\alpha_p \cos\alpha_r \\ -\frac{1}{2}u_m \sin\alpha_r - \frac{1}{2}u_t \\ \frac{1}{2}u_m \cos\alpha_p \cos\alpha_r \end{bmatrix}. \end{cases} \quad (66)$$

It is interesting to consider a few special cases of motion and how the model (66) would simplify in these special instances.

*Free fall*: Let us assume that both rotors are blocked ($\dot{\xi}_m = \dot{\xi}_t = 0$) and that they are isolated from the pilot control ($u_m = u_t = 0$). In this case, the external torque $\tau$ (19) is null. The rotational component of motion is hence described by $\{\dot{\xi}, \hat{J}_B + M_\mathcal{R}b_m b_m^\top + M_\mathcal{T}b_t b_t^\top + \hat{J}_\mathcal{R} + \hat{J}_\mathcal{T}\} = [\hat{J}_B + M_\mathcal{R}b_m b_m^\top + M_\mathcal{T}b_t b_t^\top + \hat{J}_\mathcal{R} + \hat{J}_\mathcal{T}, \xi^2]$, which represents the classical equation of a rigid body rotating freely in space under inertial forces (generally known as Euler's equation of a free rigid body).

*Constant rotor speed and negligible rotational inertia*: Assuming constant rotation speed for the principal and the tail rotors (namely, $\dot{\xi}_m = \dot{\xi}_t = 0$) and assuming that the angular momentum of the tail rotor and of the principal rotor are negligible with respect to the angular momentum of the helicopter, we obtain the simplified model $\frac{1}{2}\{\dot{\xi}, \hat{J}_B + M_\mathcal{R}b_m b_m^\top + M_\mathcal{T}b_t b_t^\top\} = \frac{1}{2}[\hat{J}_B + M_\mathcal{R}b_m b_m^\top + M_\mathcal{T}b_t b_t^\top, \xi^2] + \tau$, that is the helicopter model studied in [12].

*Hovering*: Using as reference $\mathcal{F}_E$, hovering happens when the weight $M_\mathcal{H}\bar{g}$ balances the $z$-component of the thrust. In this situation the helicopter may only translate sideways in the $x$-$y$ plane. Recalling that

$$\varphi = \varphi_m + \varphi_t = \begin{bmatrix} \frac{1}{2}u_m \sin\alpha_p \cos\alpha_r \\ -\frac{1}{2}u_m \sin\alpha_r - \frac{1}{2}u_t \\ \frac{1}{2}u_m \cos\alpha_p \cos\alpha_r \end{bmatrix},$$

defining:

$$\varphi_w := e_z^\top \begin{bmatrix} 0 \\ 0 \\ -M_\mathcal{H}\bar{g} \end{bmatrix} \text{ and } \varphi_z := e_z^\top (R\varphi)e_z, \quad (67)$$

the hovering condition reads

$$\varphi_z + \varphi_w = 0. \quad (68)$$

In fact, the scalar $\varphi_w$ denotes the (negative) intensity of gravitational pull, while the scalar term $\varphi_z$ denotes the (positive) lift thrust of the main rotor. Assuming a helicopter to be horizontal (namely, with $\mathcal{F}_B$ and $\mathcal{F}_E$'s $z$-axes aligned), the Equation (68) becomes $2M_\mathcal{H}\bar{g} = u_m \cos\alpha_p \cos\alpha_r$. As a special case, we could for simplicity consider $\alpha_p = \alpha_r = 0$. Then, by the main rotor thrust Formula (24), the hovering condition could be read as $4M_\mathcal{H}\bar{g} = C_u\rho\pi l_\mathcal{R}^4\Omega_m^2 \sin\alpha_c$. Hence, the value of the collective angle needed to maintain hovering, resulting from the hovering condition, takes the form

$$\alpha_{c,\text{hover}} = \arcsin\left(\frac{4M_\mathcal{H}\bar{g}}{\rho\pi C_u l_\mathcal{R}^4 \Omega_m^2}\right). \quad (69)$$

In general, changing the angle $\alpha_p$ or $\alpha_r$ causes a decrease in the $z$-axis thrust intensity, hence every time the *cyclic control* is operated the helicopter tends to fall. The equation below gives the value of the right collective angle with respect to $\alpha_r$ and $\alpha_p$ in order to prevent a fall condition:

$$\alpha_{c,\text{hover}} = \arcsin\left(\frac{2M_{\mathcal{H}}\bar{g}}{u_m \cos \alpha_p \cos \alpha_r}\right), \tag{70}$$

where the thrust $u_m$ comes from Equation (24).

The maximum linear velocity along the $x$-axis could be reached provided two hypothesis are met: the first is the hovering condition, in order to balance the weight force and not to decrease the helicopter height, and the second is that the horizontal component of the thrust is purely directed along the $x$-axis, namely $\alpha_r = 0$. From (68), the formula to find the corresponding pitch angle is:

$$\alpha_{p,\text{maxSpeed}} = \arccos\left(2\frac{M_{\mathcal{H}}\bar{g}}{\bar{u}_m}\right), \tag{71}$$

where $\bar{u}_m$ is a value of the thrust larger than the weight force of the helicopter. (*Remark*: As the *collective control* changes the torque exerted by the main rotor, this procedure implies a number of concurrent actions. In fact, consider the pilot wants to change the attitude of the helicopter using the *cyclic control* while keeping hovering: the *cyclic control* causes the need to boost the main rotor thrust by using the *collective control*, and the *collective control* causes an increase in the main rotor torque and hence a yaw effect that requires the *pedals control* to be managed.)

*No yawing*: The condition of no yawing is achieved when the quantity $\langle \xi, \xi_z \rangle$ stays constant to 0. Namely, the helicopter does not turn around the $z$-axis. In this case, the friction due to rotation, $\beta_r \langle \xi, \xi_z \rangle$, is 0. Assuming $\xi = 0$ at some time, it is necessary to make sure that the first derivative of the angular velocity equals zero to ensure that no yaw is present, hence $\langle \dot{\xi}, \xi_z \rangle = 0$. From (66), it follows that

$$S^{-1}\left(-2j_{\mathcal{R}}\Omega_m \xi_z + (D_t u_t - \gamma u_m)\xi_z\right)S^{-1} = 0. \tag{72}$$

As it was already underlined while discussing equations (21), in the case of constant main rotor speed $\Omega_m$, the condition (72) will become $S^{-1}((D_t u_t - \gamma u_m)\xi_z)S^{-1} = 0$ that could be reduced to $D_t u_t = \gamma u_m$.

*No drifting*: The tail thrust causes the helicopter to drift along the $y$-axis. This side effect may be compensated by choosing appropriately the roll angle $\alpha_r$ of the main rotor thrust. The equilibrium of forces along the $y$-axis is reached when $\varphi^\top e_y = 0$ (where $e_y := [0\ 1\ 0]^\top$). Since $\varphi^\top e_y = -\frac{1}{2}u_m \sin \alpha_r - \frac{1}{2}u_t$, in order not to have longitudinal forces the roll angle has to be set to:

$$\alpha_{r,\text{noDrift}} = -\arcsin\left(\frac{u_t}{u_m}\right). \tag{73}$$

With this value, the net drift force along the $y$-axis will drop to zero, meaning that no acceleration along the $y$-axis will be detected, although any pre-existing motions along the $y$-axis will not cease. Moreover, setting the angle $\alpha_r$ to this value will cause the fuselage to roll.

## 4. Numerical Methods to Simulate the Motion of a Helicopter

The principal aim of developing a mathematical model is to be able to carry out numerical simulations of a physical system through a computing platform. From this perspective, the system of differential equations (66) needs to be discretized in time in order to be implemented on a computing platform. While the equation describing the translational component of motion may be solved through a standard numerical method, the equation describing the rotational component of motion needs a specific numerical method.

An ordinary differential equation, in which the initial value is known, could be resolved numerically using the forward Euler method *fEul*. The first derivative of a function could be approximated numerically as:

$$\dot{f}_{k-1} = \frac{f_k - f_{k-1}}{h} \tag{74}$$

whereas the second derivative of a function could be approximated numerically iterating the *fEul* method as follows

$$\ddot{f}_{k-1} = \frac{\dot{f}_k - \dot{f}_{k-1}}{h} \tag{75}$$

where $k \geq 1$ denotes a discrete-time counter and $h > 0$ represents the step of resolution of the numerical method. Developing the Equations (74) and (75), the second derivative equation of a function may be approximated by $\ddot{f}_{k-2} = \frac{f_k - 2f_{k-1} + f_{k-2}}{h^2}$ with $k \geq 2$.

Using the result in Equation (66), it is possible to set up an iteration to determine numerically the trajectory of the center of mass of the helicopter, namely:

$$\frac{1}{M_{\mathcal{H}}} R_{k-2} \varphi_{k-2} - \bar{g} e_z - \frac{1}{M_{\mathcal{H}}} B\left( \frac{p_{k-1} - p_{k-2}}{h} \right) = \frac{p_k - 2p_{k-1} + p_{k-2}}{h^2},$$

which may be rewritten in explicit form as:

$$p_k = \frac{h^2}{M_{\mathcal{H}}} R_{k-2} \varphi_{k-2} - h^2 \bar{g} e_z - \frac{h}{M_{\mathcal{H}}} B(p_{k-1} - p_{k-2}) + 2p_{k-1} - p_{k-2}. \tag{76}$$

The equation $\dot{R} = R\zeta$ describes the first-order derivative of helicopter attitude. The attitude matrix $R$ belongs to the special orthogonal group SO(3). On manifolds it is not possible to perform linear operations and, as a consequence, to use directly the *fEul* method. In this case, it is necessary to use exponential map, thus:

$$R_k = \exp_{R_{k-1}}(h R_{k-1} \zeta_{k-1}). \tag{77}$$

Using the expression of exponential map tailored to the manifold SO(3) leads to the iteration

$$R_k = R_{k-1} \mathrm{Exp}(h \zeta_{k-1}). \tag{78}$$

Since the second equation in (66) describes dynamics over the Lie algebra $\mathfrak{so}(3)$, such equation may be time-descritized through the classical Euler's method: $\zeta_k = \zeta_{k-1} + h \dot{\zeta}_{k-1}$. In particular, $\dot{\zeta}_{k-1}$ represents the angular acceleration at the step $k-1$. The resulting iteration reads:

$$\begin{aligned}\zeta_k =& \zeta_{k-1} + h \cdot S^{-1}\left( [\hat{J}_{\mathcal{H}}^\star, \zeta_k^2] + 2[j_\mathcal{R} \dot{\Omega}_{\mathrm{m},k} \zeta_z - j_\mathcal{T} \dot{\Omega}_{\mathrm{t},k} \zeta_y, \zeta_k] \right. \\ & \left. -2 j_\mathcal{R} \dot{\Omega}_{\mathrm{m},k} \zeta_z + 2 j_\mathcal{T} \dot{\Omega}_{\mathrm{t},k} \zeta_y + 2\tau_k - \beta_r \langle \zeta_k, \zeta_z \rangle \zeta_z \right) S^{-1}.\end{aligned} \tag{79}$$

In summary, the complete set of iterations reads:

$$\begin{cases} p_k = \frac{h^2}{M_{\mathcal{H}}} R_{k-2} \varphi_{k-2} - h^2 \bar{g} e_z - \frac{h}{M_{\mathcal{H}}} B(p_{k-1} - p_{k-2}) + 2p_{k-1} - p_{k-2}, & k \geq 2 \\ R_k = R_{k-1} \mathrm{Exp}(h \zeta_{k-1}), & k \geq 1 \\ \zeta_k = \zeta_{k-1} + h \cdot S^{-1}([\hat{J}_{\mathcal{H}}^\star, \zeta_k^2] + 2[j_\mathcal{R} \dot{\Omega}_{\mathrm{m},k} \zeta_z - j_\mathcal{T} \dot{\Omega}_{\mathrm{t},k} \zeta_y, \zeta_k] + \\ \qquad -2 j_\mathcal{R} \dot{\Omega}_{\mathrm{m},k} \zeta_z + 2 j_\mathcal{T} \dot{\Omega}_{\mathrm{t},k} \zeta_y + 2\tau_k - \beta_r \langle \zeta_k, \zeta_z \rangle \zeta_z) S^{-1}, & k \geq 1 \\ \dot{\Omega}_{\mathrm{m},k} = (\Omega_{\mathrm{m},k} - \Omega_{\mathrm{m},k-1})/h, & k \geq 1 \\ \dot{\Omega}_{\mathrm{t},k} = (\Omega_{\mathrm{t},k} - \Omega_{\mathrm{t},k-1})/h, & k \geq 1 \\ \dot{p}_0 = 0_{3\times 1},\ p_0 = 0_{3\times 1},\ p_1 = 0_{3\times 1}, \\ R_0 = I_3,\ \zeta_0 = 0_3. \end{cases}$$

where $k = 0$ denotes the starting time and where initial conditions have been indicated as well. The quantities whose dynamics is not prescribed are either constants or externally controlled (by the pilot).

The numerical method used in the present implementation is the simplest one among the plethora of numerical methods available in the scientific literature. The Euler methods are easy to implement on a computing platform, but are the least precise ones. An analysis of the precision of the Euler method on the special orthogonal group was covered in a previous publication of the second author [22]. The precision of the numerical scheme to simulate the dynamics of a flying body be increased by accessing higher-order numerical methods such as those in the Runge–Kutta class.

## 5. Helicopter Type and Value of the Parameters

To implement the mathematical model studied, it is necessary to choose a specific helicopter model and gather values from certification sheets and data sheets. The helicopter type chosen for this study is the EC135 P2+ (also known as H135 P2+) manufactured by Airbus[TM] Corporate Helicopters. Not all parameters that appear in the equations are directly specified in the technical documentation, hence a careful usage of the equations to infer those parameters values not directly available will be illustrated. The data have been gathered from the manufacturer's flight manual [23], and other manuals [1,17–20,24–26].

The EC135 P2+ helicopter is equipped with a 4-blades bearingless main rotor and a 10-blades tail rotor and is characterized by the following features:

*Main rotor and Tail rotor*: The main characteristics of the tail rotor and of the main rotor are collected in Table 1. In particular, such table contains information about the rotors collective and cyclic angle range, rotors weight and nominal spinning velocity.

*Sizes*: For the principal dimension values, readers are referred to the manual [23]. The relevant values have been collected in Table 2, which consist in linear dimensions and weights. From the sizes of the fuselage, it is readily observed that the chosen helicopter type is relatively small, compared to larger helicopters from the army industry.

**Table 2.** Dimensions are taken from [23], page 7, and the weight of the main rotor blade from [19], page 3.

|  | Dimensions | | | Weight |
|---|---|---|---|---|
|  | [m] | | | [kg] |
| Main rotor blade |  | 5.1 |  | 33.9 [1] |
| Tail rotor blade |  | 0.5 |  | — |
| Reference axis | $x$ | $y$ | $z$ |  |
| Fuselage | 5.87 | 1.56 | 2.20 | 1134.6 [2] |

[1] The value of the height is not mentioned in any of the sources found, therefore it has been calculated from the available technical drawings. [2] The fuselage weight was computed as the weight of the empty helicopter, that is 1420 kg, removing the weight of the main rotor and of the tail rotor. The helicopter weight value was drawn from [18], page 2.

*Center of mass*: To calculate the center of mass of the helicopter it is necessary to split the helicopter's structure in three major parts, as in the development of its mathematical model:

1. The fuselage or body;
2. The main rotor;
3. The tail rotor.

It is necessary to make some assumptions to calculate the center of mass of the helicopter and to determine the values $D_m$ and $D_t$. These assumptions refer to the Figure 8. It is assumed that the center of mass of the body $c_B$ lies on the axis passing through the main rotor and perpendicular to the base. Furthermore, it is assumed that the center of mass of the main rotor $c_R$ locates on an axis tilted 5 degrees from the vertical one. In addition, the main rotor and the body may be thought of as two objects composing the system

Body − MainRotor ($_{B,R}$), and the reference system for the calculation may be thought of as having the origin located in the point $c_B$ and one axis that matches the axis tilted 5 degrees toward the point $c_R$. We can determine the value of $c_{B,R}$ by the equation

$$c_{B,R} = \frac{Weight_{MainRotor}}{Weight_{MainRotor} + Weight_{Body}} \cdot d_1. \tag{80}$$

Now, assuming $d_1 = 1.2$ m as the distance between $c_B$ and $c_R$, the above equation gives

$$c_{B,R} = \frac{277.2}{277.2 + 1134.6} \cdot 1.2 \approx 0.235614 \text{ m}. \tag{81}$$

Moreover, it is supposed that the contribution of the point $c_T$ could be disregarded since the weight of the tail rotor is negligible compared to the fuselage weight and the main rotor weight; therefore, the tail rotor does not contribute to the calculations of the helicopter center of mass, hence it results $c_{B,R} \approx c_H$. The value of $D_t \approx 6$ m has been inferred from the available data-sheets information and the structure drawings, and $D_m$, that is the distance between the center of gravity of the helicopter and the main rotor, is $D_m = d_1 − c_H \approx 0.964386$ m.

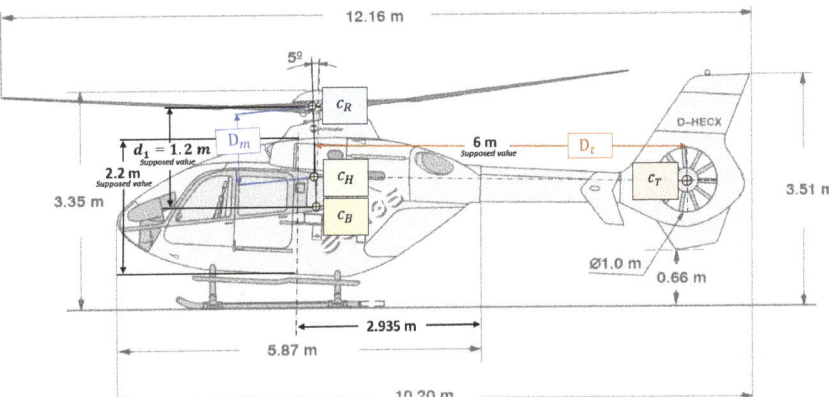

**Figure 8.** Values used to calculate the center of mass. (Figure adapted from [23], page 7.)

***Features of the engines***: The EC135 P2+ helicopter type is equipped with two PW206B2 engines from Pratt and Whitney Canada Corp$^{TM}$. To start engines there are two possibilities: manual control or automatic control. Manual control is not certificated and is normally deactivated. The automatic control is managed by the FADEC (Full Authority Digital Engine Control) that governs the starting procedure, the fuel flow and the RPMs automatically. At the start of the engines, the FADEC turns on the engines one by one until the RPMs reach the value of 98% ([1], page 437). When either the *collective control* or a manual switch are operated by the pilot, the FADEC increases the RPMs to 100% and the flight mode is engaged. When the altitude is higher than 4000 ft the speed is automatically increased to 104%, because the air density decreases. Moreover, to avoid loss of thrust when the collective angle is varied, in the main rotor (pitch) or in the tail rotor (yaw), the FADEC fixes the engine power to maintain the desired speed. The characteristics of the engines are summarized in Table 3.

**Table 3.** Values are taken from [25], pages 8 and 12. The helicopter state AEO denotes 'all engine operative', whereas the state OEI stands for 'one engine inoperative'. Typically, two possible working mode could be selected: TOP (take-off power) that has a time-limit constraint, and MCP (maximum continuous power).

| Engine Mode | Power [ kW ] | Maximum Torque [ N · m ] |
| --- | --- | --- |
| AEO TOP (max. 5 min) | $2 \times 333$ | $2 \times 519$ |
| AEO MCP | $2 \times 321$ | $2 \times 500$ |
| OEI (max. 30 s) | 547 | 851 |
| OEI (max. 2 min) | 534 | 831 |
| OEI MCP | 404 | 629 |

*Gear box*: The gear box is a complex part that transmits power, usually reducing angular velocity and increasing torque. Both helicopter engines drive the gear box that, in turn, drives the main rotor shaft and the tail rotor shaft.

*Main rotor thrust*: The Equation (26) was used to calculate the power coefficient of the main rotor, that is $C_w \approx 0.006968$, and its thrust coefficient $C_u \approx 0.045965$. It is also possible to determine the maximum thrust $u_{m,max}$ generated by the main rotor using the Equation (24), setting the throttle at 100% and the collective angle at its maximum. The obtained result is $u_{m,max} \approx 52,729 \frac{\text{kg·m·rad}^2}{\text{s}^2}$. Such numerical result was obtained by setting $\Omega_{m,max} \approx 41.364303$ rad/s, $\alpha_{c,max} = 0.541052$ rad, $\rho = 1.225$ kg/m$^3$ (which denotes, respectively, the maximum angular speed, the maximum collective angle and the air density at 15 Celsius degrees and 1 atm, from Table 1).

*Tail rotor thrust*: In the same way, it is possible to determine the power coefficient and the thrust coefficient for the tail rotor which are respectively $C_w^T \approx 0.100974$ and $C_u^T \approx 0.273201$. The maximum thrust generated by the tail rotor is $u_{t,max} \approx 2601$ kg · m · rad$^2$/s$^2$, whose value is calculated using the throttle at 100%, the angular velocity $\Omega_{t,max} \approx 375.315601$ rad/s and the maximum collective angle for the tail rotor $\alpha_{c,max}^T \approx 0.596903$ rad, from Table 1.

*Drag term*: According to Equation (28), the value of the drag term is $\gamma \approx 0.154546$ m. Such numerical result was obtained by setting the middle value of the interval of the tail rotor collective angle to $\alpha_{c,mid}^T \approx 0.151844$ rad, and the collective angle of the main rotor consistent with hovering to $\alpha_{c,hover} \approx 0.268693$ rad, from Equation (69).

*Friction terms*: Let us collect the tip velocity of the helicopter along each axis in the vector $\dot{p}_{max}$. Given the maximum velocity of the helicopter, we know that, once reached that particular value, the acceleration of the helicopter along that axis will drop to 0, because of the existence of a friction force in the opposite direction. This situation can be described as:

$$0 = R(\varphi_m + \varphi_t) - M_{\mathcal{H}}\bar{g}e_z - B\dot{p}_{max}. \tag{82}$$

Looking closely at the term $R(\varphi_m + \varphi_t)$, namely the propelling force of the helicopter, it is readily observed how it takes a special configuration when the tip speed is reached, in fact:

- To reach the tip speed along the $z$-axis, it is necessary that the $z$-axes of the inertial reference frame $\mathcal{F}_E$ and of the body-fixed reference $\mathcal{F}_B$ are aligned;
- To reach the tip speed along the $x$-axis, we consider a motion at maximum speed due to a total thrust directed along the $x$-axis while in a horizontal attitude ($R = I_3$). In this case, the thrust takes its maximum value (compatibly with the need to keep the helicopter hovering).

The Figure 9 shows the force components present in some particular helicopter attitudes. In the frame on the left-hand side of the figure, the helicopter is horizontal, namely $R = I_3$, and all the forces are directed along the $z$-axis, disregarding the force exerted by the tail rotor. In the frame on the right-hand side, the helicopter is in a hovering condition, therefore, the friction force $F_z^3 = 0$, while the friction force $F_x^2$ along the x-axis is maximum.

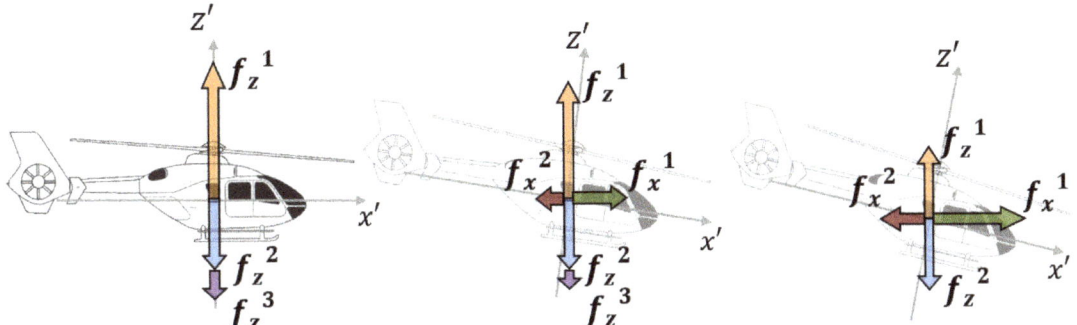

**Figure 9.** Friction terms encountered by a flying helicopter in correspondence of an increasing horizontal speed. The variables are defined as $f_z^1 = e_z^\top \varphi$, $f_z^2 = M_\mathcal{H} \bar{g}$, $f_z^3 = e_z^\top \dot{p} \beta_v$, $f_x^1 = e_x^\top \varphi$, $f_x^2 = e_x^\top \dot{p} \beta_h$.

The friction terms were calculated upon determining the tip speed of the helicopter. For the EC135 P2+ helicopter, the found values are summarized in Table 4. Since we only know the maximum linear speed along the x-axis, we consider as null the friction force along the y-axis.

**Table 4.** Airspeed value ([24], page 23). Hover turning velocity and throttle range ([23], pages 43 and 35). Rate of climbing (page 60 in [26]).

|  | Limit Values | | | Condition |
|---|---|---|---|---|
|  | [m/s] | [rad/s] | min ÷ max [%] |  |
| Airspeed | 79.7 | – | – | Sea level, +20 Celsius degrees, gross mass up to 2300 kg, TOP mode |
| Rate of climbing | 8.9 | – | – |  |
| Hover turning | – | 1.047 | – |  |
| Throttle range | – | – | 97 ÷ 104 |  |

Using the Equation (82) and the speed limit values, it is possible to infer the values of the coefficients $\beta_h$ and $\beta_v$, namely the friction term along the x-axis and the z-axis, respectively. The vector of the maximum speed reads $\dot{p}_{max} = [79.7\ ?\ 8.9]^\top$, referring to the Table 4. In addition, we considered the result from (71) and we end up with an equation depending on the unknown coefficient $\beta_h$, where the other terms are known:

$$2\beta_h e_x^\top \dot{p}_{max} = u_{m,max} \sin\left(\arccos\left(\frac{2\,M_\mathcal{H}\bar{g}}{u_{m,max}}\right)\right), \quad (83)$$

where $e_x := [1\ 0\ 0]^\top$. The Equation (83) allows to infer the value of the friction term.

The term $\arccos\left(\frac{2\,M_\mathcal{H}\bar{g}}{u_{m,max}}\right) \approx 58$ deg describes the angle with respect to the x-axis of the inertial reference frame $\mathcal{F}_E$ that the total thrust $\varphi$ must take for the helicopter to reach the maximum velocity. The orientation of the thrust $\varphi$ can be managed by the pilot by operating the *cyclic control*, which varies the angles $\alpha_p$ and $\alpha_r$, and by controlling the helicopter's overall attitude $R$.

To determine friction coefficients, the hover condition is preserved and rolling and pitching are not involved. The friction term $\beta_v$ can be determined by fixing $\alpha_p = 0$, $\alpha_r = 0$

and $R = \mathrm{diag}(1,1,1)$, which are the conditions to reach the maximum velocity along the $z$-axis. From the Equation (82), we thus obtain

$$\beta_v e_z^\top \dot{p}_{\max} = \tfrac{1}{2} u_{m,\max} - M_{\mathcal{H}} \bar{g}. \tag{84}$$

Instantiating equations (83) and (84) with known values, the friction coefficients can be easily computed. It has been found that $\beta_h \approx 281$ kg·s$^{-1}$ and $\beta_v \approx 1398$ kg·s$^{-1}$.

Using the same method, it is possible to estimate the value of $\beta_r$, the friction term linked to the yaw velocity. Let us assume the helicopter to be in the hovering condition, with $\dot{\zeta}_m = \dot{\zeta}_t = 0$. At the maximum yaw speed the angular acceleration will be null. Since we consider a hovering condition with $\alpha_r = \alpha_p = 0$, the total torque $\tau$ is equal to $\tfrac{1}{2}(D_t u_{t,\max} - \gamma u_{m,\text{hover}})\tilde{\zeta}_z$; therefore, from the second equation in (66) we obtain $0 = S^{-1}\big([\hat{J}_{\mathcal{H}}^\star, \tilde{\zeta}_{\max}^2] + (D_t u_{t,\max} - 2\gamma M_{\mathcal{H}}\bar{g} - \beta_r \langle \tilde{\zeta}_{\max}, \tilde{\zeta}_z \rangle)\tilde{\zeta}_z\big) S^{-1}$, where $\tilde{\zeta}_{\max}$ denotes the maximal yawing speed that, from the Table 4, is known to be $\tilde{\zeta}_{\max} = 1.047 \cdot \tilde{\zeta}_z$ (rad/s). Thus, isolating the friction term, this equation becomes:

$$\beta_r \langle \tilde{\zeta}_{\max}, \tilde{\zeta}_z \rangle \tilde{\zeta}_z = [\hat{J}_{\mathcal{H}}^\star, \tilde{\zeta}_{\max}^2] + (D_t u_{t,\max} - 2\gamma M_{\mathcal{H}}\bar{g})\tilde{\zeta}_z. \tag{85}$$

To determine the correct value of the friction term it is necessary to fill the Equation (85), namely the tip thrust of the tail rotor $u_{t,\max}$, the structural values, and the drag coefficient found. The computed result for this parameter is $\beta_r \approx 10797$ N·m·s·rad$^{-1}$.

### 6. Numerical Experiments and Results

A series of tests of the mathematical model were carried out by means of a MATLAB® implementation of the numerical methods explained in Section 4. In order to clarify what can be tested, and how, it could be useful to introduce the graphic control panel shown in Figure 10.

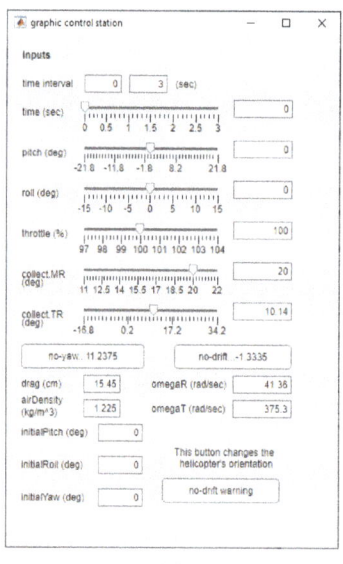

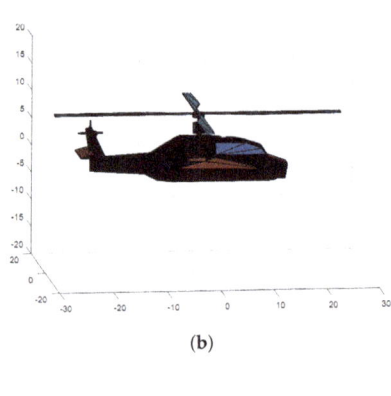

Figure 10. Graphic input interface: (a) graphic interface used to test the model; (b) graphic window to show the initial attitude of the helicopter, which is linked to the value of the matrix $R$ selected.

The cell *time interval* allows to set the time range for new experiments. The interface gives the possibility to perform series of test, therefore, the initial value of *time interval* could not be set at an instant of time $t_1$ until another experiment, which ended at $t_1$, has been completed. The slider named *Time* shows the selected instant of time in a pop-up animation window. Every slider is linked to an editable cell, hence, any value belonging to the correct range could be directly set. The five sliders *pitch, roll, throttle, collect.MR, collect.TR* are the interface to manage the value of the variables $\alpha_p$, $\alpha_r$, $\Omega_m$, $\alpha_c$, $\alpha_c^T$, respectively. The *no-yaw* button sets the angle $\alpha_c^T$ for a no-yawing condition, from Equation (72), whereas the button *no-drift* manages the value of the angle $\alpha_r$ to achieve a no-drifting condition using the Equation (73). The two editable cells *drag* and *air density* allow to input the values of the coefficients $\gamma$ and $\rho$, respectively. The three cells *initial roll, initial pitch, initial yaw* interface with the matrix $R$ forcing a value of attitude of the helicopter along the three axes $x$, $y$, $z$. The button *no-drift warning* changes the *initial roll* value in order to achieve a stationary no-drifting condition (a technique introduced in the third test).

*First test—lift response*: The first test lasts 10 s and does not involve pitch and roll angles ($\alpha_p = 0$, $\alpha_r = 0$), moreover the throttle is set at 100% and the tail rotor collective angle at $\alpha_{c,mid}^T$. About the main rotor collective angle, it has been chosen in order to produce lift along $z$-axis: the value chosen is 20 degrees. Figure 11 shows the result of this simulation. The position along $x$-axis is constant, which could seem reasonable because pitch angle is not involved. On the other hand, there is a clear decrease in the $y$-component of the positional variable $p$ due to drift effect. Notice that the direction of the tail rotor thrust is opposite to $y$-axis, as Figure 5 shows. The $z$ component of the torque $\tau$ is negative, and this is the cause of clockwise yawing. Looking closely at the entries of the position vector $p$, along the $x$-axis it can be observed a little decrease due to the combined action of the helicopter yawing and tail rotor drift. In fact, when the helicopter nose turns, the drift effect causes a slight decrease in the positional $x$-coordinate. Note that the drift force exerted by the tail rotor coincides with the $y$ component of the total thrust $\varphi$, which is $e_y^\top \varphi$. The last remarkable observation from the first test is that the $z$ component of the thrust $\varphi$ has the value of 16,191 N, which is more than the helicopter-weight force, that could be determined from Table 2 as $1420 \cdot \bar{g} \approx 13{,}925$ N. The resulting force along the $z$-axis is positive and, as described by the graph of the $z$-coordinate of the center of gravity, the helicopter lifts up.

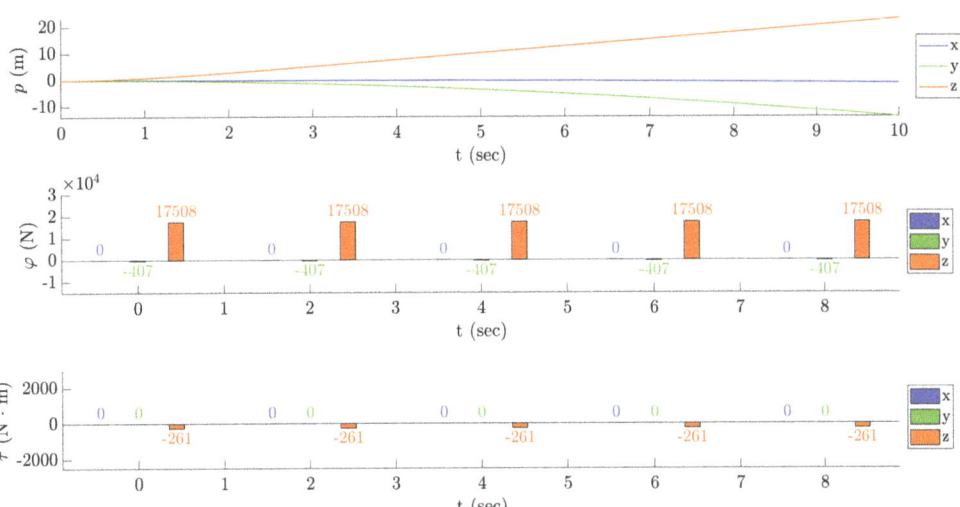

**Figure 11.** First test—lift response. Top panel: components of the position of $c_\mathcal{H}$. Middle panel: components of the thrust. Bottom panel: components of the active torque generated by rotors.

*Remark*: The $x$, $y$ and $z$ components in the torque graph have been extracted from the Equation (14) following the construction of $\tau$, namely they are calculated by $e_z^\top \tau e_y$, $e_x^\top \tau e_z$ and $e_y^\top \tau e_x$.

***Second test—no yaw***: This simulation lasts 5 s and illustrates how to select the tail rotor collective angle using the Equation (72) to achieve a no-yawing condition. From the graph of the torque $\tau$ it is clear that the torque exerted on the helicopter becomes null. Consequently, the slight decrease in the position along the *x*-axis, which was a side effect of the yawing, is no more present. The result is presented in Figure 12.

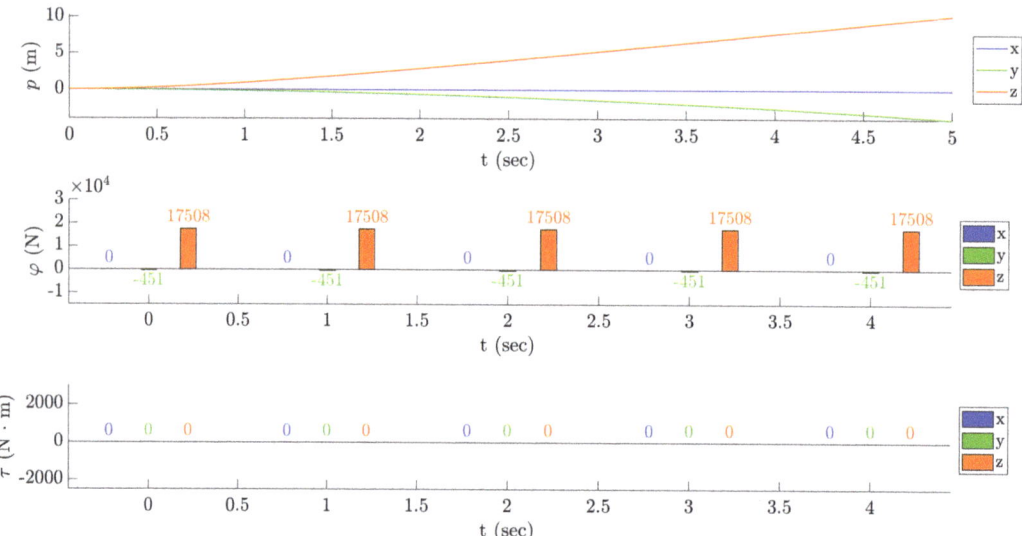

**Figure 12.** Second test—no yaw.

***Third test—neither yaw nor drift***: The third test illustrates the suppression of the drift effect due to the tail rotor. In this case, the *y* coordinate of the center of gravity of the helicopter does not vary, since the helicopter's attitude is modified by using the *no-drift warning* button in the control graphic panel. Such function does not cause a change in the angle of attack of the blades as in the previous test, but in the *initial roll* angle and, as a consequence, in the matrix *R*. In fact, the helicopter's attitude is set according to the Equation (73) along the *x*-axis, and such rotation produces an equilibrium among the drift effect and the thrust along the *y*-axis. The equilibrium among the forces causes the *y*-coordinate to stay constant, as wanted. The no-drift attitude is computed through the relation

$$R^\star := \begin{bmatrix} 1 & 0 & 0 \\ 0 & \cos(\alpha_{\text{noDrift}}) & -\sin(\alpha_{\text{noDrift}}) \\ 0 & \sin(\alpha_{\text{noDrift}}) & \cos(\alpha_{\text{noDrift}}) \end{bmatrix}.$$

The result of this experiment is shown in Figure 13. As expected, only the *z* coordinate of the helicopter varies over time.

The following tests were performed starting from the result of the third test, namely from a no yaw/no drift condition; therefore, the first 3 s of the results of each tests will be common to every execution.

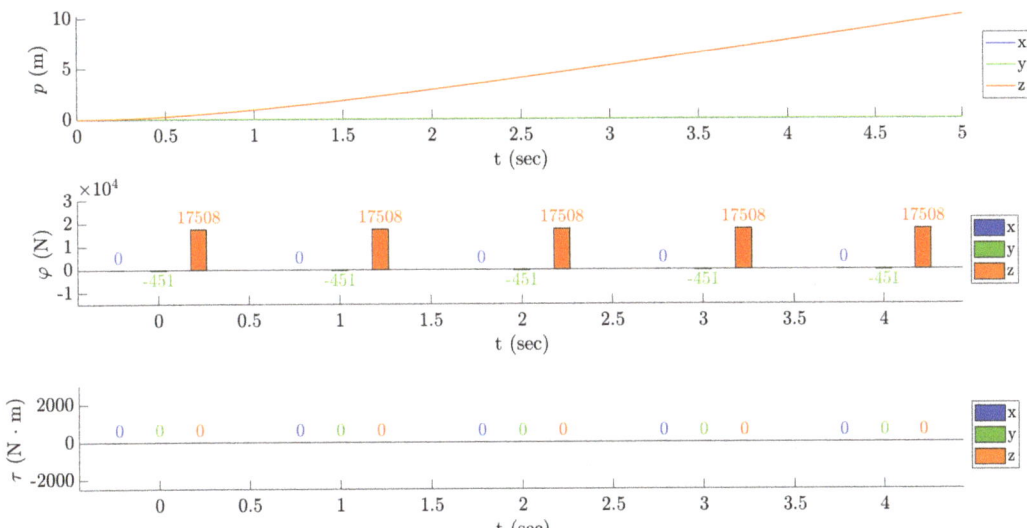

**Figure 13.** Third test—neither yaw nor drift.

*Fourth test—pitch response*: The fourth test is about pitch response. The pitch angle has been set to 5 degrees (constant) starting from the third second of the simulation to the end. In the result, illustrated in Figure 14, it can be seen an increase in the $y$-component of the total mechanical torque $\tau$. It is also possible to notice, as Figure 9 shows, that the change in the angle $\alpha_p$ causes a variation in the $\varphi$ components. Indeed, the $x$ component of the total thrust $\varphi$, that is $e_x^\top \varphi = \frac{1}{2} u_m \sin \alpha_p \cos \alpha_r$, increases as $\alpha_p$ increases, whereas the $z$ component of $\varphi$, that is $e_z^\top \varphi = \frac{1}{2} u_m \cos \alpha_p \cos \alpha_r$, decreases as $\alpha_p$ increases.

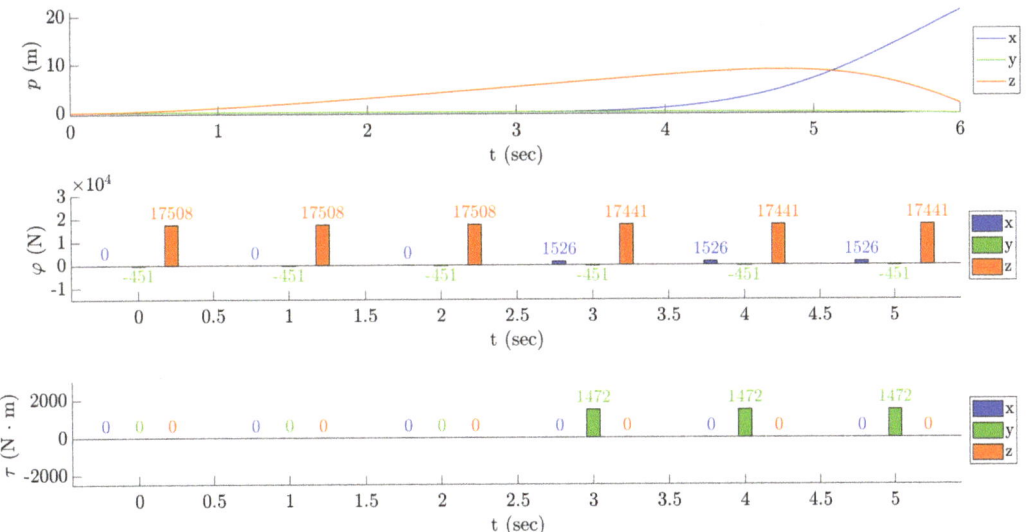

**Figure 14.** Fourth test—pitch response.

*Fifth test—positive roll response*: In this test, we set the angle $\alpha_r$ from the third second to the sixth second to 3 degrees. The obtained result is presented in Figure 15 and shows an increase in the $x$ component of the mechanical torque $\tau$. This behavior follows the Equation (66), where $\alpha_r$ is linked to the $x$ component of the mechanical torque $\tau$. In addition, using the same equation it is immediate to see that the $y$ component of $\tau$ is zero because $\alpha_p = 0$. Let us remark that instead the $z$ component of $\tau$ is zero because of the no-yaw condition.

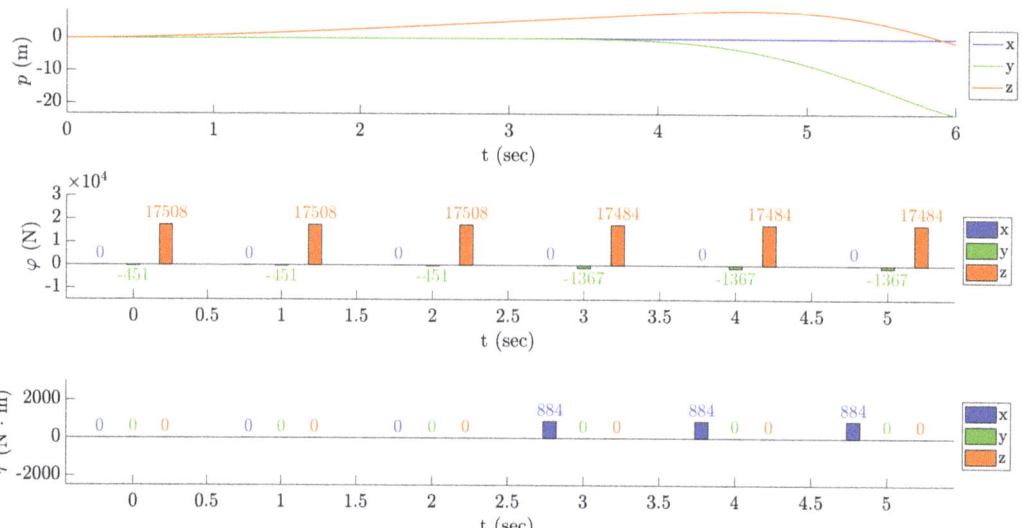

**Figure 15.** Fifth test—positive roll response.

As in the previous example, a change of the angle $\alpha_r$ causes the $z$ component of the thrust vector $\varphi$ to decrease. Moreover, the magnitude of the $y$ component of the vector $\varphi$ increases and adds up to the drifting effect of the tail rotor. It is important to point out, from the graph of the components of the positional vector $p$, that rolling causes a falling situation, as well as a shift along the $y$-axis.

*Sixth test—main rotor collective response*: The *collective control* is amply used for managing the acceleration of the helicopter. The test of this specific control system has been made increasing up to 22 degrees the main rotor collective angle starting from the third second to the tenth second. The increase in the main rotor collective angle causes a thrust boost, which increases the lifting of the helicopter. The result is shown in Figure 16. As remarked, every time *collective control* is operated the helicopter pilot also has to adjust the tail rotor collective angle, since the no-yaw flight mode depends on $u_m$, which is a function of $\alpha_c$. This side effect could be observed from the values of the $z$ component of the mechanical torque $\tau$ whose magnitude changes and needs to be adjusted through the *pedals control*.

*Seventh test—negative roll response*: The Figure 17 shows results of a test in which it has been tried to remove the drift effect using the *cyclic control*. It has already been remarked that a force control is not sufficient to remove the drift effect, since a combined helicopter's attitude control is needed. The no-drift flight mode could be achieved, for example, using a PID control on the helicopter's attitude.

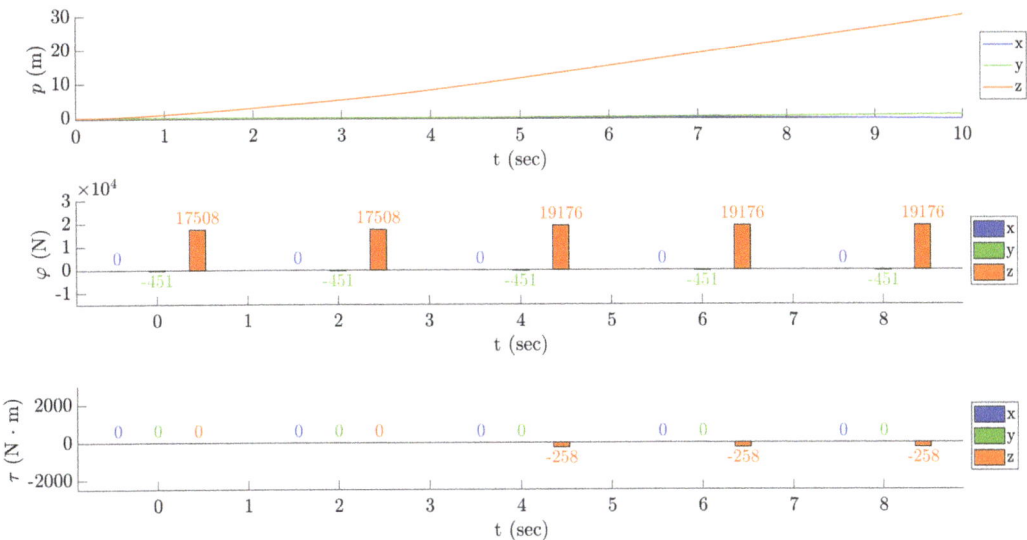

**Figure 16.** Sixth test—Main rotor collective response.

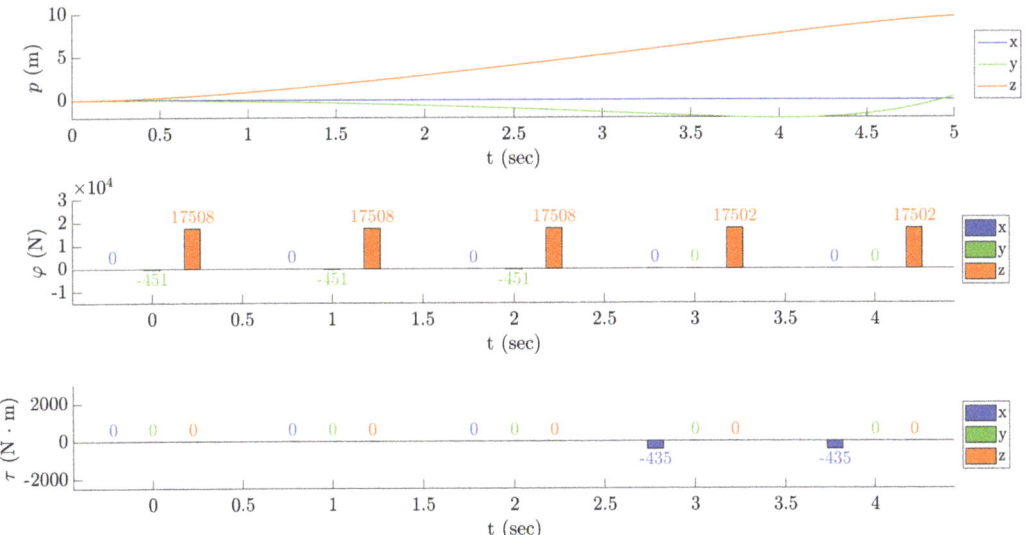

**Figure 17.** Seventh test—negative roll response.

This test was carried out using as initial setting the same setting as for the second test, whereas the last 2 s were simulated using the Equation (73) to change $\alpha_r$. Notice that an attitude controller would ensure no drifting. Ideally, a PID controller should reduce the value of the cyclic control as the roll angle of the helicopter approaches the value determined by the Equation (73).

*Eight test—free flight*: This last test consists in a simulation of a free flight achieved by setting multiple inputs for the *cyclic control* and the *collective control*. The obtained result is shown in Figure 18, while Table 5 presents the time line of the controls used. The *throttle* during the test keeps constant to 100%. The results of this simulation is exemplified by a video attached to the present paper as a supplemental material.

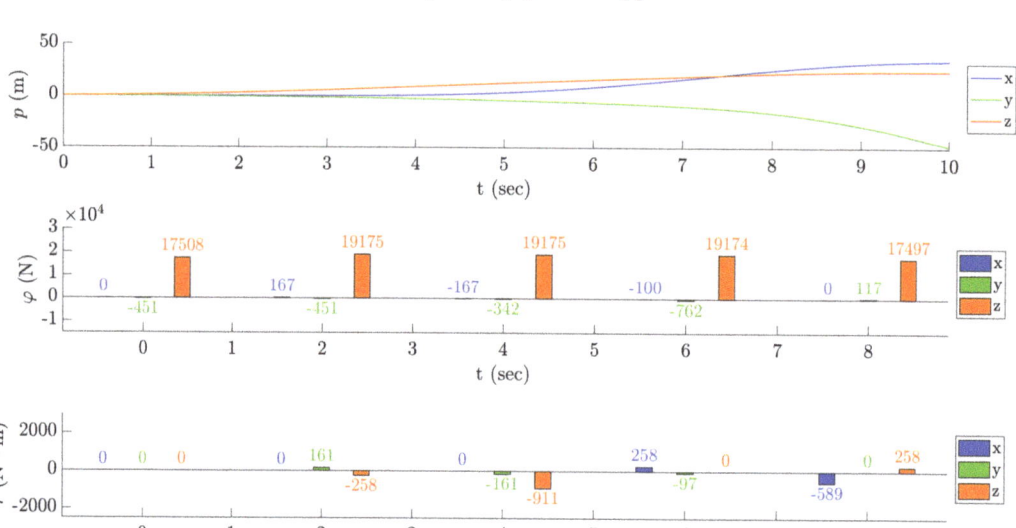

**Figure 18.** Eighth test—free flight.

A visual animation of the flight trajectory obtained in this test is available.

**Table 5.** Eighth test—free flight. The orange-colored values indicate that the *no-yaw* flight mode has been activated in that time window.

| | | Time Line of Controls | | | | |
|---|---|---|---|---|---|---|
| Time Interval | (s) | [0–2] | [2–4] | [4–6] | [6–8] | [8–10] |
| $\alpha_p$ | (deg) | 0 | 0.5 | −0.5 | −0.3 | 0 |
| $\alpha_r$ | (deg) | 0 | 0 | 0 | 0.8 | −2 |
| $\alpha_c$ | (deg) | 20 | 22 | 22 | 22 | 20 |
| $\alpha_c^T$ | (deg) | 11.24 | 11.24 | 8.5 | 12.32 | 12.32 |

## 7. Conclusions

The aim of this paper was to devise a mathematical model of a fantail helicopter in the framework of Lie-group theory. The main theoretical instrument, besides of Lie-group theory itself, is the Lagrange–d'Alembert–Pontryagin principle, which generalizes the Lagrangian formulation of dynamics to curved manifolds and to dissipative systems.

The modeling endeavor resulted in a series of differential equations, two of which describe the rotational dynamics of the helicopter body and two describe its translational dynamics. The terms in the equation have been analyzed to link the abstract mathematical notation with the physics of the real-world system under examination. In addition, a number of specific equations to calculate thrusts and power factors have been presented and merged to the Lie-group equations.

A specific section of the paper dealt with the numerical simulation of the flight of a helicopter and explained a specific numerical method to solve Lie-group-type differential equations approximately yet keeping up with the intrinsic nature of the base Lie-group (namely, the space of three-dimensional rotations).

The equations that compose the devised helicopter model include a number of parameters whose values are necessary to perform actual numerical simulations. Since most of such values are not directly available in the literature, a careful work of identification of the parameters through the devised equations has been carried out.

Numerical results have been illustrated and commented in order to elucidate some aspects of the model that were deemed of particular interest, from simple flight modes to free flight. The devised model, as the large majority of mathematical models of real-world physical phenomena, is of potential use to control engineers (who may use such mathematical model to design a state observer and an automatic control system to assist the pilot), to mechanical engineers (who may use the devised model to test a helicopter structure under stress conditions) and by pilots instructing facilities (who might use the devised model as a prototypical flight simulator).

**Supplementary Materials:** The following are available at https://www.mdpi.com/2227-7390/9/21/2682/s1.

**Author Contributions:** Conceptualization, S.F.; data curation, A.T.; writing—original draft preparation, A.T. and S.F.; writing—review and editing, S.F.; supervision, S.F. All authors have read and agreed to the published version of the manuscript.

**Funding:** This research received no external funding.

**Institutional Review Board Statement:** Not applicable.

**Informed Consent Statement:** Not applicable.

**Acknowledgments:** The authors would like to gratefully thank Toshihisa Tanaka (TUAT—Tokyo University of Agriculture and Technology) for having hosted the author A.T. during an internship in January–March 2020 and for having invited the author S.F. as a visiting professor at the TUAT in February–March 2020.

**Conflicts of Interest:** The authors declare no conflict of interest.

## References and Note

1. Eurocopter Deutschland GmbH. *EC 135—Training Manual*; Eurocopter Deutschland GmbH: Donauwörth, Germany, 2002.
2. Kim, S.; Tilbury, D. Mathematical modeling and experimental identification of an unmanned helicopter robot with flybar dynamics. *J. Robot. Syst.* **2004**, *21*, 95–116. [CrossRef]
3. Salazar, T. Mathematical model and simulation for a helicopter with tail rotor. In *Advances in Computational Intelligence, Man-Machine Systems and Cybernetics*; World Scientific and Engineering Academy and Society: 2010; pp. 27–33. Available online: http://www.wseas.us/e-library/conferences/2010/Merida/CIMMACS/CIMMACS-03.pdf (accessed on 19 October 2021).
4. Talbot, P.; Tinling, B.; Decker, W.; Chen, R. *A Mathematical Model of a Single Main Rotor Helicopter for Piloted Simulation*; Technical Report; NASA Ames Research Center: Moffett Field, CA, USA, 1982.
5. Shojaei Barjuei, E.; Caldwell, D.G.; Ortiz, J. Bond graph modeling and Kalman filter observer design for an industrial back-support exoskeleton. *Designs* **2020**, *4*, 53. [CrossRef]
6. Mizouri, W.; Najar, S.; Bouabdallah, L.; Aoun, M. Dynamic modeling of a quadrotor UAV prototype. In *New Trends in Robot Control*; Studies in Systems, Decision and Control; Ghommam, J., Derbel, N., Zhu, Q., Eds.; Springer: Singapore, 2020; Volume 270. [CrossRef]
7. Khumprom, P.; Grewell, D.; Yodo, N. Deep neural network feature selection approaches for data-driven prognostic model of aircraft engines. *Aerospace* **2020**, *7*, 132. [CrossRef]
8. Abraham, R.; Marsden, J.; Ratiu, T. *Manifolds, Tensor Analysis, and Applications*; Springer: New York, NY, USA, 1988.
9. Bullo, F.; Lewis, A. *Geometric Control of Mechanical Systems: Modeling, Analysis, and Design for Mechanical Control Systems*; Springer: New York, NY, USA, 2005.
10. Fiori, S. Nonlinear damped oscillators on Riemannian manifolds: Fundamentals. *J. Syst. Sci. Complex.* **2016**, *29*, 22–40. [CrossRef]
11. Fiori, S. Nonlinear damped oscillators on Riemannian manifolds: Numerical simulation. *Commun. Nonlinear Sci. Numer. Simul.* **2017**, *47*, 207–222. [CrossRef]
12. Kobilarov, M.; Desbrun, M.; Marsden, J.; Sukhatme, G. A discrete geometric optimal control framework for systems with symmetries. In *Robotics: Science and Systems*; MIT Press: Cambridge, UK, 2007; pp. 161–168.
13. Bloch, A.; Krishnaprasad, P.; Marsden, J.; Ratiu, T. The Euler-Poincaré equations and double bracket dissipation. *Commun. Math. Phys.* **1996**, *175*, 1–42. [CrossRef]
14. Ge, Z.M.; Lin, T.N. Chaos, chaos control and synchronization of a gyrostat system. *J. Sound Vib.* **2002**, *251*, 519–542. [CrossRef]
15. Fiori, S. Model Formulation Over Lie Groups and Numerical Methods to Simulate the Motion of Gyrostats and Quadrotors. *Mathematics* **2019**, *7*, 935. [CrossRef]

16. Rotaru, C.; Todorov, M. Helicopter Flight Physics. In *Flight Physics—Models, Techniques and Technologies*; Volkov, K., Ed.; IntechOpen Limited: London, UK, 2018.
17. Doleschel, A.; Emmerling, S. The EC135 Drive Train Analysis and Improvement of the Fatigue Strength. In Proceedings of the 33rd European Rotorcraft Forum, Kazan, Russia, 11–13 September 2007; Volume 2, pp. 1275–1319.
18. Kampa, K.; Enenkl, B.; Polz, G.; Roth, G. Aeromechanical aspects in the design of the EC135. In Proceedings of the 23rd European Rotorcraft Forum, Dresden, Germany, 16–18 September 1997; pp. 38.1–38.14.
19. Eurocopter. Eurocopter Training Service, Chapter 6—Main Rotor. 2006.
20. Axelsson, B.E.; Fulmer, J.C.; Labrie, J.P. Design of a Helicopter Hover Test Stand. Bachelor' Thesis, Worcester Polytechnic Institute, Worcester, MA, USA, 2015.
21. Yu, W.; Pan, Z. Dynamical equations of multibody systems on Lie groups. *Adv. Mech. Eng.* **2015**, *7*, 1–9. [CrossRef]
22. Fiori, S. A closed-form expression of the instantaneous rotational lurch index to valuate its numerical approximation. *Symmetry* **2019**, *11*, 1208. [CrossRef]
23. Eurocopter Deutschland GmbH. *Flight Manual EC135 P2+*; Eurocopter Deutschland GmbH: Donauwörth, Germany, 2002.
24. EASA European Aviation Safety Agency. Type Certificate Data Sheet No. EASA.R.009 for EC135. Available online: https://www.easa.europa.eu/sites/default/files/dfu/TCDS_EASA_R009_AHD_EC135_Issue_07_18Mar2015.pdf. (accessed on 13 October 2021).
25. EASA European Aviation Safety Agency. Type Certificate Data Sheet No. IM.E.017 for PW206 & PW207 Series Engines. Available online: https://www.easa.europa.eu/sites/default/files/dfu/TCDS%20IM.E.017_issue%2007_20151005_1.0.pdf (accessed on 13 October 2021).
26. Eurocopter Deutschland GmbH. *Eurocopter EC135 Technical Data*; Eurocopter Deutschland GmbH: Donauwörth, Germany, 2006.

Article

# Fracture Modelling of a Cracked Pressurized Cylindrical Structure by Using Extended Iso-Geometric Analysis (X-IGA)

Soufiane Montassir [1], Hassane Moustabchir [2], Ahmed Elkhalfi [1], Maria Luminita Scutaru [3,*] and Sorin Vlase [3,4]

[1] Faculty of Science and Technology, Sidi Mohamed Ben Abdellah University, B.P. 2202 Route d'Imouzzer, Fez 30000, Morocco; soufianemontassir@gmail.com (S.M.); aelkhalfi@gmail.com (A.E.)
[2] Laboratory of Science Engineering and Applications (LISA) National School of Applied Sciences, Sidi Mohamed Ben Abdellah University, BP 72 Route d'Imouzzer, Fez 30000, Morocco; hmoustabchir@hotmail.com
[3] Department of Mechanical Engineering, Transilvania University of Brașov, B-dul Eroilor 20, 500036 Brașov, Romania; svlase@unitbv.ro
[4] Romanian Academy of Technical Sciences, B-dul Dacia 26, 030167 Bucharest, Romania
* Correspondence: lscutaru@unitbv.ro

**Abstract:** In this study, a NURBS basis function-based extended iso-geometric analysis (X-IGA) has been implemented to simulate a two-dimensional crack in a pipe under uniform pressure using MATLAB code. Heaviside jump and asymptotic crack-tip enrichment functions are used to model the crack's behaviour. The accuracy of this investigation was ensured with the stress intensity factors (SIFs) and the J-integral. The X-IGA—based SIFs of a 2-D pipe are compared using MATLAB code with the conventional finite element method available in ABAQUS FEA, and the extended finite element method is compared with a user-defined element. Therefore, the results demonstrate the possibility of using this technique as an alternative to other existing approaches to modeling cracked pipelines.

**Keywords:** extended iso-geometric analysis; extended finite element method; crack; pipeline; ABAQUS

## 1. Introduction

The fracture phenomenon is a fundamental research topic in the field of solid mechanics, and a misunderstanding of the fracture mechanism may lead to a poor evaluation of a structure's integrity. Several experimental and theoretical studies have been conducted in this field covering this vast physical phenomenon. The cracking issue is among the problems that arise in this regard, so the study of the various kinds of discontinuities and defects that appear on the surface of structures is of major importance in the field of design and modeling for ensuring the reliability of engineering structures. These defects may appear in the form of notches, cracks, inclusions, holes, corrosion, and other types of material degradation. The existence of a defect on an engineering structure can cause material and human damage; this is caused by the propagation of cracks either in terms of direction or propagation rates. Therefore, it is necessary to predict the crack propagation rate to be able to estimate the critical load, and the acceptable length of the crack to ensure the stability of structures. The initiation and propagation of the crack requires a specific study of the defect, i.e., it is necessary to adopt a criterion of rupture in the context of fracture mechanics. Since the beginning of this research field, many methods have been developed to give a general vision of the fracture process in order to find adequate solutions and improve the strength of structures. With this technological progress, the simulation of the physical phenomena tends to become purely numerical; of course, experiments play a very important role, but due to their difficulty, and to save time and take advantage of the data-processing tools that exist today, we had to adapt to and develop in the world of digitalization. Numerical simulation and discontinuity problems still do not agree, since researchers are always looking to improve numerical solutions to be able to reflect reality.

In this sense, various numerical methods are developed that cite the classical finite element method (CFEM) [1,2], element-free Galerkin methods (EFG) [3], the extended finite element method (X-FEM) [4–6], the phase field numerical manifold method (PFNMM) [7], the boundary element method [8], and the thermo-mechanical peridynamic model (TMPM) [9]. Despite the huge use of the CFEM, however, it suffers from a certain lack of treatment of the cracking problems; it imposes that the mesh conforms with the crack or to a surface defect, and adopting a specific mesh creates difficulties at the simulation stage. The appearance of the enrichment approach [10–12] overcomes the shortcomings of the conventional method and is able to simulate all types of notches, cracks, holes, and other defects, regardless of the defect shape and mesh type. The Lagrange polynomials were used to make the interpolation with these new techniques, i.e., the geometry and the solution of the problem are approximated by these polynomials. Since the basis functions used are not the same as those used in the design, discretization errors appeared in the analysis stage [13]. The real issue occurred when we wanted to move from modeling to simulation by the process of the finite element method. Despite the integration of simulations in modeling software, such as CATIA, most software is still not good in this area, compared to software that is already simulation-integrated, such as ABAQUS. That is why there are still gaps between these two parts. This subject has become an area of interest for researchers in approximating conical shapes, as it is known that CAD software constructs this type of shape by B-spline curves. The advantage of these curves is that they are able to reproduce all types of conical shapes and free surfaces in an exact way. In this field, Hughes et al. [14] started to develop a new approach to eliminate the shortcomings of the classical method, and due to their research, iso-geometric analysis (IGA) was developed. On the other hand, industrial computational software has not yet integrated this functionality without some works that have made the implementation of these functions in specific fields of application, for example in LS-DYNA [15–17]. They were among the first in this field; precisely, they integrated IGA to study shells. In Altair RADIOSS [18], IGA was implemented and proved on industrial benchmarks. In ABAQUS [19–21], an application of IGA was implemented for linear elasticity problems. IGA has been employed in a variety of engineering fields, including the mechanics of vibration [22,23], fluid mechanics [24], electromagnetic problems [25], the medical field [26], and digital image correlation [27]. The issues addressed are diverse, including nonlinear mechanics [28], shell analysis [29–34], contact problems [35,36], fluid–structure interactions [24], the optimization of structural design [37], buckling failure [38], and crack problem analysis [39]. From this research, IGA has indicated the effectiveness of its results and it can sweep the world of numerical computation. Therefore, it can become an alternative method to the classical finite element method in the future.

Over the last few years, the mathematical formulation of the IGA method has been revised. In addition, it was expanded to fix issues concerning discontinuities, e.g., the crack, under the partition of the unity finite element method [40]. This new concept is known as the extended iso-geometric analysis (X-IGA). It has been utilized by a range of authors to address crack-growth issues in the field of fracture mechanics [41–47]. Two main functions have been added: the Heaviside jump function is used to enrich the displacement fields around the crack surfaces, while crack-tip enrichment functions are used to model the singularity at the crack tip. Most of the existing research works that use this innovative approach are limited to 2-D plate cracking problems [48,49] and cracking problems in the cantilever beam [50]. Therefore, the strategy has only been implemented with simple domain geometries. As for cracked shells, research is still in progress because of the difficulty of modeling the cracks and approximating the geometry with the NURBS functions. Furthermore, it requires a mathematical background. That is what made this kind of model interesting. In this field, X-IGA has been used for a cracked 2-D pipe. However, the method has been applied through FORTRAN language [51]. For the present investigation, we will study the efficiency and accuracy of X-IGA for cracks with 2-D pipe geometry given by CAD curves, with a special focus on ensuring that all stages of the calculation are done in MATLAB code. It is noted that mesh generation is used in MATLAB independently of another meshing software. The strategy presented in this implementation follows the philosophy used in the traditional FE codes, and also benefits from having a MATLAB routine that allows for the discretization of partial differential equations based on NURBS and B-spline. Hence, the specific aims of this paper are to formulate the X-FEM concept in the framework of NURBS and to enrich the solution in MATLAB code to adapt it for a cracked pipe under pressure. To validate this strategy, a P264GH steel gas pipe has been used with uniform inner pressure. The mechanical characteristics data of this model are taken from an experimental study [52]. Based on the above literature review, stress intensity factors (SIF) are widely used to characterize fracture mechanics. Therefore, we evaluated the SIF with this implementation and the obtained results were compared with the CFEM available in ABAQUS software and the X-FEM approach, which was implemented in the ABAQUS user-defined element (UEL) [53]. Finally, to verify the results obtained by this strategy of X-IGA application, we made a comparison with [51].

The present study begins with a brief discussion of the IGA concept, including an assessment of the B-spline and NURBS basis functions. The concept of discontinuity inside a continuum formulation is outlined to provide a background for the following discussion on the X-IGA. Later, the methodology for implementing this improved approach is introduced. The paper concludes by providing case studies of a pipe under inner uniform pressure. The obtained results of the present study were evaluated against the CFEM and X-FEM solutions.

## 2. Outline on B-splines, NURBS, and IGA Concepts

Piecewise polynomial functions with a specified degree of continuity are known as B-splines. In a knot vector, vector $\Xi = \{\xi_1, \xi_2, \ldots, \xi_{n+p+1}\}$, with $\xi_{i+1} \geq \xi_i$; $i$ is the index of knots, $n$ is the number of control points, and $p$ is the polynomial degree. A collection of coordinates in parametric space is used to create univariate B-Spline shape functions. The Cox-de Boor recursion model on the appropriate knot vectors determines the $i$th B-splines basis functions $N_{i,p}(\xi)$ [14].

For a polynomial order $p = 0$,

$$N_{i,0}(\xi) = \left\{ \begin{array}{c} 1 \text{ if } \xi_i \leq \xi < \xi_{i+1} \\ 0 \text{ otherwise} \end{array} \right\} \quad (1)$$

For $p \geq 1$,

$$N_{i,p}(\xi) = \frac{\xi - \xi_i}{\xi_{i+p} - \xi_i} N_{i,p-1}(\xi) + \frac{\xi_{i+p+1} - \xi}{\xi_{i+p+1} - \xi_{i+1}} N_{i+1,p-1}(\xi) \quad (2)$$

The B-splines shape function's first derivative with respect to the parameter is:

$$\frac{d}{d\xi} = (N_{i,p}(\xi)) = \frac{p}{\xi_{i+p} - \xi_i} N_{i,p-1}(\xi) - \frac{p}{\xi_{i+p+1} - \xi_{i+1}} N_{i+1,p-1}(\xi) \quad (3)$$

An example of a cubic B-spline basis function is illustrated in Figure 1. An open knot vector was used in this example due to the multiplicity of the first and the last knots, which are equal to $p + 1$. With IGA, the construction of the geometrical model is done with B-spline functions of open vectors. This type of vector allows for an interpolation of the basis function at the end; it is more than suitable for enforcing the boundary conditions.

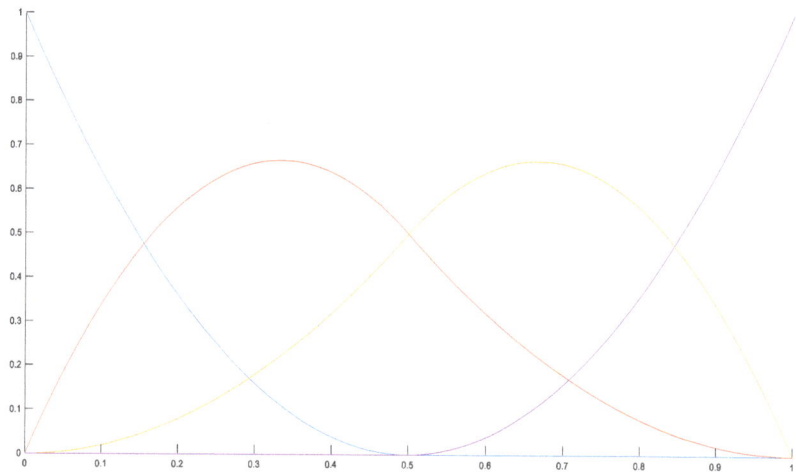

**Figure 1.** An example of cubic B-splines basis functions with knot vector $\Xi = (0, 0, 0, 0.5, 1, 1, 1)$.

*2.1. Curve and Surface Building with B-spline*

A linear combination of the shape functions and coefficients denoted by the control points constructs a B-spline curve:

$$C(\xi) = \sum_{i=1}^{n} N_{i,p}(\xi) B_i \tag{4}$$

$N_{i,p}$ and $B_i$ are the $i$th B-spline function and control points, respectively.

A B-spline surface is defined by a tensor product and parameterized by two-knot vectors, knot vector = $\{ [\xi_1, \xi_2, \ldots, \xi_{n+p+1}], [\eta_1, \eta_2, \ldots, \eta_{n+p+1}] \}$ as:

$$S(\xi, \eta) = \sum_{i=1}^{n} \sum_{j=1}^{m} N_{i,p}(\xi) M_{j,q}(\eta) B_i, \tag{5}$$

where $N_{i,p}(\xi)$ and $M_{j,q}(\eta)$ are univariate shape functions and B is the 3-vector of control point coordinates.

*2.2. Non-Uniform Rational B-splines (NURBS)*

NURBS is the generalization of the B-spline functions:

$$R_{i,p}(\xi) = \frac{N_{i,p}(\xi)\, w_i}{\sum_{\hat{i}=1}^{n} N_{\hat{i},p}(\xi) w_{\hat{i}}} \tag{6}$$

where $w_i > 0$ is the weighting parameter and $N_{i,p}(\xi)$ is the B-spline basis function. For $w_i = 1$, NURBS functions are transformed into B-splines functions. The Rhino Python Editor performs the generation of the weights.

The NURBS shape function's first derivative with respect to the parameter is:

$$\frac{d}{d\xi} R_{i,p}(\xi) = w_i \frac{W(\xi) N'_{i,p}(\xi) - W'(\xi) N_{i,p}(\xi)}{(W(\xi))^2} \tag{7}$$

where $W(\xi) = \sum_{\hat{i}=1}^{n} N_{\hat{i},p}(\xi)\, w_i$ and $W'(\xi) = \sum_{\hat{i}=1}^{n} N'_{\hat{i},p}(\xi) w_i$.

### 2.3. Curve and Surface Building with NURBS

To construct the control points for the NURBS geometry, Equation (8) is utilized:

$$(B_i)_j = \frac{(B_i^w)_j}{w_i}, \quad j = 1, \ldots, k \tag{8}$$

where $(B_i)_j$ is the $j^{th}$ element of the vector $B_i$ and $w_i$ the $i^{th}$ weight. By using Equation (6) in combination with Equation (8), the NURBS curve was constructed as:

$$C(\xi) = \sum_{i=1}^{n} R_{i,p}(\xi) B_i \tag{9}$$

By analogy with the B-spline, NURBS surfaces are constructed from a tensor product through two-knot vector arrays. It is introduced as:

$$S(\xi, \eta) = \sum_{i=1}^{n} \sum_{j=1}^{m} \frac{N_i^p(\xi) M_i^q(\eta) w_{i,j}}{\sum_{i=1}^{n} \sum_{j=1}^{m} N_i^p(\xi) M_i^q(\eta) w_{i,j}} \tag{10}$$

The reformulation (10) can be done by setting up in the following form:

$$S(\xi, \eta) = \sum_{i=1}^{n} \sum_{j=1}^{m} R_{i,j}^{p,q} B_{i,j} \tag{11}$$

where $N_i^p(\xi)$ and $M_i^p(\xi)$ are shape functions, respectively. Then, $p$ and $q$ are the order of the basis function in the two directions and m and n are the numbers of control points in the two directions.

### 2.4. Governing Equation

Let us briefly review the concept of discontinuity inside a continuum formulation. The improved displacement field term is introduced, taking into account the displacement field computation at the discontinuity. The concept of the partition of unity has been used for Lagrangian basis functions; the properties of this technique are also suitable for the B-spline and NURBS functions, which form the basis of the iso-geometric approach. The full displacement field can be expressed as the amount of two subfields by using this property:

$$u^h(x) = \sum_{I=1}^{N} \phi_I(x) \left( \alpha_I + \sum_{J=1}^{M} \beta_{IJ} K_J(x) \right) \tag{12}$$

In Equation (12), $\phi_I$ represents the standard basis function, $K_J$ is the improved basis function with m expressions, and the standard and the improved nodal degree of freedom are $\alpha_I$ and $\beta_{IJ}$, respectively. To develop the approximation, M will assume the value 2, and Equation (12) can be represented as:

$$u^h(x) = \sum_{I=1}^{N} \Phi_I(x) \left[ \alpha_I + H(x) a_I + \sum_{\beta=1}^{4} F_\beta(x) b^\beta{}_I \right] \tag{13}$$

In Equation (13), $K_J$ has been decomposed into two different enrichment terms, the Heaviside function $H(x)$ and the crack-tip function $F_\beta(x)$, in order to capture the discontinuity and the singular fields. $a_I$ and $b^\beta{}_I$ introduce the improved nodal degrees of freedom.

To follow the crack, the level set methodology has been applied. This allows for representing the discontinuity as a moving interface. The signed distance, which defines the location of an arbitrary point with respect to the interface, is still the most popular:

$$\varphi(x) = \| x - \hat{x} \| \ \text{sign}(n_{\Gamma_d} \cdot (x - \hat{x})) \tag{14}$$

where $x$ is a point in a mesh element, $\hat{x}$ is the closet point to x on the discontinuity, and $n_{\Gamma_d}$ is the interface normal.

The position of the crack in a domain is defined by the values of the level set function that is represented as follows:

$$\varphi(x) = \begin{vmatrix} < 0 \text{ if } x \in \Gamma_d^- \\ = 0 \text{ if } x \in \Gamma_d \\ > 0 \text{ if } x \in \Gamma_d^+ \end{vmatrix} \qquad (15)$$

When the body's forces are not present, the elastostatics equation, in its strongest form, is:

$$\vec{div}\,\bar{\bar{\sigma}} = \vec{0} \text{ in } \Sigma \qquad (16)$$

With the appropriate set of boundary conditions:
Essential boundary conditions:

$$\vec{u} = \vec{u_D} \text{ on } \Gamma_u \qquad (17)$$

Natural boundary conditions:

$$\bar{\bar{\sigma}}\,\vec{n} = \vec{T_D} \text{ on } \Gamma_t \qquad (18)$$

Crack surface:

$$\bar{\bar{\sigma}}\,\vec{n}_{\Gamma_d} = \vec{0} \text{ on } \Gamma_d \qquad (19)$$

where $n$ is the normal vector, $n_{\Gamma_d}$ is the normal vector regarding the interface (see Figure 2), $\sigma$ is the stress tensor, and $u$ is the displacement vector.

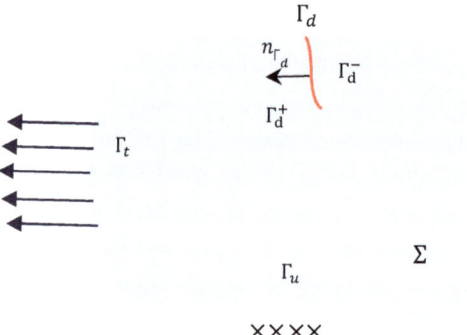

**Figure 2.** Domain $\Sigma$ with discontinuity $\Gamma_d$.

*2.5. Variational Formulation*

As illustrated in Figure 2, a 2-D domain has been considered in this work with conventional boundary conditions: the Dirichlet $\Gamma_u$ and the Neumann $\Gamma_t$ boundary. The crack face presents further traction-free boundaries $\Gamma_c$.

By using the virtual work theory, the weak form of the problem can be established from the equilibrium equation, and is represented as:

$$\mathcal{L}_{eq} = \int_\Sigma (\sigma(u):\varepsilon(q))d\Sigma - \int_\Gamma (T_D.q)d\Gamma \qquad (20)$$

The stress, the strain tensor, and the traction vector are described by $\sigma$, $\varepsilon$, and $T_D$, respectively. $u$ and $q$ denote, respectively, the displacement vector and the virtual displacement.

## 3. Extended Iso-Geometric Analysis (X-IGA)

### 3.1. X-IGA Formulation for Cracks

X-IGA aims to evaluate fractures in an engineering component without having to re-mesh it. In the framework of the X-FEM approach, the enrichment of the shape functions (B-spline, NURBS) is ensured by a Heaviside and crack-tip function since they constitute a partition unit (PU). The former function was introduced to insert a discontinuity and the latter to treat singularities at the crack tip. The approximation of the solution is given as [54]:

$$U^h(\xi) = \sum_{I \in N_{stand}} R_I(\xi) u_I + \sum_{J \in N_{CrSplit}} R_J(\xi) H(\xi) a_J + \sum_{K \in N_{CrTip}} R_K(\xi) (\sum_{\alpha=1}^{4} F_\alpha(\xi) b_K^\alpha) \quad (21)$$

where $R_I$ represents shape functions $u_I$, and $a_J$ and $b_K^\alpha$ define the standard and the further DOFs, respectively. $N_{stand}$ includes the standard nodes of the mesh, $N_{CrSplit}$ includes the nodes of elements that have been split by the crack faces, and $N_{CrTip}$ includes the nodes of the crack-tip elements. The parameter coordinates are represented by $\xi$. The Heaviside function, wherein the quantities on both parts of the split element are different, is denoted by $H(\xi)$. The purpose of the crack-tip function, among others, is to increase the accuracy of the results, and it is denoted by $F_\alpha$. These enrichment functions are characterized by the following equations [55]:

$$H(\xi) = \text{sign}(\varphi(\xi)) = \begin{cases} -1 \text{ if } \varphi(\xi) < 0 \\ +1 \text{ if } \varphi(\xi) > 0 \end{cases}, \quad (22)$$

$$F_\alpha(\xi) = F_\alpha(r, \theta) = \left\{ r^{\frac{1}{2}} \left[ \sin\frac{\theta}{2}, \cos\frac{\theta}{2}, \sin\theta\sin\frac{\theta}{2}, \cos\frac{\theta}{2}\sin\theta \right] \right\} \quad (23)$$

The polar coordinates of the crack tip are represented by $r, \theta$.

After substituting Equation (19) in the equilibrium equation, the final phase of the processing stage is to solve the linear algebra system:

$$K^{enr} U^{enr} = F^{enr}, \quad (24)$$

where $K^{enr}$ denotes an enriched stiffness matrix, $F^{enr}$ denotes a force vector, and $U^{enr}$ denotes an enriched displacement vector as:

$$U^{enr} = \{U \ d \ b_1 \ b_2 \ b_3 \ b_4\}^T, \quad (25)$$

where $U$ is the DOF vector for the IGA normal, $d$ is the DOF for Heaviside enrichment, and $b_1, b_2, b_3$, and $b_4$ are the DOF vectors for the crack-tip enrichment functions. $K^{enr}$ and $F^{enr}$ are constructed as follows from the element stiffness matrix:

$$K^{enr} = \begin{bmatrix} K^{uu} & K^{ua} & K^{ub_1} & K^{ub_2} & K^{ub_3} & K^{ub_4} \\ K^{au} & K^{aa} & K^{ab_1} & K^{ab_2} & K^{ab_3} & K^{ab_4} \\ K^{b_1 u} & K^{b_1 a} & K^{b_1 b_1} & K^{b_1 b_2} & K^{b_1 b_3} & K^{b_1 b_4} \\ K^{b_2 u} & K^{b_2 a} & K^{b_2 b_1} & K^{b_2 b_2} & K^{b_2 b_3} & K^{b_2 b_4} \\ K^{b_3 u} & K^{b_3 a} & K^{b_3 b_1} & K^{b_3 b_2} & K^{b_3 b_3} & K^{b_3 b_4} \\ K^{b_4 u} & K^{b_4 a} & K^{b_4 b_1} & K^{b_4 b_2} & K^{b_4 b_3} & K^{b_4 b_4} \end{bmatrix} \quad (26)$$

$$F^{enr} = \{F^u \ F^a \ F^{b_1} \ F^{b_2} \ F^{b_3} \ F^{b_4}\}^T. \quad (27)$$

The number of control points $N_{stand}$, $N_{CrSplit}$, $N_{CrTip}$, and the number of enrichment basis functions $N_{ef}$ determine the size of each $K^{enr}$ and $F^{enr}$. Each component can be written as follows:

$$K_{ij}^{rs} = \int_{\Sigma_e} (B_i^r)^T C B_j^s \, d\Sigma \ (r, s = u, d, b_1, b_2, b_3, b_4) \quad (28)$$

$$F_i^u = \int_{\Gamma_u} R_i^T T_D d\Gamma, \tag{29}$$

$$F_i^a = \int_{\Gamma_t} R_i^T H T_D d\Gamma, \tag{30}$$

$$F_i^{b_\alpha} = \int_{\Gamma_t} R_i^T f_{b_\alpha} T_D d\Gamma \ (\alpha = 1, 2, 3, 4) \tag{31}$$

$$B_i^u = \begin{bmatrix} \frac{\partial R_i}{\partial x} & 0 \\ 0 & \frac{\partial R_i}{\partial y} \\ \frac{\partial R_i}{\partial y} & \frac{\partial R_i}{\partial x} \end{bmatrix} \tag{32}$$

$$B_i^a = \begin{bmatrix} \frac{\partial R_i}{\partial x} H & 0 \\ 0 & \frac{\partial R_i}{\partial y} H \\ \frac{\partial R_i}{\partial y} H & \frac{\partial R_i}{\partial x} H \end{bmatrix} \tag{33}$$

$$B_i^{b_\alpha} = \begin{bmatrix} \frac{\partial R_i f_{b_\alpha}}{\partial x} & 0 \\ 0 & \frac{\partial R_i f_{b_\alpha}}{\partial y} \\ \frac{\partial R_i f_{b_\alpha}}{\partial y} & \frac{\partial R_i f_{b_\alpha}}{\partial x} \end{bmatrix} (\alpha = 1, 2, 3, 4) \tag{34}$$

The Heaviside function (Equation (22)) is represented by H, while the crack-tip enrichment function (Equation(23)) is represented by $f_{b_\alpha}$.

### 3.2. Construction of 2-D Pipe with X-IGA Concept

A patch with an internal interface is used to model a 2-D pipe problem (Figure 3); this interface is the result of coinciding control points in the circumferential direction at the beginning and end of the patch. To solve the problem correctly, we must confine the control values of these overlapping control points so that each pair of control values for the corresponding coincident control points is the same. The master–slave approach is a simple solution to this problem. This method is implemented in the present study by the global renumbering of DOFs in the physical space, where coincident DOFs are numbered by the same indices. The global numbering is saved in an array and given as an additional input argument.

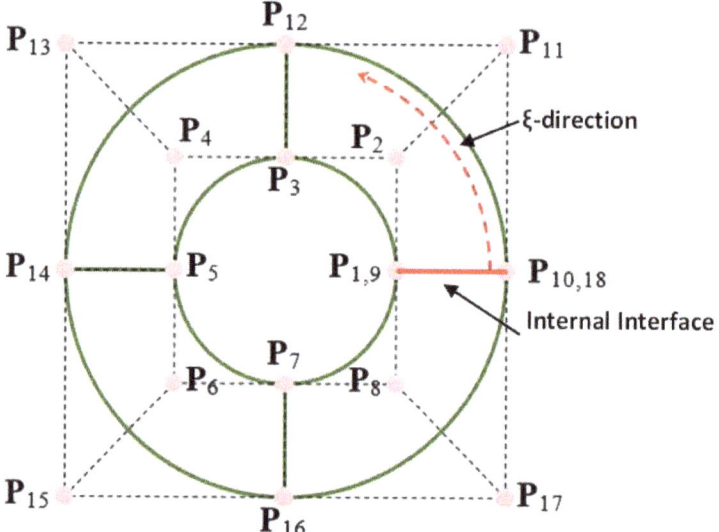

**Figure 3.** 2-D pipe with the iso-geometric analysis approach.

## 3.3. Enrichment Topology for Control Points

In IGA, every basis function is linked to its corresponding control point in a specific way. The intersection of each supported domain with the crack face leads to enrichment by the Heaviside function. The domain support, which contains the crack tip, will be enriched by the singular function. According to [56], there are two ways to enrich the crack tip: with either topological or geometrical enrichment. In this work, topological enrichment has been employed. Figure 4 shows a schematic representation of this concept.

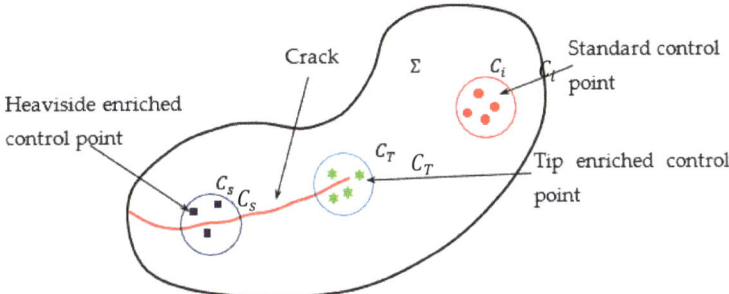

**Figure 4.** The enrichment concept used in NURBS modelling [57].

According to the Figure 4, $C_s$ denotes the crack-face control points, while $C_T$ denotes the crack-tip-enriched control points, and $C_i$ denotes the standard control point. For the purpose of selecting enriched control points, the level set method has been used. We applied the procedure that has been used by [58]. Initially, the level set values of the crack at the mesh's vertices are computed according to these level sets, and the formulation determines the elements intersected by the crack and the crack-tip element.

## 4. Numerical Integration in the Elastic Field

For numerical integration, the standard Gaussian quadrature method cannot be applied explicitly to the XIGA since it contains various discontinuous elements. To assess the integration rule of the crack's split and tip elements in this study, the triangular sub-domain methodology is used (Figure 5). This technique has been successfully implemented by X-FEM [42], as seen in Figure 6. Each sub-triangle element has a defined number of integration points.

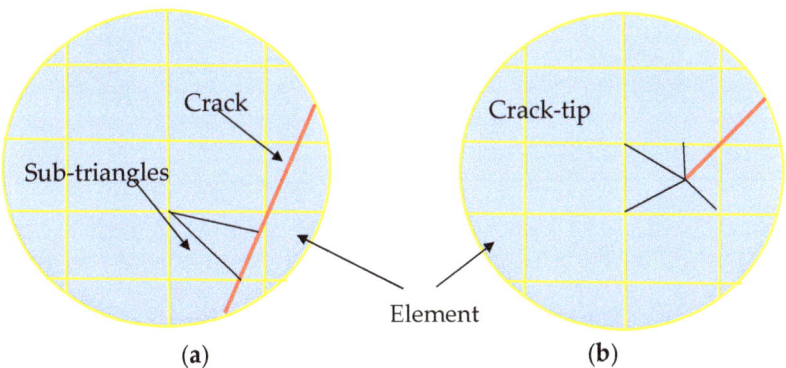

**Figure 5.** Partitioning the crack face (**a**) and crack tip (**b**) using the sub-triangulation technique [42].

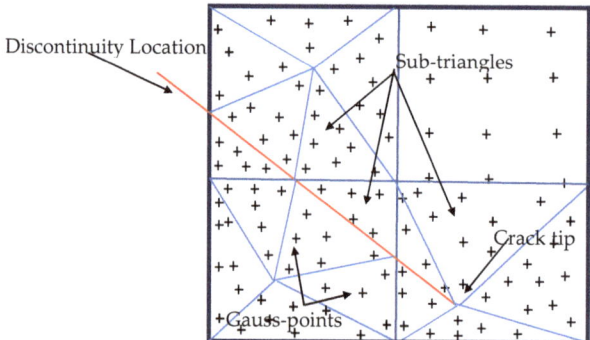

**Figure 6.** Sub-triangles technique around the crack and the distribution of the Gauss point.

## 5. Fracture Parameter Evaluation

The stress intensity factor (SIF) is a crucial criterion in crack formation and growth research. As a result, when a numerical simulation is used to solve fracture mechanics issues, one of the objectives is to quantify the SIF. There are a range of ways to calculate this parameter, which is used in so much of the interaction integral method [55]:

$$M = \int_\Gamma \left[ \sigma_{ij}^{act} \frac{\partial u_i^{aux}}{\partial x_1} + \sigma_{ij}^{aux} \frac{\partial u_i^{act}}{\partial x_1} - W^M \delta_{1j} \right] \frac{\partial q}{\partial x_j} d\Gamma, \quad (35)$$

where $M$ is the interaction integral with the aux and act index defining the auxiliary and the actual state, respectively. The stress and the displacement are represented by $\sigma_{ij}$ and $u_i$, respectively. $W^M$ represents the interaction work; it can be described in the following form:

$$W^M = \frac{1}{2}\left(\sigma_{ij}^{act}\varepsilon_{ij}^{aux} + \sigma_{ij}^{aux}\varepsilon_{ij}^{act}\right), \quad (36)$$

where $\varepsilon_{ij}$ and $q$ define the strain and an arbitrary function, whose values are defined as:

$$q = \begin{cases} 1, |at\ the\ crack\ tip \\ 0, |along\ the\ contour \end{cases} \quad (37)$$

The stress intensity factor with these two modes (I, II) and the J-integral are related by:

$$J = \frac{1}{E'}\left(K_I^2 + K_{II}^2\right). \quad (38)$$

Based on this correlation, the following equation has been derived:

$$M = \frac{2}{E'}\left(K_I^{act}K_I^{aux} + K_{II}^{act}K_{II}^{aux}\right) \quad (39)$$

where

$$E' = \begin{cases} E\ for\ plane\ stress \\ \frac{E}{1-\nu^2}\ for\ plane\ strain \end{cases} \quad (40)$$

$E$ represents the Young's Modulus, $v$ is the Poisson's ratio, and $M$ represents the interaction integral.

The crack opening is the field of interest of this study. Therefore, the SIF in mode $I$ can be obtained by choosing $K_I^{aux}=1$ and $K_{II}^{aux} = 0$. Finally, the SIF is defined as:

$$K_I = \frac{E'}{2}M \quad (41)$$

## 6. Process of Implementing a X-FEM Code in ABAQUS

The idea of implementing the X-FEM technique is based on the fact that the ABAQUS software does not include the stress intensity factor computation for a 2-D crack. The two enriched functions that correspond to the Heaviside and the crack-tip functions that appeared in Equation (13) were implemented through a user-defined element (UEL). The implementation process requires three phases: pre-processing, processing, and post-processing. In this study, to make the implementation easier, the input file (XFEM.inp) has been generated by ABAQUS, and then the interaction between the crack and the mesh was constructed to apply Equation (14). Therefore, the enriched elements and nodes have been determined. The user-defined element (UEL) in ABAQUS is used to program the processing stage (UEL.for) to compute the stresses and strains. The last stage consists of calculating the Mode-I SIF, KI, by a post-processing code; indeed, the interaction integral method explained in section five (Fracture Parameter Evaluation) has been programmed in FORTRAN. The details for the file descriptions are based on [53], who introduce the X-FEM implementation in ABAQUS. Thus, we have adopted the principle of this implementation for our problem. The ABAQUS command is used to run the main file (XFEM.inp) and the user subroutine file (UEL.for): Abaqus job = X-FEM user = UEL.for.

The resulting information of this simulation is stored in (XFEM.fil), and the ABAQUS output conventions are included in this file. Then, we used an external subroutine for computing the stress intensity factors. This subroutine is compiled through the ABAQUS command "ABAQUS make job = interaction_integral", and then run with the command "ABAQUS interaction_integral". The implementation process is described in Figure 7.

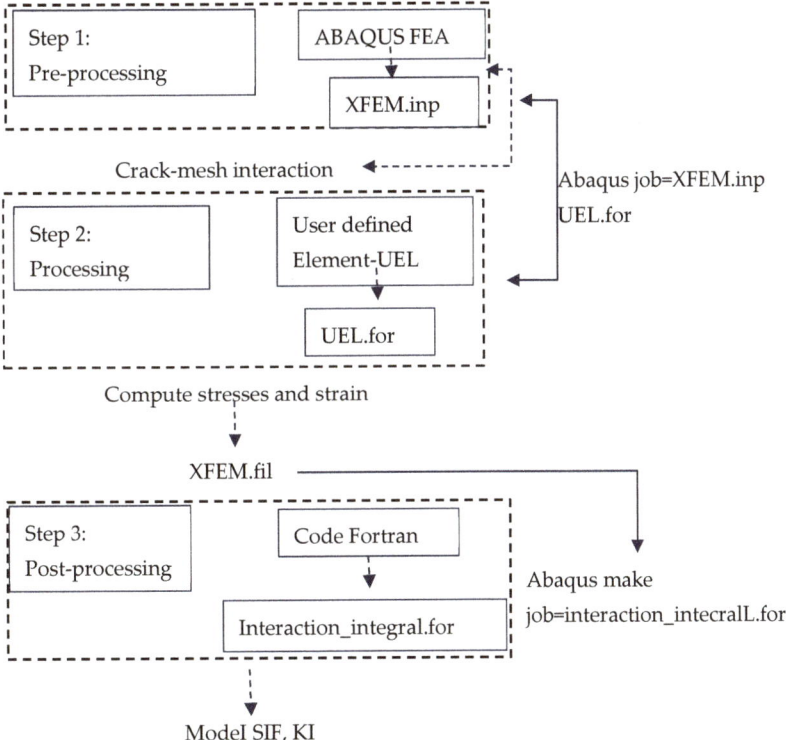

**Figure 7.** X-FEM implementation process in a Finite Element Code.

## 7. Process of Implementing a X-IGA Code in MATLAB

In this section, the main steps of the XIGA implementation code to numerically model a structure with a pre-existing discontinuity (a crack) have been summarized in Figure 8. To understand the implementation, it is necessary to know two major things: the three steps of a finite element code (pre-processing, processing, and post-processing), and the identification of a crack using the level set method (LSM). The flowchart starts with the introduction of the input parameters, which contain the geometrical data and the material properties. Then, the elasticity matrix is integrated. In addition, the data required by the NURBS functions, such as the polynomial order, control points, and node vectors, are introduced to build the NURBS model. The next step is to introduce the crack data, length, and coordinates of the crack points with the level set method to determine the position of the crack and to select the enrichment points. Then, the Heaviside and crack-tip functions are imposed on the nodes according to the technique that precedes this step. Then, the boundary conditions, nodal force vector, and stiffness matrix are calculated. The stress, strain, and displacement values are the output of this process. These data can be put to use in the interaction integral to compute the stress intensity factor.

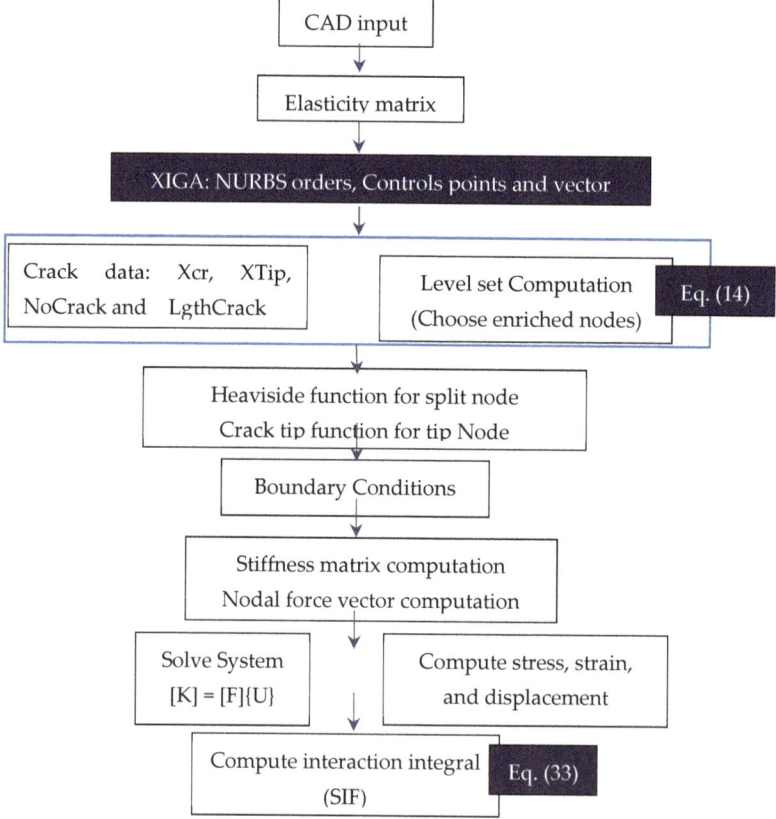

**Figure 8.** Flowchart of the implementation.

## 8. Numerical Results and Discussions

The iso-geometric analysis is extended and used in this section to analyze the cracked model under mechanical loading; a specimen from the industrial field is chosen. Pressure pipes are commonly used and the performance of these systems is still under investigation. In a 2-D linear static analysis case, an isotropic and homogeneous pipe has been studied. The plane stress is thought to be the stress distribution state. For this analysis, The P264GH material is used and its mechanical and chemical characteristics are described in Tables 1 and 2, respectively.

**Table 1.** Mechanical characteristics for the P264GH [59].

| Properties | Values |
|---|---|
| Young's Modulus | 207 GPa |
| Poisson ratio | 0.3 |
| Yield Stress | 340 MPa |
| Ultimate tensile strength | 440 MPa |
| Elongation to fracture | 35% |
| Fracture Toughness | 95 MPa$\sqrt{m}$ |

**Table 2.** Chemical characteristic of P264GH—% by weight [60].

| Material | C | P | Al | Mn | S | Si | Fe |
|---|---|---|---|---|---|---|---|
| Tested steel | 0.135 | 0.013 | 0.027 | 0.665 | 0.002 | 0.195 | Bal. |
| P264GH steel according to the Standard EN10028.2-92 | 0.18 | 0.025 | 0.02 | 1 | 0.015 | 0.4 | Bal. |

To examine the reliability, accuracy, and efficiency of this approach, an external axial crack was studied. The results of the analysis were compared with the results of standard ABAQUS software using the classical finite element method (CFEM), which is based on the integral contour and the X-FEM method using a subroutine UEL code.

In each of these parametric directions $\xi$ and $\eta$, the degree of the NURBS polynomial is two ($p = q = 2$). In the first direction $\xi$, the knot vector is open and with interior duplication, and in the second direction $\eta$, the knot vector is open and without internal duplication.

The integration is done along each Gauss point direction $(p + 1) \times (q + 1)$, and as the sub-triangles approach was used in this case, 13 Gauss quadrature points were imposed for each sub-triangle. For each numerical method, different mesh sizes were examined. It is important to be aware that the crack was represented as a straight segment.

In order to figure out the fracture parameter, i.e., to estimate the stress state near a crack tip, the stress intensity factor (SIF), KI, and the J-Integral are extracted. For the three techniques used in this work, the interaction integral method was implemented. The X-IGA technique was implemented in MATLAB code. The ABAQUS software was used for the CFEM and X-FEM to extract the KI, but for the X-FEM, it is important to mention that the software does not support the computation of this parameter in the 2-D domain, which led to the use of a subroutine UEL.

*8.1. Two-Dimensional Pipe with an Axial Crack under Uniform Pressure*

In the present study, an isotropic and homogenous 2-D pipe including a 2 mm straight edge crack with uniform pressure distribution and an outer and inner radius of Ro=20 mm and Ri=10 mm, respectively, is examined. Three models were used in this study, which included 320, 480, and 640 element numbers for the X-IGA method, and 1297, 3844, and 85,942 element numbers for the CFEM, with a step size of 1, 0.5, and 0.1, respectively, and 470, 1265, and 2747 element numbers for X-FEM with a step size of 1, 0.45, and 0.25 around the crack, respectively. The NURBS geometry is represented by using several patches (subdomains) with an internal interface. In the circumferential direction, the control points coincide with each other on the patch, and this implies the creation of interface terms. The NURBS geometry is illustrated in Figure 9a.

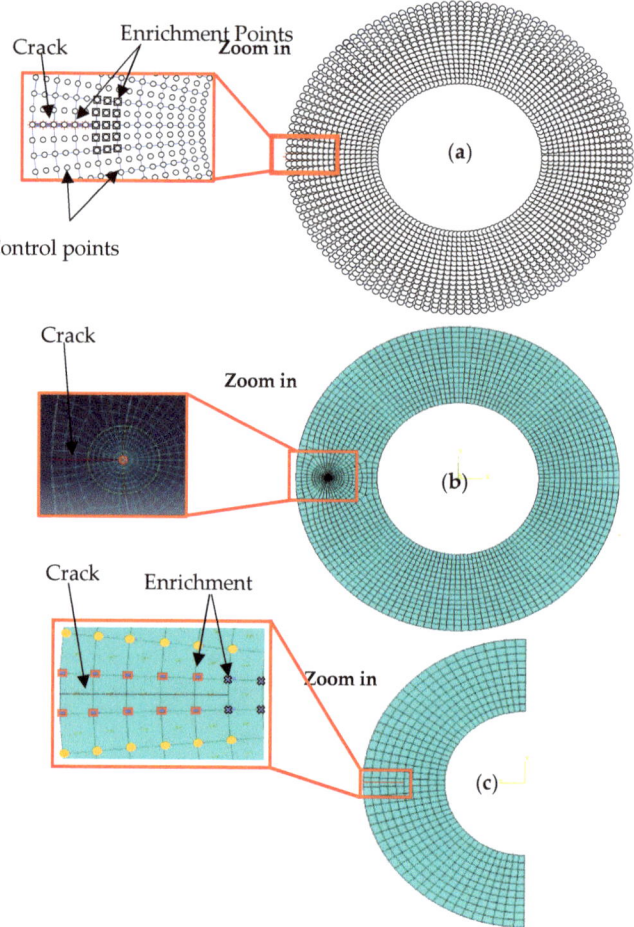

**Figure 9.** Geometry construction: (**a**) the NURBS model, (**b**) the CFEM model, (**c**) X-FEM model.

The FEM model has been represented by a finite element mesh on ABAQUS software. The most important thing is that the modeling of the crack requires a specific treatment, i.e., it requires building a particular mesh around the crack. A ring of a triangle, as represented in Figure 9b, is formed at the crack tip, along with concentric layers of structured quads [61].

For the X-FEM method (Figure 9c), the meshing was done in a simple way since the crack was modeled independently of the meshing. It should be noted that, for this technique, a half tube was treated because the implementation is heavy when using UEL subroutines, so to overcome this problem, a half tube was analyzed with the symmetry boundary conditions.

To perform the numerical simulation, we present Table 3, which summarizes the crack length, crack position, and pressures applied.

**Table 3.** Data details for the numerical simulation.

| Crack Length Ratios (a/t) | Crack Position $(x_1\ y_1;\ x_2\ y_2)$ (mm) | Applied Pressure (MPa) |
|---|---|---|
| 0.2, 0.3, 0.4, 0.5, 0.6, 0.7, 0.8 | $(-20\ 0;\ (-20 + a)\ 0)$ | 2.5 |

Figure 10 illustrates the distribution of the Von Mises stress by the three techniques for $\frac{a}{t} = 0.3$, and a pressure of 2.5 MPa [52] has been applied. The accuracy of the stress and strain distribution in a geometry takes a very important place in fracture mechanics, especially when it concerns cracking problems.

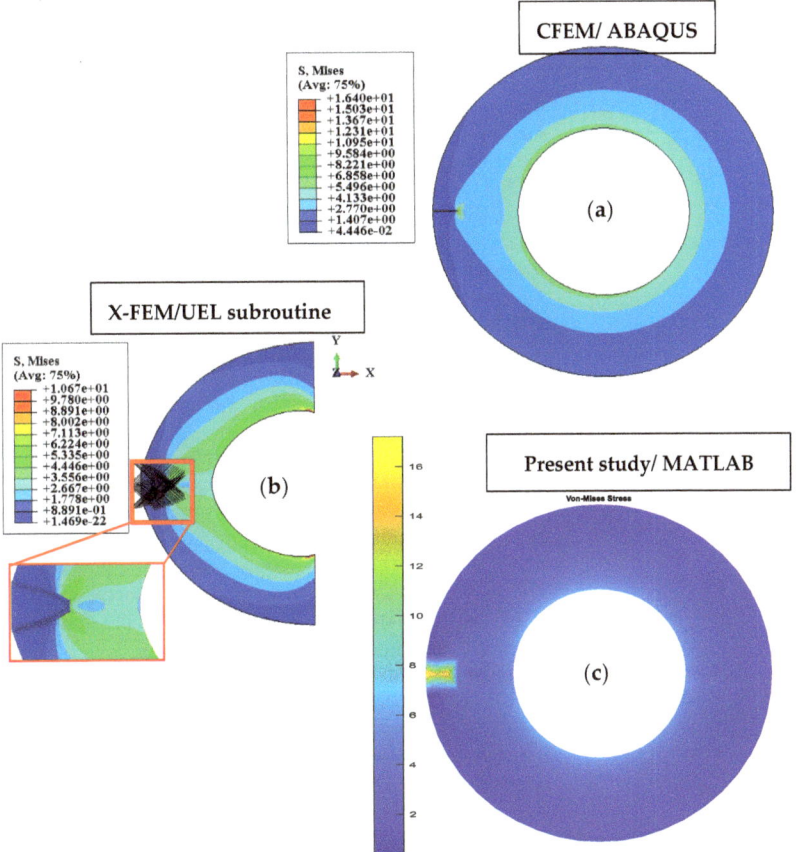

**Figure 10.** The distribution of Von Mises stress: (**a**) with FEM analysis, (**b**) with X-FEM analysis, (**c**) with the improved technique (X-IGA).

The zone of interest is the crack tip where the degree of damage of the defect has been known. Obtaining a regularity of stresses in this region is a priority for numerical methods since the calculation of fracture parameters, such as the stress intensity factor, as well as the angle of deviation of the crack propagation, is based on the value of stresses at the crack tip.

By comparison of the three results of Figure 10, there is a similarity in the results of the numerical simulation. However, some errors can appear when using these numerical techniques. The errors' origins are detailed as follows: with the CFEM analysis, the field of strains and stresses become singular at the crack tip, even with the integration of the singularity in the model to improve the accuracy of the results. With the implementation of the X-FEM method via user subroutine UEL in ABAQUS, a mesh sensitivity affects the simulation results. For the present study, there is no singularity at the crack tip and there is a minor sensitivity of mesh. The next section of this work shows a calculation of different meshes for all the different methods. The X-IGA method implemented in MATLAB can be an alternative to these numerical methods. The results presented in Figure 10 can support this conclusion because 4112 elements were used for the full tube with the CFEM method, 2747 elements for the half tube with the X-FEM method, and only 384 elements for the XIGA method. For this reason, the evaluation of the stresses around an existing crack on a pipe can be made by the present study with large elements and with a weak error, which appears in the solution discretization.

*8.2. Evaluation of the Fracture Parameter*

In order to present the accuracy of the X-IGA technique, as well as the regularity of the stress distribution around the crack tip, the SIFs were extracted and the value of the mode I ($K_I$) was calculated in this study, since the degree of damage that corresponds to the opening of the crack is more severe than with the other modes [62]. To ensure that this study is inscribed in linear elasticity, a pressure of 2.5 MPa [52] was applied. Three models are evaluated for different mesh sizes. Therefore, the result of CFEM, X-FEM, and X-IGA are compared. Moreover, the interaction integral approach, which is known as M-integral, was used for calculating the SIFs.

For this comparison, the depth of the crack was varied from the thickness of the model a/t; the thickness is $t = 10$ mm and $a = 2, 3, \ldots, 8$ mm. Figures 11 and 12 illustrate the results of the computation. The comparison between the results obtained by the X-IGA and CFEM methods shows a good similarity. It is observed that fine meshes must be used to discretize the areas around the crack tips for the CFEM, whereas the meshes are coarser for the XIGA. It is clear to observe that only 640 elements are required by the present study to obtain the result of 1297 elements with the CFEM method. Results obtained by implementing X-IGA in MATLAB are in good agreement with the CFEM method implemented in ABAQUS. The mentioned technique did not require much effort in dealing with the mesh in the crack tip, whereas using ABAQUS required a specific meshing technique, due to the singularity at the crack tip, and this can affect the numerical results.

As for the comparison between the X-IGA and X-FEM, the results show a good similarity. It is also observed that, with X-FEM, a fine mesh must be used without a particular treatment around the crack, since the crack is modelled independently of the mesh. It is clear to observe that the present study requires just 640 elements for obtaining the result of 2747 elements using X-FEM. The same enrichment functions for overcoming the problem of singularity are used for both methods and only the shape functions are modified. Therefore, there is a similarity of results between the two numerical methods. When the depth of the crack approaches the inside of the pipe, the SIFs values become maximized. It is clear to see that whatever technique is employed to simulate the crack, the SIFs gradually increase with the crack length.

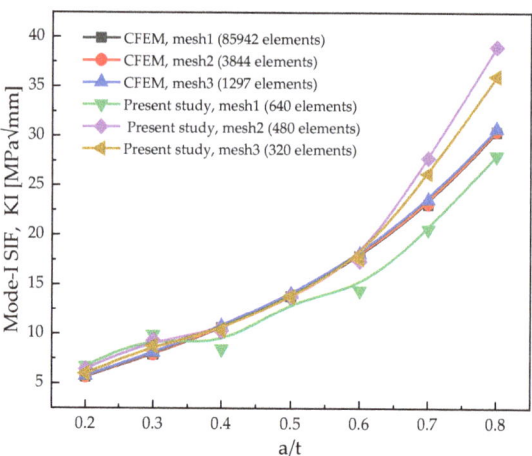

**Figure 11.** The comparison between the CFEM and X-IGA method for a 2-D cracked pipe.

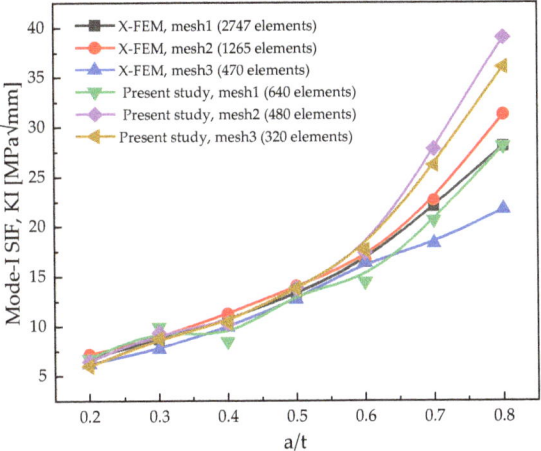

**Figure 12.** The comparison between the X-FEM and X-IGA method for a 2-D cracked pipe.

By these results, implementing X-IGA in MATLAB shows the possibility to calculate the fracture mechanics parameters for a cracked pipe under uniform pressure with a large mesh size, compared to the other approaches (Figure 12). Therefore, it can be an alternative way to evaluate the damage of a cracked pipe.

In addition to the SIF, the J-integral has specific importance when it comes to the numerical stress analysis of cracks. To give more physical meaning to the analysis, and to validate the strategy used in the application of X-IGA to address the cracking pipe problem, we evaluated the J-integral with various crack lengths, as shown in Figure 13. Here, we used 640 elements for the present study, and 2747 and 85,642 elements for X-FEM and CFEM, respectively. The J-integral value increased gradually with the crack length and the results obtained are similar to other numerical results.

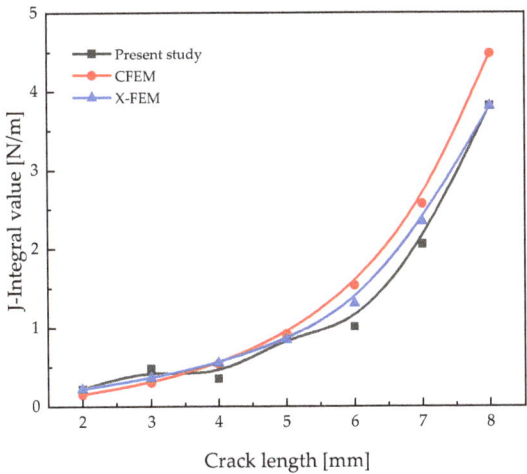

**Figure 13.** J-integral value obtained by the present study, X-FEM, and CFEM.

This problem is also analyzed by [51]; they used a UEL subroutine in the ABAQUS software to evaluate the stress intensity factor. With the same element number and for $a = 5$ mm and $p = 2.5$ MPa, we evaluated the performance of the present study and the results of both implementations are compared with Folias solutions [63], as illustrated in Table 4.

**Table 4.** Comparative study of the present investigation and implementation using Fortran for $a$ = 5 mm.

| Method/ Implementation | X-IGA/ Fortran [51] | XIGA/ Fortran [51] | Present Study MATLAB | Present Study MATLAB | Folias Solution [63] |
|---|---|---|---|---|---|
| Element Number | 470 | 767 | 470 | 767 | — |
| KI (MPa$\sqrt{mm}$) | 13.18 | 13.015 | 13.85 | 13.51 | 14.37 |
| Error (%) | 8.306 | 9.454 | 3.645 | 6.01 | |

Table 3 proves the significance of the present study implemented in MATLAB. It is interesting to note that, in the present study, the error for a model with 470 elements is 3.645%, while for the study implemented in Fortran the error is 8.306%. Additionally, with 767 elements, the error for the present study is 6.01%, while for the study implemented in Fortran the error is 9.454%. It is observed that the X-IGA implementation strategy that was followed in the present study had a higher accuracy than the strategy followed by [51].

In addition, the accuracy of the present study can be realized by computing the error of SIFs (%) with various crack lengths, as illustrated in Figure 14. It is noted that the maximum error of the present study is 12.25%, while the study implemented in Fortran [51] is 27.83%.

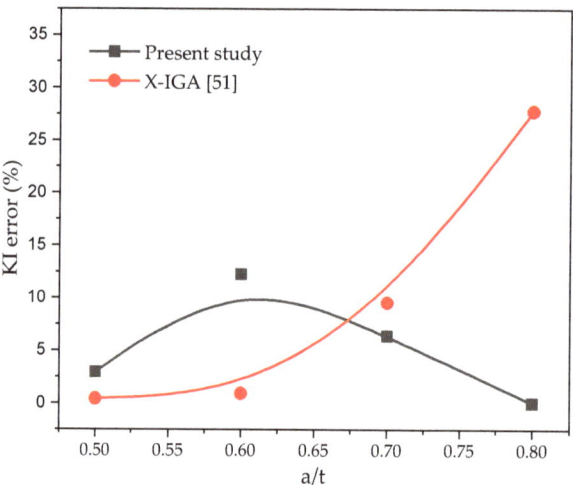

**Figure 14.** Error of SIFs with various crack length for the present study and the reference [51].

### 8.3. Effect of Pressure on the Fracture Parameter Calculation

Finally, in order to check the efficiency of the present study for the calculation of the fracture parameter in the 2-D pipe domain, the inner pressure was varied with the initial crack length $a$ = 4 mm, so that the pressure did not exceed 4.5 MPa [64], keeping the analysis in the elastic domain. The results of the SIFs and the J-Integral of CFEM, X-FEM, and X-IGA are presented in Figure 15. It is observed that for the three-analyses technique, the stress intensity factor increases with the increase of inner pressure. It is also observed that the results for both the X-FEM and X-IGA methods became similar each time the pressure was increased; this is due to the use of enrichment functions at the crack tip. The singularity at the crack tip and the mesh dependency for the CFEM method makes their results inaccurate.

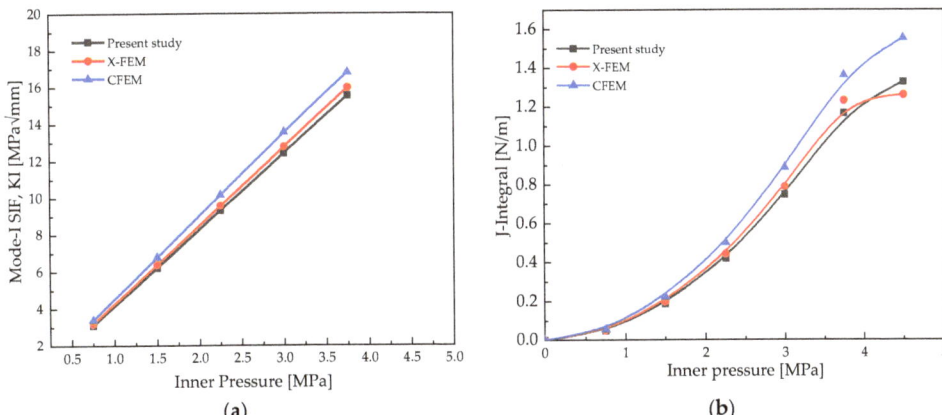

**Figure 15.** (a) SIFs and (b) J-Integral values for different pressures by the CFEM, X-FEM, and EX-IGA methods.

From these results, the X-IGA method can be used in the fracture parameters computation on cracked cylindrical structures under uniform pressure, the X-IGA method has been well validated for the calculation of cracked plates [48], and this study is an extension of the research work that has already been done in this field. The comparison between the most-known methods in the field of numerical computation will allow us to justify the role of this new technique: it can replace the existing methods in the current calculation codes in the future, and the validation of the efficiency of this technique by the different research works will convince the users in the industrial field. This is due to several reasons; the most important reason, for the industrial sector, is the cost of calculation. Instead of designing the model and approximating it by several meshing processes, it is enough to use the geometry directly by the same shape functions, named B-spline or NURBS; after that, the model will be reproduced precisely. It has been observed that X-IGA just needs a small element for a crack in a pipe, compared to other techniques, and that with these elements, the same results have been obtained, so if there is a complex geometry, the calculation procedure will be very fast compared to the current method.

## 9. Conclusions

The aim of this investigation was to implement the X-IGA technique into a MATLAB code for modeling cracked pipelines. The main theoretical approach was the NURBS principle, which provided a higher-order continuity in numerical modeling.

An external crack in a two-dimensional pipe subjected to a uniform pressure has been studied. The accuracy of this technique has been examined by deriving the stress intensity factors and the J-integral. For several mesh sizes and for different inner pressures, SIFs and the J-integral were extracted by X-IGA analysis using MATLAB code and its accuracy was validated with the enrichment technique (X-FEM) using FORTRAN language and the conventional finite element method (X-FEM) using ABAQUS software. It has been shown that, when using the X-IGA analysis:

- The cracked pipe modeling does not need a finer mesh than other numerical techniques. Therefore, the cost of computational will be reduced;
- The regularity of the stress and strain at the crack tip is obtained;
- The geometry was constructed exactly with using NURBS, which avoids the discretization error. Therefore, confident results can be achieved;
- The error on the SIFs is minimal compared to X-IGA implemented by FORTRAN.

Other problems can be addressed by the provided technique, such as crack-growth problems and dynamic fracture analyses in pipelines, which are planned for future research.

**Author Contributions:** All the authors conceived the framework and structured the whole manuscript, checked the results, and completed the revision of the paper. The authors have equally contributed to the elaboration of this manuscript. All authors have read and agreed to the published version of the manuscript.

**Funding:** This research received no external funding.

**Institutional Review Board Statement:** Not applicable.

**Informed Consent Statement:** Not applicable.

**Data Availability Statement:** Not applicable.

**Acknowledgments:** The authors thank Catalin Iulian Pruncu from the University of Strathclyde, UK, for proofreading assistance and overall guidance, which made drafting this work possible.

**Conflicts of Interest:** The authors declare no conflict of interest.

## References

1. Barsoum, R.S. On the Use of Isoparametric Finite Elements in Linear Fracture Mechanics. *Int. J. Numer. Methods Eng.* **1976**, *10*, 25–37. [CrossRef]
2. Cheung, S.; Luxmoore, A.R. A Finite Element Analysis of Stable Crack Growth in an Aluminium Alloy. *Eng. Fract. Mech.* **2003**, *70*, 1153–1169. [CrossRef]
3. Pasetto, M.; Baek, J.; Chen, J.-S.; Wei, H.; Sherburn, J.A.; Roth, M.J. A Lagrangian/Semi-Lagrangian Coupling Approach for Accelerated Meshfree Modelling of Extreme Deformation Problems. *Comput. Methods Appl. Mech. Eng.* **2021**, *381*, 113827. [CrossRef]
4. Moës, N.; Belytschko, T. Extended Finite Element Method for Cohesive Crack Growth. *Eng. Fract. Mech.* **2002**, *69*, 813–833. [CrossRef]
5. Montassir, S.; Yakoubi, K.; Moustabchir, H.; Elkhalfi, A.; Rajak, D.K.; Pruncu, C.I. Analysis of Crack Behaviour in Pipeline System Using FAD Diagram Based on Numerical Simulation under XFEM. *Appl. Sci.* **2020**, *10*, 6129. [CrossRef]
6. Yakoubi, K.; Montassir, S.; Moustabchir, H.; Elkhalfi, A.; Pruncu, C.I.; Arbaoui, J.; Farooq, M.U. An Extended Finite Element Method (XFEM) Study on the Elastic T-Stress Evaluations for a Notch in a Pipe Steel Exposed to Internal Pressure. *Mathematics* **2021**, *9*, 507. [CrossRef]
7. Yang, L.; Yang, Y.; Zheng, H. A Phase Field Numerical Manifold Method for Crack Propagation in Quasi-Brittle Materials. *Eng. Fract. Mech.* **2021**, *241*, 107427. [CrossRef]
8. Partridge, P.W.; Brebbia, C.A. *Dual Reciprocity Boundary Element Method*; Springer Science & Business Media: Berlin/Heidelberg, Germany, 2012; ISBN 94-011-3690-4.
9. Bazazzadeh, S.; Mossaiby, F.; Shojaei, A. An Adaptive Thermo-Mechanical Peridynamic Model for Fracture Analysis in Ceramics. *Eng. Fract. Mech.* **2020**, *223*, 106708. [CrossRef]
10. Daux, C.; Moës, N.; Dolbow, J.; Sukumar, N.; Belytschko, T. Arbitrary Branched and Intersecting Cracks with the Extended Finite Element Method. *Int. J. Numer. Methods Eng.* **2000**, *48*, 1741–1760. [CrossRef]
11. Belytschko, T.; Black, T. Elastic Crack Growth in Finite Elements with Minimal Remeshing. *Int. J. Numer. Methods Eng.* **1999**, *45*, 601–620. [CrossRef]
12. Andrade, H.; Leonel, E. An Enriched Dual Boundary Element Method Formulation for Linear Elastic Crack Propagation. *Eng. Anal. Bound. Elem.* **2020**, *121*, 158–179. [CrossRef]
13. Bhardwaj, G.; Singh, I.; Mishra, B.; Bui, T. Numerical Simulation of Functionally Graded Cracked Plates Using NURBS Based XIGA under Different Loads and Boundary Conditions. *Compos. Struct.* **2015**, *126*, 347–359. [CrossRef]
14. Hughes, T.J.; Cottrell, J.A.; Bazilevs, Y. Isogeometric Analysis: CAD, Finite Elements, NURBS, Exact Geometry and Mesh Refinement. *Comput. Methods Appl. Mech. Eng.* **2005**, *194*, 4135–4195. [CrossRef]
15. Benson, D.; Bazilevs, Y.; Hsu, M.-C.; Hughes, T. A Large Deformation, Rotation-Free, Isogeometric Shell. *Comput. Methods Appl. Mech. Eng.* **2011**, *200*, 1367–1378. [CrossRef]
16. Chen, Y.; Lin, S.; Faruque, O.; Alanoly, J.; El-Essawi, M.; Baskaran, R. Current Status of Lsdyna Isogeometric Analysis in Crash Simulation. In Proceedings of the 14th International LS-DYNA Conference, Detroit, MI, USA, 12–14 June 2016.
17. Latimer, C.; Kópházi, J.; Eaton, M.; McClarren, R. Spatial Adaptivity of the SAAF and Weighted Least Squares (WLS) Forms of the Neutron Transport Equation Using Constraint Based, Locally Refined, Isogeometric Analysis (IGA) with Dual Weighted Residual (DWR) Error Measures. *J. Comput. Phys.* **2021**, *426*, 109941. [CrossRef]
18. Occelli, M.; Elguedj, T.; Bouabdallah, S.; Morançay, L. LR B-Splines Implementation in the Altair RadiossTM Solver for Explicit Dynamics IsoGeometric Analysis. *Adv. Eng. Softw.* **2019**, *131*, 166–185. [CrossRef]
19. Elguedj, T.; Duval, A.; Maurin, F.; Al-Akhras, H. Abaqus User Element Implementation of NURBS Based Isogeometric Analysis. In Proceedings of the 6th European Congress on Computational Methods in Applied Sciences and Engineering, Vienna, Austria, 10–14 September 2012; pp. 10–14.

20. Duval, A.; Elguedj, T.; Al-Akhras, H.; Maurin, F. AbqNURBS: Implémentation D'éléments Isogéométriques Dans Abaqus et Outils de Pré-et Post-Traitement Dédiés; CSMA: Giens, France, 2015.
21. Lai, Y.; Zhang, Y.J.; Liu, L.; Wei, X.; Fang, E.; Lua, J. Integrating CAD with Abaqus: A Practical Isogeometric Analysis Software Platform for Industrial Applications. *Comput. Math. Appl.* **2017**, *74*, 1648–1660. [CrossRef]
22. Xue, Y.; Jin, G.; Ding, H.; Chen, M. Free Vibration Analysis of In-Plane Functionally Graded Plates Using a Refined Plate Theory and Isogeometric Approach. *Compos. Struct.* **2018**, *192*, 193–205. [CrossRef]
23. Chen, D.; Zheng, S.; Wang, Y.; Yang, L.; Li, Z. Nonlinear Free Vibration Analysis of a Rotating Two-Dimensional Functionally Graded Porous Micro-Beam Using Isogeometric Analysis. *Eur. J. Mech. -A/Solids* **2020**, *84*, 104083. [CrossRef]
24. Kamensky, D. Open-Source Immersogeometric Analysis of Fluid–Structure Interaction Using FEniCS and TIGAr. *Comput. Math. Appl.* **2021**, *81*, 634–648. [CrossRef]
25. Simona, A.; Bonaventura, L.; de Falco, C.; Schöps, S. IsoGeometric Approximations for Electromagnetic Problems in Axisymmetric Domains. *Comput. Methods Appl. Mech. Eng.* **2020**, *369*, 113211. [CrossRef]
26. Bucelli, M.; Salvador, M.; Quarteroni, A. Multipatch Isogeometric Analysis for Electrophysiology: Simulation in a Human Heart. *Comput. Methods Appl. Mech. Eng.* **2021**, *376*, 113666. [CrossRef]
27. Rouwane, A.; Bouclier, R.; Passieux, J.-C.; Périé, J.-N. Adjusting Fictitious Domain Parameters for Fairly Priced Image-Based Modeling: Application to the Regularization of Digital Image Correlation. *Comput. Methods Appl. Mech. Eng.* **2021**, *373*, 113507. [CrossRef]
28. Du, X.; Zhao, G.; Wang, W.; Guo, M.; Zhang, R.; Yang, J. NLIGA: A MATLAB Framework for Nonlinear Isogeometric Analysis. *Comput. Aided Geom. Des.* **2020**, *80*, 101869. [CrossRef]
29. Benson, D.; Bazilevs, Y.; Hsu, M.-C.; Hughes, T. Isogeometric Shell Analysis: The Reissner–Mindlin Shell. *Comput. Methods Appl. Mech. Eng.* **2010**, *199*, 276–289. [CrossRef]
30. Mi, Y.; Yu, X. Isogeometric MITC Shell. *Comput. Methods Appl. Mech. Eng.* **2021**, *377*, 113693. [CrossRef]
31. Sobhani, E.; Masoodi, A.R.; Ahmadi-Pari, A.R. Vibration of FG-CNT and FG-GNP Sandwich Composite Coupled Conical-Cylindrical-Conical Shell. *Compos. Struct.* **2021**, *273*, 114281. [CrossRef]
32. Sobhani, E.; Moradi-Dastjerdi, R.; Behdinan, K.; Masoodi, A.R.; Ahmadi-Pari, A.R. Multifunctional Trace of Various Reinforcements on Vibrations of Three-Phase Nanocomposite Combined Hemispherical-Cylindrical Shells. *Compos. Struct.* **2021**, *279*, 114798. [CrossRef]
33. Rezaiee-Pajand, M.; Masoodi, A.R. Analyzing FG Shells with Large Deformations and Finite Rotations. *World J. Eng.* **2019**, *16*, 636–647. [CrossRef]
34. Marathe, S.P.; Raval, H.K. Numerical Investigation on Forming Behavior of Friction Stir Tailor Welded Blanks (FSTWBs) during Single-Point Incremental Forming (SPIF) Process. *J. Braz. Soc. Mech. Sci. Eng.* **2019**, *41*, 424. [CrossRef]
35. Temizer, I.; Wriggers, P.; Hughes, T. Contact Treatment in Isogeometric Analysis with NURBS. *Comput. Methods Appl. Mech. Eng.* **2011**, *200*, 1100–1112. [CrossRef]
36. Dimitri, R.; Zavarise, G. Isogeometric Treatment of Frictional Contact and Mixed Mode Debonding Problems. *Comput. Mech.* **2017**, *60*, 315–332. [CrossRef]
37. Wang, Y.; Wang, Z.; Xia, Z.; Poh, L.H. Structural Design Optimization Using Isogeometric Analysis: A Comprehensive Review. *Comput. Modeling Eng. Sci.* **2018**, *117*, 455–507. [CrossRef]
38. Yu, T.; Yin, S.; Bui, T.Q.; Liu, C.; Wattanasakulpong, N. Buckling Isogeometric Analysis of Functionally Graded Plates under Combined Thermal and Mechanical Loads. *Compos. Struct.* **2017**, *162*, 54–69. [CrossRef]
39. Kaushik, V.; Ghosh, A. Experimental and XIGA-CZM Based Mode-II and Mixed-Mode Interlaminar Fracture Model for Unidirectional Aerospace-Grade Composites. *Mech. Mater.* **2021**, *154*, 103722. [CrossRef]
40. Melenk, J.M.; Babuška, I. The Partition of Unity Finite Element Method: Basic Theory and Applications. *Comput. Methods Appl. Mech. Eng.* **1996**, *139*, 289–314. [CrossRef]
41. De Luycker, E.; Benson, D.J.; Belytschko, T.; Bazilevs, Y.; Hsu, M.C. X-FEM in Isogeometric Analysis for Linear Fracture Mechanics. *Int. J. Numer. Methods Eng.* **2011**, *87*, 541–565. [CrossRef]
42. Ghorashi, S.S.; Valizadeh, N.; Mohammadi, S. Extended Isogeometric Analysis for Simulation of Stationary and Propagating Cracks. *Int. J. Numer. Methods Eng.* **2012**, *89*, 1069–1101. [CrossRef]
43. Bhardwaj, G.; Singh, I.; Mishra, B. Numerical Simulation of Plane Crack Problems Using Extended Isogeometric Analysis. *Procedia Eng.* **2013**, *64*, 661–670. [CrossRef]
44. Singh, S.; Singh, I.V.; Bhardwaj, G.; Mishra, B. A Bézier Extraction Based XIGA Approach for Three-Dimensional Crack Simulations. *Adv. Eng. Softw.* **2018**, *125*, 55–93. [CrossRef]
45. Khatir, S.; Wahab, M.A. A Computational Approach for Crack Identification in Plate Structures Using XFEM, XIGA, PSO and Jaya Algorithm. *Theor. Appl. Fract. Mech.* **2019**, *103*, 102240. [CrossRef]
46. Yadav, A.; Patil, R.; Singh, S.; Godara, R.; Bhardwaj, G. A Thermo-Mechanical Fracture Analysis of Linear Elastic Materials Using XIGA. *Mech. Adv. Mater. Struct.* **2020**, 1–26. [CrossRef]
47. Nguyen, V.P.; Anitescu, C.; Bordas, S.P.; Rabczuk, T. Isogeometric Analysis: An Overview and Computer Implementation Aspects. *Math. Comput. Simul.* **2015**, *117*, 89–116. [CrossRef]
48. Li, K.; Yu, T.; Bui, T.Q. Adaptive Extended Isogeometric Upper-Bound Limit Analysis of Cracked Structures. *Eng. Fract. Mech.* **2020**, *235*, 107131. [CrossRef]

49. Khatir, S.; Boutchicha, D.; Le Thanh, C.; Tran-Ngoc, H.; Nguyen, T.; Abdel-Wahab, M. Improved ANN Technique Combined with Jaya Algorithm for Crack Identification in Plates Using XIGA and Experimental Analysis. *Theor. Appl. Fract. Mech.* **2020**, *107*, 102554. [CrossRef]
50. Gu, J.; Yu, T.; Tanaka, S.; Yuan, H.; Bui, T.Q. Crack Growth Adaptive XIGA Simulation in Isotropic and Orthotropic Materials. *Comput. Methods Appl. Mech. Eng.* **2020**, *365*, 113016. [CrossRef]
51. El Fakkoussi, S.; Moustabchir, H.; Elkhalfi, A.; Pruncu, C. Application of the Extended Isogeometric Analysis (X-IGA) to Evaluate a Pipeline Structure Containing an External Crack. *J. Eng.* **2018**, *2018*. [CrossRef]
52. Moustabchir, H.; Arbaoui, J.; Azari, Z.; Hariri, S.; Pruncu, C.I. Experimental/Numerical Investigation of Mechanical Behaviour of Internally Pressurized Cylindrical Shells with External Longitudinal and Circumferential Semi-Elliptical Defects. *Alex. Eng. J.* **2018**, *57*, 1339–1347. [CrossRef]
53. Giner, E.; Sukumar, N.; Tarancón, J.; Fuenmayor, F. An Abaqus Implementation of the Extended Finite Element Method. *Eng. Fract. Mech.* **2009**, *76*, 347–368. [CrossRef]
54. Hou, W.; Jiang, K.; Zhu, X.; Shen, Y.; Hu, P. Extended Isogeometric Analysis Using B++ Splines for Strong Discontinuous Problems. *Comput. Methods Appl. Mech. Eng.* **2021**, *381*, 113779. [CrossRef]
55. Mohammadi, S. *Extended Finite Element Method: For Fracture Analysis of Structures*; John Wiley & Sons: Hoboken, NJ, USA, 2008; ISBN 0-470-69799-7.
56. Béchet, É.; Minnebo, H.; Moës, N.; Burgardt, B. Improved Implementation and Robustness Study of the X-FEM for Stress Analysis around Cracks. *Int. J. Numer. Methods Eng.* **2005**, *64*, 1033–1056. [CrossRef]
57. Yadav, A.; Godara, R.; Bhardwaj, G. A Review on XIGA Method for Computational Fracture Mechanics Applications. *Eng. Fract. Mech.* **2020**, *230*, 107001. [CrossRef]
58. Nguyen, V.P.; Bordas, S. Extended Isogeometric Analysis for Strong and Weak Discontinuities. In *Isogeometric Methods for Numerical Simulation*; Springer: Berlin/Heidelberg, Germany, 2015; pp. 21–120.
59. Moustabchir, H.; Zitouni, A.; Hariri, S.; Gilgert, J.; Pruncu, C. Experimental–Numerical Characterization of the Fracture Behaviour of P264GH Steel Notched Pipes Subject to Internal Pressure. *Iran. J. Sci. Technol. Trans. Mech. Eng.* **2018**, *42*, 107–115. [CrossRef]
60. Moustabchir, H.; Pruncu, C.; Azari, Z.; Hariri, S.; Dmytrakh, I. Fracture Mechanics Defect Assessment Diagram on Pipe from Steel P264GH with a Notch. *Int. J. Mech. Mater. Des.* **2016**, *12*, 273–284. [CrossRef]
61. Creating a Contour Integral Crack. Available online: https://abaqus-docs.mit.edu/2017/English/SIMACAECAERefMap/simacae-t-enghelpcrack.htm (accessed on 6 May 2021).
62. Suresh Kumar, S.; Naren Balaji, V. Mode-I, Mode-II, and Mode-III Stress Intensity Factor Estimation of Regular-and Irregular-Shaped Surface Cracks in Circular Pipes. *J. Fail. Anal. Prev.* **2020**, *20*, 853–867. [CrossRef]
63. Gajdoš, L'.; Šperl, M. Evaluating the Integrity of Pressure Pipelines by Fracture Mechanics. *Appl. Fract. Mech.* **2012**, 283.
64. Moustabchir, H. *Etude Des Défauts Présents Dans Des Tuyaux Soumis à Une Pression Interne*; University of Lorraine: Metz, French, 2008.

Article

# Assessment of Complex System Dynamics via Harmonic Mapping in a Multifractal Paradigm

Gabriel Gavriluț [1], Liliana Topliceanu [2], Manuela Gîrțu [3], Ana Maria Rotundu [1], Stefan Andrei Irimiciuc [4,5,*] and Maricel Agop [5,6]

[1] Faculty of Phisics, Alexandru Ioan Cuza University, Bulevardul Carol I nr. 11, 700506 Iași, Romania; gavrilutgabriel@yahoo.com (G.G.); anabotezatu82@gmail.com (A.M.R.)
[2] Faculty of Engineering, Vasile Alecsandri University of Bacau, 600115 Bacau, Romania; lili@ub.ro
[3] Department of Mathematics and Informatics, Vasile Alecsandri University of Bacau, 600115 Bacau, Romania; girtum@yahoo.com
[4] National Institute for Laser, Plasma and Radiation Physics, 409 Atomistilor Street, 077125 Bucharest, Romania
[5] Romanian Scientists Academy, 54 Splaiul Independentei, 050094 Bucharest, Romania; magop@tuiasi.ro
[6] Department of Physics, "Gh. Asachi" Technical University of Iasi, 700050 Iasi, Romania
* Correspondence: stefan.irimiciuc@inflpr.ro

**Citation:** Gavriluț, G.; Topliceanu, L.; Gîrțu, M.; Rotundu, A.M.; Irimiciuc, S.A.; Agop, M. Assessment of Complex System Dynamics via Harmonic Mapping in a Multifractal Paradigm. *Mathematics* 2021, *9*, 3298. https://doi.org/10.3390/math 9243298

Academic Editors: Catalin I. Pruncu and Maria Luminița Scutaru

Received: 30 November 2021
Accepted: 17 December 2021
Published: 18 December 2021

**Publisher's Note:** MDPI stays neutral with regard to jurisdictional claims in published maps and institutional affiliations.

**Copyright:** © 2021 by the authors. Licensee MDPI, Basel, Switzerland. This article is an open access article distributed under the terms and conditions of the Creative Commons Attribution (CC BY) license (https:// creativecommons.org/licenses/by/ 4.0/).

**Abstract:** In the present paper, nonlinear behaviors of complex system dynamics from a multifractal perspective of motion are analyzed. In the framework of scale relativity theory, by analyzing the dynamics of complex system entities based on continuous but non-differentiable curves (multifractal curves), both the Schrödinger and Madelung scenarios on the holographic implementations of dynamics are functional and complementary. In the Madelung scenario, the holographic implementation of dynamics (i.e., free of any external or internal constraints) has some important consequences explicated by means of various operational procedures. The selected procedures involve synchronous modes through SL (2R) transformation group based on a hidden symmetry, coherence domains through Riemann manifold embedded with a Poincaré metric based on a parallel transport of direction (in a Levi Civita sense). Other procedures used here relate to the stationary-non-stationary dynamics transition through harmonic mapping from the usual space to the hyperbolic one manifested as cellular and channel type self-structuring. Finally, the Madelung scenario on the holographic implementations of dynamics are discussed with respect to laser-produced plasma dynamics.

**Keywords:** harmonic mapping; complex system dynamics; SL (2R) group; hidden symmetries

## 1. Introduction

Nonlinearity is accepted as one of the most fundamental properties of any complex system dynamics. Interactions between the structural units of any complex system imply mutual constraints at different scale resolution. Then, the universality of the dynamics laws for any complex system becomes natural and must be reflected in various theoretical models that could describe their dynamics. Some of the usual theoretical models are based on the hypothesis that the variables characterizing the complex system dynamics are differentiable, which can be otherwise unjustified. In such a perspective, the validations of the previously described type of models need to be seen as sequential and applicable on restricted domains for which integrability and differentiability are respected. Since nonlinearity implies predominantly non-differentiable behaviors in the description of complex system dynamics, it is necessary to explicitly introduce the scale resolution in the equations defining the dynamics-governing variables. It implies that any variables used in the description of any complex system now have a dual dependence on the space-time coordinates and the scale resolution. In this new perspective, for instance, instead of using variables defined by non-differentiable functions, approximations of these complex functions that will be used at various scale resolutions are becoming available and operational. Therefore, all variables used to define the complex system dynamics will work as a limit of families of functions, which for a null scale resolution are non-differentiable but for non-null scale resolution are differentiable. The previous mathematical procedure involves the development of suitable geometrical structures and a class of models for which the motion laws are integrated with the scale laws. Such geometrical structures are built on the concept of multifractality, and the equivalent theoretical models are based on the scale relativity theory, either with the fractal dimension $D_f = 2$ (standard model) or in an arbitrary and constant dimension (multifractal theory of motion). In this class of models (non-differentiable), the complex system's structural unit's dynamics can be described by continuous but non-differentiable movement curves (multifractal motion curves). These curves exhibit self-similarity as their main property at any of the points forming the curve, which translates into behaviors of holographic type (every part reflects the global system). Such a complex approach suggests that only holographic implementations can offer complete descriptions of the complex system dynamics [1–3].

According to our previous report from [4], by assimilating any complex fluid with a mathematic object of fractal type in the framework of scale relativity theory (SRT) [5], various non-linear behaviors through a fractal hydrodynamic-type description as well as through a fractal Schrodinger-type description, were established. Thus, the fractal hydrodynamic -type description implies holographic implementations of dynamics through velocity fields at non-differentiable scale resolution, via fractal soliton, fractal soliton-kink, and fractal minimal vortex. The fractal Schrodinger-type description thus implies holographic implementation of complex system dynamics though in-phase coherences of fractal state fields via Airy fractal functions. In this last description, various operational procedures can become functional. We can mention the fractal cubes with fractal SL(2R) group invariance through in-phase coherence of the structural unit dynamics of any complex fluid, fractal SL(2R) groups through dynamic synchronization among the complex system structural units, fractal Riemann manifolds induced by fractal cubics and embedded with a Poincaré metric through apolar transport of cubes, and harmonic mapping from the usual space to the hyperbolic one. These procedures become operational so that several possible scenarios towards chaos (fractal periodic doubling scenario), but without fully transitioning into chaos, (non-manifest chaos) can be obtained.

In this work, we will analyze from a multifractal perspective the nonlinear dynamics of complex systems, generalizing the results from [4]. In such context, exploring a hidden symmetry under the form of synchronization groups of complex system entities leads to the generation of a Riemann manifold with a hyperbolic type metric via parallel transport of direction. Then, accessing complex systems' nonstationary dynamics is performed thorough harmonic mapping from the usual space to the hyperbolic one.

## 2. Mathematical Model

### 2.1. Motion Equation

In the following, any complex system can be assimilated with a multifractal object. Then, since in the framework of scale relativity theory [6–9], the dynamics of complex system entities are described through continuous and non-differentiable curves (multifractal curves), the motion equation (with geodesics equation status) becomes (for detail see [6–9]):

$$\frac{d\hat{V}^i}{dt} = \partial_t \hat{V}^i + \hat{V}^l \partial_l \hat{V}^i + \frac{1}{4}(dt)^{[\frac{2}{f(\alpha)}]-1} D^{lk} \partial_l \partial_k \hat{V}^i = 0, \qquad (1)$$

where

$$\hat{V}^l = V_D^l - V_F^l$$
$$D^{lk} = d^{lk} - i\hat{d}^{lk}$$
$$d^{lk} = \lambda_+^l \lambda_+^k - \lambda_-^l \lambda_-^k$$
$$\hat{d}^{lk} = \lambda_+^l \lambda_+^k + \lambda_-^l \lambda_-^k$$

$$\partial_t = \frac{\partial}{\partial t}, \ \partial_l = \frac{\partial}{\partial x^l}, \ \partial_l \partial_k = \frac{\partial}{\partial x^l} \frac{\partial}{\partial x^k}, \ i = \sqrt{-1}, \ l, k = 1, 2, 3.$$

In relation (1), the meanings of the variables and parameters are as follows:
- $x^l$ is the multifractal spatial coordinate;
- $t$ is the non-multifractal time having the role of an affine parameter of the motion curves;
- $\hat{V}^l$ is the multifractal complex velocity;
- $V_D^l$ is the differentiable velocity independent of the scale resolution;
- $V_F^l$ is the non-differentiable velocity dependent on the scale resolution;
- $dt$ is the scale resolution;
- $f(\alpha)$ is the singularity spectrum of order $\alpha$;
- $\alpha$ is the singularity index and is a function of fractal dimension $D_f$;
- $D^{lk}$ is the constant tensor associated with the differentiable–non-differentiable transition;
- $\lambda_+^l \left( \lambda_+^k \right)$ is the constant vector associated with the backward differentiable–non-differentiable dynamic processes;
- $\lambda_-^l \left( \lambda_-^k \right)$ is the constant vector associated with the forward differentiable–non-differentiable dynamic processes.

The relation (1) shows that in the most general case of complex system structural unit dynamics, regardless of the fractalization type, the multifractal inertial, $\partial_t \hat{V}^i$, the multifractal convective, $\hat{V}^l \partial_l \hat{V}^i$, and the multifractal dissipative effects, $\frac{1}{4}(dt)^{[\frac{2}{f(\alpha)}]-1} D^{lk} \partial_l \partial_k \hat{V}^i$, are achieving balance at any point of the movement curve.

By using the singularity spectrum, the following patterns in the complex system dynamics can be distinguished: monofractal patterns that imply dynamics in homogenous complex systems characterized though a single fractal dimension and having the same scaling properties in any time interval; multifractal patterns that include dynamics in inhomogeneous and anisotropic complex systems characterized simultaneously by a wide variety of fractal dimensions. Thus, $f(\alpha)$ allows the identification of the universality classes in the dynamics of any complex system even when the strange attractors associated with these dynamics have different aspects. For details on the singularity spectrum and its implication for the dynamics of complex systems, see [10–12].

### 2.2. Schrodinger and Madelung Scenarios in the Description of Complex System Dynamics

For a large temporal scale resolution with respect to the inverse of the highest Lyapunov exponent [7–9], the class of deterministic trajectories of any complex system entity can be substituted by the class of virtual trajectories. Then, the concept of definite trajectories is replaced by the one of density of probability. The multifractality is then expressed by means of multi-stochasticity and becomes functional when describing the dynamic of any complex system in the form of multifractal fluid dynamics (for details see [5–9]).

Many modes of multifractalization through stochasticization processes can be defined. Among the most utilized processes, the Markovian and non-Markovian stochastic processes are found [10–12]. In the following description of complex system dynamics, only multifractalizations by means of Markovian stochastic processes will be discussed, i.e., those specified by constraints [10–12]:

$$\lambda_+^i \lambda_+^l = \lambda_-^i \lambda_-^l = 2\lambda \delta^{il}, \tag{2}$$

where $\lambda$ is a constant associated with the differentiable–non-differentiable transitions and $\delta^{il}$ is the Kronecker pseudo-tensor. Based on (3), the motion Equation (1) become (for details on the mathematical procedure see [7–9]):

$$\frac{d\hat{V}^i}{dt} = \partial_t \hat{V}^i + \hat{V}^l \partial_l \hat{V}^i - i\lambda(dt)^{[\frac{2}{f(\alpha)}]-1} \partial_l \partial^l \hat{V}^i = 0. \tag{3}$$

The relation (4) shows that for the case of complex system structural unit dynamics, only for multifractalization by means of Markovian stochastic processes (for the case of Brownian or Levy type motions) in any point of the motion curves, the local multifractal complex acceleration, $\partial_t \hat{V}^i$, the multifractal complex convection, $\hat{V}^l \partial_l \hat{V}^i$, and the multifractal complex dissipation $i\lambda(dt)^{[\frac{2}{f(\alpha)}]-1} \partial_l \partial^l \hat{V}^i$ are in equilibrium.

In the following, let it be allowed that the motions of the entities belonging to any complex system are irrotational. Then, the multifractal complex velocity fields from (2) become:

$$\hat{V}^i = -2i\lambda(dt)^{[\frac{2}{f(\alpha)}]-1} \partial^i \ln \Psi, \tag{4}$$

where

$$\chi = -2i\lambda(dt)^{[\frac{2}{f(\alpha)}]-1} \ln \Psi \tag{5}$$

is the multifractal complex scalar potential of the complex velocity fields from (5) and $\Psi$ is the function of states (on the significance of $\Psi$, see [5–10]). In these conditions, substituting (5) in (4) and using the mathematical procedures from [6–9], the motion Equation (4) takes the form of the multifractal Schrödinger equation:

$$\lambda^2(dt)^{[\frac{4}{f(\alpha)}]-2} \partial^l \partial_l \Psi + i\lambda(dt)^{[\frac{2}{f(\alpha)}]-1} \partial_t \Psi = 0. \tag{6}$$

Therefore, for the complex velocity fields (5), the dynamics of any complex system entity are described through Schrödinger type "regimes" at various scale resolutions (Schrödinger's multifractal description). Equation (7) defines the Schrödinger scenario on the holographic implementation of complex system dynamics.

Moreover, if $\Psi$ is chosen in the form (Madelung's type choice):

$$\Psi = \sqrt{\rho} e^{is}, \tag{7}$$

where $\sqrt{\rho}$ is the amplitude and $s$ is the phase, then the multifractal complex velocity fields (5) take the explicit form:

$$\hat{V}^i = 2\lambda(dt)^{[\frac{2}{f(\alpha)}]-1} \partial^i s - i\lambda(dt)^{[\frac{2}{f(\alpha)}]-1} \partial^i \ln \rho, \tag{8}$$

which implies the real multifractal velocity fields:

$$V_D^i = 2\lambda(dt)^{[\frac{2}{f(\alpha)}]-1} \partial^i s \tag{9}$$

$$V_F^i = \lambda(dt)^{[\frac{2}{f(\alpha)}]-1} \partial^i \ln \rho. \tag{10}$$

In (10), $V_D^i$ is the differential velocity field, while in (11), $V_F^i$ is the multifractal velocity field.

By (9)–(11) and using the mathematical procedure from [6–10], the motion Equation (4) reduces to the multifractal Madelung equations:

$$\partial_t V_D^i + V_D^l \partial_l V_D^i = -\partial^i Q \tag{11}$$

$$\partial_t \rho + \partial_l \left( \rho V_D^l \right) = 0, \tag{12}$$

with $Q$ the multifractal specific potential:

$$Q = -2\lambda^2(dt)^{[\frac{4}{f(\alpha)}]-2} \frac{\partial^l \partial_l \sqrt{\rho}}{\sqrt{\rho}} = -V_F^i V_F^i - \frac{1}{2}\lambda(dt)^{[\frac{2}{f(\alpha)}]-1} \partial_l V_F^l. \tag{13}$$

Equation (12) corresponds to the multifractal specific momentum conservation law, while Equation (13) corresponds to the multifractal states density conservation law. The multifractal specific potential (14) implies the multifractal specific force:

$$F^i = -\partial^i Q = -2\lambda^2 (dt)^{[\frac{4}{f(\alpha)}]-2} \partial^i \frac{\partial^l \partial_l \sqrt{\rho}}{\sqrt{\rho}}, \qquad (14)$$

which is a measure of the multifractality of the motion curves.

Therefore, for the multifractal complex velocity fields (9), the dynamics of any complex system are described through Madelung-type "regimes" at various scale resolutions (Madelung's multifractal description). Equations (12)–(14) define the Madelung scenario on the holographic implementation for complex system dynamics. In this context, any complex system entity is in a permanent interaction with a multifractal medium through the multifractal specific force (15). All complex systems can be identified with a multifractal fluid, the dynamics of which are described by the multifractal Madelung equations (see (12)–(14)). The velocity field $V_F^i$ does not represent the contemporary dynamics. Since $V_F^i$ is missing from (13), this velocity field contributes to the transfer of the multifractal specific momentum and to the multifractal energy focus. Any analysis of $Q$ should consider the "self" nature of the specific momentum transfer of multifractal type. Then, the conservation of the multifractal energy and the multifractal momentum ensure the reversibility and the existence of the multifractal eigenstates.

If the multifractal tensor is considered:

$$\hat{\tau}^{il} = 2\lambda^2 (dt)^{[\frac{4}{f(\alpha)}]-2} \rho \partial^i \partial^l \ln \rho, \qquad (15)$$

the equation defining the multifractal forces that derive from the multifractal specific potential $Q$ can be written in the form of a multifractal equilibrium equation:

$$\rho \partial^i Q = \partial_l \hat{\tau}^{il}. \qquad (16)$$

Since $\hat{\tau}^{il}$ can be also written in the form:

$$\hat{\tau}^{il} = \eta \left( \partial_l V_F^i + \partial_i V_F^l \right), \qquad (17)$$

with

$$\eta = \lambda (dt)^{[\frac{2}{f(\alpha)}]-1} \rho \qquad (18)$$

a multifractal linear constitutive equation for a multifractal "viscous fluid" can be highlighted. In such a context, the coefficient $\eta$ can be interpreted as a multifractal dynamic viscosity coefficient of the multifractal fluid.

### 2.3. Synchronization Modes in Complex System Dynamics through a "Hidden" Symmetry

The existence of multifractal specific force (15) and the multifractal viscosity tensor (16) will be considered as the "trigger" of the complex system processes that lead both to instabilities and to self-structuring. If the multifractal specific potential is constant, through (15) for the one-dimensional case, the following condition is satisfied:

$$\frac{\partial^2 \sqrt{\rho}}{\partial x^2} + k_0^2 \sqrt{\rho} = 0, \qquad (19)$$

with

$$k_0^2 = \frac{E}{2\lambda^2 (dt)^{[\frac{4}{f(\alpha)}]-2}}.$$

In the above relation, $E$ is the multifractal energy of the complex system's entity and $m_0$ is the rest mass. The solution of (20) can be written in the form

$$\sqrt{\rho} = z e^{i(k_0 x + \theta)} + \bar{z} e^{-i(k_0 x + \theta)}, \tag{20}$$

where $z$ is a complex amplitude, $\bar{z}$ is its complex conjugate, $\theta$ is a specific phase and $x$ is the multifractal spatial coordinate. In such a context, $z$ and $\theta$ "scan" each entity of the complex system, which has as a general characteristic Equation (20), and thus the same $k_0$.

Equation (20) has a multifractal hidden symmetry by means of a homographic group. Indeed, the ratio $\varepsilon$ of two independent linear solutions of Equation (20) is a solution of multifractal Schwartz's differential equation (for the classical case, see [6–9]):

$$\{\varepsilon, x\} = \left(\frac{\varepsilon''}{\varepsilon'}\right)' - \frac{1}{2}\left(\frac{\varepsilon''}{\varepsilon'}\right)^2 = 2k_0^2 \tag{21}$$

$$\varepsilon' = \frac{d\varepsilon}{dx}, \quad \varepsilon'' = \frac{d^2\varepsilon}{dx^2}. \tag{22}$$

The left part of (22) is invariant with respect to the multifractal homographic transformation

$$\varepsilon \leftrightarrow \varepsilon' = \frac{a\varepsilon + b}{c\varepsilon + d} \tag{23}$$

with $a, b, c, d$ multifractal real parameters. The relation (24) corresponding to all possible values of these parameters defines the multifractal group SL(2R) (for the classical case, see [13,14]).

Thus, all of the complex system entities having the same $k_0$ are in biunivocal correspondence with the transformation of the multifractal group SL(2R). This allows the construction of a personal parameter $\varepsilon$ for each individual complex system entity. Indeed, as a guide, it is chosen in the general form of solution of (22), which is written as

$$\varepsilon' = l + m \tan(k_0 x + \theta) \tag{24}$$

Thus, through $l$, $m$, and $\theta$, it is possible to characterize any complex systems entity. In such conjecture, identifying the phase from (25) with the one from (21), the personal parameter becomes:

$$\varepsilon(x) = \frac{z + \bar{z}\varepsilon}{1 + z}, \quad z = l + im, \quad \bar{z} = l - im, \quad \varepsilon \equiv e^{2i(k_0 x + \theta)}. \tag{25}$$

The fact that (25) is also a solution of (22) implies, by explicitly solving (24), that the multifractal group SL(2R):

$$z' = \frac{az + b}{cz + d}, \quad \bar{z}' = \frac{a\bar{z} + b}{c\bar{z} + d}, \quad \varepsilon' = \frac{c\bar{z} + d}{cz + d}\varepsilon. \tag{26}$$

Therefore, the multifractal group (27) works as a synchronization mode among various entities of any complex system process to which the amplitudes and the phases are also connected. More precisely, through (27) the phase of $\varepsilon$ is only moved with a quantity depending on the amplitude of the complex system at the transition among various complex system entities. Moreover, the amplitude of the movement is also affected from a multifractal homographic perspective. The usual synchronization modes manifested through delay of the amplitudes and phases of the complex system entities must describe here only a particular case.

## 2.4. Riemann's Manifold Generated through Synchronization Processes

According to the mathematical procedures from [6–9,15–17], the space of multifractal group (27) can be structured by means of $(z, \bar{z}, \varepsilon)$ parameters, as a multifractal Riemann's manifold. Indeed, the structure of multifractal group (27) is typical of an $SL(2R)$ one, which is taken in the standard form

$$[A_1, A_2] = A_1, \quad [A_2, A_3] = A_3, \quad [A_3, A_1] = -2A_2 \tag{27}$$

where $A_k, k = 1, 2, 3$ are the multifractal infinitesimal generators of the group. Since the multifractal group is simple transitive, these multifractal generators can be found as components of the multifractal Cartan coframe (for the classical case, see Cartan [15]) from the relation.

$$d(f) = \sum \frac{\partial f}{\partial x^k} dx^k = \left\{ \omega^1 \left[ z^2 \frac{\partial}{\partial z} + \bar{z}^2 \frac{\partial}{\partial \bar{z}} + (z - \bar{z})\varepsilon \frac{\partial}{\partial \varepsilon} \right] + 2\omega^2 \left( z \frac{\partial}{\partial z} + \bar{z} \frac{\partial}{\partial \bar{z}} \right) + \omega^3 \left( \frac{\partial}{\partial z} + \frac{\partial}{\partial \bar{z}} \right) \right\} (f) \tag{28}$$

where $\omega^k$ are the components of the multifractal Cartan coframe which can be found from the system:

$$dz = \omega^1 z^2 + 2\omega^2 z + \omega^3, \quad d\bar{z} = \omega^1 \bar{z}^2 + 2\omega^2 \bar{z} + \omega^3, \quad d\varepsilon = \omega^1 \varepsilon (z - \bar{z}) \tag{29}$$

Thus, both the multifractal infinitesimal generators and the multifractal coframe are obtained by identifying the right-hand side of (29) with the standard dot product of multifractal algebra $SL(2R)$

$$\omega^1 A_3 + \omega^3 A_1 - 2\omega^2 A_2, \tag{30}$$

so that

$$A_1 = \frac{\partial}{\partial z} + \frac{\partial}{\partial \bar{z}}, \quad A_2 = z \frac{\partial}{\partial z} + \bar{z} \frac{\partial}{\partial \bar{z}}, \quad A_3 = z^2 \frac{\partial}{\partial z} + \bar{z}^2 \frac{\partial}{\partial \bar{z}} + (z - \bar{z})\varepsilon \frac{\partial}{\partial \varepsilon} \tag{31}$$

and

$$\omega^1 = \frac{d\varepsilon}{(z - \bar{z})\varepsilon}, \quad 2\omega^2 = \frac{dz - d\bar{z}}{z - \bar{z}} - \frac{z + \bar{z}}{z - \bar{z}} \frac{d\varepsilon}{\varepsilon}, \quad \omega^3 = \frac{z d\bar{z} - \bar{z} dz}{z - \bar{z}} + \frac{z \bar{z} d\varepsilon}{(z - \bar{z})\varepsilon}. \tag{32}$$

In real terms from (26), these last multifractal equations can be written as

$$A_1 = \frac{\partial}{\partial l}, \quad A_2 = l \frac{\partial}{\partial l} + m \frac{\partial}{\partial m}, \quad A_3 = (l^2 - m^2) \frac{\partial}{\partial l} + 2lm \frac{\partial}{\partial m} + 2m \frac{\partial}{\partial \theta} \tag{33}$$

$$\omega^1 = \frac{d\theta}{2m}, \quad \omega^2 = \frac{dm}{m} - \frac{l}{m} d\theta, \quad \omega^2 = \frac{l^2 + m^2}{2m} d\theta + \frac{m dl - l dm}{m}. \tag{34}$$

It should be mentioned that in [6–9], it does not work with the previous multifractal differential forms, but with the multifractal absolute invariant differentials:

$$\omega^1 = \frac{dz}{(z - \bar{z})\varepsilon}, \quad \omega^2 = -i \left( \frac{d\varepsilon}{\varepsilon} - \frac{dz + d\bar{z}}{z - \bar{z}} \right), \quad \omega^3 = \frac{-\varepsilon d\bar{z}}{z - \bar{z}} \tag{35}$$

or, in real terms, exhibiting a three-dimensional Lorentz structure of this multifractal space

$$\Omega^1 = \omega^1 = d\theta + \frac{dl}{m}, \quad \Omega^2 = \cos\theta \frac{dl}{m} + \sin\theta \frac{dm}{m}, \quad \Omega^3 = -\sin\theta \frac{dl}{m} + \cos\theta \frac{dm}{m} \tag{36}$$

The advantage of this representation is that it makes obvious the multifractal connection with the multifractal Poincaré representation of the multifractal Lobachevsky plane. Indeed, the multifractal metric is:

$$\frac{ds^2}{g} = \left( \omega^2 \right)^2 - 4\omega^1 \omega^2 = \left( \frac{d\varepsilon}{\varepsilon} - \frac{dz + d\bar{z}}{z - \bar{z}} \right)^2 + 4 \frac{dz d\bar{z}}{(z - \bar{z})^2}, \tag{37}$$

or in real terms

$$-\frac{ds^2}{g} = -\left( \Omega^1 \right)^2 + \left( \Omega^2 \right)^2 + \left( \Omega^3 \right)^2 = -\left( d\theta + \frac{dl}{m} \right)^2 + \frac{dl^2 + dm^2}{m^2},$$

where $g$ is a multifractal constant.

This multifractal metric reduces to that of Poincaré

$$\frac{ds^2}{g} = -4\frac{dzd\bar{z}}{(z-\bar{z})^2} = -4\frac{dl^2+dm^2}{m^2}. \tag{38}$$

in the case when $\omega^2 = 0$ or $\Omega^1 = 0$, which defines the variable $\theta$ as the "angle of parallelism" (in Levi-Civita sense) of the multifractal hyperbolic plane (the multifractal connection). The multifractal Riemann manifold can further be associated with particular coherence domains induced by the parallel transport of direction. In fact, if in modern terms $\frac{dl}{m}$ represents the multifractal connection form of the multifractal hyperbolic plane, the relations in (37) then represent a general multifractal Bäcklung transformation in that multifractal plane. For the classical case, see [16].

*2.5. Complex System Dynamics via Harmonic Mapping*

In the following, we will generate non-stationary dynamics in complex systems through harmonic map generation. Indeed, let us assume that the complex system dynamics are described by the variables $(Y^j)$, for which the following multifractal metric was discovered:

$$h_{ij}dY^idY^j \tag{39}$$

in an ambient space of multifractal metric:

$$\gamma_{\alpha\beta}dX^\alpha dX^\beta. \tag{40}$$

In this situation, the field equations of the complex system dynamics are derived from a variational principle, connected to the multifractal Lagrangian:

$$L = \gamma^{\alpha\beta}h_{ij}\frac{dY^idY^j}{\partial X^\alpha \partial X^\beta}. \tag{41}$$

In the current case, (40) is given by (39) with the constraint $\omega^2 = 0$, the field variables being $z$ and $\bar{z}$ or, equivalently, the real and imaginary part of $z$. Therefore, if the variational principle:

$$\delta \int L\sqrt{\gamma}d^3x, \tag{42}$$

is accepted as a starting point, where $\gamma = |\gamma_{\alpha\beta}|$, the main purpose of the complex system dynamics research would be to produce multifractal metrics of the multifractal Lobachevski plane (or related to it). In such a context, the multifractal Euler equations corresponding to the variational principle (43) are:

$$(z-\bar{z})\nabla(\nabla z) = 2(\nabla z)^2 \tag{43}$$

$$(z-\bar{z})\nabla(\nabla \bar{z}) = 2(\nabla \bar{z})^2,$$

which allows the solution:

$$h = \frac{\cosh\left(\frac{\Phi}{2}\right) - \sinh\left(\frac{\Phi}{2}\right)e^{-i\alpha}}{\cosh\left(\frac{\Phi}{2}\right) + \sinh\left(\frac{\Phi}{2}\right)e^{-i\alpha}}, \ \alpha \in \mathbb{R}, \tag{44}$$

with $\alpha$ real and arbitrary, as long as $\left(\frac{\Phi}{2}\right)$ is the solution of a Laplace-type equation for the free space, such that $\nabla^2\left(\frac{\Phi}{2}\right) = 0$. For a choice of the form $\alpha = 2\Omega t$, in which case a temporal dependency was introduced in the complex system dynamics, (45) becomes:

$$h = \frac{i\left[e^{2\Phi}\sin(2\Omega t) - \sin(2\Omega t) - 2ie^{\Phi}\right]}{e^{2\Phi}[\cos(2\Omega t) + 1] - \cos(2\Omega t) + 1}. \tag{45}$$

In Figures 1–3, multiple nonlinear behaviors of complex dynamics at scale resolutions in dimensionless coordinates are presented via Python simulations: (i) nonlinear behaviors at a global scale resolution (Figure 1a,b); (ii) nonlinear behaviors at a differentiable scale resolution (Figure 2a,b); (iii) nonlinear behaviors at a non-differentiable scale resolution (Figure 3a,b). Let it be noted that, whatever the scale resolution, complex system dynamics prove themselves to be reducible to self-structuring patterns. The structures are present in pairs of two large patterns that are intercommunicated in an intermittent way. In the 0–20 range for $\Omega$ and $t$, the resulting structures are communicating with each other via a channel created along the symmetry axis for $t \sim 10$. This channel is also seen for different $(\Omega; t)$ coordinates, which is interpreted as an intermittency in the structure bonding. Based on the properties of the studied system, there are some associations with real physical phenomena that can be made. The self-structuring process is a well-known aspect of low-temperature plasmas [18,19]. In recent years there have been some reports on structuring of the laser-produced plasmas [19–21], with impacts in pulsed laser deposition technology. The data presented here can be correlated with the plasma structuring (into a fast structure and a slow structure, also named Coulomb and thermal structure, respectively, after the dominant ejection mechanism) during expansion based on the ablation mechanism and ionization state [21–23]. In recent years, a change in the understanding of this structure has been reported, and a separation based on the ionization state was more plausible for the energetic structuring of the plasma [21,24–26]. In a series of papers [26–30], it was shown that each structure can be correlated with certain properties of the target. For this reason, the use of a multifractal model would be suitable for understanding plasma structuring and exploring the relation between the structure, which, as of now, is outside the reach of any of the tools used [31,32]. The model shows that for the structuring process, a communication channel is formed that will automatically appear. If the same rational treatment is applied to the study of plasma structuring, we identify that these channels are the double layer forming at the interface between the two structures. Comments on the effect of the double layer separating the two-plasma structure were made in [28–32], and it was shown that it plays an important role in controlling the kinetics of laser-produced plasmas. Our model highlights an important aspect of the plasma double layers: they are 3-dimensional objects with different properties seen at different investigation scales that transcend the planar expansion. This is seen from Figures 1–3, where we see that the channel is present for different $(\Omega; t)$ coordinates. The transcendence of the plasma double layer over several resolution scales is understandable, as the average value is of a few tens of Debye lengths [32], which is the core resolution scale in plasma physics. The presence of a transient double layer driving the dynamics of a laser-produced plasma is relatively novel and has been investigated through other modeling approaches and experimental investigations.

Let it be noted that the mathematical formalism of the multifractal theory of motion naturally implies various operational procedures (invariance groups, harmonic mappings, group isomorphisms, embedding manifolds, etc.) with quite a number of applications in complex systems and plasma physics dynamics [32]. Plotting $h$, once again in dimensionless parameters, also highlights certain temporal self-similar properties, with the multifractal structures being contained into similar multifractal structures at much higher scales (Figure 4a–c). Let us also note that the structure's communication channel has an exponential decrease in the $(\Omega; t)$ plane, which reflects the dissipation processes [32] occurring during laser-produced plasma expansion. When they expand, laser-produced plasmas lose particles and energy through collisional/radiative processes. This will be reflected in the weakening of the plasma double layer and limiting of the reach to a small plasma volume in the proximity of the double layer. The model manages to express the dissipation of the plasma through the reduction of the channel amplitude on the $\Omega$ axis as the time variable is increased. This result represents an important step forward in understating the dynamics at the front of the plume. Most of plasma diagnostics and even modeling are concerned with late-time interactions mostly occurring in the core of the plasma. Our model manages to capture, albeit in a multifractal picture, dissipation processes and possible recombination occurring at the front of the plume. Expanding the reach of our results, we could find future implementation for pulsed-laser deposition, where the front of a subsequent plasma always interacts with the already-deposited film.

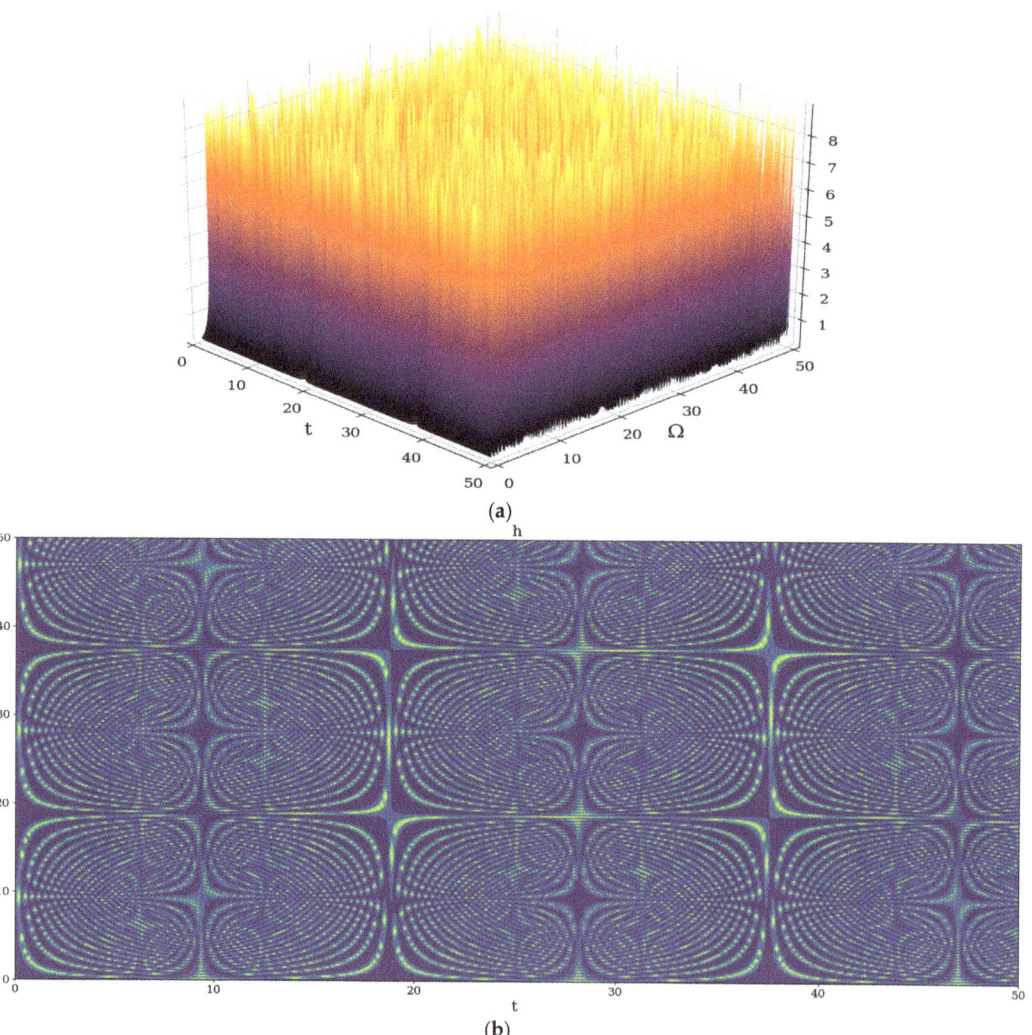

**Figure 1.** (**a**): 3D dynamics at global scale resolution of $h(\Omega, t)$ with $\Phi = 2.35$. (**b**): 2D dynamics at global scale resolution of $h(\Omega, t)$ with $\Phi = 2.35$.

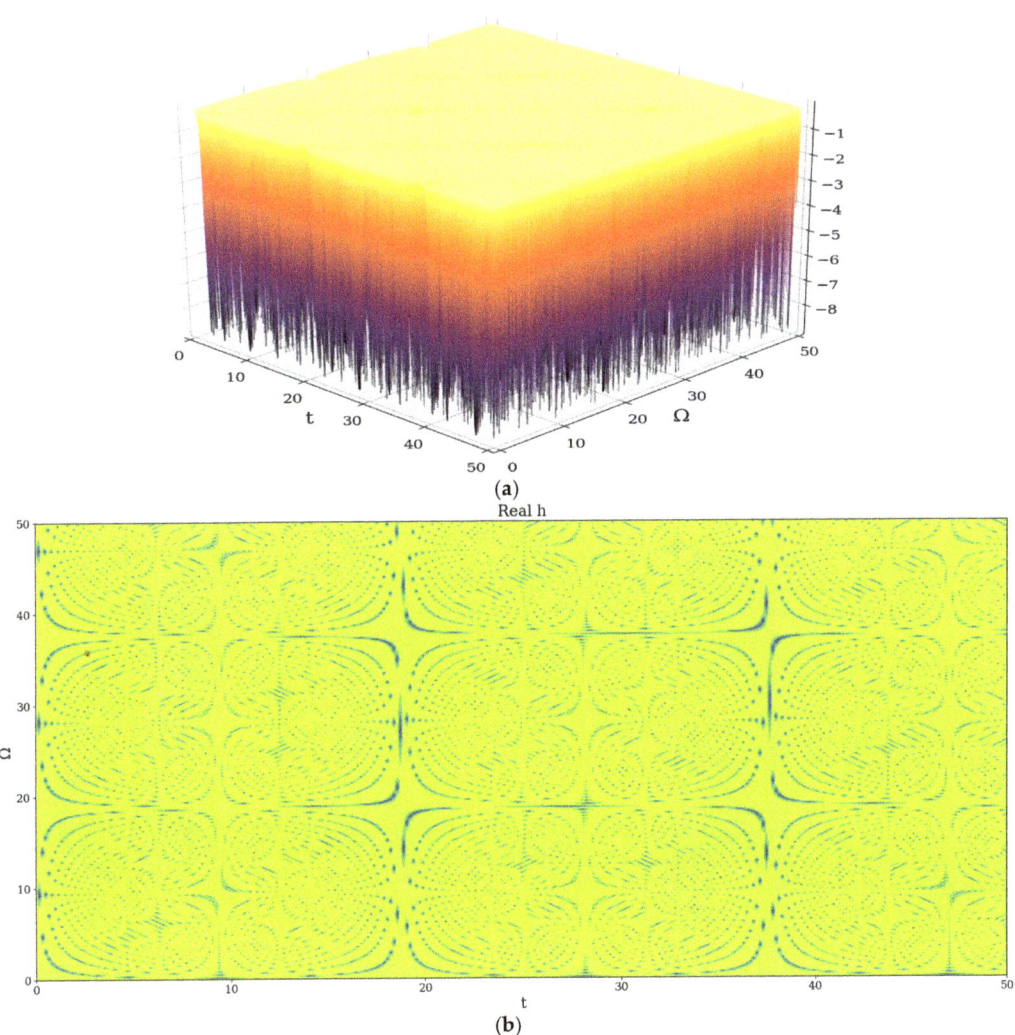

**Figure 2.** (**a**): 3D dynamics at differentiable scale resolution of $Re[h(\Omega, t)]$ with $\Phi = 2.35$. (**b**): 2D dynamics at differentiable scale resolution of $Re[h(\Omega, t)]$ with $\Phi = 2.35$.

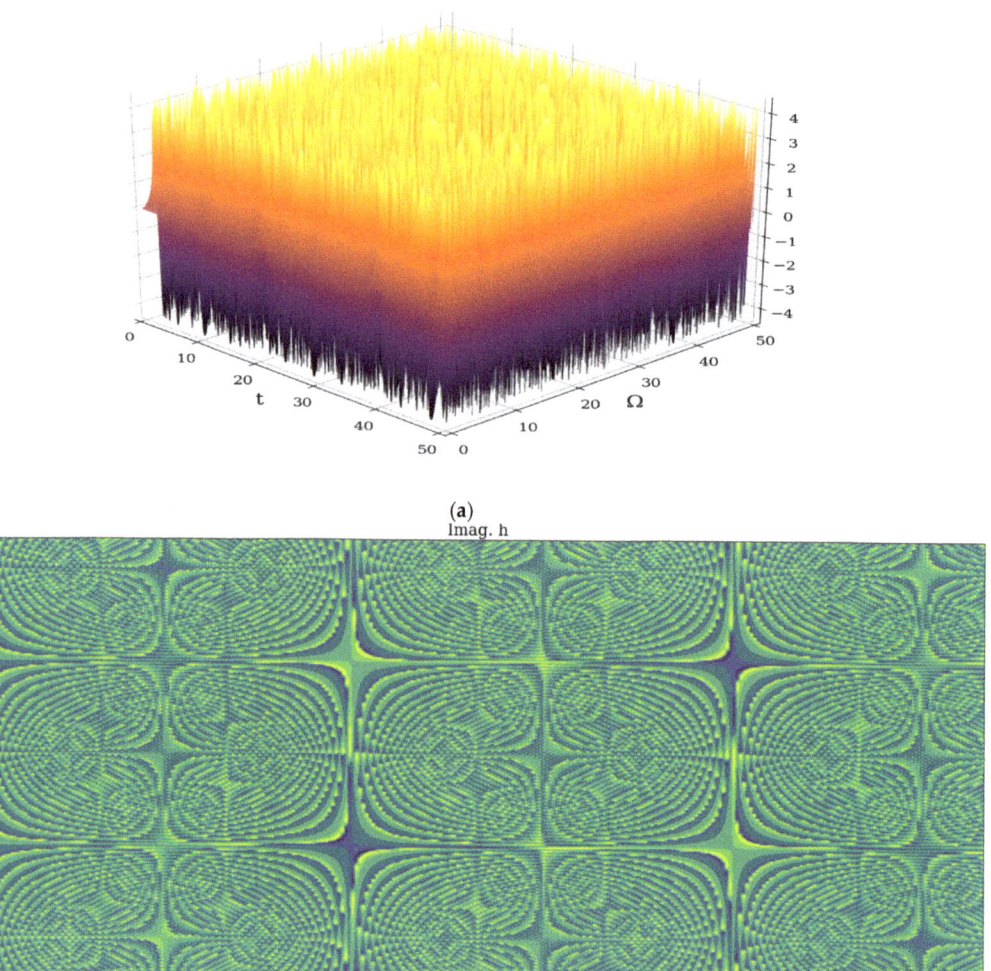

**Figure 3.** (**a**): 3D dynamics at non-differentiable scale resolution of $Im[h(\Omega, t)]$ with $\Phi = 2.35$. (**b**): 2D dynamics at non-differentiable scale resolution of $Im[h(\Omega, t)]$ with $\Phi = 2.35$.

The results presented in Figure 4a–c also specify that, through self-structuring of the complex system entities, channel-type patterns can also be observed.

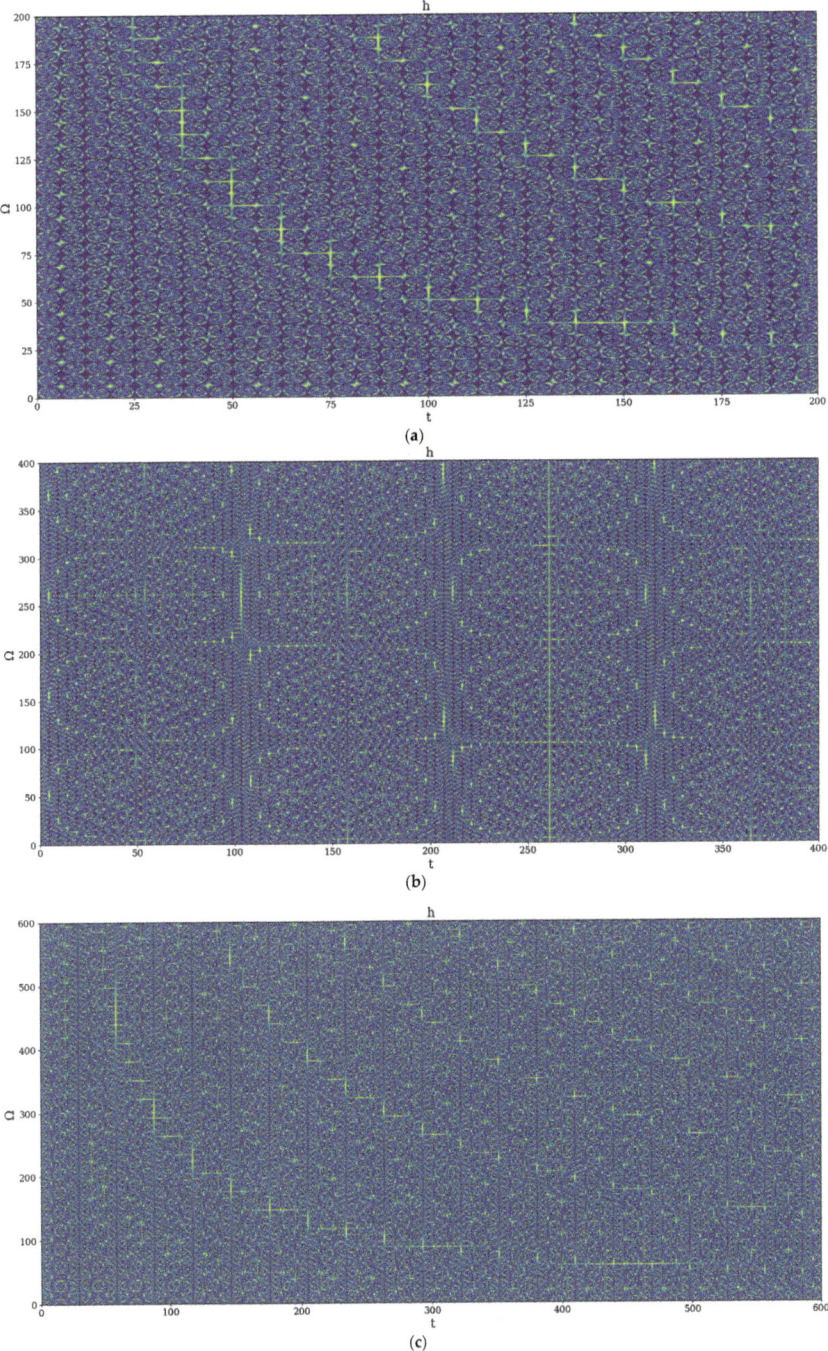

**Figure 4.** (**a**): 2D dynamics at global scale resolution of $h(\Omega = 0 - 200, t = 0 - 200)$; $h = 5$. (**b**): 2D dynamics at global scale resolution of $h(\Omega = 0 - 400, t = 0 - 400)$; $h = 5$. (**c**): 2D dynamics at global scale resolution of $h(\Omega = 0 - 600, t = 0 - 600)$; $h = 5$.

## 3. Conclusions

By considering that any complex system dynamics can be assimilated with a mathematical object of multifractal type, various non-linear behaviors in the framework of the scale relativity theory of motion are developed. In such a context, Schrödinger's and Madelung's holographic implementation scenarios for any complex system dynamics become operational through the multifractal motion curves. Exploring at various scale resolutions a hidden symmetry of stationary dynamics in the Madelung description, synchronization modes are seen forming through the SL (2R) group between the complex system entities. In the synchronization process, the amplitudes and phase of the motions of any complex system entity are shown to be connected, while the amplitude attributed to each motion can be tailored from a multifractal homographic perspective. The usual synchronization modes were proved to be manifested through the delay of the amplitude and phases of the complex system entities, and are here a particular case. The space induced by means of SL(2R) group parameters was structured at various scale resolutions as a Riemann manifold (multifractal Riemann manifold). The generators of a special Cartan coframe and their associated metrics were found. When a parallel transport of direction in the Levi-Civita sense became functional, the metric was reduced to that of Poincare, with the angle of parallelism of the hyperbolic plane defining the connections. Riemann manifolds were associated with coherence domains, with the coherence on each domain being induced by parallel transport of direction. Access to non-stationary dynamics at various scale resolutions became possible via harmonic mapping from the usual space to the hyperbolic one. Then, self-structuring of cellular and channel types were produced. The results are discussed with possible interpretations for the dynamics of laser-produced plasmas.

**Author Contributions:** Conceptualization, M.A. and S.A.I.; methodology, G.G. and L.T.; investigation, M.G., A.M.R. and S.A.I.; resources, G.G. and M.G.; writing—original draft preparation, S.A.I., and M.A.; writing—review and editing, M.A. and S.A.I. All authors have read and agreed to the published version of the manuscript.

**Funding:** This research received no external funding.

**Institutional Review Board Statement:** Not applicable.

**Informed Consent Statement:** Not applicable.

**Data Availability Statement:** Data will be available on request.

**Acknowledgments:** We would like to acknowledge the anonymous reviewers for their insightful comments that helped improve the manuscript.

**Conflicts of Interest:** The authors declare no conflict of interest.

## References

1. Djebali, R.; Mebarek-Oudina, F.; Rajashekhar, C. Similarity solution analysis of dynamic and thermal boundary layers: Further formulation along a vertical flat plate. *Phys. Scr.* **2021**, *96*, 085206. [CrossRef]
2. Hamrelaine, S.; Mebarek-Oudina, F.; Sari, M.R. Analysis of MHD Jeffery Hamel Flow with Suction/Injection by Homotopy Analysis Method. *J. Adv. Res. Fluid Mech. Ther. Sci.* **2020**, *58*, 173–186.
3. Alkasassbeh, M.; Omar, Z.; Mebarek-Oudina, F.; Raza, J.; Chamkha, A. Heat transfer study of convective fin with temperature-dependent internal heat generation by hybrid block method. *Heat Transf. Asian Res.* **2019**, *48*, 1225–1244. [CrossRef]
4. Saviuc, A.; Gîrțu, M.; Topliceanu, L.; Petrescu, T.-C.; Agop, M. "Holographic Implementations" in the Complex Fluid Dynamics through a Fractal Paradigm. *Mathematics* **2021**, *9*, 2273. [CrossRef]
5. Nottale, L. *Scale Relativity and Fractal Space-Time: A New Approach to Unifying Relativity and Quantum Mechanics*; Imperial College: London, UK, 2011.
6. Merches, I.; Agop, M. *Differentiability and Fractality in Dynamics of Physical Systems*; World Scientific: Hackensack, NJ, USA, 2016.
7. Agop, M.; Paun, V.P. On the new perspectives of fractal theory. In *Fundaments and Applications*; Romanian Academy Publishing House: Bucharest, Romania, 2017.
8. Mazilu, N.; Agop, M.; Merches, I. *Scale Transitions as Foundations of Physics*; World Scientific: Singapore, 2021.

9. Maziulu, N.; Agop, M.; Merches, I. The mathematical principles of scale relativity theory. In *The Concept of Interpretation*; CRC Press, Taylor and Francis Group: Boca Raton, FL, USA, 2020.
10. Strogatz, S.H. *Nonlinear Dynamics and Chaos*, 2nd ed.; CRC Press: Boca Raton, FL, USA, 2015.
11. Cristescu, C.P. Nonlinear dynamics and chaos. In *Theoretical Fundaments and Applications*; Romanian Academy Publishing House: Bucharest, Romania, 2008.
12. Mandelbrot, B.B. *Fractal and Chaos*; Springer: Berlin/Heidelberg, Germany, 2004.
13. Isaacs, I.M. *Finite Group Theory American Mathematical Society*; Providence: Rhode Island, RI, USA, 2008.
14. Ramadevi, P.; Dubey, V. *Group Theory for Physicists with Applications*; Cambridge University Press: Cambridge, UK, 2019.
15. Cartan, I. *Riemannian Geometry in an Orthogonal Frame*; World Scientific: Singapore, 2001.
16. Flanders, H. *Differential Forms with Applications to the Physical Sciences*; Dover Publication, Inc.: New York, NY, USA, 2012.
17. Felsager, B. *Geometry, Particle and Fields*; Springer: New York, NY, USA, 1998.
18. Dimitriu, D.G.; Irimiciuc, S.A.; Popescu, S.; Agop, M.; Ionita, C.; Schrittwieser, R.W. On the interaction between two fireballs in low-temperature plasma. *Phys. Plasmas* **2015**, *22*, 113511. [CrossRef]
19. Volkov, N.A. Splitting of laser-induced neutral and plasma plumes: Hydrodynamic origin of bimodal distributions of vapor density and plasma emission intensity. *J. Phys. D Appl. Phys.* **2021**, *54*, 37LT01. [CrossRef]
20. Irimiciuc, S.A.; Hodoroaba, B.C.; Bulai, G.; Gurlui, S.; Craciun, V. Multiple structure formation and molecule dynamics in transient plasmas generated by laser ablation of graphite. *Spectrochim. Acta-Part B At. Spectrosc.* **2020**, *165*, 105774. [CrossRef]
21. Kumar, R. Self-structuring in Laser-Blow-Off Plasma Plume. *Int. J. Sci. Eng. Res.* **2012**, *3*, 1–9.
22. Irimiciuc, S.A.; Chertopalov, S.; Craciun, V.; Novotný, M.; Lancok, J. Investigation of laser-produced plasma multistructuring by floating probe measurements and optical emission spectroscopy. *Plasma Process. Polym.* **2020**, *11*, 2000136. [CrossRef]
23. Morozov, A.A.; Evtushenko, A.B.; Bulgakov, A.V. Gas-dynamic acceleration of laser-ablation plumes: Hyperthermal particle energies under thermal vaporization. *Appl. Phys. Lett.* **2015**, *106*, 054107. [CrossRef]
24. Baraldi, G.; Perea, A.; Afonso, C.N. Dynamics of ions produced by laser ablation of several metals at 193 nm. *J. Appl. Phys.* **2011**, *109*, 043302. [CrossRef]
25. Leitz, K.H.; Redlingshofer, B.; Reg, Y.; Otto, A.; Schmidt, M. Metal Ablation with Short and Ultrashort Laser Pulses. *Phys. Procedia* **2011**, *12*, 230–238. [CrossRef]
26. Anoop, K.K.; Polek, M.P.; Bruzzese, R.; Amoruso, S.; Harilal, S.S. Multidiagnostic analysis of ion dynamics in ultrafast laser ablation of metals over a large fluence range. *J. Appl. Phys.* **2015**, *117*, 083108. [CrossRef]
27. Irimiciuc, S.A.A.; Gurlui, S.; Nica, P.; Focsa, C.; Agop, M. A compact non-differential approach for modeling laser ablation plasma dynamics. *J. Appl. Phys.* **2017**, *121*, 083301. [CrossRef]
28. Williams, G.O.; O'Connor, G.M.; Mannion, P.T.; Glynn, T.J. Langmuir probe investigation of surface contamination effects on metals during femtosecond laser ablation. *Appl. Surf. Sci.* **2008**, *254*, 5921–5926. [CrossRef]
29. Skočić, M.; Dojić, D.; Bukvić, S. Formation of double-layer in the early stage of nanosecond laser ablation. *J. Quant. Spectrosc. Radiat. Transf.* **2019**, *227*, 57–62. [CrossRef]
30. Eliezer, S.; Nissim, N.; Martínez Val, J.M.; Mima, K.; Hora, H. Double layer acceleration by laser radiation. *Laser Part. Beams* **2014**, *32*, 211–216. [CrossRef]
31. Beilis, I. Modeling of the plasma produced by moderate energy laser beam interaction with metallic targets: Physics of the phenomena. *Laser Part. Beams* **2012**, *30*, 341–356. [CrossRef]
32. Kokai, F.; Takahashi, K.; Shimizu, K.; Yudasaka, M.; Iijima, S. Shadowgraphic and emission imaging spectroscopic studies of the laser ablation of graphite in an Ar gas atmosphere. *Appl. Phys. A Mater. Sci. Process.* **1999**, *69*, 223–227. [CrossRef]

Article

# Damage Detection and Isolation from Limited Experimental Data Using Simple Simulations and Knowledge Transfer

Asif Khan [1], Jun-Sik Kim [2] and Heung Soo Kim [1,*]

[1] Department of Mechanical, Robotics and Energy Engineering, Dongguk University-Seoul, 30 Pildong-ro 1 Gil, Jung-gu, Seoul 04620, Korea; asif_dgu@dgu.edu
[2] Department of Mechanical System Engineering, Kumoh National Institute of Technology, Gumi-si 39177, Korea; junsik.kim@kumoh.ac.kr
* Correspondence: heungsoo@dgu.edu; Tel.: +82-2-2260-8577; Fax: +82-2-2263-9379

**Abstract:** A simulation model can provide insight into the characteristic behaviors of different health states of an actual system; however, such a simulation cannot account for all complexities in the system. This work proposes a transfer learning strategy that employs simple computer simulations for fault diagnosis in an actual system. A simple shaft-disk system was used to generate a substantial set of source data for three health states of a rotor system, and that data was used to train, validate, and test a customized deep neural network. The deep learning model, pretrained on simulation data, was used as a domain and class invariant generalized feature extractor, and the extracted features were processed with traditional machine learning algorithms. The experimental data sets of an RK4 rotor kit and a machinery fault simulator (MFS) were employed to assess the effectiveness of the proposed approach. The proposed method was also validated by comparing its performance with the pre-existing deep learning models of GoogleNet, VGG16, ResNet18, AlexNet, and SqueezeNet in terms of feature extraction, generalizability, computational cost, and size and parameters of the networks.

**Keywords:** computer simulations; actual systems; deep learning; transfer learning; autonomous feature extraction; machine learning

**Citation:** Khan, A.; Kim, J.-S.; Kim, H.S. Damage Detection and Isolation from Limited Experimental Data Using Simple Simulations and Knowledge Transfer. *Mathematics* **2022**, *10*, 80. https://doi.org/10.3390/math10010080

Academic Editors: Maria Luminița Scutaru and Catalin I. Pruncu

Received: 29 November 2021
Accepted: 24 December 2021
Published: 27 December 2021

**Publisher's Note:** MDPI stays neutral with regard to jurisdictional claims in published maps and institutional affiliations.

**Copyright:** © 2021 by the authors. Licensee MDPI, Basel, Switzerland. This article is an open access article distributed under the terms and conditions of the Creative Commons Attribution (CC BY) license (https:// creativecommons.org/licenses/by/ 4.0/).

## 1. Introduction

Rotating machinery is a common and critical type of mechanical equipment used in a wide variety of modern industrial applications. Catastrophic failure of rotating machinery may result in substantial economic loss and injury to personnel. Turbines are key rotating parts of power plants and are susceptible to mechanical defects, such as unbalance [1,2], misalignment [3,4] rubbing [5,6], oil whirl [7], and oil whip [8,9], during operation. The presence of defects in turbines may cause performance degradation or even collapse of the entire system if not rectified in a timely manner. To ensure safe and reliable operation of rotating machinery, it is imperative that operators be able to promptly detect, isolate, and quantify different faults using vibration signals obtained through accelerometers or proximity sensors.

The most commonly used methods of fault diagnosis include model-based methods and data-driven methods [10–13]. In model-based methods, the physics underlying the system's behavior are modeled and used for fault diagnosis. It is difficult or even impossible to precisely model the behavior of complex systems, owing to the wide range of structural complexities and environmental uncertainties that affect such systems [14]. Data-driven methods use data obtained from sensors in the system to carry out fault diagnosis; these methods do not require much knowledge about the underlying kinematics and physics of the failure of the system [15,16]. In traditional data-driven fault diagnosis methods using machine learning, the signals from sensors are usually subjected to preprocessing (e.g., noise removal, domain transformation (time to frequency), signal decomposition (empirical mode decomposition)), extraction of discriminative features (e.g., time and frequency domain statistical features), selection of features that are more sensitive to damage (e.g., feature ranking), and processing of the selected features with supervised or unsupervised machine learning algorithms [17]. The performance of machine learning algorithms for fault diagnosis is heavily dependent on the set of discriminative features that is selected [18]. A set of statistical features may work well for one problem and may fail completely for another problem in the same domain but on a different scale [19]. In general, there is no optimized set of processing steps for fault diagnosis using handcrafted statistical features from sensor data and machine learning algorithms. For instance, a data-driven diagnostic strategy that uses simulation data may not be generalized to a dataset from an experimental setup of the same problem without a complex process of model updating [20,21]. Moreover, the extraction of damage-sensitive features is labor-intensive and requires considerable diagnostic skills and domain expertise [22,23]. In addition, even an experienced diagnostic expert may spend a long time optimizing the set of discriminative features to diagnose a certain problem.

Deep learning has been successfully implemented for a variety of applications, such as image classification, speech recognition, computer vision, medical diagnosis, finance, marketing, and a multitude of other applications [23–26]. The inherent capability of deep learning models to automatically extract features from raw data to describe the underlying problem is one of their most celebrated benefits. Additionally, deep learning models can deal with unstructured data in different formats (e.g., texts, images, pdf files, etc.) to uncover the latent relationships between different data types and make important predictions [27,28]. In general, data-driven methods that use deep learning algorithms require sufficient data on the healthy and faulty states of the system for the development of robust and effective fault diagnosis strategies. Although data on the healthy state of a system is generally available in sufficient amounts, data on different defective states can be limited or even completely unavailable due to the high cost associated with running the machinery in the presence of defects. To make up for the dearth of failure data from different defective states of expensive machinery, computer simulations can be employed to generate a sufficient amount of healthy and faulty data using simplified mathematical models of the actual machinery [29]. However, there are gaps between the data from simulations and actual systems and a labor-intensive process is required to identify the parameters of the actual system and tune those parameters to bring the response characteristics of the simulation model closer to those of the actual system [30]. Additionally, despite the process of parameter identification from the actual system, it is not guaranteed that a diagnostic strategy developed from a simulation model will perform equally well for the detection, isolation, and quantification of different defects in the actual system. In general, the better a computer simulation represents the response behavior of an actual system, the greater its computational cost, and vice versa.

One way to bridge the gap between computer simulation and actual systems while keeping the simulation as simple as possible involves transfer leaning or cross-domain knowledge transfer [31,32]. The fundamental idea of transfer learning is to leverage the knowledge from a semantically related problem to solve a new problem with a different domain distribution. In the general framework of transfer learning, a learning body learns the required properties and parameters from a source task with a substantial amount of labeled data and transfers/tunes those parameters to a target task with a limited amount of labeled or unlabeled data [12,33]. Generally, the source and target data have different statistical distributions [31,34]. Cao et al. [35] proposed transfer learning from a pretrained deep convolutional neural network (CNN) for fault diagnosis of a gearbox with limited data. The source domain consisted of a large number of labeled natural images and the target domain comprised graphical images from the gearbox vibration signals. Xu et al. [36] presented transfer CNNs for online fault diagnosis of bearings and pumps. In their work, related datasets were used to train several offline CNNs, then their shallow layers were transferred to an online CNN to improve its diagnostic performance. Yan et al. [37]

studied the application of knowledge transfer for fault diagnosis in rotary machines while considering the variation of working conditions, fault locations, types of machines, and different faults. Hasan and Kim [38] studied transfer learning for fault diagnosis of bearings under different working conditions. The only difference between the source and target tasks was the speed of rotation. Li et al. [39] studied the fault diagnosis of rolling bearings via deep convolution domain adversarial transfer learning. Zhang et al. [40] proposed a fault diagnosis strategy for bearings under different working conditions using transfer learning with neural networks. Huang et al. [41] presented a boosted algorithm (SharedBoost) to explore transfer learning for multiple data sources and compared its results with those of other transfer learning methods.

This paper attempted to employ simple simulation models of a rotor system to devise a robust and autonomous diagnostic strategy for actual rotating machinery. First principles were used to develop a simple two degree of freedom model of a rotor system with three types of defects (unbalance, parallel misalignment, and point rubbing). The simple rotor model was used to generate a large amount of vibration data by considering different operation speeds and different defect severity levels. For robustness, the simulation data was also contaminated with different levels of white Gaussian noise. The vibration signals from the simulation models were transformed into scalograms [42,43], which were then used to obtain a pretrained customized deep neural network. The pretrained network was employed as a generalized autonomous feature extractor from the experimental data sets of an RK4 rotor kit and a machinery fault simulator (MFS). The extracted features were processed with several conventional machine learning algorithms and an optimum classifier was identified. The performance of the customized deep learning network for autonomous feature extraction is also compared to that of other existing pretrained models, such as AlexNet [44], GoogleNet [45], ResNet [46], Vgg16 [47], etc. The proposed approach is invariant to the number of health states in the simulation and experimental domains, while no attempts are made to minimize the gap between the two domains.

## 2. Proposed Methodology

The limited nature of data from different defective states of actual machinery prohibits the use of deep learning models for autonomous feature extraction and diagnostics. Development of exact simulation models that replicate the response characteristics of the actual machinery is often computationally expensive. Although simple simulation models can provide insights into the characteristic behaviors of actual machinery in the presence of defects and are less computationally expensive, they do not account for all the uncertainties in the actual system. Transfer learning or cross-domain knowledge transfer could help to leverage the advantages of simple simulation models for fault diagnosis of the actual system. A schematic illustration of the basic idea of transfer learning in the context of the current problem is shown in Figure 1.

A large amount of source data is required to train, validate, and test a deep learning model with the highest possible degree of accuracy. That pretrained model can be used for automatic feature extraction from a limited dataset for a target task, or its weights and bias can be transferred to a limited target dataset using the concept of fine tuning [48,49]. In our case, the parameters of the model trained on simulation data are employed to extract high-level discriminative features from the target task of the experimental data. In transfer learning, the types of defects in the source data and target data are not necessarily the same [35,37,50]. A schematic illustration of the general workflow of the current work, which involves employing simple simulation models to detect, isolate, and quantify different types of defects in actual mechanical systems, is depicted in Figure 2.

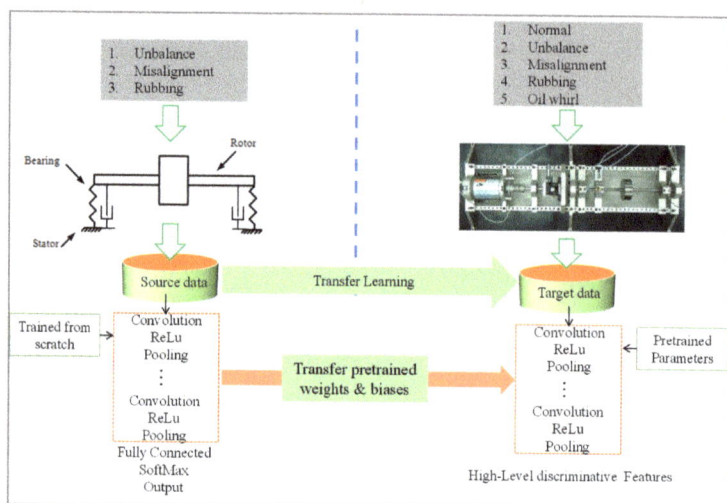

**Figure 1.** Fundamental idea of transfer learning.

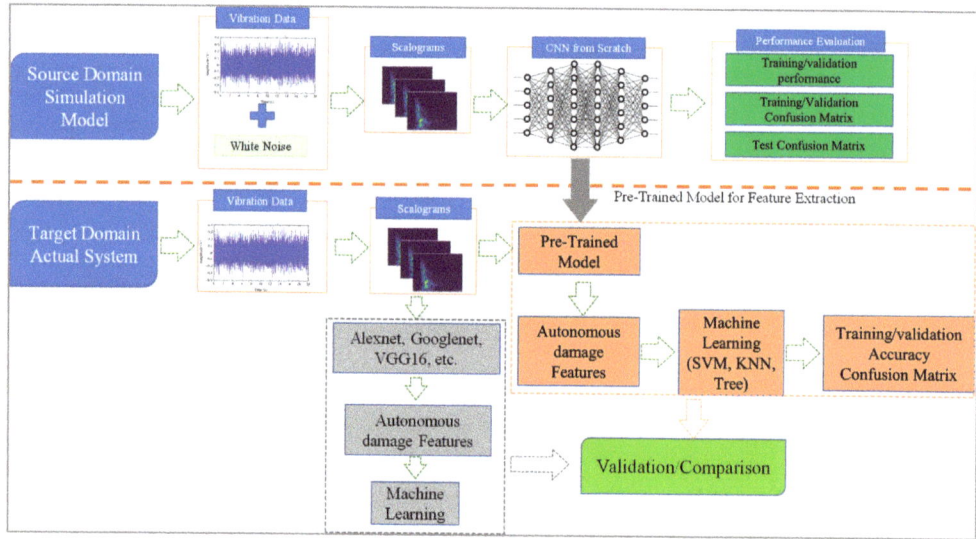

**Figure 2.** Overall workflow of the proposed methodology.

Herein, a simple simulation model was employed to generate a large amount of source data for the representative health states of the source task. For robustness, the simulation data was contaminated with different levels of white Gaussian noise. The noisy source data was transformed into scalograms via MATLAB and the scalograms were used to train, validate, and test a customized CNN.

The neural network trained on simulation data was used to automatically extract discriminative features from the response scalograms of the experimental data of real machines. The discriminative features were processed using traditional machine learning algorithms, such as support vector machine (SVM), tree classifier, K-nearest neighbor (KNN), etc. The results of the autonomous feature extraction via a customized neural network trained on simulation data are also compared with the results of feature extraction via available pretrained deep learning models (e.g., Alexnet, GoogLeNet, VGG16) in terms of classification accuracy, generalization, computational cost, hardware requirement, etc. The proposed approach was validated for two datasets from an RK4 rotor kit by GE Bently Nevada (1631 Bently Parkway South, Minden, Nevada USA 89423) and a machinery fault simulator (MFS) by SpectraQuest (8227 Hermitage Road, Richmond, VA 23228 USA). Although, in the current work, the proposed approach was employed for the diagnosis of rotating machinery, this approach could be extended to the damage assessment of laminated composites, civil infrastructures, industrial robots, gearboxes, and others, where simple simulations could be developed to gain insights into the fault characteristics of the actual systems.

### 2.1. Simulation Model and Source Data Generation

As described in the previous sections, developing a simulation model that precisely matches the response characteristics of an actual system in the absence and presence of defects is either too computationally expensive or completely impossible for complex systems. Although simple simulation models of different types of actual machinery (e.g., a turbine simplified as a shaft-disk system) have been used to gain insight into the characteristics of various defects in the actual system, it is never a guarantee that the simulation models can be employed to assess damage in the actual system using a conventional approach. To bridge the gap between the actual systems and their simulated counterparts, transfer learning or cross-domain knowledge transfer provides a natural solution. However, transfer learning requires a large amount of data from the source task. This section describes the simple mathematical models of a shaft-disk system with different defects that were used to generate a large dataset for the source task. The simple rotor system considered in this work consists of a single disk of mass m mounted at the center of a shaft with length L, as shown in Figure 3.

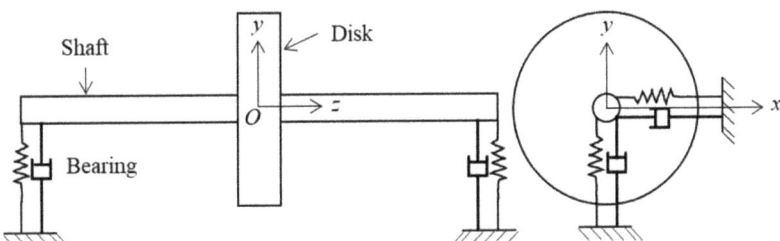

Figure 3. Simple rotor-disk system for the generation of a large amount of source data.

The shaft is supported by two bearings at the ends; the bearings are linearized, ideally with stiffness and damping. The support and/or foundation are assumed to be rigid. The dynamic response of the shaft-disk system is represented by a fixed coordinate system at the center of the disk. The system is characterized in terms of transverse displacements, and the vibration along the axis of the shaft is ignored. For the isotropic properties of the bearings at the two ends, and the disk mounted at the center of the shaft, the dynamics of the system in Figure 3 can be defined by a time-dependent equation, as follows:

$$m\ddot{x}(t) + c_{xT}\dot{x}(t) + k_{xT}x(t) = F_x(t) \\ m\ddot{y}(t) + c_{yT}\dot{y}(t) + k_{yT}y(t) = F_y(t) \qquad (1)$$

where $x$ and $y$ are the displacements at the disk along the $x$ and $y$ axes, respectively, $m$ is the mass of the disk, $c_{xT}$ and $c_{yT}$, respectively, denote the total damping at the two bearings along the $x$ and $y$ axes, and $k_{xT}$ and $k_{yT}$ denote the total stiffness at the two bearings along the $x$ and $y$ axes, respectively. The terms $F_x$ and $F_y$ refer to the general forces acting on the system along the $x$ and $y$ direction, respectively. The anisotropic supports can be modeled using the approach proposed by Filippi et al. [51]. The characteristic behavior of the forcing functions acting on the system depends on the type of defect in the system. In this work, three defects (unbalance, misalignment, and rubbing) of different magnitudes were considered in the system shown in Figure 3 to generate a large amount of source data. In practice, the pristine or healthy state of the rotating machinery has a small amount of residual unbalance that cannot be completely removed despite efforts at balancing. This small amount of residual unbalance is considered to be within the acceptable range (i.e., the system is considered to be healthy) if the amplitude of the vibration signals is within a certain level of the root mean square (rms) as prescribed by the standards of ISO 7919-2 [52–54]. Residual unbalance in the system exists when the center of mass is not coincident with the center of rotation. The motion equation used to simulate the residual unbalance in the rotor system is shown as follows in Equation (2):

$$m\ddot{x}(t) + c_{xT}\dot{x}(t) + k_{xT}x(t) = m e_r \Omega^2 \cos(\Omega t + \alpha)$$
$$m\ddot{y}(t) + c_{yT}\dot{y}(t) + k_{yT}y(t) = m e_r \Omega^2 \sin(\Omega t + \alpha)$$
(2)

where $e_r$ is the eccentricity between the center of mass and center of rotation, $\alpha$ is the phase angle of residual unbalance, and $\Omega$ is the rotational speed of the shaft.

The presence of unbalance, misalignment, and rubbing can be simulated as additional forces along with the residual unbalance. The forcing functions for the three defects are shown by Equations (3)–(5), respectively, as follows:

$$F_{x\_unb} = m_a e_a \Omega^2 \cos(\Omega t + \beta)$$
$$F_{y\_unb} = m_a e_a \Omega^2 \sin(\Omega t + \beta)$$
(3)

$$F_{x\_mis} = FX_2 \cos(\Omega t + \psi) + FX_2 \cos(2\Omega t + \psi)$$
$$F_{y\_mis} = FY_2 \sin(\Omega t + \psi) + FY_2 \sin(2\Omega t + \psi)$$
(4)

$$F_{x\_rub} = -k_r(x - \delta_0) H(x - \delta_0)$$
$$F_{y\_rub} = f k_r(x - \delta_0) H(x - \delta_0)$$
(5)

where $m_a$ is the added unbalance to the disk with an eccentricity of $e_a$ and phase angle of $\beta$. The term $\Omega$ denotes the speed of rotation. The terms $FX_i$ and $FY_i$ ($i$ = 1, 2) are the external forces due to parallel misalignment with a phase angle of $\psi$. The term $k_r$ is the stiffness of the axial rub-impact rod, $f$ is the friction coefficient of between the two parts, $H$ is the Heaviside function, and $\delta_0$ is the gap between the rotor and stator. Further details on the mathematical modeling can be found in Appendix A.

The mathematical models of unbalance (Equation (3)), misalignment (Equation (4)), and rubbing (Equation (5)) were employed to generate the large amount of source data necessary for the transfer learning strategy shown in Figure 2. The basic parameters of the three simulation models are given in Table 1.

Herein, the added unbalance was simulated with a fixed value of eccentricity ($e_a$) by varying the value of the added mass from 1 to 20 g at increments of 2 g, misalignment was simulated with a parallel misalignment along the $y$-bending angular flexibility rate axis from 8 to 26 mm at increments of 2 mm, and rubbing was simulated by reducing the values of clearance between the rotor and stator from $9.2 \times 10^{-8}$ to $4.7 \times 10^{-8}$ m at decrements of $0.5 \times 10^{-8}$ m. For the three defects (unbalance, misalignment, and rubbing) of the shaft-disk system, ten different levels of severity were considered, and each defective case of the system was operated at 50 different speeds of rotation from 300 to 6810 rpm at increments of 120 rpm. The steady-state vibration responses of the system were obtained along the $x$ and $y$ axes at the disk location by solving the differential equation (Equations (3)–(5)) of each defect via Newmark's time integration algorithm [55]. The number of steady-state responses for each defect with all severity levels was $20 \times 50 = 1000$ samples.

Table 1. Material properties and parameters of the simulation models.

| Model | Parameter | Value |
|---|---|---|
| General | Length of shaft ($L$) | 1 m |
| | Modulus of elasticity ($E$) | $211 \times 10^9$ Pa |
| | Modulus of Rigidity ($G$) | $81.1 \times 10^9$ Pa |
| | Diameter of shaft ($d_s$) | 0.01 m |
| | Diameter of disk ($d$) | 0.075 m |
| | Thickness of disk ($h$) | 0.0254 m |
| | Density of shaft and disk ($\rho$) | 7810 kg/m$^3$ |
| | Mass of disk $m = \rho h \pi d2/4$ | 0.8764 kg |
| | Stiffness at bearing 1 along $x$-axis ($k_{x1}$) | $1.0 \times 10^6$ N/m |
| | Stiffness at bearing 1 along $y$-axis ($k_{y1}$) | $1.0 \times 10^6$ N/m |
| | Stiffness at bearing 2 along $x$-axis ($k_{x2}$) | $1.0 \times 10^6$ N/m |
| | Stiffness at bearing 2 along $y$-axis ($k_{y2}$) | $1.0 \times 10^6$ N/m |
| | Damping at bearing 1 along $x$-axis ($c_{x1}$) | 1000 Ns/m |
| | Damping at bearing 1 along $y$-axis ($c_{y1}$) | 1000 Ns/m |
| | Damping at bearing 2 along $x$-axis ($c_{x2}$) | 1000 Ns/m |
| | Damping at bearing 2 along $y$-axis ($c_{y2}$) | 1000 Ns/m |
| Residual Unbalance | Mass eccentricity ($e_r$) | 0.000015 m |
| | Phase angle ($\alpha$) | 0° |
| Unbalance | Added masses ($m_a$) | (1:2:20) g |
| | Phase angle ($\beta$) | 0° |
| Misalignment | Misalignment along $x$-axis ($\Delta X_1 = -\Delta X_2$) | 0 m |
| | Misalignment along $y$-axis ($\Delta Y_1 = -\Delta Y_2$) | (8:2:26) mm |
| | Center of articulation ($Z_3$) | 0.024 m |
| | Bending angular flexibility rate ($K_b$) | 0.35 degree/Nm |
| Rubbing | Power of Motor ($P$) | 700 Watt |
| | Clearance between rotor and stator ($\delta_0$) | $(9.2: -0.5:4.7) \times 10^{-8}$ m |
| | Stiffness of axial rub-impact rod ($k_r$) | $1.2 \times 10^7$ Pa |
| | Coefficient of friction ($f$) | 0.7 |

To account for noise in the signals from actual systems, all steady-state responses of the three defects were added to white Gaussian noise with a signal-to-noise ratio (SNR) of 31 to 40 using the MATLAB function *awgn* (add white Gaussian noise to signal). The basic mathematical form of the *awgn* function is shown by Equation (6).

$$S_{noise} = S + Z \\ Z \sim N(0, \mu) \tag{6}$$

where $S$ is the original signal without noise and $Z$ refers to the random noise having normal/Gaussian distribution with zero mean and $\mu$ variance. $S_{noise}$ is the output signal contaminated with noise. Additional mathematical details of adding white Gaussian noise can be found in the MATLAB documentation and the published literature, as shown in the references [56–58]. The decision to use a range of SNR from 31 to 40 was made after looking at the effect of different SNR values on the original signals obtained from simulations. The effect of different values of SNR on the original signal of 9 g unbalance at 300 rpm is shown in Figure 4.

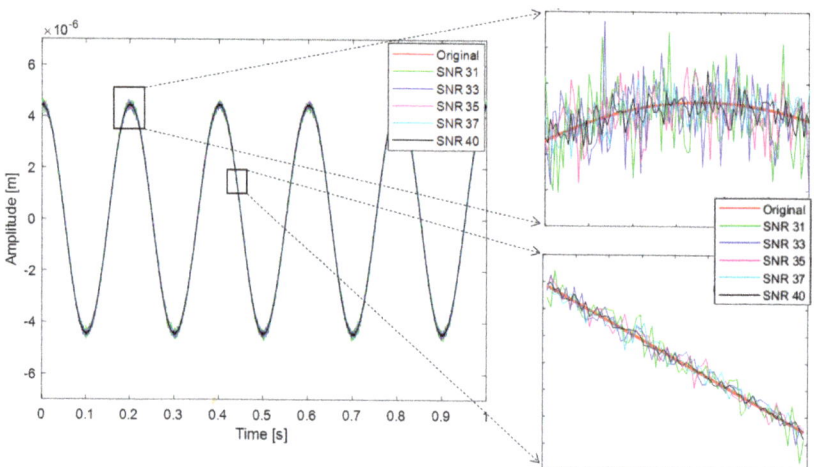

**Figure 4.** Effect of different values of sound-to-noise-ratio (SNR) on the signals obtained from the simulation model.

As shown in Figure 4, it was observed that the SNR range of 31 to 40 accounts for the higher and lower levels of noise in the source simulation data.

This noise contamination of the 1000 steady-state signals of each defect resulted in $20 \times 50 \times 10 = 10{,}000$ samples for each defect. The steady-state response signals of the three defects with and without noise were combined, resulting in 33,000 samples (11,000 samples for each defect) that served as the source data from the simulation model.

### 2.2. Deep Learning Model for Simulation Data

The 33,000 response signals from the simulation model were transformed into scalograms using continuous wavelet transform (CWT). A scalogram is essentially a time-frequency representation of a time domain signal that is generated from the absolute value of the CWT coefficients of that signal. The mathematical details regarding the transformation of a time series to a scalogram using wavelet analysis can be found in the references [59,60]. In this work, MATLAB was used to design a CWT filter bank with a sampling frequency of 8500 Hz (the same as the signal acquisition frequency) and the default number of voices per octave (10 wavelet bandpass filters per octave) [61]. The analytic Morse wavelet with the default values of the symmetry parameter and time-bandwidth product was used in the filter bank [62–64]. More details on the parametric study of the effect of the parameters of wavelet transform can be found in the references [65–67]. The filter bank was used to transform all the time series from the simulation models to scalograms. Figure 5 depicts some samples of unbalance, misalignment, and rubbing scalograms for a given speed of rotation out of 33,000 scalograms from the simulation data.

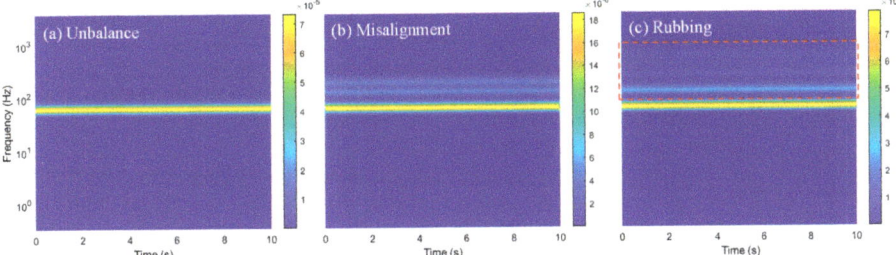

**Figure 5.** Sample unbalance, misalignment, and rubbing scalograms in the simulation model at a steady state of 3660 rpm with $y$-axis on a logarithmic scale; (**a**) Unbalance; (**b**) Misalignment; (**c**) Rubbing.

The scalograms of the three defects have distinct characteristics in the time-frequency domain. In general, the presence of unbalance, misalignment, and rubbing in a rotating system are characterized by the presence of distinct frequency spectra at 1X (speed of rotation) [68], frequency spectra at 1X and the integral multiples thereof (2X, 3X), and frequency spectra at 1X and its sub- and super-harmonics depending on the speed of rotation [69,70], respectively. The general characteristics of unbalance, misalignment, and rubbing can be observed in the scalograms in Figure 5, where the presence of unbalance, misalignment, and rubbing are shown by a distinct frequency component at the speed of rotation, the speed of rotation and its integral multiples, and by super harmonics (dashed red rectangle), respectively. In addition, note that the $y$-axis is a logarithmic scale.

The scalograms of the source data from the simulation were used to train, validate, and test a convolutional neural network (CNN). Figure 6 depicts the detailed architecture of the CNN used in the current work.

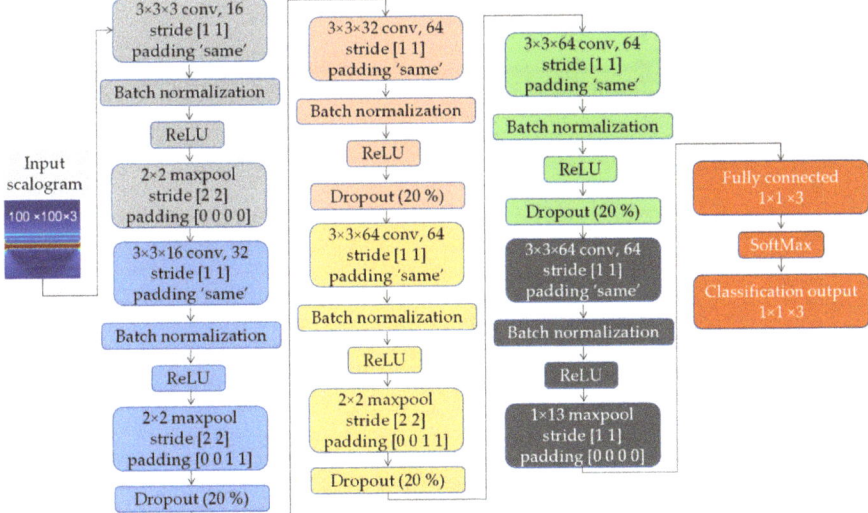

**Figure 6.** Architecture of the convolutional neural network used in the current study.

In the CNN architecture, convolutional, batch normalization, and ReLU layers are used to extract high-level features from the input scalograms, and max pooling layers are employed to down-sample those features [35]. A dropout layer is inserted to minimize the chances of overfitting during the training process [71]. The classification layer adopts a SoftMax function [72,73] to classify the extracted features into three different classes:

unbalance, misalignment, and rubbing in the shaft-disk system. In the current architecture, the max pooling layers in the first, second, and fourth hidden layers were used to account for invariances in the simulation scalograms. Since the pretrained model was to be employed as a generalized feature extractor from the experimental data, the max pooling layers in the third and fifth hidden layers were excluded to accommodate the local variations in the autonomous features of the target task.

To train the CNN, the weights were randomly initialized and tuned from scratch using Adam optimizer as an optimization function. The data set of 33,000 scalograms was split into 80% training, 10% validation during the training, and 10% independent test datasets. To avoid memory problems, the scalograms were loaded in the form of an image data store using the function "imageDatastore" in MATLAB. Figure 7 shows the accuracy and loss for the training and validation of the network.

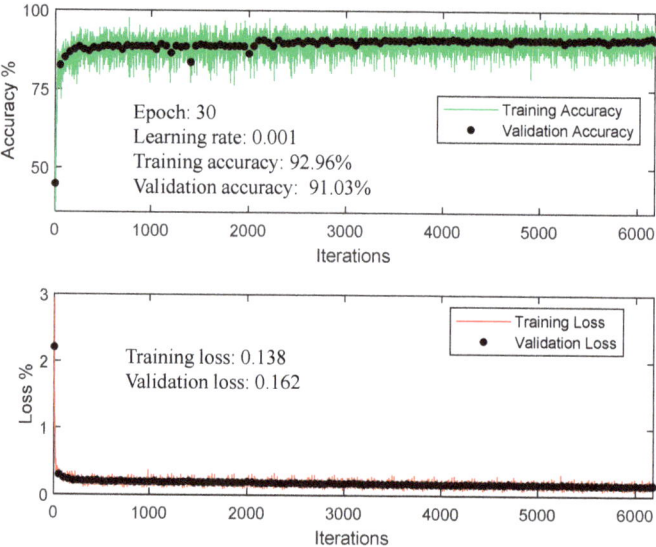

**Figure 7.** Training and validation of the customized deep learning model using simulation data.

Here, 80% of the data (training data) was used to train the CNN, while 10% of the data (validation data) was used to evaluate the performance of the model at each iteration of the training process. The training/validation accuracy refers to the classification/validation accuracy for each mini batch of the training/validation dataset. The training/validation loss indicates the performance of the model after each iteration of optimization and denotes the sum of errors for each example of the training/validation data. From Figure 7, the overlap between the training and validation accuracies and losses as well as the validation accuracy of 91.5% imply that the network is optimally learning from the training data and could be generalized to unseen instances.

To verify the generalization of the pretrained CNN to an unseen data set, the model was tested on the remaining 10% of the dataset (testing data). Figure 8 shows the test confusion matrix.

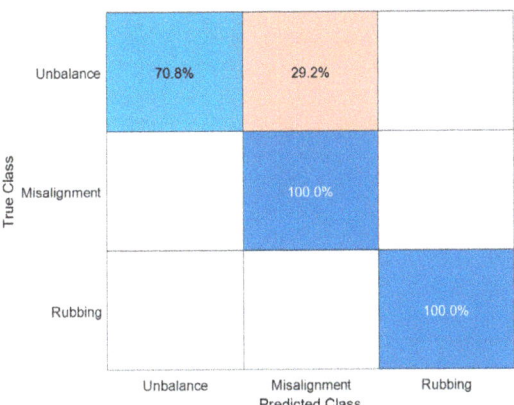

**Figure 8.** Test confusion matrix showing the performance of the pretrained deep learning model on unseen simulation test data (90.27% accuracy on test data).

As shown in Figure 8, the pretrained network successfully identified the presence of misalignment and rubbing with 100% accuracy; however, it confused 29.2% instances (321 observations) of unbalance with misalignment. The reason for the confusion between unbalance and misalignment is that the misalignment model in Equation (4) only simulates parallel misalignment along the $y$-axis, resulting in misaligned response characteristics along the $y$-axis and unbalance response characteristics along the $x$-axis. For lower values of added unbalance, the unbalance response from the misalignment model along the $x$-axis and the actual added unbalance will be confused. Thus, 29.2% instances of unbalance were confused with misalignment.

### 2.3. Experimental Data

Two experimental data sets were employed to validate the effectiveness of the proposed approach. The first experimental vibration data for different health states of the shaft-disk system was obtained from an RK4 rotor kit, a product of GE Bently Nevada (1631 Bently Parkway South, Minden, Nevada USA 89423). The vibration signals were obtained via proximity sensors for the following health states: normal (residual unbalance), unbalance, rubbing, misalignment, and oil whirl. The experimental configuration of the different health states of the rotor system is shown in Figure 9.

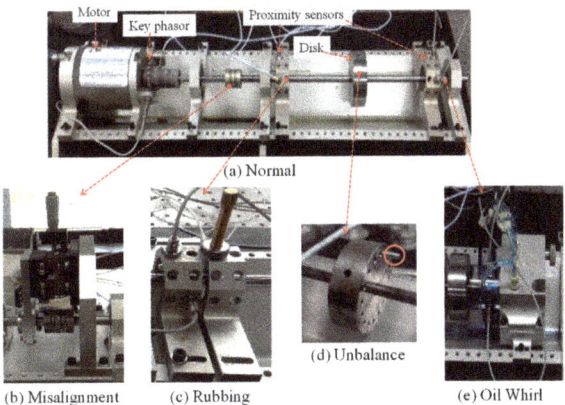

**Figure 9.** RK4 Rotor kit with different health states; (**a**) Normal; (**b**) Misalignment; (**c**) Rubbing; (**d**) Unbalance; (**e**) Oil Whirl.

Despite efforts to perfectly balance the system in the normal state, there existed a small amount of unbalance in the system; the amplitude of the resulting vibration signal was within the acceptable range of 10 μm of the root mean square (rms) level, as determined by the ISO standard 7919-2.

The unbalance state was induced by attaching a 15 g screw to the disk (Figure 9d). A special jig (Figure 9b) was employed to induce a parallel misalignment of 20 μm along the y-axis at the coupling location. The rubbing state was simulated with a rubbing screw (Figure 9c) that contacted the shaft when a 15 g mass was attached to the disk. The position of the rubbing screw was adjusted such that the shaft contacted the rubbing screw once per revolution at 3600 rpm (steady-state condition for all health conditions). An additional tool kit (Figure 9e) was used to induce the oil whirl phenomenon at an oil pressure of 35 kPa.

Two sets of proximity sensors placed near each bearing were used to acquire the vibration response signals for all the health states of the rotor kit; for each set of proximity sensors, the two were installed at right angles along the $x$ and $y$ axes. To ensure repeatability and account for experimental uncertainty, each health state was executed five times, and the rotor kit was reassembled before each experiment. All five health states of RK4 were studied at a steady-state condition of 3600 rpm. The dataset for all health states consisted of 100 signals, with 20 signals for each case (4 signals for each health state × 5 executions of each experiment).

The CWT filter bank designed for the simulation data was used to transform the vibration signals from the RK4 rotor kit for all health states to scalograms without any preprocessing. We aimed to gain insight into the characteristics of different defects and compare the results with the outcomes of the simple simulations. Figure 10 depicts sample scalograms of the experimental data for the different health states.

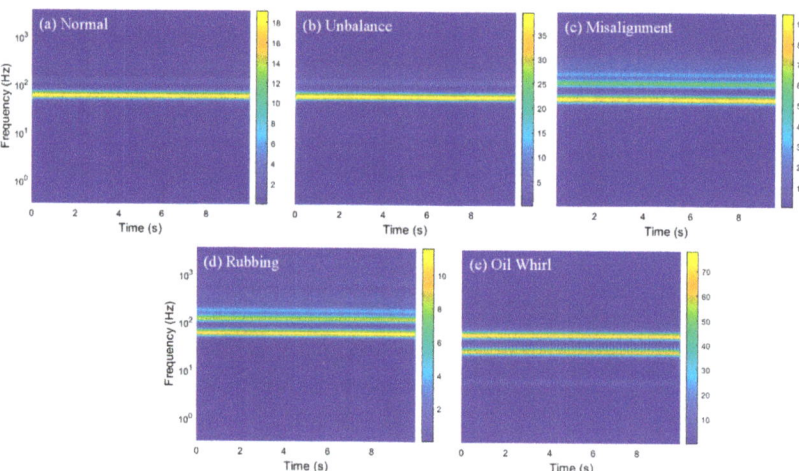

**Figure 10.** Sample scalograms of different health states in the RK4 Rotor kit at steady state of 3600 rpm: (**a**) normal; (**b**) unbalance; (**c**) misalignment; (**d**) rubbing; and (**e**) oil whirl (y-axis is a logarithmic scale).

In Figure 10, some high-frequency contents are observed in the scalogram of the normal state, implying either the presence of noise or some other small unavoidable defects alongside the small unbalance in the system. Additionally, comparing the scalograms from the experimental data with their simulated counterparts shows that the scalograms of the experimental data demonstrate more complex behavior in terms of time-frequency content, which confirms that extremely simple mathematical models cannot replicate the exact dynamic response behavior of an actual system with and without defects. Furthermore, the health states of normal and oil whirl were not considered in the source simulation data.

In the next section, the CNN model pretrained on simulation data is used to automatically extract discriminative features from the scalograms of experimental data.

### 2.4. Autonomous Feature Extraction Using Pretrained Models

In transfer learning, a model developed and trained for one task is reused as a starting point for another related task, without expending much time or computational resources [33]. As stated previously, the inner layers of a CNN autonomously extract high-level features from the input images and use those features in the last fully connected and classification layers to distinguish between different classes of input images. In the architecture of a pretrained network, some layers can be eliminated to retrieve the high-level features from layer activation, and those features can be processed with traditional machine learning algorithms, as depicted in Figure 11.

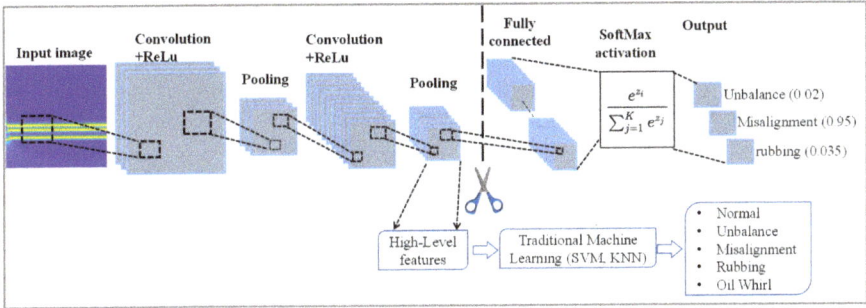

**Figure 11.** Autonomous feature extraction via a pretrained deep learning model.

The automatically extracted high-level features can be used to train, validate, and test traditional machine learning algorithms, such as SVM, tree classifier, KNN, etc. In this work, the activations from the last max pooling layer of the CNN trained on simulation data were used as discriminative features for the scalograms of the experimental data from the RK4 rotor kit. The autonomously extracted features were processed with several different machine learning classifiers; Figure 12 shows a comparison of the different classifiers in terms of overall classification accuracy and area under the ROC (receiver operating characteristic) curve. The ROC area is obtained by graphing the true and false positive rates and its value implies a tradeoff between recall and fallout. An ROC area close to 1 indicates that the model is able to achieve a high recall (true positive rate) while maintaining a low fallout (false positive rate) [74].

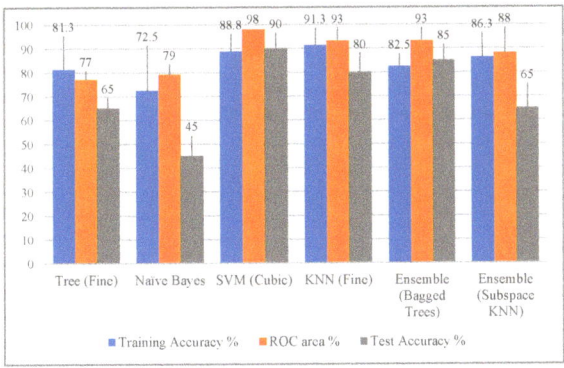

**Figure 12.** Performance of different classifiers using the automatically extracted features.

As shown in Figure 12, the minimum and maximum training accuracies were 72.5% and 91.3%, for the naïve Bayes and KNN classifiers, respectively. However, the overall training accuracy could be deceiving, and the model may have overfitted the training data. The matrices of ROC area and prediction results on an independent test set would help to fully explore the behavior of the supervised learning classifiers.

In Figure 12, SVM stands out as the optimum classifier in terms of training accuracy (88.8%), ROC area (98%), and test accuracy (90%). Figure 13 shows the confusion matrix of SVM on the 80% training dataset, created to gain further insight into the classification performance of SVM.

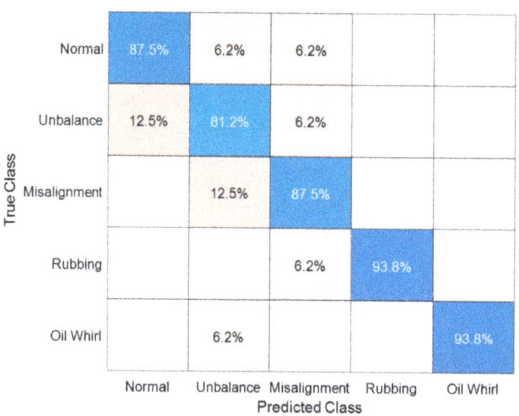

**Figure 13.** Training/validation confusion matrix of cubic SVM on the features automatically extracted by the pretrained deep learning model from the original RK4 data.

During the training process, the classifier confused 6.2% of the instances of normal as unbalance and misalignment, 12.5% of the instances of unbalance as normal, 6.2% of the instances of unbalance as misalignment, 12.5% of the instances of misalignment as unbalance, 6.2% of the instances of rubbing as misalignment, and 6.2% of the instances of oil whirl as unbalance. Here, 6.2% and 12.5% instances refer to one and two observations, respectively. According to the training confusion matrix, the loss of accuracy was mainly due to the confusion of 12.5% instances of misalignment with unbalance and 12.5% instances of unbalance with the normal state. The physical reason for this confusion is that only parallel misalignment was induced along the $y$-axis, while the response along the $x$-axis is purely due to residual unbalance that may coincide with the added unbalance. Similarly, a possible explanation for confusing unbalance with the normal state is that the two share the same response characteristics and only differ in amplitude. The domain and class invariance of the proposed approach is verified from the high classification accuracy of the health states of normal and oil whirl, which were not considered in the source simulation domain.

The results of the pretrained cubic SVM on the unseen test dataset are shown in the form of a confusion matrix in Figure 14.

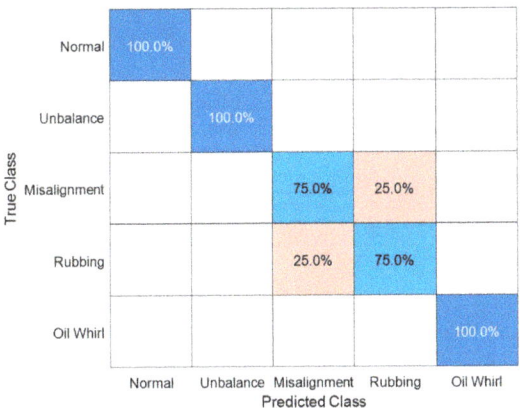

**Figure 14.** Test confusion matrix of cubic SVM on features automatically extracted by the pretrained deep learning model from the original RK4 data.

As shown in Figure 14, 25% of instances of misalignment were confused with rubbing and 25% of the instances of rubbing were confused with misalignment. The results of the test confusion matrix are within an acceptable range, as 25% of instances is equivalent to one observation out of four from the 20% test data.

Given the above discussion, the results of autonomous feature extraction via a CNN that was pretrained on simulation data are physically reasonable. However, the limited size of the training and the test datasets make it difficult to draw a general conclusion. One option is to obtain more data from the testbed by repeating the experiments; however, the experiments have already been repeated five times.

Another option is to employ the concept of virtual sensors around the shaft, as introduced by Jung et al. [75], to artificially augment the data without performing any further experiments. In this work, the concept of virtual sensors is adopted to synthetically augment the experimental data. The main idea of virtual sensors is to obtain synthetic vibration signals from the vibration signals of the actual orthogonal proximity sensors by rotating the cartesian coordinate system with respect to the z-axis, as depicted in Figure 15.

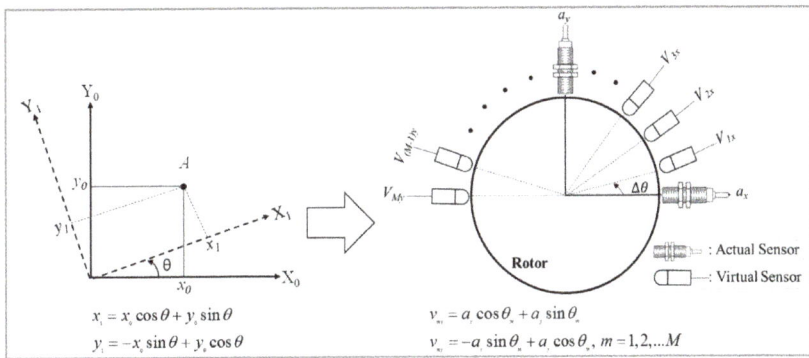

**Figure 15.** Concept of virtual sensors based on simple transformation of coordinates.

The virtual signals are obtained from the actual signals using the following coordinate transformation:

$$\begin{aligned} x_{Vm} &= \cos(m\Delta\theta)x_a + \sin(m\Delta\theta)y_a \\ y_{Vm} &= -\sin(m\Delta\theta)x_a + \cos(m\Delta\theta)y_a \\ &(m = 1, 2, \ldots, M) \end{aligned} \quad (7)$$

where $x_{Vm}$ and $y_{Vm}$ are virtual signals along the rotated $x$ and $y$ axes, respectively. The terms $a_x$ and $a_y$ refer to the actual signals obtained via the proximity sensors along the original $x$ and $y$ axes, respectively, $\Delta\theta$ is the angle of rotation for the coordinate system of virtual signals, and $M$ denotes the number of virtual signals. Owing to symmetry around the shaft, the maximum number of virtual sensors is $M = \pi/\Delta\theta$. As shown in a previous paper [75], $x_{Vm}$ is equal to $y_{Vm+M/2}$; hence, in this work only $x_{Vm}$ was retained from Equation (6) for synthetic data augmentation. To identify the optimum number of virtual sensors for the current task, a parametric study was carried out for different numbers of virtual sensors and the effect was evaluated terms of training/validation accuracy, ROC area, and the number of instances per class as a result of data augmentation, as shown in Figure 16.

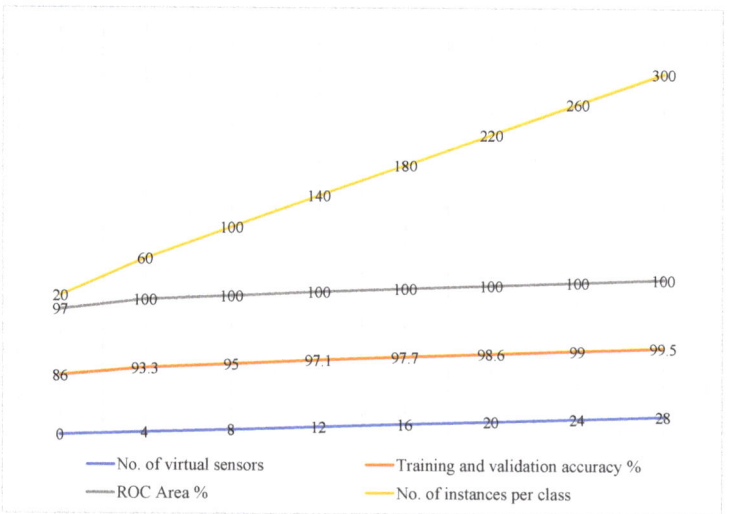

**Figure 16.** Different numbers of virtual sensors and their effect on the classification performance and size of dataset.

To obtain the results shown in Figure 16, the original and augmented datasets were transformed into scalograms and processed via the pretrained CNN to extract discriminative features. A cubic SVM was employed to classify the extracted features into different classes using 10-fold cross-validation. The results showed that the training and validation accuracy could be increased to 99.5% by synthetic data augmentation using virtual sensors; however, the increase in the evaluation matrices of training/validation accuracy and ROC are relatively small compared with the increase in the size of the augmented dataset. Thus, because of this tradeoff between the size of the augmented dataset and classification accuracy, the number of virtual sensors was set at 12 for further analysis. To provide more insight into the problem, the augmented data was split into 80% training and 20% test data. Figure 17 shows the per class training/validation performance of the cubic SVM on the training data in the form of a confusion matrix with 97.3% accuracy.

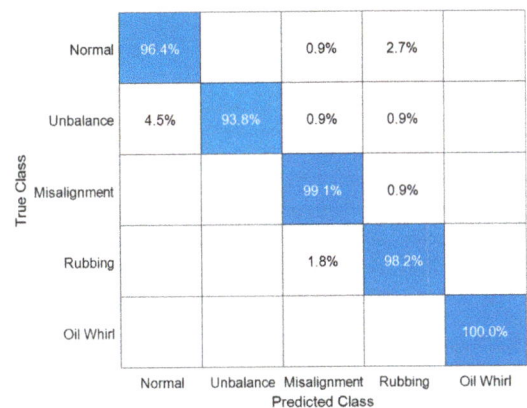

**Figure 17.** Training/validation confusion matrix of SVM (cubic) on automatically extracted features from the pretrained deep learning model with 12 virtual sensors on RK4.

The cubic SVM was trained via 10-fold cross-validation on the synthetically augmented data using 12 virtual sensors. In Figure 17, the higher true positive rate and the lower false positive rate for each class demonstrate the optimum performance of the proposed methodology on the experimental data set. Additionally, note that the per class classification performance increased compared with the results from the data without augmentation in Figure 13. To show that the vibration signals synthesized through virtual sensors did not cause overfitting of the machine learning model, the cubic SVM pretrained on the augmented data (synthesized and measured) was employed to make predictions on the 20% unseen test data. Here, the unseen data describes a data set not seen by the network during the training/validation process. The pretrained model showed a test accuracy of 97.14%, with the confusion matrix shown in Figure 18.

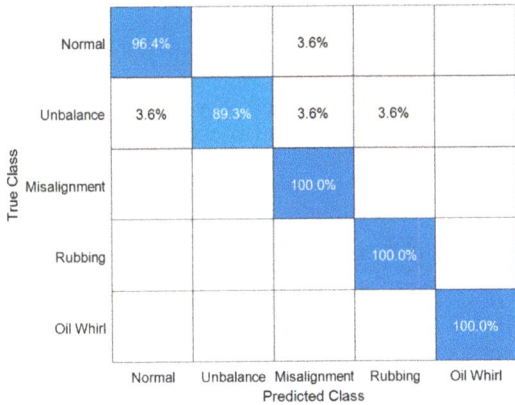

**Figure 18.** Test confusion matrix of SVM (cubic) on automatically extracted features from the pretrained deep learning model with 12 virtual sensors on RK4.

As shown in Figure 18, the test accuracy on the augmented data increased from 90% to 97.14% in comparison with the performance on the measured data, which would not have been possible in the case of overfitting due to synthesized signals.

To further explore the robustness of the proposed approach and its ability to bridge the gap between simple computer simulations and actual experiments, the deep learning model trained on simulation data was compared with pre-existing deep learning models of GoogleNet, Vgg16, ResNet18, AlexNet, and SqueezeNet [76] in terms of feature extraction. The pre-existing deep learning models are trained and optimized on natural images and have fixed network architectures [31]. The image dataset that is commonly employed to train the existing pretrained networks is usually a subset of the ImageNet database [77]. For instance, Vgg16 is pretrained on approximately 1.5 million images with 41 layers, and Alexnet is pretrained on approximately 1.2 million images with eight layers and 60 million parameters. To verify the performance of the customized deep learning model for autonomous feature extraction from a limited amount of experimental data, the performance of the cubic SVM on autonomously extracted features from the simulation model was compared with the performance of Googlenet, Vgg16, Resnet18, Alexnet, and Squeezenet, as shown in Table 2.

**Table 2.** Comparison of the customized pretrained simulation model with existing pretrained deep learning models (12 virtual sensors and cubic SVM).

|  | Training/Validation Accuracy % | ROC Area% | Testing Accuracy % |
| --- | --- | --- | --- |
| Simulation Model | 97.5 | 100 | 97.14 |
| GoogleNet | 93.8 | 99 | 83.57 |
| Vgg16 | 95 | 100 | 83.5 |
| Resnet18 | 97.1 | 100 | 86.43 |
| Alexnet | 95.2 | 99 | 85.71 |
| SqueezeNet | 90.2 | 99 | 81.43 |

For the results in Table 2, the features extracted by all the deep learning models were split into 80:20 for training and testing, respectively. The 80% training data was used to train a cubic SVM through 10-fold cross-validation, and the resulting trained model was employed to make predictions on the 20% test dataset. According to the results shown in Table 2, all the deep learning models performed reasonably well in terms of training/validation accuracy, ROC area, and test accuracy, which validates the performance of the customized deep learning model.

The results shown in Table 2 bring up an obvious question: if the existing pretrained models perform equally well on the limited experimental dataset, then why bother using a simulation dataset and a customized deep learning model?

The motivation behind the customized deep learning model is that the existing pretrained networks (AlexNet, VGG16) have fixed architectures, a fixed number of parameters, and limited flexibility for controlling the dimensions of the extracted discriminative features, whereas the customized deep learning model offers more flexibility in terms of network size, number of parameters, and dimensions of the extracted discriminative features. Furthermore, as seen from the test classification accuracy of the simulation model in Table 2, a deep learning model pretrained on source data that resembles the target data of the transfer learning scheme would provide better generalizability. To further support the effectiveness of the proposed approach, the autonomous feature extraction through a customized deep learning model for the data of 12 virtual sensors was compared with the feature extraction through the pre-existing deep learning models in terms of size of the network, parameters of the network, and computational time, as shown in Table 3.

Table 3. Comparison of customized deep learning model with pre-existing deep learning models.

| Name | Size of Network (Bytes) | Number of Network Parameters | Computational Time (sec) | |
|---|---|---|---|---|
| | | | CPU * | GPU ** |
| Simulation Model | 631,037 | 139,587 | 4.542 | 1.65 |
| GoogleNet | 29,670,809 (191.7%) | 6,698,552 (192.2%) | 25.66 (139.8%) | 2.55 (42.7%) |
| Vgg16 | 554,895,306 (199.5%) | 138,357,54 (199.6%) | 149.47 (188.2%) | 6.33 (117.2%) |
| Resnet18 | 47,156,446 (194.7%) | 11,694,312 (195.2%) | 23.56 (135.3%) | 2.33 (34.1%) |
| Alexnet | 245,283,524 (198.9%) | 60,965,224 (199.0%) | 10.14 (76.3%) | 1.86 (12.12) |
| SqueezeNet | 5,232,394 (156.9%) | 1,235,496 (159.4%) | 17.59 (117.9%) | 1.89 (13.3%) |

* CPU: Intel i7-4790 with 32 GB RAM, ** GPU: NVIDIA GeForce RTX 2080 Ti.

In Table 3, the percentage value in each cell is the percentage of the difference between the value for the customized deep learning model and that of a pre-existing deep learning model. One can observe that the customized deep learning model with relatively simple architecture outperformed the pre-existing deep models developed and trained by experts with a massive amount of training data.

In addition, as seen from the computation time, the problem-specific customized deep learning model has more potential for practical implementation with less hardware requirements than pre-existing deep learning models. Furthermore, in the framework of transfer learning, the input data to the pretrained models must be of the same size as that of the original data used during the pretraining of the network (e.g., image size, number of channels etc.); resizing a data set as per the requirement of pre-existing deep learning models may remove significant information in terms of the image size reduction or image size increment. However, such issues could be easily handled with a customized deep learning model specifically designed, trained, and transferred for a given engineering problem as achieved in the current work.

In the previous discussion, the experimental data set from RK4 rotor kit consisted of five health states at a steady state speed of 3600 rpm and a single severity level of each health state. To further verify the robustness of the proposed approach, a more extensive data set from SpectraQuest's machinery fault simulator (MFS) [78] kit was employed. In this work, five health states (normal, horizontal misalignment, unbalance, outer race fault in bearing, and rolling element fault in bearing) with different speeds of operation (49 speeds for each health state) and different severity levels (two severity levels) of each health state were considered. Furthermore, the bearing defects were studied in the presence of a 6 g and 35 g unbalanced mass. A detailed discussion of the data set can be referred to in reference [79] and it is available for download at the website in reference [80]. The vibration signals from the MFS were transformed into scalograms using the same filter bank as used for the data from the simulation models and RK4 rotor kit. The deep learning model pretrained on simulation data was employed to extract discriminative features from the scalograms of the experimental data from the MFS kit and SVM was employed to classify those features into different classes. The SVM classifier was trained through 10-fold cross-validation and Figure 19 shows its training/validation confusion matrix on the 90% training data with a classification accuracy of 97.8%.

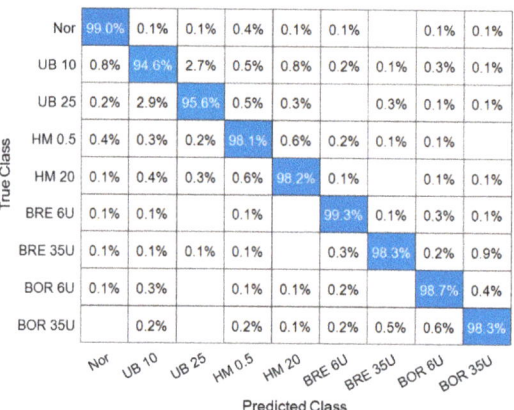

**Figure 19.** Training/validation confusion matrix of SVM on automatically extracted features through the pretrained deep learning model from the data of the MFS kit.

To verify that the SVM did not overfit the training data, Figure 20 shows the testing performance in the form of a confusion matrix on the 10% independent test data with 97.31% accuracy.

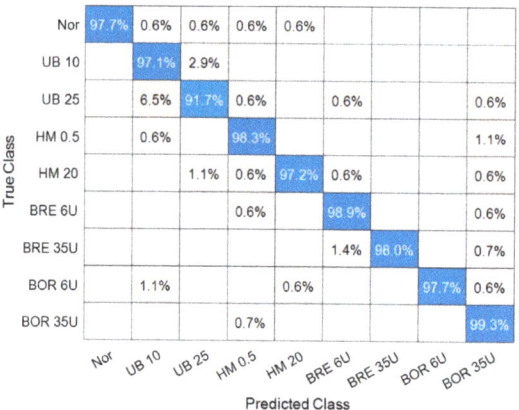

**Figure 20.** Test confusion matrix of SVM on features automatically extracted by the pretrained deep learning model from the data of the MFS kit.

In Figure 20, the tags should be interpreted as follows: HM 0.5: horizontal misalignment of 0.5 mm; HM 20: horizontal misalignment of 20 mm; UB 10: unbalance with 10 g mass; UB 25: unbalance with 25 g mass; Nor: normal; BRE 35U: bearing with rolling element fault and 35 g unbalance mass; BRE 6U: bearing with rolling element fault and 6 g unbalance mass; BOR 35U: bearing with outer race fault and 35 g unbalance mass; and BOR 6U: bearing with outer race fault and 6 g unbalance mass.

The results show that the model can distinguish different health states and their severity levels with a minimum accuracy of 94.6% for UB 10 and a maximum accuracy of 99.3% for BRE 6U. The misclassification results are within the acceptable range. The high accuracy of the model on the features extracted through the simulation model shows that the model is robust to different speeds of operations (49 different speeds for each health state) and that the extracted features are only sensitive to the presence of defects in the rotor system. Additionally, the high accuracy on the bearing faults in the presence of different unbalance loads confirms the robustness of the model to different loads. Furthermore, the target domain class invariance of the customized deep learning model is verified from the high accuracy on the bearing faults, which were not considered in the source simulation data.

From the test confusion matrix of Figure 20, the high accuracy of 97.31% on the 10% independent test data shows that the SVM model pretrained on the discriminative features of the deep learning model did not overfit the training data. The essence of the current work is that a domain invariant generalized feature extractor developed from simple simulations can accommodate the gap in the response characteristics and new health states in the target domain in a supervised learning framework.

## 3. Conclusions

This work proposed a domain and class invariant generalized feature extractor using a supervised learning framework of transfer learning. A source simulation domain with three health states was employed to detect, isolate, and quantify five health states in the target experimental domain without minimizing the gap in the response characteristics of the two domains. The source domain was comprised of the simulation model of a few representative health states of the target domain, and simulation models were not required for all prospective health states of the actual target system. The proposed methodology relies on transfer learning, where a customized deep learning model is trained, validated, and tested on a substantial set of simulation data, and then the pretrained model is employed to autonomously extract discriminative features from a small experimental target dataset. This work also discussed the synthetic augmentation of the limited experimental data using virtual sensors, where the output from the virtual sensors was defined in terms of the actual sensors using the concept of coordinate transformation. Synthetic augmentation of the experimental data enhanced the performance of the proposed approach in terms of training/validation accuracy (from 88.8% to 99.5%), test accuracy (90% to 97.14%), and ROC area (from 97% to 100%). The effectiveness of the proposed approach was validated by comparing its results with the pre-existing deep learning models of GoogleNet, VGG16, ResNet18, AlexNet, and SqueezeNet in terms of training, testing, generalization, size of the network, parameters of the network, and computational time. The current approach was found to perform relatively better in terms of generalizability and computation cost with more flexibility for a given engineering problem.

The proposed approach autonomously extracts discriminative features from the vibration-based scalograms of a limited experimental dataset and eliminates the need for labor-intensive hand-crafted statistical features. In addition, the source simulation signals and target experimental signals are directly transformed into scalograms using a single filter bank that eliminates the need for complex preprocessing. The generalized autonomous discriminative features are robust to variations in the operating conditions, severity levels of different health states, and scale of the source and target domains. This work could be extended to assess faults in laminated composites, gearboxes, industrial robots, civil infrastructures, etc.

**Author Contributions:** Conceptualization, H.S.K., J.-S.K. and A.K.; methodology, A.K. and H.S.K.; software, A.K.; validation, A.K. and H.S.K.; formal analysis, A.K.; investigation, A.K.; resources, H.S.K.; writing—original draft preparation, A.K.; writing—review and editing, A.K., H.S.K. and J.-S.K.; supervision, H.S.K.; project administration, H.S.K.; funding acquisition, H.S.K. All authors have read and agreed to the published version of the manuscript.

**Funding:** This research was supported by the Basic Science Research Program through the National Research Foundation of Korea (NRF-2020R1A2C1006613), funded by the Ministry of Education, and was also conducted as part of a research project (R17GA08) of the Korea Electric Power Corporation.

**Institutional Review Board Statement:** Not applicable.

**Informed Consent Statement:** Not applicable.

**Data Availability Statement:** The data that support the findings of this study are available from the corresponding author upon reasonable request.

**Conflicts of Interest:** The authors declare no conflict of interest.

## Appendix A

To simulate the added unbalance in the system, which represents unbalance that is commonly encountered in practice, a small mass of magnitude $m_a$ is attached at an eccentricity of $e_a$ and phase angle of $\beta$ to the disk, causing a harmonic centrifugal force of magnitude $m_a \times e_a \times \Omega^2$ along the $x$ and $y$ axes of the system when the system rotates at speed $\Omega$. The motion of the system in the presence of added unbalance is expressed as follows:

$$m\ddot{x}(t) + c_{xT}\dot{x}(t) + k_{xT}x(t) = m e_r \Omega^2 \cos(\Omega t + \alpha) + m_a e_a \Omega^2 \cos(\Omega t + \beta)$$
$$m\ddot{y}(t) + c_{yT}\dot{y}(t) + k_{yT}y(t) = m e_r \Omega^2 \sin(\Omega t + \alpha) + m_a e_a \Omega^2 \sin(\Omega t + \beta) \quad (A1)$$

Misalignment is another common defect in rotating machinery. Misalignment in the coupled machine shafts generates reaction forces in the coupling [4]. In this work, the Gibbons [5] model was adopted to simulate the presence of misalignment in the rotor-disk system of Figure 3. A schematic of the Gibbons model of parallel misalignment is shown in Figure A1 [81,82].

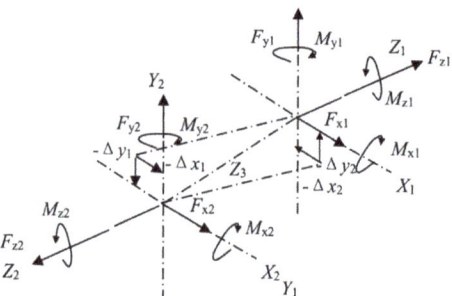

**Figure A1.** Schematic of the Gibbons parallel misalignment model. Reprinted with permission from ref. [82]. Copyright 2021 Elsevier.

Here, $Z_1$ and $Z_2$, respectively, denote the centerlines of the driver and driven shafts, which are offset by $\Delta Y$ along the vertical direction and by $\Delta X$ along the horizontal direction. The term $Z_3$ denotes the coupling center of articulation; $M_X$, $M_Y$, and $M_Z$ are the three moments; and $F_X$, $F_Y$, and $F_Z$ are the three reaction forces. The moments and forces exerted by coupling on the driver and driven shafts are shown by Equation (A2).

$$\theta_1 = \sin^{-1}(\Delta X_1/Z_3), \ \theta_2 = \sin^{-1}(\Delta X_2/Z_3)$$
$$\phi_1 = \sin^{-1}(\Delta Y_1/Z_3), \ \phi_2 = \sin^{-1}(\Delta Y_2/Z_3)$$
$$M_{X1} = T_q \sin \theta_1 + K_b \phi_1, \ MX_2 = T_q \sin \theta_2 - K_b \phi_2$$
$$M_{Y1} = T_q \sin \phi_1 - K_b \theta_1, \ MY_2 = T_q \sin \phi_2 + K_b \theta_2, \quad (A2)$$
$$F_{X1} = (-MY_1 - MY_2)/Z_3, \ FX_2 = -FX_1$$
$$F_{Y1} = (MX_1 + MX_2)/Z_3, \ FY_2 = -FY_1$$

where $K_b$ is the bending angular flexibility rate of the flexible coupling and $T_q$ is the torque of the rotor shaft, which is calculated in terms of motor power $P$ and speed of rotation $\Omega$, as given by Equation (A3).

$$P = T_q \times \Omega \quad (A3)$$

The moments and forces of Equation (A2) appear as periodic forces with $1\Omega$ and $2\Omega$ components, and the equation of motion in the presence of parallel misalignment is modified as follows:

$$\begin{array}{rl} m\ddot{x}(t) + c_{xT}\dot{x}(t) + k_{xT}x(t) = & m e_r \Omega^2 \cos(\Omega t + \alpha) + FX_2 \cos(\Omega t + \psi) \\ & + FX_2 \cos(2\Omega t + \psi) \\ m\ddot{y}(t) + c_{yT}\dot{y}(t) + k_{yT}y(t) = & m e_r \Omega^2 \sin(\Omega t + \alpha) + FY_2 \sin(\Omega t + \psi) \\ & + FY_2 \sin(2\Omega t + \psi) \end{array} \quad (A4)$$

where $\psi$ is the phase angle. In addition, note that, besides the misalignment forces, residual unbalance is present in the system, as shown by the first term on the right side of Equation (A4).

To simulate the rubbing phenomenon between the rotor and stator, it is assumed that a single rub-impact occurs at the disk location, as shown by the schematic in Figure A2.

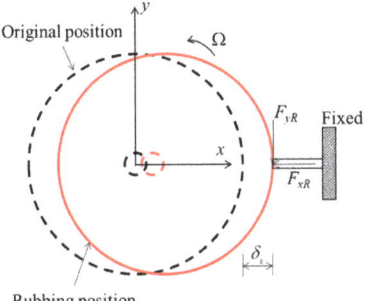

**Figure A2.** Schematic of a single rub-impact between the rotor and stator.

It is assumed that there is a small gap of $\delta_0$ between the rotor and stator. The rub-impact occurs when the axial displacement of the shaft due to unbalance is larger than $\delta_0$. The equation of motion in the presence of a single-rub impact is given by Equation (A5):

$$\begin{array}{l} m\ddot{x}(t) + c_{xT}\dot{x}(t) + k_{xT}x(t) = m e_r \Omega^2 \cos(\Omega t + \alpha) + F_{xR}(x,y) \\ m\ddot{y}(t) + c_{yT}\dot{y}(t) + k_{yT}y(t) = m e_r \Omega^2 \sin(\Omega t + \alpha) + F_{yR}(x,y) \end{array} \quad (A5)$$

where the terms $F_{xR}$ and $F_{yR}$ denote the nonlinear forces along the $x$ and $y$ axes, respectively, due to the single rub-impact between the rotor and stator and are expressed as follows [83]:

$$\begin{array}{l} F_{xR} = -k_r(x - \delta_0)H(x - \delta_0) \\ F_{yR} = fk_r(x - \delta_0)H(x - \delta_0) \end{array} \quad (A6)$$

where $k_r$ is the stiffness of the axial rub-impact rod, $f$ is the coefficient of friction between the rotor and stator, and $H$ is the Heaviside function, which is expressed as follows:

$$H(x - \delta_0) = \begin{cases} 0 & if \ x < \delta_0 \\ 1 & if \ x \geq \delta_0 \end{cases} \quad (A7)$$

## References

1. Sudhakar, G.; Sekhar, A.S. Identification of Unbalance in a Rotor Bearing System. *J. Sound Vib.* **2011**, *330*, 2299–2313. [CrossRef]
2. Jain, J.R.; Kundra, T.K. Model Based Online Diagnosis of Unbalance and Transverse Fatigue Crack in Rotor Systems. *Mech. Res. Commun.* **2004**, *31*, 557–568. [CrossRef]
3. Tonks, O.; Wang, Q. The Detection of Wind Turbine Shaft Misalignment Using Temperature Monitoring. *CIRP J. Manuf. Sci. Technol.* **2017**, *17*, 71–79. [CrossRef]
4. Verma, A.K.; Sarangi, S.; Kolekar, M.H. Experimental Investigation of Misalignment Effects on Rotor Shaft Vibration and on Stator Current Signature. *J. Fail. Anal. Prev.* **2014**, *14*, 125–138. [CrossRef]

5. Al-bedoor, B.O. Transient Torsional and Lateral Vibrations of Unbalanced Rotors with Rotor-to-Stator Rubbing. *J. Sound Vib.* **2000**, *229*, 627–645. [CrossRef]
6. Jiang, J.; Ulbrich, H.; Chavez, A. Improvement of Rotor Performance under Rubbing Conditions through Active Auxiliary Bearings. *Int. J. Non-Linear Mech.* **2006**, *41*, 949–957. [CrossRef]
7. Luo, L.-Y.; Fan, Y.-H.; Tang, J.-H.; Chen, T.-Y.; Zhong, N.-R.; Feng, P.-C.; Kao, Y.-C. Frequency Enhancement of Oil Whip and Oil Whirl in a Ferrofluid–Lubricated Hydrodynamic Bearing–Rotor System by Magnetic Field with Permanent Magnets. *Appl. Sci.* **2018**, *8*, 1687. [CrossRef]
8. El-Shafei, A.; Tawfick, S.H.; Raafat, M.S.; Aziz, G.M. Some Experiments on Oil Whirl and Oil Whip. *J. Eng. Gas Turbines Power* **2007**, *129*, 144–153. [CrossRef]
9. Schweizer, B. Oil Whirl, Oil Whip and Whirl/Whip Synchronization Occurring in Rotor Systems with Full-Floating Ring Bearings. *Nonlinear Dyn.* **2009**, *57*, 509–532. [CrossRef]
10. Gao, Z.; Cecati, C.; Ding, S.X. A Survey of Fault Diagnosis and Fault-Tolerant Techniques—Part I: Fault Diagnosis with Model-Based and Signal-Based Approaches. *IEEE Trans. Ind. Electron.* **2015**, *62*, 3757–3767. [CrossRef]
11. Mansouri, M.; Harkat, M.-F.; Nounou, H.N.; Nounou, M.N. *Data-Driven and Model-Based Methods for Fault Detection and Diagnosis*; Elsevier: Amsterdam, The Netherlands, 2020.
12. Wang, X.; Shen, C.; Xia, M.; Wang, D.; Zhu, J.; Zhu, Z. Multi-Scale Deep Intra-Class Transfer Learning for Bearing Fault Diagnosis. *Reliab. Eng. Syst. Saf.* **2020**, *202*, 107050. [CrossRef]
13. Xu, Z.; Saleh, J.H. Machine Learning for Reliability Engineering and Safety Applications: Review of Current Status and Future Opportunities. *Reliab. Eng. Syst. Saf.* **2021**, *211*, 107530. [CrossRef]
14. Lu, B.; Li, Y.; Wu, X.; Yang, Z. A Review of Recent Advances in Wind Turbine Condition Monitoring and Fault Diagnosis. In Proceedings of the 2009 IEEE power electronics and machines in wind applications, Lincoln, NE, USA, 24–26 June 2009; IEEE: Piscataway Township, NJ, USA, 2009; pp. 1–7.
15. Yin, S.; Ding, S.X.; Xie, X.; Luo, H. A Review on Basic Data-Driven Approaches for Industrial Process Monitoring. *IEEE Trans. Ind. Electron.* **2014**, *61*, 6418–6428. [CrossRef]
16. Gómez, M.J.; Castejón, C.; García-Prada, J.C. Automatic Condition Monitoring System for Crack Detection in Rotating Machinery. *Reliab. Eng. Syst. Saf.* **2016**, *152*, 239–247. [CrossRef]
17. Khan, S.; Yairi, T. A Review on the Application of Deep Learning in System Health Management. *Mech. Syst. Signal Process.* **2018**, *107*, 241–265. [CrossRef]
18. Shukla, S.; Yadav, R.N.; Sharma, J.; Khare, S. Analysis of Statistical Features for Fault Detection in Ball Bearing. In Proceedings of the 2015 IEEE International Conference on Computational Intelligence and Computing Research (ICCIC), Madurai, India, 10–12 December 2015; pp. 1–7.
19. Oh, H.; Jung, J.H.; Jeon, B.C.; Youn, B.D. Scalable and Unsupervised Feature Engineering Using Vibration-Imaging and Deep Learning for Rotor System Diagnosis. *IEEE Trans. Ind. Electron.* **2017**, *65*, 3539–3549. [CrossRef]
20. Mottershead, J.E.; Friswell, M.I. Model Updating in Structural Dynamics: A Survey. *J. Sound Vib.* **1993**, *167*, 347–375. [CrossRef]
21. Cavalini, A.A., Jr.; Lobato, F.S.; Koroishi, E.H.; Steffen, V., Jr. Model Updating of a Rotating Machine Using the Self-Adaptive Differential Evolution Algorithm. *Inverse Probl. Sci. Eng.* **2016**, *24*, 504–523. [CrossRef]
22. Immovilli, F.; Bianchini, C.; Cocconcelli, M.; Bellini, A.; Rubini, R. Bearing Fault Model for Induction Motor with Externally Induced Vibration. *IEEE Trans. Ind. Electron.* **2012**, *60*, 3408–3418. [CrossRef]
23. LeCun, Y.; Bengio, Y.; Hinton, G. Deep Learning. *Nature* **2015**, *521*, 436–444. [CrossRef]
24. Bakator, M.; Radosav, D. Deep Learning and Medical Diagnosis: A Review of Literature. *Multimodal Technol. Interact.* **2018**, *2*, 47. [CrossRef]
25. Voulodimos, A.; Doulamis, N.; Doulamis, A.; Protopapadakis, E. Deep Learning for Computer Vision: A Brief Review. *Comput. Intell. Neurosci.* **2018**, *2018*, 7068349. [CrossRef]
26. Pierson, H.A.; Gashler, M.S. Deep Learning in Robotics: A Review of Recent Research. *Adv. Robot.* **2017**, *31*, 821–835. [CrossRef]
27. Morozov, V.; Petrovskiy, M. An Approach for Complex Event Streams Processing and Forecasting. In Proceedings of the 2020 26th Conference of Open Innovations Association (FRUCT), Yaroslavl, Russia, 20–24 April 2020; IEEE: Piscataway Township, NJ, USA, 2020; pp. 305–313.
28. Dey, L.; Meisheri, H.; Verma, I. Predictive Analytics with Structured and Unstructured Data-a Deep Learning Based Approach. *IEEE Intell. Inform. Bull.* **2017**, *18*, 27–34.
29. Gecgel, O.; Ekwaro-Osire, S.; Dias, J.P.; Serwadda, A.; Alemayehu, F.M.; Nispel, A. Gearbox Fault Diagnostics Using Deep Learning with Simulated Data. In Proceedings of the 2019 IEEE International Conference on Prognostics and Health Management (ICPHM), San Francisco, CA, USA, 17–20 June 2019; IEEE: Piscataway Township, NJ, USA, 2019; pp. 1–8.
30. Tiwari, R. *Rotor Systems: Analysis and Identification*; CRC Press: Boca Raton, FL, USA, 2017.
31. Zheng, H.; Wang, R.; Yang, Y.; Yin, J.; Li, Y.; Li, Y.; Xu, M. Cross-Domain Fault Diagnosis Using Knowledge Transfer Strategy: A Review. *IEEE Access* **2019**, *7*, 129260–129290. [CrossRef]
32. Li, C.; Zhang, S.; Qin, Y.; Estupinan, E. A Systematic Review of Deep Transfer Learning for Machinery Fault Diagnosis. *Neurocomputing* **2020**, *407*, 121–135. [CrossRef]
33. Pan, S.J.; Yang, Q. A Survey on Transfer Learning. *IEEE Trans. Knowl. Data Eng.* **2010**, *22*, 1345–1359. [CrossRef]

34. Zhang, W.; Li, X.; Ma, H.; Luo, Z.; Li, X. Transfer Learning Using Deep Representation Regularization in Remaining Useful Life Prediction across Operating Conditions. *Reliab. Eng. Syst. Saf.* **2021**, *211*, 107556. [CrossRef]
35. Cao, P.; Zhang, S.; Tang, J. Preprocessing-Free Gear Fault Diagnosis Using Small Datasets with Deep Convolutional Neural Network-Based Transfer Learning. *IEEE Access* **2018**, *6*, 26241–26253. [CrossRef]
36. Xu, G.; Liu, M.; Jiang, Z.; Shen, W.; Huang, C. Online Fault Diagnosis Method Based on Transfer Convolutional Neural Networks. *IEEE Trans. Instrum. Meas.* **2020**, *69*, 509–520. [CrossRef]
37. Yan, R.; Shen, F.; Sun, C.; Chen, X. Knowledge Transfer for Rotary Machine Fault Diagnosis. *IEEE Sens. J.* **2020**, *20*, 8374–8393. [CrossRef]
38. Hasan, M.J.; Kim, J.-M. Bearing Fault Diagnosis under Variable Rotational Speeds Using Stockwell Transform-Based Vibration Imaging and Transfer Learning. *Appl. Sci.* **2018**, *8*, 2357. [CrossRef]
39. Li, F.; Tang, T.; Tang, B.; He, Q. Deep Convolution Domain-Adversarial Transfer Learning for Fault Diagnosis of Rolling Bearings. *Measurement* **2021**, *169*, 108339. [CrossRef]
40. Zhang, R.; Tao, H.; Wu, L.; Guan, Y. Transfer Learning with Neural Networks for Bearing Fault Diagnosis in Changing Working Conditions. *IEEE Access* **2017**, *5*, 14347–14357. [CrossRef]
41. Huang, P.; Wang, G.; Qin, S. Boosting for Transfer Learning from Multiple Data Sources. *Pattern Recognit. Lett.* **2012**, *33*, 568–579. [CrossRef]
42. Byeon, Y.-H.; Pan, S.-B.; Kwak, K.-C. Intelligent Deep Models Based on Scalograms of Electrocardiogram Signals for Biometrics. *Sensors* **2019**, *19*, 935. [CrossRef] [PubMed]
43. Peng, Z.K.; Chu, F.L.; Tse, P.W. Detection of the Rubbing-Caused Impacts for Rotor–Stator Fault Diagnosis Using Reassigned Scalogram. *Mech. Syst. Signal Process.* **2005**, *19*, 391–409. [CrossRef]
44. Krizhevsky, A.; Sutskever, I.; Hinton, G.E. ImageNet Classification with Deep Convolutional Neural Networks. In *Advances in Neural Information Processing Systems 25*; Pereira, F., Burges, C.J.C., Bottou, L., Weinberger, K.Q., Eds.; Curran Associates, Inc.: New York, NJ, USA, 2012; pp. 1097–1105.
45. Szegedy, C.; Liu, W.; Jia, Y.; Sermanet, P.; Reed, S.; Anguelov, D.; Erhan, D.; Vanhoucke, V.; Rabinovich, A. Going Deeper with Convolutions. In Proceedings of the 2015 IEEE Conference on Computer Vision and Pattern Recognition (CVPR), Boston, MA, USA, 7–12 June 2015; IEEE: Piscataway Township, NJ, USA, 2015; pp. 1–9.
46. He, K.; Zhang, X.; Ren, S.; Sun, J. Deep Residual Learning for Image Recognition. In Proceedings of the 2016 IEEE Conference on Computer Vision and Pattern Recognition (CVPR), Las Vegas, NV, USA, 27–30 June 2016; IEEE: Piscataway Township, NJ, USA, 2016; pp. 770–778.
47. Simonyan, K.; Zisserman, A. Very Deep Convolutional Networks for Large-Scale Image Recognition. *arXiv* **2015**, arXiv:1409.1556.
48. Yosinski, J.; Clune, J.; Bengio, Y.; Lipson, H. How Transferable Are Features in Deep Neural Networks? In *Advances in Neural Information Processing Systems 27*; Ghahramani, Z., Welling, M., Cortes, C., Lawrence, N.D., Weinberger, K.Q., Eds.; Curran Associates, Inc.: New York, NK, USA, 2014; pp. 3320–3328.
49. Guo, Y.; Shi, H.; Kumar, A.; Grauman, K.; Rosing, T.; Feris, R. Spottune: Transfer Learning through Adaptive Fine-Tuning. In Proceedings of the Proceedings of the IEEE Conference on Computer Vision and Pattern Recognition, Long Beach, CA, USA, 16–20 June 2019; pp. 4805–4814.
50. Shao, S.; McAleer, S.; Yan, R.; Baldi, P. Highly Accurate Machine Fault Diagnosis Using Deep Transfer Learning. *IEEE Trans. Ind. Inform.* **2018**, *15*, 2446–2455. [CrossRef]
51. Filippi, M.; Carrera, E. Stability and Transient Analyses of Asymmetric Rotors on Anisotropic Supports. *J. Sound Vib.* **2021**, *500*, 116006. [CrossRef]
52. Standard, I.S.O. *Mechanical Vibration-Evaluation of Machine Vibration by Measurements on Non-Rotating Parts*; ISO/IS: 1996; p. 10816. Available online: https://www.iso.org/obp/ui/#iso:std:iso:7919:-1:ed-2:v1:en (accessed on 18 February 2021).
53. Standardization, I.O. for ISO 20816-1: 2016 (En), Mechanical Vibration—Measurement and Evaluation of Machine Vibration—Part 1: General Guidelines. 2016. Available online: https://www.iso.org/obp/ui/#iso:std:iso:20816:-1:ed-1:v1:en (accessed on 18 February 2021).
54. ISO 16084:2017(En), Balancing of Rotating Tools and Tool Systems. Available online: https://www.iso.org/obp/ui/#iso:std:iso:16084:ed-1:v1:en (accessed on 15 January 2021).
55. Hashamdar, H.; Ibrahim, Z.; Jameel, M. Finite Element Analysis of Nonlinear Structures with Newmark Method. *IJPS* **2011**, *6*, 1395–1403. [CrossRef]
56. Jondral, F.K. White Gaussian Noise–Models for Engineers. *Frequenz* **2018**, *72*, 293–299. [CrossRef]
57. Liu, N.; Schumacher, T. Improved Denoising of Structural Vibration Data Employing Bilateral Filtering. *Sensors* **2020**, *20*, 1423. [CrossRef] [PubMed]
58. Add White Gaussian Noise to Signal—MATLAB Awgn. Available online: https://www.mathworks.com/help/comm/ref/awgn.html (accessed on 23 December 2021).
59. Benítez, R.; Bolós, V.J.; Ramírez, M.E. A Wavelet-Based Tool for Studying Non-Periodicity. *Comput. Math. Appl.* **2010**, *60*, 634–641. [CrossRef]
60. Türk, Ö.; Özerdem, M.S. Epilepsy Detection by Using Scalogram Based Convolutional Neural Network from EEG Signals. *Brain Sci.* **2019**, *9*, 115. [CrossRef]
61. Lee, D.T.; Yamamoto, A. Wavelet Analysis: Theory and Applications. *Hewlett Packard J.* **1994**, *45*, 44–52.

62. Olhede, S.C.; Walden, A.T. Generalized Morse Wavelets. *IEEE Trans. Signal Process.* **2002**, *50*, 2661–2670. [CrossRef]
63. Lilly, J.M.; Olhede, S.C. Higher-Order Properties of Analytic Wavelets. *IEEE Trans. Signal Process.* **2008**, *57*, 146–160. [CrossRef]
64. Lilly, J.M.; Olhede, S.C. Generalized Morse Wavelets as a Superfamily of Analytic Wavelets. *IEEE Trans. Signal Process.* **2012**, *60*, 6036–6041. [CrossRef]
65. Silik, A.; Noori, M.; Altabey, W.A.; Ghiasi, R.; Wu, Z. Comparative Analysis of Wavelet Transform for Time-Frequency Analysis and Transient Localization in Structural Health Monitoring. *Struct. Durab. Health Monit.* **2021**, *15*, 1–22. [CrossRef]
66. De Moortel, I.; Munday, S.A.; Hood, A.W. Wavelet Analysis: The Effect of Varying Basic Wavelet Parameters. *Sol. Phys.* **2004**, *222*, 203–228. [CrossRef]
67. Bolós, V.J.; Benítez, R. The Wavelet Scalogram in the Study of Time Series. In *Advances in Differential Equations and Applications*; Springer: Berlin/Heidelberg, Germany, 2014; pp. 147–154.
68. Yamamoto, G.K.; da Costa, C.; da Silva Sousa, J.S. A Smart Experimental Setup for Vibration Measurement and Imbalance Fault Detection in Rotating Machinery. *Case Stud. Mech. Syst. Signal Process.* **2016**, *4*, 8–18. [CrossRef]
69. Chen, S.; Yang, Y.; Peng, Z.; Wang, S.; Zhang, W.; Chen, X. Detection of Rub-Impact Fault for Rotor-Stator Systems: A Novel Method Based on Adaptive Chirp Mode Decomposition. *J. Sound Vib.* **2019**, *440*, 83–99. [CrossRef]
70. Ma, H.; Tai, X.; Sun, J.; Wen, B. Analysis of Dynamic Characteristics for a Dual-Disk Rotor System with Single Rub-Impact. *Adv. Sci. Lett.* **2011**, *4*, 2782–2789. [CrossRef]
71. Srivastava, N.; Hinton, G.; Krizhevsky, A.; Sutskever, I.; Salakhutdinov, R. Dropout: A Simple Way to Prevent Neural Networks from Overfitting. *J. Mach. Learn. Res.* **2014**, *15*, 1929–1958.
72. Nwankpa, C.; Ijomah, W.; Gachagan, A.; Marshall, S. Activation Functions: Comparison of Trends in Practice and Research for Deep Learning. *arXiv* **2018**, arXiv:1811.03378.
73. Khan, A.; Sohail, A.; Zahoora, U.; Qureshi, A.S. A Survey of the Recent Architectures of Deep Convolutional Neural Networks. *Artif Intell Rev* **2020**, *53*, 5455–5516. [CrossRef]
74. Hanley, J.A.; McNeil, B.J. The Meaning and Use of the Area under a Receiver Operating Characteristic (ROC) Curve. *Radiology* **1982**, *143*, 29–36. [CrossRef]
75. Jung, J.H.; Jeon, B.C.; Youn, B.D.; Kim, M.; Kim, D.; Kim, Y. Omnidirectional Regeneration (ODR) of Proximity Sensor Signals for Robust Diagnosis of Journal Bearing Systems. *Mech. Syst. Signal Process.* **2017**, *90*, 189–207. [CrossRef]
76. Iandola, F.N.; Han, S.; Moskewicz, M.W.; Ashraf, K.; Dally, W.J.; Keutzer, K. SqueezeNet: AlexNet-Level Accuracy with 50x Fewer Parameters and <0.5MB Model Size. *arXiv* **2016**, arXiv:1602.07360.
77. Russakovsky, O.; Deng, J.; Su, H.; Krause, J.; Satheesh, S.; Ma, S.; Huang, Z.; Karpathy, A.; Khosla, A.; Bernstein, M.; et al. ImageNet Large Scale Visual Recognition Challenge. *Int. J. Comput. Vis.* **2015**, *115*, 211–252. [CrossRef]
78. SpectraQuest Inc. Available online: https://spectraquest.com/ (accessed on 18 February 2021).
79. Marins, M.A.; Ribeiro, F.M.; Netto, S.L.; da Silva, E.A. Improved Similarity-Based Modeling for the Classification of Rotating-Machine Failures. *J. Frankl. Inst.* **2018**, *355*, 1913–1930. [CrossRef]
80. MAFAULDA: Machinery Fault Database. Available online: http://www02.smt.ufrj.br/~{}offshore/mfs/page_01.html (accessed on 24 November 2021).
81. Gibbons, C.B. Coupling Misalignment Forces. In Proceedings of the Proceedings of the 5th Turbomachinery Symposium, College Station, TX, USA, 12–14 October 1976; Texas A&M University, Gas Turbine Laboratories: College Station, TX, USA, 1976.
82. Jalan, A.K.; Mohanty, A.R. Model Based Fault Diagnosis of a Rotor–Bearing System for Misalignment and Unbalance under Steady-State Condition. *J. Sound Vib.* **2009**, *327*, 604–622. [CrossRef]
83. Tai, X.; Ma, H.; Liu, F.; Liu, Y.; Wen, B. Stability and Steady-State Response Analysis of a Single Rub-Impact Rotor System. *Arch. App. Mech.* **2015**, *85*, 133–148. [CrossRef]

MDPI  
St. Alban-Anlage 66  
4052 Basel  
Switzerland  
Tel. +41 61 683 77 34  
Fax +41 61 302 89 18  
www.mdpi.com  

*Mathematics* Editorial Office  
E-mail: mathematics@mdpi.com  
www.mdpi.com/journal/mathematics

www.ingramcontent.com/pod-product-compliance
Lightning Source LLC
LaVergne TN
LVHW070221100526
838202LV00015B/2072